Geostatistics for Natural Resources Characterization

Part 1

NATO ASI Series

Advanced Science Institutes Series

A series presenting the results of activities sponsored by the NATO Science Committee, which aims at the dissemination of advanced scientific and technological knowledge, with a view to strengthening links between scientific communities.

The series is published by an international board of publishers in conjunction with the NATO Scientific Affairs Division

A	Life Sciences	Plenum Publishing Corporation
B	Physics	London and New York
C	Mathematical and Physical Sciences	D. Reidel Publishing Company Dordrecht, Boston and Lancaster
D	Behavioural and Social Sciences	Martinus Nijhoff Publishers
E	Engineering and Materials Sciences	The Hague, Boston and Lancaster
F	Computer and Systems Sciences	Springer-Verlag
G	Ecological Sciences	Berlin, Heidelberg, New York and Tokyo

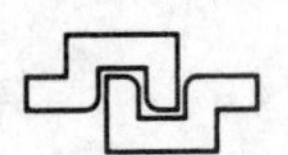

Series C: Mathematical and Physical Sciences Vol. 122 – Part 1

Geostatistics for Natural Resources Characterization

Part 1

edited by

G. Verly

Department of Applied Earth Sciences, Stanford University, Stanford, U.S.A.

M. David

Departement de Chimie Mineral, Ecole Polytechnique, Montreal, Canada

A. G. Journel

Department of Applied Earth Sciences, Stanford University, Stanford, U.S.A.

and

A. Marechal

S.N. Elf Aquitaine (Production), Pau, France

D. Reidel Publishing Company

Dordrecht / Boston / Lancaster

Published in cooperation with NATO Scientific Affairs Division

Proceedings of the NATO Advanced Study Institute on
Geostatistics for Natural Resources Characterization
Stanford Sierra Lodge, South Lake Tahoe, California, U.S.A.
September 6 – 17, 1983

Library of Congress Cataloging in Publication Data

NATO Advanced Study Institute on Geostatistics for Natural Resources Characterization (2nd : 1983 : South Lake Tahoe, Calif.)
Geostatistics for natural resources characterization.
(NATO ASI series. Series C, Mathematical and physical sciences; vol. 122)
"Proceedings of the NATO Advanced Study Institute on Geostatistics for Natural Resources Characterization, Stanford Sierra Lodge, South Lake Tahoe, California, U.S.A., September 6–17, 1983"—T.p. verso.
"Published in cooperation with NATO Scientific Affairs Division."
Includes bibliographical references and index.
1. Mine valuation—Statistical methods—Congresses. 2. Geology—Statistical methods—Congresses. I. Verly, G. (Georges) II. North Atlantic Treaty Organization. Scientific Affairs Division. III. Title. IV. Series: NATO ASI Series. Series C, Mathematical and physical sciences; Vol. 122.
TN272.N38 1983 553'.0724 84-2155
ISBN 90-277-1746-X (pt. 1)
ISBN 90-277-1747-8 (pt. 2)

Published by D. Reidel Publishing Company
P.O. Box 17, 3300 AA Dordrecht, Holland

Sold and distributed in the U.S.A. and Canada
by Kluwer Academic Publishers,
190 Old Derby Street, Hingham, MA 02043, U.S.A.

In all other countries, sold and distributed
by Kluwer Academic Publishers Group,
P.O. Box 322, 3300 AH Dordrecht, Holland

D. Reidel Publishing Company is a member of the Kluwer Academic Publishers Group

---

Printed in The Netherlands.

TABLE OF CONTENTS

FOREWORD

Since October 1975 and the first NATO ASI "Geostat - Roma 1975", there has not been an advanced workshop on geostatistics where experts from throughout the world could meet, teach and debate without the pressure of an ordinary professional conference. This second NATO ASI "Geostat - Tahoe 1983" was intended as a high-level teaching activity yet opened to all new ideas and contributions to the discipline of geostatistics. It was expected that the institute would bridge the gap since "Geostat - Roma 1975" and establish the state of the art of the discipline as of 1983.

"Geostat - Tahoe 1983" fulfilled all expectations. The institute, held in September 1983 at the Stanford Sierra Lodge near Lake Tahoe, California, was attended by all major players in the field, representing evenly the academy and the industry of 23 different countries. Twelve guest lecturers were backed by some 50 original contributions. Particularly important for the future was the active participation of graduate students, giving evidence of the dynamism of a still very young discipline.

Applications of geostatistics are no longer limited to the mining industry, and the original scope of the ASI had to be widened to witness the progress made in such fields as hydrology, soil sciences, pollution control and geotechnical engineering. Also important was the cross-fertilization from other statistical branches such as spectral analysis and robust statistics.

The NATO ASI "Geostat - Tahoe 1983" will be remembered as a major milestone in the development of the discipline. I invite all of you who did not make it to the ASI to dive into these proceedings and share a bit of the excitement that pervaded this summer's end at Lake Tahoe.

A.G. Journel
Director of the ASI

ACKNOWLEDGEMENTS

The Scientific Affairs Division of NATO is gratefully acknowledged for having sponsored this Advanced Study Institute. Additional support was provided by Stanford University, Ecole Polytechnique de Montreal, and Ecole des Mines de Paris.

We thank the group of following companies which helped finance the Institute:

BRGM, France
CERCHAR, France
COGEMA, France
Elf Aquitaine (Production), France
Schlumberger E.P.S., France
TECMIN, France
Shell, Netherlands

We are especially grateful to the team of Stanford students who made possible the organization of the ASI.

The overwhelming success of this Institute would not have been possible without the dedication of our lecturers and contributors.

The Organizing Committee

George Verly
Michel David
Andre Journel
Alain Marechal

Stanford, October 1983

LIST OF PARTICIPANTS

ALFARO, Marco: Professor, Departemento Minas, Universidad de Chile, Casilla 2777,Santiago, Chile.

AMEGEE, Kodjo: Student, Departement of Agricultural Engineering, Oregon State University, Corvallis, Oregon 97331, U.S.A.

ARMSTRONG, Margaret: Research Engineer,Centre de Geostatistique, ENSMP, 35 rue St. Honore, 77305 Fontainebleau, France.

BAECHER Gregory B.: Professor, Departement of Civil Engineering, Masachusset Institute of Technology, Cambridge, Massachusetts 02139, U.S.A.

BAELE Steve M.: Mining Engineer, U.S. Borax Corp., 3075 Wishire Blvd., Los Angeles, California 90010, U.S.A.

BALIA, Lobo : Mineral Technology Development Center, Jl. Jendral Sudirman No.623, Bandung, Indonesia.

BARNES, Randal J.: Instructor, Colorado School of Mines, Golden, Colorado 80401, U.S.A.

BELISLE, Jean Marc: Project Manager, Geostat Systems International, 5180 Queen Mary, Suite 370, Montreal, Quebec H3X 3E7, Canada

BERLANGA, Juan M.: —Professor, Division de Posgrado, Fac. Ingenieria , Universidad de Mexico, Mexico D.F. 04510, Mexico

BILONICK, Richard A.: Statistician, Consolidation Coal Co., Scientific Systems, Consol Plaza , Pittsburgh, Pennsylvania 15241, U.S.A.

BONHAM-CARTER, Graeme F.: Research Scientist, Geological Survey of Canada, 601 Booth Street, Ottawa, Ontario K1A 0E8, Canada

BORGMAN, Leon E.: Prof. of Geology and Statistics, Dept. of Statistics, University of Wyoming, Laramie, Wyoming 82071, U.S.A.

BRYAN, Rex, C.: Director, Geostat Systems Inc., P.O. Box 1193, Golden, Colorado 80402, U.S.A.

BUXTON, Bruce: Student, Dept. of Applied Earth Sciences, Stanford University, Stanford, California 94305, U.S.A.

CAMISANI-CALZOLARI, F.A.M.G.: Project Head, Nuclear Development Corp., Uranium Evaluation Group, Private Bag X256 Pretoria 0001, South Africa

CARR, James R.: Professor, Dept. of Geological Engineering, University of Missouri, Rolla, Missouri 65401, U.S.A.

CHICA-OLMO, Marion: Student, Centre de Geostatistique, ENSMP, 35 rue St. Honore, 77305 Fontainebleau, France

CHILES, Jean-Paul: Mining Engineer, B.R.G.M., B.P.6009, 45018 Orleans, France

CHU, Darong: Student, 1126 C San Pablo Av., #13, Albany, California 94706, U.S.A.

CHUNG, Chang-Jo F.: Research Scientist, Geological Survey of Canada, 601 Booth Street, Ottawa, Ontario K1A 0E8, Canada

CLARK, Isobel: Managing Director, Geostokos Ltd., 8a. Lower Grosvenor Place, London SW1W 0EN, United Kingdom

CRESSIE, Noel A.C.: Senior Lecturer, Dept.of Math. Sciences, The Flinders University, Bedford Park, S.A. 5042, Australia.

CULHANE, Patrick, G.: Manager, Applications Service, Sohio Petroleum Company, 100 Pine Street, San Francisco, California 94111, U.S.A.

DAGBERT, Michel: Manager, Geostat Systems International, 5180 Queen Mary, Suite 370, Montreal, Quebec H3X 3E7, Canada

DAMAY, Jacques: Director Minas-Exploracion, S.M.M.P., Alfonso XII-30, Madrid (14), Spain

DAVID, Michel: Professor, Dept. de Genie Mineral, Ecole Polytechnique, C.P. 6079 Succ. A, Montreal, Quebec H3C 3A7, Canada.

DAVIS, Bruce M.: Chief Geostatistician, St. Joe American Corp., 2002 N. Forbes Blvd, Tucson, Arizona 85705, U.S.A.

DAVIS, Michael W.: Principal Geostatistician, Fluor Mining and Metal Inc., 10 Twin Dolphin Drive, Redwood City, California 94065, U.S.A.

DERAISME, Jacques: Research Engineer, Centre de Geostatistique, ENSMP, 35 rue St. Honore, 77305 Fontainebleau, France

DESBARATS, Alexandre: Student, Dept. of Applied Earth Sciences, Stanford University, Stanford, California 94305, U.S.A.

DEVARY, Joseph L.: Research Scientist, Pacific Northwest Laboratory , P.O. Box 999, Richland, Washington 99352, U.S.A.

DIEHL, Peter: Student, Franckerstr. 2, 2300 Kiel 1, West Germany

DOWD, Peter A.: Lecturer, University of Leeds, Leeds LS2 9JT, England, United Kingdom

DUBRULE, Olivier: Senior Technical Analyst, Sohio Petroleum Company, 100 Pine Street, San Francisco, California 94111, U.S.A.

DUNN, Mark R.: Research Mathematician, Shell Development Company, P.O. Box 1380, Houston, Texas 77001, U.S.A.

ENGLUND, Evan J.: Sr. Minerals Geologist, Exxon Minerals Company, P.O. Box 4508, Houston, Texas 77210, U.S.A.

ERWIN, Les: Geologist, Texaco USA, Bellaire Research Lab., P.O. Box 425, Bellaire, Texas 77401, U.S.A.

FLATMAN, George T.: Statistician, U.S. EPA EMSL-LV, Station EAD, P.O. Box 15027, Las Vegas, Las Vegas, Nevada 89114, U.S.A.

FRANCOIS-BONGARCON, Dominique: Vice-president, Geomath Inc., 4891 Independence St., Suite 250, Wheat Ridge, Colorado 80033, U.S.A.

FROIDEVAUX, Roland: Senior Geostatistician, David Robertson and Associates, 145 King St. West, 24th floor, Toronto, Ontario M5H 1V8, Canada.

GALLEGO, Alejandrino: Ingenierio de Minas, S.M.M. Penarroya, Alfonso XII-30, Madrid (14), Spain

GALLI, Alain: Research Engineer, Centre de Geostatistique, ENSMP, 35 rue St. Honore, 77305 Fontainebleau, France

GASCOINE, Chris E.: Technical Specialist Geostatistics, Esso Resources Canada Ltd., 339 - 50 Avenue S.E., Calgary, Alberta T2G 2B3, Canada

GERDIL-NEUILLET, Francoise: Engineer,TOTAL-CFP, 39-43 Quai Andre Citroen, 75015 Paris, France

GERKEN, Dennis J.: Senior Software Manager, Geomath Inc., 4891 Independence St., Suite 250, Wheatridge , Colorado 80033, U.S.A.

GIRARDI, Jorge P.: Researcher, Inst. de Investigaciones Mineras, Universidad Nacional de San Juan, Avda. Libertador Gral. San Martin 1109 (oeste), 5400 San Juan, Argentina

GIROUX, Gary H.: Partner, Montgomery Consultants Ltd., 203-2786 W. 16th Ave., Vancouver, British Colombia V6K 3C3, Canada

GLASS, Charles E.: Professor, Dept. of Minning & Geological Engineering, University of Arizona, Tucson, Arizona 85721, U.S.A.

GUARASCIO, Massimo: Professor, Inst. di Geologia Applicata, Universita di Roma, via Endessiana 18, Roma 00184, Italy

GUERTIN, Kateri: Student, Dept. of Applied Earth Sciences, Stanford University, Stanford, California 94305, U.S.A.

GUIBAL, Daniel: Senior Geostatistician, Siromines, 71 York Street, Sydney, NSW 2000, Australia

HAGAN, Randy: Geomathematician, Anaconda Minerals Co., 555 17th Street, Denver, Colorado 80202, U.S.A.

HELWICK, Sterling, J. Jr.: Researcher, Exxon Production Research Co., P.O. Box 2189, Houston, Texas 77001, U.S.A.

HERNANDEZ-GARCIA, Gustavo: Engineer, Secretaria de Energia, Minas e Industria Paraestatal, San Luis Potosi 211-1er. Piso, Mexico D.F. 06700, Mexico

HODOS, Ellen F.: Mining Consultant, Placer Service Corp., 12050 Charles Drive Suite A, Nevada City, California 95959

ISAAKS, Edward H.: Student, Dept of Applied Earth Sciences, Stanford University, Stanford, California 94305, U.S.A.

JARVIS, Merigon M.: Processing Geophysicist, Compagnie Generale de Geophysique, 47/55 The Vale, Acton, London W3 7RR, England

JONES, Thomas A.: Sr. Research Associate, Exxon Production Research Co., P.O. Box 2189, Houston, Texas 77001, U.S.A.

JOURNEL, Andre G.: Professor, Dept. of Applied Earth Sciences, Stanford University, Stanford, California 94305, U.S.A.

KIM, Young C.: Professor, Dept. of Mining and Geolological Engineering., University of Arizona, Tucson, Arizona 85721, U.S.A.

KLEINGELD, W.J.: Ore Evaluation Manager, De Beers Consolidated Mines Ltd., P.O. Box 616, Kimberley 8300, South Africa

KNUDSEN, Pete H.: Professor, Mining Engineering Dept., Montana College of Mineral Sciences and Technology, Butte, Montana 59701, U.S.A.

KULKARNI, Ram B.: Head Systems Engineering Group, Woodward-Clyde Cons., One Walnut Creek Center, 100 Pringle Ave., Walnut Creek, California 94596, U.S.A.

KWA, Leong B.: Student, Mackay School of Mines, University of Nevada, Reno, Nevada 89557, U.S.A.

LAILLE, Jean-Paul: Research Engineer, Centre de Geostatistique, ENSMP, 35 rue St. Honore, 77305 Fontainebleau, France

LAJAUNIE, Christian: Research Engineer, Centre de Geostatistique, ENSMP, 35 rue St. Honore, 77305 Fontainebleau, France

LEMMER, Carina: Senior Analyst, Technical Systems, Gold Field SA, P.O. Box 1167, Johannesburg 2000, South Africa

LONERGAN E.T.: Project Supervisor, Placer Development Ltd., P.O. Box 49330, Bentall Postal Station, Vancouver, British Colombia V7X 1P1, Canada

LUDVIGSEN, Erik: Head of Divison, Division of Mining Engineering, The Norwegian Institute of Technology, 7034 Trondheim-NTH, Norway

LUSTER, Gordon R.: Student, Dept. of Applied Earth Sciences, Stanford University, Stanford, California 94305, U.S.A.

LYEW-AYEE, Parris A.: Director Bauxite Division, Jamaica Bauxite Institute, P.O. Box 355, Kingston 6, Jamaica, W.I.

MAINVILLE, Alain: Student, Dept. de Genie Mineral, Ecole Polytechnique, C.P. 6079 Succ. A, Montreal, Quebec H3C 3A, Canada

MALLET, Jean Laurent: Professor, Ecole Nationale Superieure de Geologie, B.P. 452, 54001 Nancy, France

MAO, Nai-hsien: Geophysicist, U.C. Lawrence Livermore National Lab., P.O. Box 808, Mail Stop L224, Livermore, California 94550, U.S.A.

MARBEAU, Jean-Paul: President, Geomath Inc., 4891 Independence St., Suite 250, Wheatridge, Colorado 80033, U.S.A.

MARCOTTE Denis: Research Associate, Dept. de Genie Mineral, Ecole Polytechnique, C.P. 6079 Succ. A, Montreal, Quebec H3C 3A, Canada

MARECHAL, Alain: Engineer, Societe Elf Aquitaine Prod., DOTI-Developpement, 64018 Pau, France

de MARSILY, Ghislain: Centre d'informatique Geologique, ENSMP, 35 rue St. Honore, 77305 Fontainebleau, France

MATHERON, Georges: Director, Centre de Geostatistique, ENSMP, 35 rue St. Honore, 77305 Fontainebleau, France

MILIORIS, George: Student, Dept. of Mining and Metallurgical Engineering, Mc Gill University, 3480 University St., Montreal, Quebec H3A 2A7, Canada

MOREAU, Alain: Student, Dept. de Genie Mineral, Ecole Polytechnique, C.P. 6079 Succ. A, Montreal, Quebec H3C 3A, Canada

MOUSSET-JONES, Pierre: Professor, Mackay School of Mines, University of Nevada, Reno, Nevada 89557, U.S.A.

MUGE, Fernando: Professor, CVRM - Instituto Superior Technico, Av. Rovisco Pais, 1096 Lisboa, Portugal

MYERS, Jeffrey C.: Student, Colorado School of Mines, Golden, Colorado 80401, U.S.A.

MYERS, Donald E.: Professor, Dept. of Mathematics, University of Arizona, Tucson, Arizona 85721, U.S.A.

NEUMAN, Schlomo P.: Professor, Dept. of Hydrology and Water Resources, University of Arizona, Tucson, Arizona 85721, U.S.A.

OMRE, Henning: Student, Dept. of Applied Earth Sciences, Stanford University, Stanford, California 94305, U.S.A.

OOSTERVELD, M.M.: Ore Evaluation Consultant, Anglo American Corp., 14 Kekewitch Drive, Kimberley 8301, South Africa

PARKER, Harry M.: Manager, Geology and Geostatistics, Fluor Mining and Metal Inc., 10 Twin Dolphin Drive, Redwood City, California 94065, U.S.A.

PAULSON, Gary D.: Sr. Minerals Geologist, Exxon Minerals Company, P.O. Box 4508, Houston, Texas 77210, U.S.A.

PEREIRA, Henrique G.: Professor, CVRM - Instituto Superior Technico, Av. Rovisco Pais, 1096 Lisboa, Portugal

RAO, A.R.: Professor, School of Civil Engineering, Purdue University, West Lafayette, Indiana 47907, U.S.A.

RAYMOND, Gary F.: Sr. Mine Project Engineer, Cominco, Ltd., P.O. Box 2000, Kimberley, British Colombia V1A 2G3, Canada

REMACRE, Armando: Student, Centre de Geostatistique, ENSMP, 35 rue St. Honore, 77305 Fontainebleau, France

RENARD, Didier: Research Engineer, Centre de Geostatistique, ENSMP, 35 rue St. Honore, 77305 Fontainebleau, France

RIVOIRARD, Jacques: Research Engineer, Centre de Geostatistique, ENSMP, 35 rue St. Honore, 77305 Fontainebleau, France

ROGADO, Jose Q.: Professor, CVRM - Instituto Superior Technico, Av. Rovisco Pais, 1096 Lisboa, Portugal

ROYER, Jean-Jacques: Research Engineer, CRPG-CNRS, 15 rue N.-D. des Pauvres, 54501 Vandoeuvre les Nancy, France

RUTLEDGE, Robert W.: 172 Burns Road, Turramurra, N.S.W. 2074, Australia

SABOURIN, Raymond L.: Mining Geologist and Geostatistician, Federal Government, 555 Booth St., Ottawa, Ontario K1A 0E8, Canada

SANDJIVY, Luc: Research Engineer, Centre de Geostatistique, ENSMP, 35 rue St. Honore, 77305 Fontainebleau, France

SANS, Henri: Geostatistician, COGEMA, B.P. 4, 78141 Velizy-Villacoublay, France

de Sarquis, Maria A.M.: Researcher, Inst. de Investigaciones Mineras, Universidad Nacional de San Juan, Avda. Libertador Gral. San Martin 1109 (oeste), 5400 San Juan, Argentina

SCHAEBEN, Helmut: Visiting Scholar, Dept. of Geology and Geophysics, University of California at Berkeley, Berkeley, California 94707, U.S.A.

SCHECK, Donald E.: Professor, Dept. of Industrial and Systems Engineering, 313 Engineering Bldg., Ohio University, Athens, Ohio 45701, U.S.A.

SINCLAIR, Alistair J.: Professor, Dept. Geological Sciences, University of British Columbia, Vancouver, British Colombia V6T 2B4, Canada

SINDING-LARSEN, Richard: Professor, Dept. of Geology, Trondheim University, 7001 Trondheim, Norway

SOARES, Amilcar O.: Research Fellow, CVRM - Instituto Superior Technico, Av. Rovisco Pais, 1096 Lisboa, Portugal

SOLOW, Andrew R.: Student, Dept. of Applied Earth Sciences, Stanford University, Stanford, California 94305, U.S.A.

SOULIE, Michel: Professor, Dept. de Genie Civil, Ecole Polytechnique, C.P. 6079 Succ. A, Montreal, Quebec H3C 3A7, Canada.

STEFFENS, Francois E.: Professor, Dept. of Statistics, UNISA, P.O. Box 392, Pretoria, South Africa

SULLIVAN Jeffrey: Student, Dept. of Applied Earth Sciences, Stanford University, Stanford, California 94305, U.S.A.

SWITZER, Paul: Professor, Dept. of Statistics, Stanford University, Stanford, California 94305, U.S.A.

TAHERI, Mojtaba S.: Principal Investigator - Geostatistician, Dept. de Ciencias de la Tierra, INTEVEP, Apartado 76343, Caracas - 1070A, Venezuela

TROCHU, Francois: Student, Dept. de Genie Civil, Ecole Polytechnique, C.P. 6079 Succ. A, Montreal, Quebec H3C 3A7, Canada.

UNAL, Ahmet: Student, Virginia Polytechnic Institute and State University, 213 Holden Hall, Blacksburst, Virginia 24061, U.S.A.

UYTTENDAELE, Michel: Student, Dept. de Geologie, Universite de Liege, 45 Avenue des Tilleuls, B-4000 Liege, Belgium

VALENTE, Jorge: Head of Mining Division, Paulo Abib Eng. S.A., R. Martim Francisco 331/402, CEP 30.000 - Bello Horizonte, Brazil

VERLY, Georges: Student, Dept. of Applied Earth Sciences, Stanford University, Stanford, California 94305, U.S.A.

VIERA-BRAGA, Luiz P.: Student, Dept. d'Informatique et de Recherche Operationnelle, Universite de Montreal, C.P. 6128, Succursale A,Montreal, Quebec H3C 3J, Canada

WEBSTER, Richard: Sr. Principal Scientific Officer, Rothamsted Experimental Station, Harpenden, Herts AL5 2JQ, England

YANCEY, James D.: Sr. Research Associate, Arco Oil & Gas Co., P.O. Box 2819, Dallas, Texas 75221, U.S.A.

YOUNG, Dae S.: Professor, Mining Engineering Dept., Michigan Technological University, Houghton, Michigan 49931, U.S.A.

IMPROVING THE ESTIMATION AND MODELLING OF THE VARIOGRAM

Margaret ARMSTRONG
Centre de Géostatistique et Morphologie Mathématique
Ecole Nationale Supérieure des Mines de Paris
35 rue Saint Honoré, 77305 FONTAINEBLEAU (France)

ABSTRACT - The efficacy of geostatistics depends, to a large extent, on the quality of the estimate obtained for the variogram. Several robust estimators of the variogram have been proposed recently. A close examination of many examples of non robust variograms suggests that this work has been going in the wrong direction. The variogram cloud approach used by Chauvet (5) seems a more fruitful starting point. Several ways of fitting variogram models using the variogram cloud will be discussed. The sensitivity of the ultimate kriging weights and the kriging variance to small changes in the variogram model or in the location of sample points will also be studied.

1. INTRODUCTION

Over the past few years, the geostatistical community has begun to feel the need for "robust" geostatistics. This is not to say that the problems are new. As early as 1965 Matheron warned of the variability of the experimental variogram for large values of h. More recently, Alfaro (1) investigated the variability of the experimental variogram for the normal distribution and for a few other distributions. So the question is by no means new. What has changed is that practicing geostatisticians have passed the first stage of learning how geostatistics works and have begun perfecting and refining the techniques they use. The second factor that has brought about the change is that professional statisticians familiar with "robust" statistical methods have begun applying their knowledge to the problem.

To date this has led to a rather piece-meal approach to the

G. Verly et al. (eds.), Geostatistics for Natural Resources Characterization, Part 1, 1–19.

question. Improvements have been made here and there but an overall approach has been lacking. It is to be hoped that the papers presented in this section of the workshop and the ensuing discussion will lead to a more unified view of the problems. With this in mind, the objective of this paper is to provide a schematic overview of the question, to indicate the main work that has already been done and to suggest points where improvements are needed. Some specific proposals will be made.

2. THE FIRST STEP : THE ROBUST VARIOGRAM

The main steps in a geostatistical study are represented schematically in Figure 1.

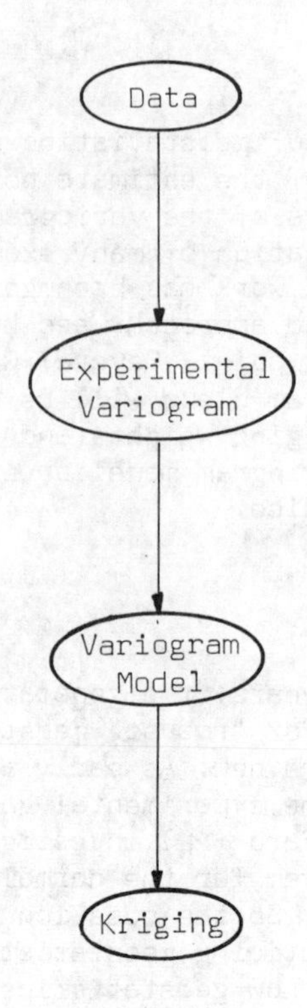

Figure 1. Steps involved in geostatistical studies

Before this conference the main thrust of the work on "robustness" was on various types of "robust" variograms. They therefore covered only the first stage of a study. To mention just a few, Cressie and Hawkins (7) advocated a fourth root transform variogram ; Armstrong and Delfiner (4) studied quantile variograms and "Huberized" variograms (i.e. using Huber's estimator of scale instead of the sample variance). Henley (11) proposed several non-parametric estimators. Other suggestions will be made in the course of this conference.

The methods proposed prior to the conference have one point in common : they all calculate a more robust estimator of the quantity $E\,(Z(x+h) - Z(x))^2$ from the experimental values of $(Z(x_i+h) - Z(x))^2$ in fixed distance classes.

Except for this point of similarity the "robust" variogram estimators differ widely in their choice of the estimator of $E(Z(x+h) - Z(x))^2$. As Cressie and Hawkins pointed out, this estimation problem can be considered from several points of view : that of estimating the mean or central tendency of the population of square differences or that of estimating the variance or scale parameter of the population of differences. The first stand-point leads to estimators such as the fourth root estimator proposed by Cressie and Hawkins, the quantile estimators (notably the median) of Armstrong and Delfiner, the non-parametric estimators (Henley) and the median absolute difference (Dowd, this volume). The second approach leads to estimators like the "Huberized" variogram.

One disadvantage of the estimators in the first category is that their measure of central tendency coincides with the mean only in the case of a symmetric distribution, and the distribution of square differences is far from this. The only way to determine the relationship between the mean and the measure of central tendency adopted is by supposing the distribution known. Some authors (e.g. Cressie and Hawkins) make an explicit choice of distribution which can then be judged on its merits. However, most simply ignore this point. And when the underlying distribution is not stated, two problems arise. Firstly, it is more difficult to evaluate the estimator. Secondly, and more importantly, there is no way to make a change of support.

The second approach avoids the problems coming from trying to apply robust methods to skew distributions by treating the distribution of differences rather than square differences, which can be considered to be symmetric. See Armstrong and Delfiner for details. However it is still interesting to know what sort of non-robust behaviour this has been designed for.

So when comparing the different proposals for robust variograms we can ask

1) Is the underlying distributional assumption stated explicitly ? If not, what distributional assumptions are needed to make this estimator optimal ?

2) Is this choice of distribution appropriate ? (This point will be considered in more detail in the next section).

3) What properties does the estimator have for this distribution ? How will these properties hold for other distributions ?

3. COMMON CAUSES OF NON ROBUST BEHAVIOUR

Whether they are explicitly stated or not, robust methods are designed to counter-act certain types of non-robust behaviour and not others. Clearly they cannot be expected to deal with every kind. In a previous section we emphasized the need to know what type of "non-robust" behaviour the estimator had been designed to handle. The other side of the question is to know what types of non-robust behaviour are commonly encountered.

This question has received much less attention than it deserves. Krige and Magri (13) studied the impact of a small percentage of outliers in gold data (which are approximately lognormal. Armstrong (2) encountered similar problems with outliers in coal data. Fang and Stark (18) considered the influence of a trend on the calculation of the variogram. Interesting as these studies are, they show only the tip of the iceberg.

To get a more comprehensive idea of the range of problems with variograms and some idea of their relative importance, the author collected all the troublesome cases that colleagues at Fontainebleau or elsewhere had encountered over a period of a couple of years. The results are presented in full elsewhere (Armstrong (3) . Table 1 summarizes the results, and suggests some solutions.

Although the table is fairly self-explanatory, a few comments may be helpful. The first cause cited is the presence of artefacts, that is, oddities introduced by the user. These include such silly mistakes as forgetting to test for missing values or setting all the zero values to 1.0000 E-31 before taking logs. These have to be dealt with case by case. Another common problem is an unsuitable choice of the distance classes. When samples are on an almost regular grid, this can lead to highly erratic variograms. This can be remedied by changing the distance class parameter. Another

Cause	Suggested Treatment
1. Artefacts	To be decided case by case
2. Poor choice of distance classes	Change variogram step length
3. Geographically distinct populations (or outliers)	Treat zones separately
4. Outliers, skew distributions and mixed populations	a) Treat separately if possible and if appropriate (e.g. eliminate outliers manually) b) Use "robust" variogram methods, or other methods of downweighting extreme values c) For data ranging over different orders of magnitude : change of scale methods (e.g. log-normal kriging, etc.)

TABLE 1

easy to treat problem is that of geographically distinct populations or outliers. While these are easily overcome (once they have been identified) others are not so easy to deal with. Outliers, skew distributions and mixed populations fall into this category. In many ways the distinction between "outliers", "skew distributions" and "mixed populations" is rather tenuous. Values which might appear as outliers in a relatively small sample, might well seem like part of the tail of a skew distribution or as a separate population if more data become available. So it is hardly surprising to find similar methods used in all these categories.

Before any action is taken, one has to decide whether the outliers (etc) are an essential (inseparable) part of the phenomenon under study or whether they are extraneous and should be removed. Examples of the latter are outliers due to typographical errors, or mixed populations related to different sampling campaigns.

However, in cases like gold or uranium where small quantities of very high grade material are mixed in with a matrix of waste, the ultimate estimation procedure must take account of both types of material. The few percent of high grade material ought not be treated as a contaminant which is polluting the data sets, as is done in the contaminated normal model. While this model is appropriate in cases where a few aberrant values have slipped in among the population to be estimated, as can happen in physical sciences, this is not the case here. For one thing, the miners are far more interested by that small proportion of ore that is "contaminating" the waste. But more importantly, the essential question is to estimate some parameters of these inextricably mixed populations rather than to estimate one separately from the influence of the other. So the author of this article would question the appropriateness of the contaminated normal model.

But it is nevertheless true that the few high values completely dominate the variogram and make it numerically unstable. Something has to be done. Three broad types of solution exist :

(1) downweighting the high values

(2) robust variograms

(3) change of scale methods such as lognormal kriging

The first two types both work by downweighting the extreme values by more or less sophisticated methods, while the third variety diminishes the effect of large values by transforming them.

The objective of collecting these troublesome cases together had been to see what type of problems produce non-robust behaviour in variograms and hence to see what types of solutions were required, more particularly to see what types of non-robust behaviour need to be covered by the model. A close study of these cases suggested the following conclusions :

- Given the wide variety of types of non-robust behaviour, a careful study of the data using quite ordinary data cleaning methods is indispensable since this clears up the vast majority of problems. In particular, the idea that robust geostatistics might one day eliminate the need for a critical structural analysis should be opposed in the firmest terms possible.

- Some types of non robust behaviour (notably the presence of outliers) which are the same from the statistical point of view have to be treated in very different ways depending on their nature. Outliers due to typographical errors should be eliminated completely, not merely down-weighted as is done by robust techniques whereas extremely high gold or uranium grades should be retained. Clearly, robust techniques should only be applied when the cause of the non-robust behaviour is known and should be chosen in function of this knowledge.

- The sensitivity of the variogram to the choice of the distance classes suggests that the present approach to "robust" variogram estimators based on fixed distance classes may be misguided. It would be better to work on the whole variogram cloud (i.e. the cross plot of square differences against the distance) as was suggested by Chauvet (5).

- Since the final aim of a geostatistical study is kriging (estimation) and not obtaining the experimental variogram, one could ask whether a more overall definition of robustness is required. The idea of "robust kriging" follows logically from this. Since this will be dealt with by other speakers it will not be treated here. While "robust kriging" represents a step towards a more overall view of "robustness", it is still not general enough. In particular it ignores the important question of the sensitivity of the ultimate results of kriging to the choice of the variogram model and to the location of the data. This will be discussed in a later section.

Before concluding this section, some questions which, in the author's opinion, need additional research will be mentioned .

1) More work is needed on the causes of erratic experi-

mental variograms. This would be doubly helpful : firstly students of geostatistics can learn a lot from other people's problems and secondly research into robust geostatistics needs this information to progress.

2) The only model for non-robust behaviour proposed to date is the contaminated normal. Even those who are in favour of this model would have to agree that this is a rather limited range. More models are needed and they have to be based on geostatistical experience rather then being drawn from classical statistical literature.

3) Looking at Figure 1 we see that there has been some work on the first step in a geostatistical study (in the form of the robust variogram) and more recently on the third step. But even "robust" kriging presupposes that the variogram model has been chosen. Unfortunately little work seems to have been done on the second step. A tentative suggestion on this subject will be made in a later section.

We now go back to study the sensitivity of the results of the kriging to the location of the data points and to the choice of the variogram model.

4. THE SENSIVITY OF KRIGING, AND THE KRIGING SYSTEM

In his book on robust statistical procedures Huber [12] defined "robustness" as signifying insensitivity against small deviations from the a ssumptions and went on to comment that statisticians are usually interested in distributional robustness (i.e. when the shape of the true underlying distribution deviates slightly from the assumed model, usually the Gaussian law) The existing proposals for "robust" variograms and "robust" kriging are natural developments from this statistical vision of robustness.

However this view of "robustness" is too narrow for geostatistics since it overlooks several equally important aspects of the question such as the sensitivity of the kriging system to the choice of the variogram model and to the location of samples.

Going further one could ask to what extent the kriging weights and the kriging variance depend on these two factors. Most of the time, the results of kriging are fairly stable. However since this is not always so, it would be better to be able to quantify the robustness of kriging to small perturbations. This would help identify potentially unstable cases and allow the user to take appropriate action.

Before launching into the mathematics, two examples will be presented to show how bad things can be. (These were taken from Diamond and Armstrong (9)).

EXAMPLE 1 : Kriging a 1.0 x 1.0 square centred on the origin from three sample points ((- .4, 0) , (.4, 0) , (.39, .1)) using a Gaussian variogram with parameter a = 1, with and without a nugget effect.

Kriging Weight for	No Nugget Effect Sill = 1.0- Nugget effect = 0	1% Nugget Effect Sill = 0.99 Nugget Effect = 0.01
(-.4, 0)	4982	.4954
(.4, 0)	.3970	.3249
(.39,.1)	.1048	.1797
Kriging Variance	.0184	.0222

TABLE 2 : Kriging weights and kriging variances for two variogram models.

EXAMPLE 2 : Kriging the same 1.0 x 1.0 square using the same Gaussian model (with no nugget effect) using three sample points. Two sets of data locations considered.

Data Points	Kriging Weights	Data Points	Kriging Weights
(-.4, 0)	.4982	(-.4, 0)	.4998
(.4, 0)	.3970	(.4, 0)	.4797
(.39,.1)	.1048	(.41,.1)	.0245
Kriging Variance	.0184	Kriging Variance	0186

TABLE 3 : Kriging weights and kriging variances on moving one sample point.

These examples have been "cooked" to produce dramatic results. But the fact still remains : even a small system (4 x 4) can be as ill-conditioned as this. The mind boggles at what might happen with large data sets of unfavourably located samples.

The problem is to find a mathematical way of characterizing the change in the kriging weights and the kriging variance provoked by a small change in the variogram model, and later by a change in the location of the data. This begs the question of what is meant by a "small" change in a variogram model.

Definition of a δ neighbourhood of a variogram : a valid (isotropic) variogram g(h) lies in the δ-neighbourhood $N_\delta(\gamma)$ of the variogram $\gamma(h)$ if

$$1 - \delta \leq g(h)/\gamma(h) \leq 1 + \delta \quad \begin{cases} 0 < \delta < 1 \\ 0 \leqslant h < \infty \end{cases} \tag{1}$$

Suppose that a block V is to be kriged from N point samples located at x_i, i = 1, 2,...,N. The kriging system corresponding to the variogram γ is

$$\Gamma X_\gamma = B_\gamma \tag{2}$$

where $X_\gamma^T = (\lambda_1, \ldots, \lambda_N, \mu)$

$B_\gamma^T = (\overline{\gamma}(x_1, V), \ldots, \overline{\gamma}(x_N, V), 1)$

and
$$\Gamma = \begin{bmatrix} \gamma(x_1,x_1) & \ldots\ldots\ldots & \gamma(x_1,x_N), & 1 \\ & & & \\ \gamma(x_1,x_N) & \ldots\ldots\ldots & \gamma(x_N,x_N) & 1 \\ 1 & \ldots\ldots\ldots & 1 & 0 \end{bmatrix}$$

Similarly the system corresponding to g can be written as $G X_g = B_g$.

From (1), integration over V gives, for i = 1, 2,...,N,

$$(1 - \delta)\ \overline{\gamma}(x_i, V) < \overline{g}(x_i, V) < (1 + \delta)\ \overline{\gamma}(x_i, V) \tag{3}$$

whence

$$(1 - \delta)\ \|B_\gamma\| < \|B_g\| < (1 + \ )\ \|B_\gamma\| \tag{4}$$

where $\|B\|$ is some L(p) vector norm (usually with p = 1,2 or ∞).

A similar argument gives

$$(1 - \delta) \|\Gamma\| < \|G\| < (1 + \delta) \|\Gamma\| \qquad (5)$$

where $\|G\|$ is the matrix norm consistent with $\|Y\|$ defined by $\|G\| = \sup_{\|Y\|=1} \|GY\|$.

On letting $\Delta X = X_\gamma - X_g$, $\Delta B = B_\gamma - B_g$ and $\Delta\Gamma = \Gamma - G$, it can be shown that

$$\|\Delta X\| / \|X\| \leq 2\delta \ K(\Gamma)/(1 - \delta K(\Gamma)) \qquad (6)$$

where $K(\Gamma)$ is the conditioning number of the matrix Γ. The details of the matrix algebra are given in Diamond and Armstrong (9) (Note that $\delta K(\Gamma)$ must be less than 1 to ensure the non singularity of one of the intermediate matrices).

The difference between the kriging variances $|\sigma_\gamma^2 - \sigma_g^2|$ can also be expressed in terms of δ , $K(\Gamma)$ and $\overline{\gamma}(V,V)$. Diamond and Armstrong also obtained similar formulae for the case of a perturbation in the location of the sample points.

This means that the relative change in the kriging weights, $\|\Delta X\| / \|X\|$, and the change in the kriging variance can now be quantified for all the variograms belonging to a δ-neighbourhood. The fact that the upper bounds obtained are given in terms of the conditioning number of the kriging matrix Γ is hardly surprising. Since the conditioning number increases with the size of the system and when rows are very similar (as happens with clustered data points), one can expect problems in either of these cases. This merely provides a mathematical justification of the standard geostatistical practices of limiting the size of kriging systems and regrouping clusters of points.

One important conclusion from the theoretical work is that it suggests that it ought to be standard practice to calculate the conditioning number of kriging matrices. In cases where the value found is too high, the geostatistician would then be able to take suitable action. This might involve re-thinking the choice of the variogram model (in particular avoiding troublesome ones like the Gaussian), or increasing the nugget effect, or limiting the number of points taken into consideration.

One negative aspect of the work to date is that it is limited to variograms belonging to the same δ - neighbourhood. This definition of similarity in variograms regroups variograms which look alike, but it also means that they must have the same sort of behaviour at the origin. This is an unfortunate limitation, since it completely excludes models with a non-zero nugget effect from the neighbourhood of models without one. A more flexible choice of the definition of a δ neighbourhood might allow this case to be included as well, but it would probably result in unacceptable large limits for $\| \Delta X \| / \| X \|$ and for $| \sigma_g^2 - \sigma_\gamma^2 |$.

As was mentioned in section 3, two specific aspects of "robustness" in the wider sense of the word will be discussed in this article. The first; the sensitivity of the kriging system, covers the whole of the geostatistical study from step 1 right through to step 3 (Figure 1) but it assumes that the variogram model has been chosen - or at least lies within certain bounds. The next section discusses how this model can be chosen.

5. WHY BOTHER ABOUT ALTERNATIVE WAYS OF FITTING VARIOGRAMS ?

Although some work has been done on improved ways of calculating the experimental variogram and on making kriging more robust once the variogram is known, not much has been done on alternative ways of fitting a model to the experimental variogram. It is true that various types of cross validation procedures have been developed (Delfiner (8), Chung (6)) and that others will be presented at this conference. But these tend to be fairly heavy from the computational point of view.

Clearly the main reason for this lack of interest in alternative fitting methods is that users are satisfied with the traditional "by eye" approach. This does have the tremendous advantage over statistical methods of allowing the user to include a lot of information that is difficult to quantify. Although this lack of interest can be justified for stationary (or intrinsic) data, the situation is quite different with IRF-k. At present the degree of the drift and the variogram (or generalized covariance) are fitted by a combination of weighted least squares and cross-validation (jack-knife) techniques. It is therefore interesting to have a closer look at alternative fitting methods in the simpler stationary case, if only so that they can be applied later in the more complicated case. This may make it possible to enlarge the range of covariances that can be used, to include other models, notably those with a sill.

The Problems with Least Squares.

Since the least squares method is undoubtedly the best known curve fitting method among non-statisticians, it is a good starting point in the search for alternative methods. Restating its disadvantages, even though they are well-known, will help to clarify what is required of a good variogram fitting technique.

1) The first problem with least squares is that if it is applied blindly to fit arbitrarily chosen functions such as polynomials, it can lead to the selection of an inadmissible model (i.e. one which does not have the appropriate properties of positive definiteness). This can be overcome either by working within a family of acceptable models (e.g. exponential or spherical models) or alternatively by eliminating inadmissible models afterwards, as is done in BLUEPACK.

2) Least squares assumes that the errors (i.e. the differences between the function being fitted and its observed value) are independent. Since the observed values here are square differences, this assumption is untenable. It is worth noting that although least squares does not make any explicit distributional assumptions (hence its intuitive appeal) it is optimal only when the errors are normally distributed.

3) In the simplest form of least squares, all the errors are given equal weighting. Clearly this is far from ideal. For a start, since the behaviour of the variogram near the origin has a decisive influence on the results of the kriging (as was shown by Example 1 in Section 3), the quality fit is particularly important for short distances. This is reinforced by the fact that the kriging configuration (usually) only includes points up to a certain distance and so the fit of the variogram for large distances is irrelevant. As well as this, once a reasonable estimate of the sill has been obtained, additional values at large distances do not add much extra information. From this it is clear that the variogram values near the origin should receive a much heavier weighting than distant values.

4) It is difficult to incorporate geological information into the fit. This criticism applies equally to other statistical methods as well.

It would be possible to overcome all these objections and still use least squares. This can be done by specifying the type of model to be fitted (e.g. the sum of 1 or 2 exponential models plus a nugget effect), by choosing an appropriate weighting function and applying the least squares approach to the whole variogram cloud after it has been thoroughly cleaned and checked. The equations are not difficult to derive.

Let y_i denote the ith square difference corresponding to two points distant h_i apart. Suppose that the model to be fitted is an exponential plus a nugget effect. That is $\gamma(h) = C_o + C - C \exp(-h/a)$ for $h > 0$. On putting $C_o + C = C'$ and $1/a = a'$, the sum of squares Q becomes

$$Q = \sum w_i (y_i - C' + C \exp(-h_i a'))^2 \tag{7}$$

where w_i is the weighting factor for a distance h_i. On differentiating and setting the derivatives to zero, the following equations are obtained :

$$\begin{aligned}
&\sum w_i y_i - C' \sum w_i + C \sum w_i e^{-h_j a'} = 0 \\
&\sum w_i y_i e^{-h_i a'} - C' \sum w_i e^{-h_i a'} + C \sum w_i e^{-2h_i a'} = 0 \\
&\sum w_i y_i h_i e^{-h_i a'} - C' \sum w_i h_i e^{h_i a'} + C \sum w_i h_i e^{-2h_i a'} = 0
\end{aligned} \tag{8}$$

The fact that the parameter a (or a') appears in the exponential makes these difficult to solve. A similar but even nastier problem arises with the spherical model. In that case, the contribution of any value to the sum of squares depends on whether the distance h_i is greater than the range or not. Consequently in both these cases it is possible to minimize the sum of squares in a two stage procedure by solving the equations for a fixed value of "a" and then letting "a" vary, but this a very unwieldy procedure. If two or more nested structures are used one would have great difficulty finding the minimum.

Clearly, least squares looked at from this point of view, is utterly unworkable. However a suitable transformation may facilitate things.

Weighted Area Methods.

The function that geostatisticians call an exponential variogram occurs in biology under the name of the Von Bertalanffy growth curve. Of the many methods proposed for fitting this curve, the weighted area technique of Leedow and Tweedie (14) seems the most interesting for geostatistics. The essential idea is to fit the model $\gamma(h)$ by equating the weighted area under the curve

$$\int_0^\infty \gamma(h)\, s e^{-sh}\, dh$$

with the observed quantity. In other words, the theoretical Laplace transform is equated with the observed one. From the geostatistical point of view the weighting function se^{-sh} is ideal as it downweights the tail of the variogram.

The technique can best be illustrated by an example. Suppose that the exponential variogram $\gamma(h) = C(1 - \exp(-ha'))$ is to be fitted to the variogram cloud (Note a' replaces 1/a as before). Its theoretical Laplace transform is

$$\mathcal{L}(s) = \frac{Ca}{s+a} \tag{9}$$

The observed Laplace transforms can be calculated for a fixed value of s using the formula

$$\mathcal{L}^*(s) = \sum_{i=1} y_i \left(e^{-sh_{i-1}} - e^{-sh_i}\right) \tag{10}$$

where y_i is, as before, the square difference between two points distant h_i apart. The indices i must be arranged in order of h increasing. (Note : this formula can also be applied to the ordinary experimental variogram in which case y_i should be replaced by the average variogram value for all the couples in the ith distance class (h_{i-1}, h_i). The end corrections mentioned later would then be advisable).

If the observed transforms $\mathcal{L}^*(s)$ corresponding to two values of s (s_1 and s_2) are denoted by L_1 and L_2 then the estimates of C and a can be found from the equations

$$L_1 = \frac{C\,a}{s_1 + a} \qquad L_2 = \frac{Ca}{s_2 + a} \tag{11}$$

Hence

$$\hat{a} = -\frac{L_2\, s_2 - L_1\, s_1}{L_1 - L_2} \qquad \hat{C} = \frac{L_1\, L_2\, (s_1 - s_2)}{L_1\, s_1 - L_2\, s_2} \tag{12}$$

These equations are clearly much simpler to solve than the preceding least squares ones. The point is that a suitably chosen transformation can convert a non-linear problem into a much simpler linear one.

Leedow and Tweedie also consider the 3 parameter growth curve which is equivalent to the sum of an exponential model plus a nugget effect. The resulting three equations are also linear in s and L. Furthermore they suggest some end corrections which would be very useful when the first value is far from zero or when the last value is below the sill.

As the solution to (12) depends on the values chosen for s_1 and s_2 , one can ask how they should be chosen. Morgan and Tweedie (16) suggest choosing s by calculating the likelihood corresponding to the parameters calculated for each set of s and choosing the set that maximizes the likelihood. They found that in many models it was best to choose the values of s close together. In this case presented here, maximizing the likelihood reduces to a one dimensional search along the diagonal of the s-space.

It is interesting to compare this approach with least squares. The simplest way to do this is by assuming the errors to be independent, normally distributed variables, in which case maximum likelihood is equivalent to least squares. After extensive simulation trials on both the two and three parameter models, Leedow and Tweedie found that the weighted area technique (with end corrections) had a high relative efficiency (from 75% to 85%). So it can certainly be recommended if it is felt that the errors are normally distributed.

Unfortunately in geostatistics, as in fish growth curves, this assumption seems implausible. However, as Leedow and Tweedie noted, "In that situation, the distribution-free or moment-like properties of the weighted area method are an advantage, in that it restricts parameter estimates to a sensible range".

Additional information on weighted area techniques (notably their statistical properties) is given in (10), (17).

From this discussion, even though the research on this in geostatistics is still at an early stage, weighted area techniques seem to be a promising method for fitting variogram models. Certainly the use of an exponential weighting function (i.e. using Laplace transforms) looks feasible for the exponential model. However the situation is not so rosy where the spherical is concerned. Unfortunately the Laplace transform of the spherical model

$$\gamma(h) = C\left(\frac{3h}{2a} - \frac{h^3}{2a^3}\right) \tag{13}$$

is

$$\mathcal{L}(s) = C\left\{\frac{3}{2\,as} - \frac{1}{2(as)^3} + 3\frac{e^{-as}}{(as)^2}\left(1 + \frac{1}{as}\right)\right\} \tag{14}$$

The non-linearity of the Laplace transform makes it much more difficult to solve and hence , for our purposes quite uninteresting. However a more suitably chosen weighting factor may overcome this problem.

6. DISCUSSION AND CONCLUSIONS

Geostatisticians are well aware of the crucial role of the variogram model in the overall geostatistical estimation procedure. The concept of a δ neighbourhood around a variogram merely provides a theoretical framework for the idea that similar-looking variograms ought to lead to similar results from kriging. It also gives specific bounds to the relative change in the kriging weights and the kriging variance resulting from a slight perturbation in either the variogram model or the location of the data points. Implicitly it indicates how fussy we need to be when fitting a model. Once the model is known up to a δ neighbourhood, the results of the kriging are not going to change much. However this implicitly requires a good knowledge (or rather, a suitable choice) of the nugget effect, particularly if the latter is small relative to the total sill. In the long run, these ideas also have implications for any goodness of fit criterion proposed for fitting variogram models. One would want any such criterion to be consistent in the sense that two different models belonging to the same $\hat{\delta}$ neighbourhood ought to have similar "goodness of fit" statistics since they would produce comparable results in the kriging.

This notion of the sensitivity of the kriging system to minor perturbations is just one aspect of the wider question of "robustness". Unfortunately most of the work to date has taken a narrower vision of the problem. Suggestions have been made to "robustify" particular steps in the geostatistical estimation procedure, generally by inserting robust statistical procedures, but an overall approach to the question has been lacking.

In the author's opinion, it is time to rethink the whole question of robustness from the start. We need to ask ourselves several questions :

1) Do we need more sophisticated methods ? The author feels that at least 80% of the problems encountered in getting a sensible estimate of the experimental variogram can be overcome using common sense and ordinary data cleaning techniques. Moreover she fears that the undiscerning use of "robust" geostatistics would lead to "papering over the cracks" to quote Tukey (19). In cases where our answer to the first question is affirmative we should then ask :

2) What do we need ?

(i) a robust estimate of the experimental variogram

(ii) a robust way of fitting the variogram model to either the experimental variogram or the variogram cloud

(iii) a more robust type of kriging

Then before developing any new procedures, we have to decide

3) What kind of non-robust behaviour should this new procedure be designed to counter-act ? To do this, a model of non-robustness is needed. In choosing one, we should take heed of Tukey's advice (19) : "To serve us well, the model has to adequately portray the behaviour of the measurements as they really are. It is not enough to represent how we wish the measurements had been (but were not)". A harsh message, but a vital one.

REFERENCES

(1) Alfaro, M., 1979 : "Etude de la robustesse des simulations de fonctions aléatoires". Doc. Ing. Thesis, Ecole des Mines de Paris, 161 p.

(2) Armstrong, M., 1980 : "Application de la géostatistique aux problèmes de l'estimation du charbon". Doc. Ing. Thesis, Ecole des Mines de Paris, 180 p.

(3) Armstrong, M., 1982 : "Commoner problems seen in variograms". Internal note n° 774, C.G.M.M. (in press).

(4) Armstrong, M. and Delfiner, P., 1980 : "Towards a more robust variogram : a case study on coal". Internal note n° N-671, C.G.M.M., Fontainebleau.

(5) Chauvet, P., 1982 : "The variogram cloud". 17th APCOM Symposium, Colorado School of Mines, Golden, Colorado.

(6) Chung, C.F., 1984 : "Use of the Jack-knife Method to Estimate Auto-correlation Functions (or variograms). Lake Tahoe NATO A.S.I. Series, Reidel Publ. Corp., Dordrecht.p. 55.

(7) Cressie, N. and Hawkins, J., 1980 : "Robust estimation of the variogram". Int. J. of Math. Geol., 12, 2, pp. 115-126.

(8) Delfiner, P., 1976 : "Linear estimation of non-stationary spatial phenomena". In "Advanced Geostatistics in the Mining Industry", M. Guarascio (Ed.), NATO ASI Series, Reidel Publ. Co., Dordrecht, Holland, pp. 49-68.

(9) Diamond, P. and Armstrong, M., 1983 : "Robustness of Variograms and conditioning of kriging matrices". Internal Note n° 804, CGMM, Fontainebleau (Submitted for publication).

(10) Fergin, P.D., Tweedie, R.L.* and Belyea, C., 1981 : "Explicit parameter estimation in hierarchical models using time-weighted linear combinations of observations". (Submitted for publication).

(11) Henley, S., 1981 : "Non-Parametric Geostatistics", Applied Science Publ., Ltd., London.

(12) Huber, P.J., 1977 : "Robust statistical procedures" CBMS-NSF, S.I.A.M., 56 p.

(13) Krige, D.G., Magri, E.J., 1982 : "Studies of the effect of outliers and data transformation on variogram estimates for a base metal and a gold orebody". Int. J. of Math. Geol. 14, 6, pp. 557-564.
(14) Leedow, M.I. and Tweedie, R.L. , 1981 : "Weighted area techniques for the estimation of von Bertalanffy growth curve parameters". (Submitted for publication).
(15) Matheron, G., 1965 : "Les Variables régionalisées et leur estimation".Doc. Ing. Thesis, Masson, Paris, 306 p.
(16) Morgan, B.J.T. and Tweedie, R.L , 1982 : "Stable parameter estimation using Laplace transforms in non-linear regression". (Submitted for publication).
(17) Schuh, H.J., and Tweedie, R.L. , 1979 : "Parameter estimation using transform estimation in time-evolving models". Math. Biosci., 45, pp. 37-67.
(18) Stark, T.H., and Fang, J.H., 1982 : "The effect of drift on the experimental semi-variogram". Int. J. of Math. Geol. 14, 4, pp. 309-320.
(19) Tukey, J.W., 1973 : "Introduction to today's data analysis". Technical Report n° 40, Series D, Statistics Dept., Princeton University. (Presented at the Conference on Critical Engineering of Chemical and Physical Structural Information, Dartmouth College).

TOWARDS RESISTANT GEOSTATISTICS

Noel Cressie

The Flinders University of South Australia

ABSTRACT
Implicit in many of the geostatistical techniques developed, is faith that the data are Gaussian (normal) or can be conveniently transformed to Gaussianity. A mining engineer knows however that contamination is present all too often. This paper will make an exploratory analysis of spatial data, graphing and summarizing in a way that is resistant to that contamination. The ultimate goal is variogram estimation and kriging, and the data analytic techniques used reflect this. The pocket plot is a new way of looking at contributions of small regions of points to variogram estimation. The gridded data are thought of as a higher way table and analysed by median polish. The residual table is shown to contain useful information on the spatial relationships between data. Coal ash measurements in Pennsylvania are analysed from this resistant point of view.

Keywords: box plot, EDA, inter-quartile range, kriging, median polish, plus-one-fit, pocket plot, resistance, spatial data, square root differences, stationarity, stem and leaf plot, variogram.

INTRODUCTION

For some years now we have held the view that geostatistics could profit considerably from the philosophy and methods of modern data analysis, as it currently pervades the mainstream of applied statistics. A lot of time, effort and money is put into geological exploration, but very little into *statistical exploration*. Of course the data are always looked at and summarized,

G. Verly et al. (eds.), Geostatistics for Natural Resources Characterization, Part 1, 21–44.

but mostly in rather classical ways, as if they were the result of a random sample or a time series, rather than as a spatial collection of dependent random variables whose dependence is strongly tied to relative spatial locations. The goal of this paper is to present an *exploratory data analysis* for *spatial data*, suited to the needs of mining engineers and geostatisticians.

By its very nature, exploratory data analysis must not trust every observation equally, but rather isolate for further perusal observations which look suspiciously atypical. Having said this, there is the implication that we have some underlying stochastic model in mind, from which departures need to be clearly identified. Indeed we have; any of us who has drawn a histogram (or stem and leaf plot) of a batch of numbers must know how much simpler our statistical analysis is when the histogram appears bell-shaped with no unusually large or small observations to concern us. In the dependent spatial context, our underlying stochastic model is that all the data (which are observed in a comparable way) come from a joint normal (or Gaussian) distribution whose correlation structure depends only on the *spatial relationships* between the data. When one allows for the possibility of an initial (normalizing) transformation on the observations, this underlying model is quite general; it is also the classical one from which all geostatistical inference has developed. But most importantly and most realistically, one will expect that some observations will be atypical due to an anomaly or a recording error or ···. This we will model, by adding to the underlying Gaussian model a small component of "contamination".

Data analytic methods that graph and summarize the way the observations act and interact, should first of all be relatively unaffected by the presence of contamination, and secondly be able to highlight the contaminant from the rest. Such methods are called *resistant*, and that part of any geostatistical analysis which uses them, can safely be called *resistant geostatistics*. There is a very important philosophical point to be made here, namely that it is the underlying Gaussian component which is of primary interest, and ultimately it is *its* parameters which need to be estimated.

In any geostatistical study, there should ideally be inputs from many different areas of expertise. The team should at the very least include a geologist, a mining engineer, a metallurgist, a statistician, and a financial manager. This paper will be written from a statistician's point of view, but in no way do we wish to detract from the role of the others in the complete study. The statistician can typically be expected to lead the team through the following five stages.

1. Graphing and summarizing the data.

2. Detecting and allowing for non-stationarity.
3. Estimating spatial relationships, usually through the variogram (sometimes called structural analysis).
4. Estimating the *in situ* resources, usually through kriging.
5. Assessing the recoverable reserves.

When we started researching this paper, we had hoped to pass quickly to stage 3, but we found those crucial first two stages of exploratory data analysis somewhat lacking. Most of what we are to present involves a closer look at stages 1 and 2, as it pertains to the central problem of variogram estimation. Our research *per se* on variogram estimation and kriging has yielded a number of interesting results, which unfortunately, due to lack of space, will have to be published elsewhere. Those who would like an introduction to these ideas should refer to Cressie (1979), Cressie and Hawkins (1980) and Hawkins and Cressie (1984).

The present paper will use coal ash data from Gomez and Hazen (1970, Tables 19 and 20) to illustrate the techniques. The data are discussed, graphed and summarized from an exploratory data analysis (EDA) point of view (Tukey, 1977) in Section 2. Section 3 introduces more sophisticated data analytic techniques, such as transformations and a new plot which detects small pockets of non-stationarity (called a *pocket plot*). Further exploitation of the spatial nature of the data allows simple adjustments to be made for non-stationarity. Usual objections that the bias of the classical variogram estimator becomes prohibitive, are countered by appropriate first moment calculations. These two aspects are considered in Section 4. Section 5 compares the residual variogram to the original variogram, and indicates what conclusions are possible from the coal ash data.

2. THE DATA SET

Data on coal ash obtained from Gomez and Hazen (1970, Tables 19 and 20), for the Robena Mine Property in Greene County, Pennsylvania, serve to illustrate the methods of this paper. Other data sets have received equal treatment with similar success; the only requirement really is that the observations occur on a regular or near to regular grid. We have used the 208 coal ash measurements at core locations with west co-ordinates greater than 64,000 feet; spatially this defines an approximately square grid, with 2,500 feet spacing, running roughly SW-NE and NW-SE. Figure 1 displays the core locations. In order to talk easily and sensibly about directions, we decided to reorient the grid so that it appears to run in an E-W, N-S direction. Figure 2 shows the coal ash measurements and their spatial location; comparison of Figures 1 and 2 will show how the grid was reoriented.

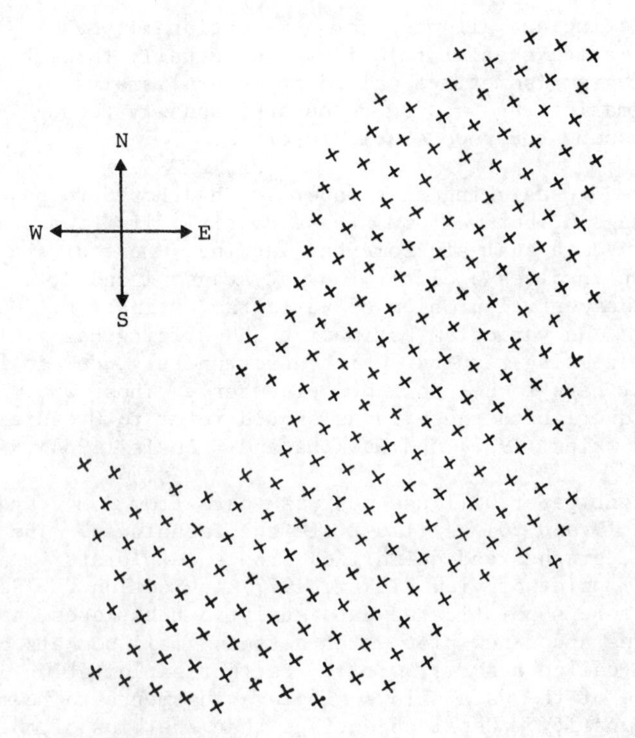

Figure 1: Core locations of coal ash data.

From now on we will refer to the locations of the cores according to the orientation of Figure 2. What follows is an exploratory analysis of the spatial data using techniques of EDA (Tukey, 1977), as well as new techniques very much in the spirit of EDA.

If we ignore the spatial nature of the observations, and treat them simply as a batch of numbers, then the stem and leaf plot gives some idea of distribution shape. (Implicit in doing this is the belief that the data are stationary, something which is often not the case); see Figure 3. Clearly there is one unusually large observation, and several others on the high side and on the low side which are suspect. But as analysers of *spatial* data, we can only be suspicious of observations when they are unusual *with respect to their neighbours*. Implicitly this is saying that we believe the data to be governed by some sort of *local* stationarity (which will in general not guarantee a global stationarity across the ore body). It has been observed that a proportional effect variogram model (Journel and Huijbregts, 1978) is sometimes adequate to deal with local stationarity. Our starting point is more general.

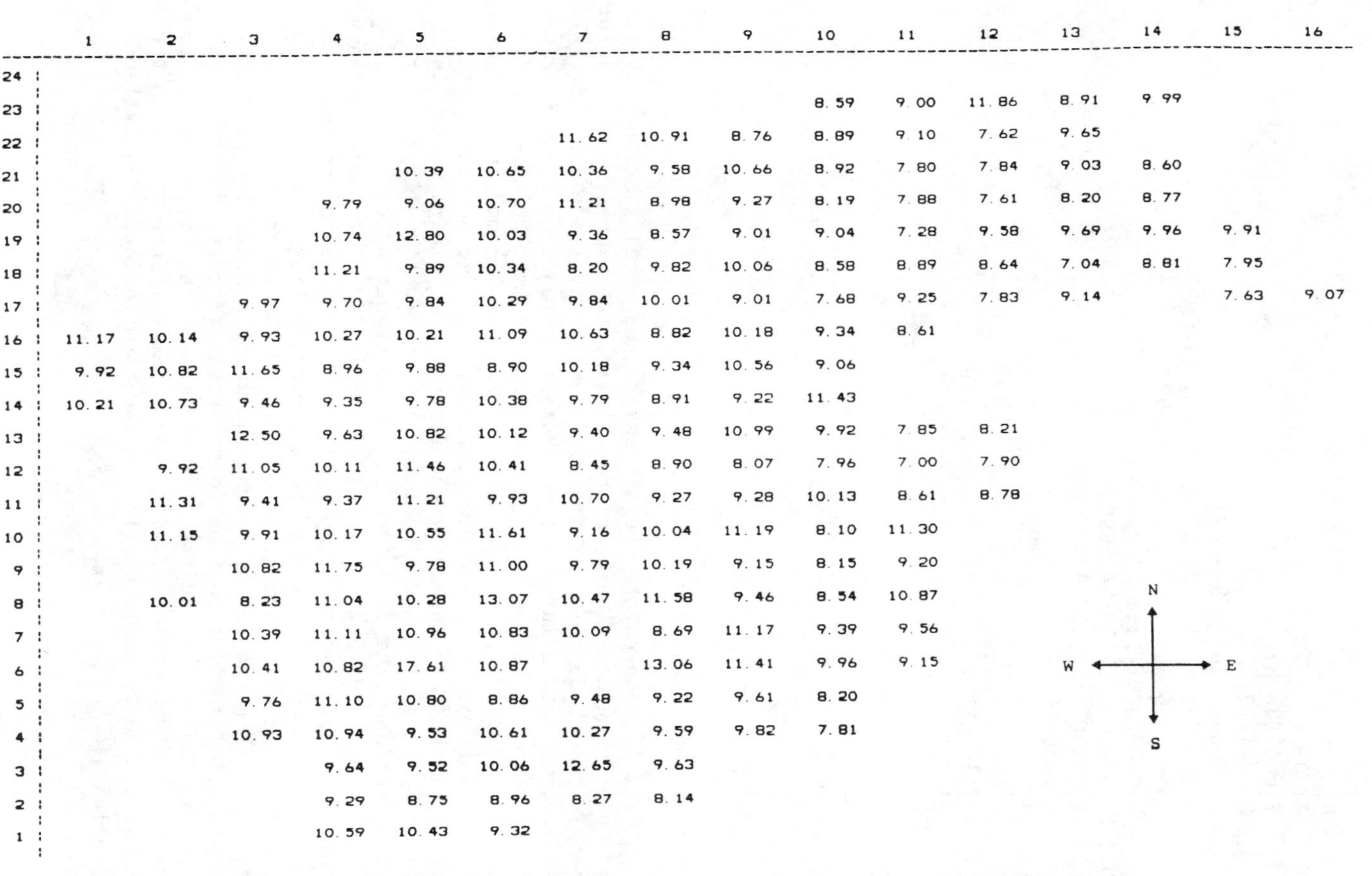

	1	2	3	4	5	6	7	8	9	10	11	12	13	14	15	16
24																
23										8.59	9.00	11.86	8.91	9.99		
22							11.62	10.91	8.76	8.89	9.10	7.62	9.65			
21					10.39	10.65	10.36	9.58	10.66	8.92	7.80	7.84	9.03	8.60		
20				9.79	9.06	10.70	11.21	8.98	9.27	8.19	7.88	7.61	8.20	8.77		
19				10.74	12.80	10.03	9.36	8.57	9.01	9.04	7.28	9.58	9.69	9.96	9.91	
18				11.21	9.89	10.34	8.20	9.82	10.06	8.58	8.89	8.64	7.04	8.81	7.95	
17			9.97	9.70	9.84	10.29	9.84	10.01	9.01	7.68	9.25	7.83	9.14		7.63	9.07
16	11.17	10.14	9.93	10.27	10.21	11.09	10.63	8.82	10.18	9.34	8.61					
15	9.92	10.82	11.65	8.96	9.88	8.90	10.18	9.34	10.56	9.06						
14	10.21	10.73	9.46	9.35	9.78	10.38	9.79	8.91	9.22	11.43						
13			12.50	9.63	10.82	10.12	9.40	9.48	10.99	9.92	7.85	8.21				
12		9.92	11.05	10.11	11.46	10.41	8.45	8.90	8.07	7.96	7.00	7.90				
11		11.31	9.41	9.37	11.21	9.93	10.70	9.27	9.28	10.13	8.61	8.78				
10		11.15	9.91	10.17	10.55	11.61	9.16	10.04	11.19	8.10	11.30					
9			10.82	11.75	9.78	11.00	9.79	10.19	9.15	8.15	9.20					
8		10.01	8.23	11.04	10.28	13.07	10.47	11.58	9.46	8.54	10.87					
7			10.39	11.11	10.96	10.83	10.09	8.69	11.17	9.39	9.56					
6			10.41	10.82	17.61	10.87		13.06	11.41	9.96	9.15					
5			9.76	11.10	10.80	8.86	9.48	9.22	9.61	8.20						
4			10.93	10.94	9.53	10.61	10.27	9.59	9.82	7.81						
3				9.64	9.52	10.06	12.65	9.63								
2				9.29	8.75	8.96	8.27	8.14								
1				10.59	10.43	9.32										

Figure 2: Reoriented core locations and core values of coal ash data.

```
 6 |
 6 |
 7 | 003
 7 | 66678888899
 8 | 00111222222234
 8 | 56666666788888899999999
 9 | 000000001111122222223333333444444
 9 | 555556666666666778888888888999999999
10 | 000000001111111222222333334444444
10 | 56666667777788888899999
11 | 0000011122222222 3344
11 | 5666689
12 |
12 | 568
13 | 11
13 |
14 |
14 |
15 |
15 |
16 |
16 |
17 |
17 | 6
```

Figure 3: Stem and leaf plot of coal ash data.

We will incorporate the abovementioned local stationarity into the underlying Gaussian model by allowing the *mean* of a data point to be a smooth function of its *position*, however the *covariance* between two data points will remain only a function of their *relative position*. Figure 4 is an attempt to summarize this mean non-stationarity, using the two measures of (sample) median and (sample) mean across rows and down columns. Thus we see the spatial aspect of the data enter for the first time. Clearly there is a marked trend in the E-W direction, although not really so in the N-S direction. The use of median and mean serves two purposes. First from an EDA point of view the median is a resistant summary statistic, but also its comparison to the non-resistant mean summary has the additional function of highlighting rows or columns which may contain atypical observations. If (mean - median) is "too big" then the row or column should be scanned for possible outliers. A small amount of formal statistics tells us what is meant by "too big". Suppose for the moment that $Y_1, Y_2, \ldots, Y_n$ are i.i.d. from symmetric f with mean μ and variance σ^2. Then the median can be formally written in the asymptotically equivalent form (provided $f(\mu) \neq 0$):

$$\tilde{Y} = \mu + \frac{1}{n} \cdot \sum_{i=1}^{n} \frac{\text{sgn}(Y_i - \mu)}{2f(\mu)}, \tag{2.1}$$

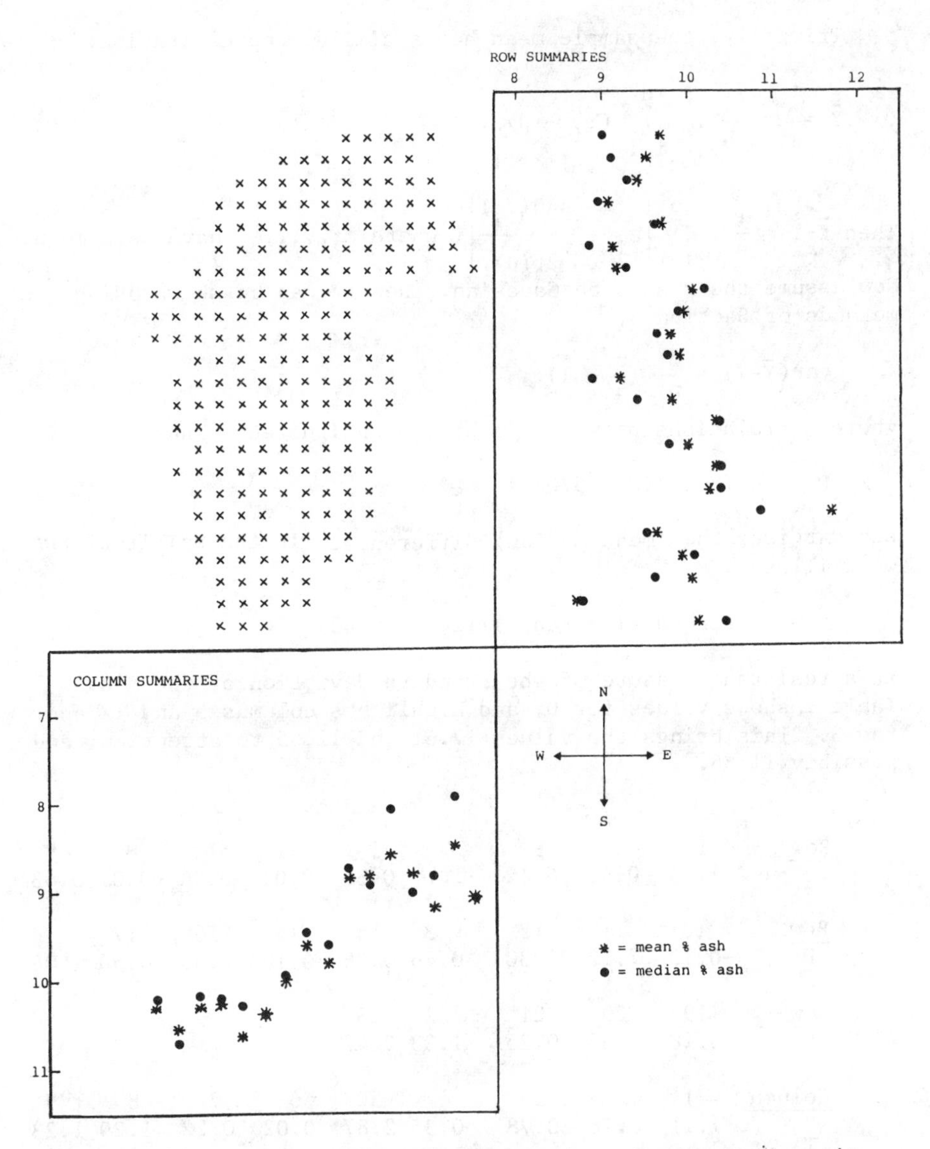

Figure 4: Mean and median summaries of non-stationarity.

where $\text{sgn}(x) = 1$ if $x > 0$, $= 0$ if $x = 0$, $= -1$ if $x < 0$.

Trivially, the sample mean has a similar representation:

$$\bar{Y} = \mu + \frac{1}{n} \cdot \sum_{i=1}^{n} (Y_i - \mu) . \tag{2.2}$$

Then $\bar{Y}-\tilde{Y} = \frac{1}{n} \cdot \sum_{i=1}^{n} \left(\varepsilon_i - \frac{\text{sgn}(\varepsilon_i)}{2f(\mu)} \right)$, where $\varepsilon_1,\ldots,\varepsilon_n$ have zero mean. Now assume the ε's to be Gaussian. Then it can be shown using the methods of Section 4,

$$\text{var}(\bar{Y}-\tilde{Y}) = \frac{\sigma^2}{n} \left(\frac{\pi}{2} - 1\right) , \tag{2.3}$$

where correlations between ε's have been ignored. Thus,

$$D = \sqrt{n}\ (\bar{Y}-\tilde{Y})/(\hat{\sigma}\sqrt{0.5708}\ ) \tag{2.4}$$

standardises the (mean-median) difference. In the spirit of EDA, we will use

$$\hat{\sigma} = (\text{interquartile range})/(2 \times 0.6745)$$

as a resistant measure of the standard deviation of the Y's. Table 1 shows values for D, and highlights columns 5 and 12 and row 3. This brings the values 17.61 and 12.65 to attention, and possibly 11.86.

Row	1	2	3	4	5	6	7	8	9
D	-1.54	-0.40	6.12	-0.45	0.35	2.01	-0.56	-0.07	0.63

Row	10	11	12	13	14	15	16	17	18
D	-0.18	2.12	0.80	0.46	0.78	0.10	-1.05	-0.60	1.05

Row	19	20	21	22	23
D	0.18	0.35	0.25	1.33	2.47

Column	1	2	3	4	5	6	7	8	9
D	1.11	-0.76	0.78	0.35	2.87	0.02	0.22	1.29	1.23

Column	10	11	12	13	14	15
D	1.03	-0.58	3.17	-1.24	1.39	1.48

Table 1: D is the standardized (mean-median) difference.

It is easy to see that if there are covariances between the Y's, and they are positive, then (2.3) is an under-estimate of the true variance of $\bar{Y}-\tilde{Y}$. Hence (2.4) is really a liberal diagnostic, in that it highlights *more* atypical rows and columns than it should.

Another more usual way of detecting atypical observations, also relies on the belief of local stationarity. Figures 5 and 6 show graphs of Z_{t+h} versus Z_t, as t varies over the core locations; Figure 5 shows h = 1 in an E-W direction, and Figure 6 shows h = 1 in a N-S direction. The outlier Z(6,5) = 17.61 is blatantly obvious, and other values such as Z(3,7) = 12.65, Z(6,8)= 13.06, Z(8,6) = 13.07, Z(13,3) = 12.50, Z(19,5) = 12.80 seem to be atypical, given their surrounding values.

We are gradually accumulating a set of possible troublesome observations, although an explanation of their "non-compliance" may still be the global non-stationarity (allowed for in the mean of the underlying Gaussian distribution). Whether they do in fact need special attention will ultimately be determined by their influence in variogram estimation.

3. TRANSFORMING TO DETECT ATYPICAL VARIOGRAM CONTRIBUTIONS

3.1 The Intrinsic Hypothesis

We have gone about as far as we can without introducing formally a model and its parameters. Suppose that the grade of an ore body at a point t (here in $\mathbb{R}^2$) is the realization of a random function $\{Z_t\}$, and that this is observed at certain points $\{t_i\}$ (here an approximately square grid) of the ore body. Then the *intrinsic hypothesis* assumes second order stationarity of first differences, namely

$$E(Z_{t+h} - Z_t) = 0 \tag{3.1}$$

$$E(Z_{t+h} - Z_t)^2 = 2\gamma(h). \tag{3.2}$$

The quantity $2\gamma(h)$ is known as the variogram, and is the crucial parameter of geostatistics; see Matheron (1963) and Journel and Huijbregts (1978). This is a more general model than that of second order stationarity of the $\{Z_t\}$, but when the latter is appropriate,

$$\mathrm{cov}(Z_t, Z_{t+h}) = C(h), \tag{3.3}$$

and $\gamma(h) = C(0) - C(h)$.

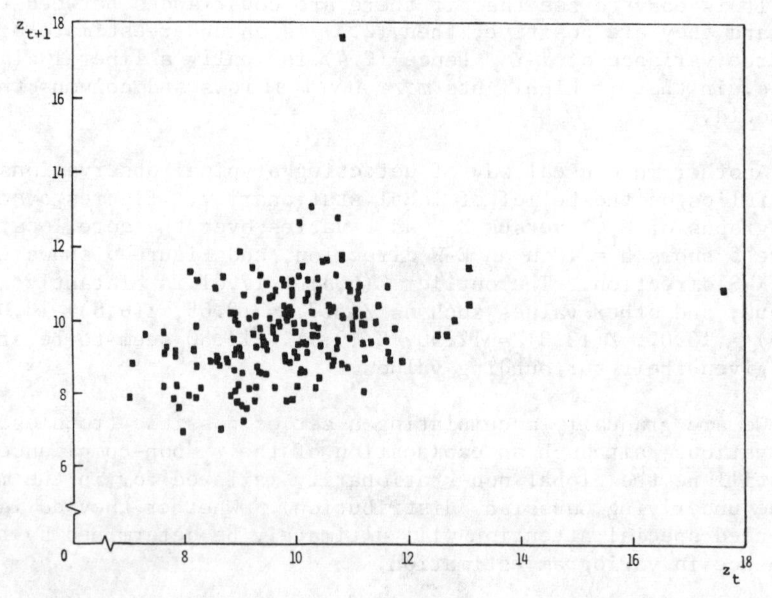

Figure 5: Plot of z_{t+1} versus z_t in the E-W direction.

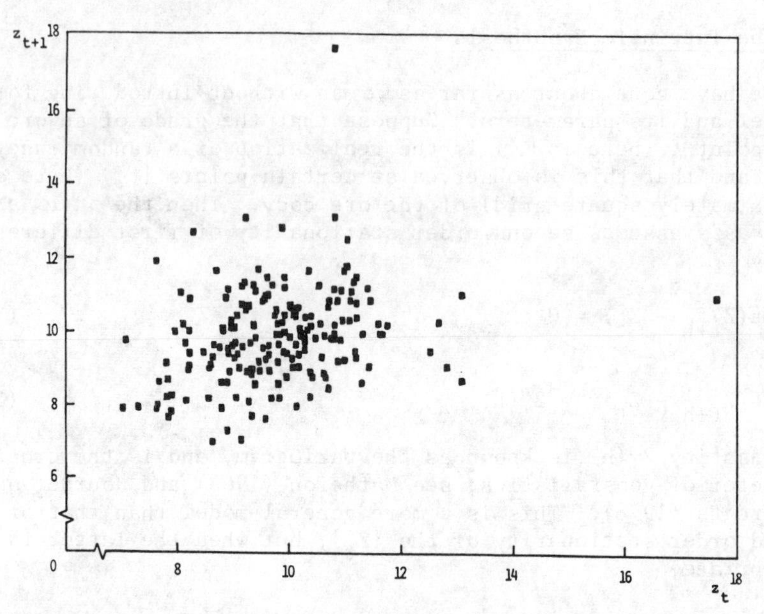

Figure 6: Plot of z_{t+1} versus z_t in the N-S direction.

Most of the time we will assume the $C(\cdot)$ function exists, and indeed use it instead of $\gamma(\cdot)$ when convenient. The classical estimator of the variogram proposed by Matheron (1963) is,

$$2\hat{\gamma}(h) = \frac{1}{N_h} \sum_{i=1}^{N_h} (Z_{t_i+h} - Z_{t_i})^2. \tag{3.4}$$

It *is* unbiased, however it possesses very poor resistance properties. It is badly affected by outliers due to the $(\cdot)^2$ term, and its "cousin", the variogram cloud, is not efficacious in highlighting these atypical observations.

3.2 Transforming the Differences

Cressie and Hawkins (1980) argued that a more resistant approach to the estimation of the variogram was to transform the problem to location estimation from a symmetric distribution. Now for Gaussian Z's, $(Z_{t+h} - Z_t)^2$ is distributed as $2\gamma(h) \times \chi_1^2$, where χ_1^2 is a chi-squared random variable on one degree of freedom. Thus $2\hat{\gamma}(h)$ is the first moment of a *highly skewed* random variable. Using the set of power transformations proposed by Box and Cox (1964), Cressie and Hawkins (1980) found that $|Z_{t+h} - Z_t|^{\frac{1}{2}}$ has a skewness of 0.08 and a kurtosis of 2.48 (compared with 0 and 3 for the normal distribution). Estimates of location such as the mean and the median were then applied to the *transformed differenced data* $\{|Z_{t_i+h} - Z_{t_i}|^{\frac{1}{2}}\}$. These estimates were then raised to the 4th power to bring them back to the correct scale, and adjusted for bias. This results in the two variogram estimators:

$$2\bar{\gamma}(h) = \left\{ \frac{1}{N_h} \sum_{i=1}^{N_h} |Z_{t_i+h} - Z_{t_i}|^{\frac{1}{2}} \right\}^4 / (0.457 + 0.494/N_h) \tag{3.5}$$

$$2\tilde{\gamma}(h) = (\text{med}\{|Z_{t_i+h} - Z_{t_i}|^{\frac{1}{2}}\})^4 / (0.457 + 0.494/N_h). \tag{3.6}$$

The superior properties of these estimators over the classical estimator (3.4) are presented in Hawkins and Cressie (1984). Put aside for the moment these theoretical considerations, and return to the realms of data analysis.

Consider the coal ash data, and for illustrative purposes suppose differences are taken in the E-W direction. It has been proposed by Chauvet (1982) that the variogram cloud, "cousin" to (3.4), is a useful diagnostic tool. I propose that a *square root differences cloud*, "cousin" to (3.5) and (3.6), is a much more powerful data analytic technique than Chauvet's. The variogram cloud is simply an x-y plot of the y-values $\{(Z_{t_i+h} - Z_{t_i})^2;$ $i = 1,\ldots,N_h\}$ at the x-value h, as h varies from 1,2,.... Figure 7 shows box plots (see Tukey, 1977) of the "cloud" as h varies. Figure 8 shows the square root differences cloud, obtained by using $\{|Z_{t_i+h} - Z_{t_i}|^{\frac{1}{2}}; i = 1,\ldots,N_h\}$ as y-values at the x-value h; h = 1,2,....

Let us compare these two plots. We all know what the box plot of a roughly *symmetric* batch should look like (and besides the mean should roughly be equal to the median). Therefore any departures from this are easy to detect by eye. On the contrary when the batches are inherently skewed, as they are in the variogram cloud, it is difficult to gauge whether the large value is the result of the skewness or the result of an atypical observation. It is also clear now when comparing Figures 7 and 8, how the square root differences do indeed lead to a more resistant variogram estimator; look at the influence of the outlier 17.61 for h = 6 in the variogram cloud (Figure 7) and for h = 6 in the square root differences cloud (Figure 8).

3.3 The Pocket Plot

The data analytic methods presented so far, have been good at detecting either gross trends or isolated outliers. A method is needed which will identify a localised area as being atypical, when the data should be exhibiting stationarity. Again this is done by exploiting the spatial nature of the data, as before through row and column co-ordinates.

The immediate aim of the geostatistician is to estimate the spatial relationship (i.e. variogram) between data points, and then to use this estimate to predict the grade of mining blocks and to provide an estimate of the predictor's variability. Although (3.5) offers a resistant approach to variogram estimation, there is still the concern that a possibly significant fraction of differences $Z_{t_i+h} - Z_{t_i}$ may be inappropriate in the estimation of $2\gamma(h)$. Locations on the grid which exhibit measurements "different" from the rest, need to be identified. These pockets of non-stationarity, once discovered, may be initially ignored but must eventually be modelled and incorporated into final resource appraisals. The *pocket plot* (so called because of its use in detecting pockets of non-stationarity, and because it evolved from a thick wad of many plots down to just one which could be carried around in a pocket) is a simple idea which I will illustrate on N-S differences of the coal ash data. Concentrate on row j of the grid. For any other row, k say, there are a certain number (N_{jk}) of differences defined; let $\bar{Y}_{jk}$ denote the mean of these

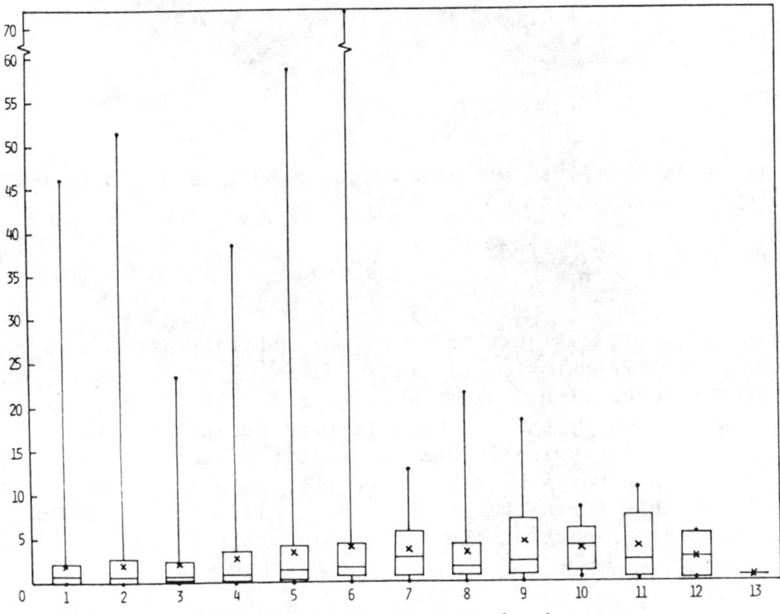

Figure 7: Variogram cloud.

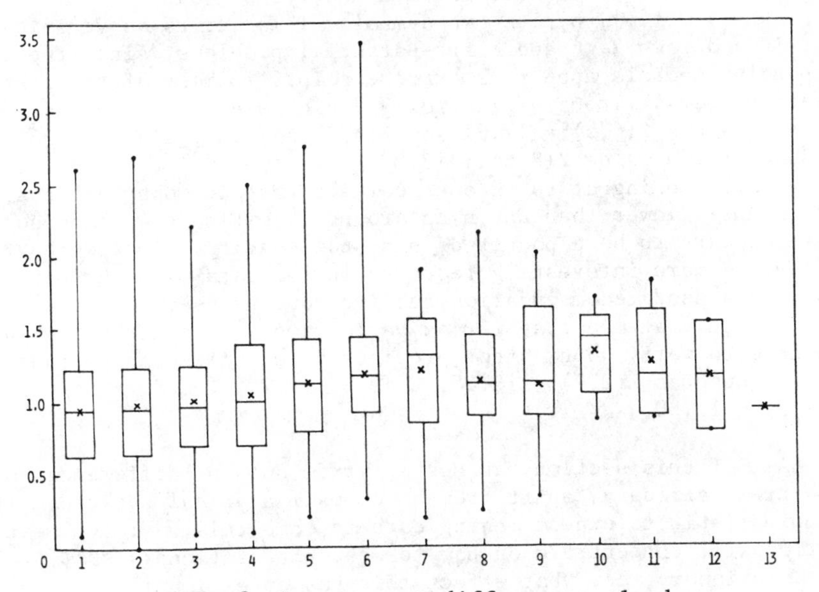

Figure 8: Square root differences cloud.

$|\text{difference}|^{\frac{1}{2}}$, averaged over the N_{jk} terms. Let $h = |j-k|$, and (see (3.5))

$$\bar{\bar{Y}}_h = \frac{1}{N_h} \sum_{i=1}^{N_h} |Z_{t_i+h} - Z_{t_i}|^{\frac{1}{2}},$$

or alternatively $\bar{\bar{Y}}_h$ is just the weighted mean of $\bar{Y}_{jk}$'s such that $|j-k| = h$. Then define

$$P_{jk} = \bar{Y}_{jk} - \bar{\bar{Y}}_h; \tag{3.7}$$

$\{P_{jk};\ k = 1,2,\ldots\}$ is just the residual contribution of row j to the variogram estimator at differing lags. Ideally, these points will be scattered either side of zero, but if there is something unusual about row j, then it will give an unusual contribution at all lags and will typically show a scatter of points *above* the zero level. Now vary the row j, and put the point scatters beside each other; this forms the *pocket plot* as illustrated in Figure 9 where the central part of the scatter is replaced by the box of the box plot.

Clearly rows 2, 6 and 8 are atypical, in that their $\{P_{jk}\}$ points are scattered about a level above zero. Notice that the kth row has been identified in each scatter of points $\{P_{jk};\ k = 1,2,\ldots\}$, by its own symbol. This serves as verification that indeed rows 2, 6 and 8 are potential problems, since the corresponding symbols appear as extreme points in most of the other point scatters. Trouble with row 6 could be expected because of the two values Z(6,5) = 17.61 and Z(6,8) = 13.06, and with row 8 because of the value Z(8,6) = 13.07. *But trouble with row 2 is a surprise!* Looking at row 2 one realises that its observations are consistently lower than anything around (see Figure 4), and so there appears to be a pocket of non-stationarity. That we have discovered more interesting features in the data with the pocket plot, is a good recommendation for its routine use. But the added bonus is that at least as far as variograms are concerned, we do not have to worry about other previously identified suspicious values, such as Z(2,7) = 12.65, Z(13,3) = 12.50, and Z(19,5) = 12.80.

All of this section has assumed that data or differences of data are stationary, apart from some small atypical pockets. It is unrealistic to expect mining data to come this way; the next section will concentrate on how to make non-stationary *spatial* data stationary, and what effect this has on estimating the spatial relationships.

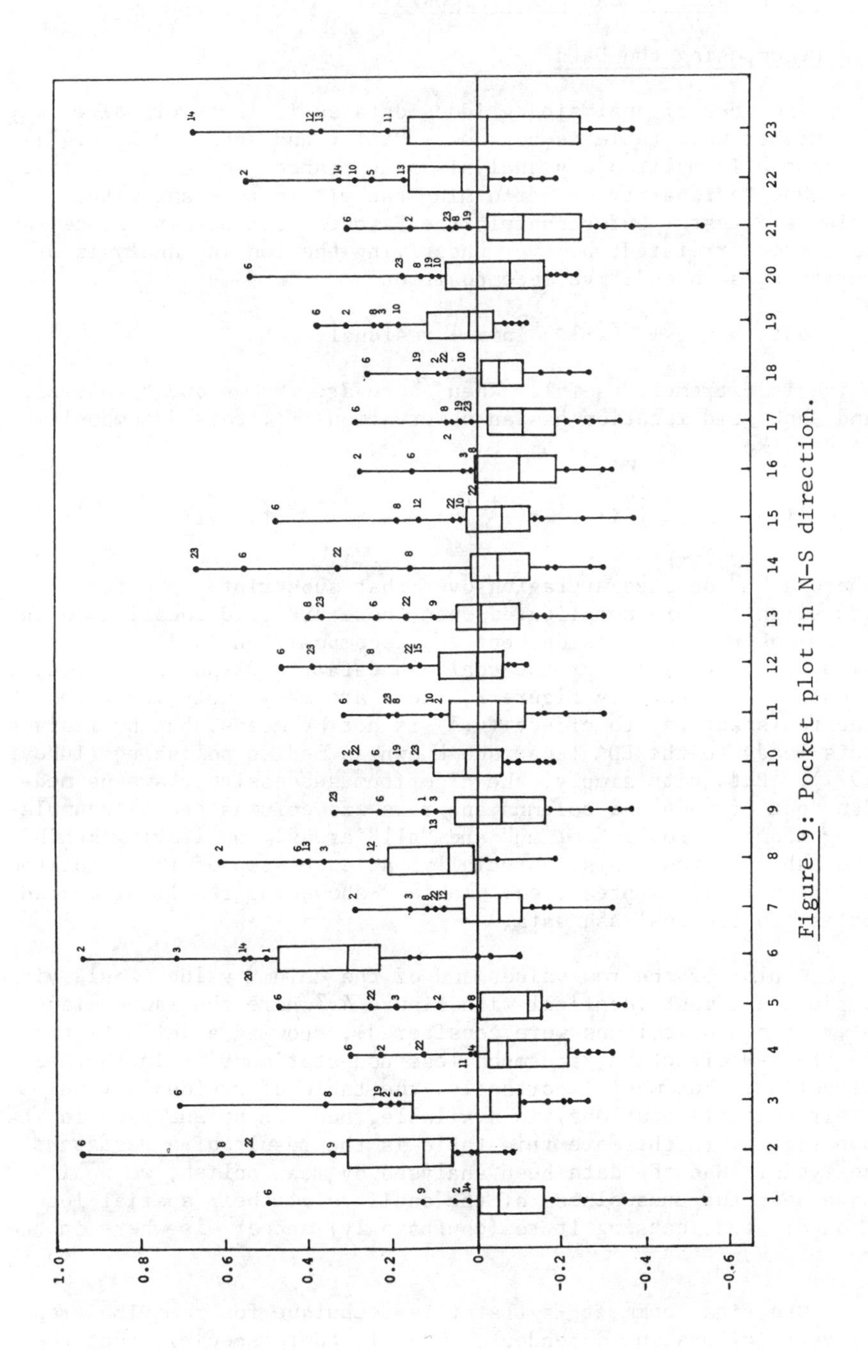

Figure 9: Pocket plot in N-S direction.

4. DEALING WITH GLOBAL NON-STATIONARITY

4.1 Decomposing the Data

The idea of analysing gridded data as if they were a two way (or higher way) table is not new. Miller and Kahn (1962, P.411) propose a formal two way analysis of variance, and claim to test for non-stationarity by performing the within rows and within columns F tests. Unfortunately the F tests mean nothing since the data are correlated, however underlying the two way analysis of variance is an additive decomposition:

$$\text{data} = \text{all} + \text{row} + \text{column} + \text{residual}, \tag{4.1}$$

which is extremely useful. When there are r rows and c columns, and each grid location has an observation Y_{ij}, this decomposition looks like

$$Y_{ij} = Y_{\cdot\cdot} + (Y_{i\cdot} - Y_{\cdot\cdot}) + (Y_{\cdot j} - Y_{\cdot\cdot}) + (Y_{ij} - Y_{i\cdot} - Y_{\cdot j} + Y_{\cdot\cdot}),$$

where a "•" denotes averaging over that subscript. The formulae are slightly more complicated when there are grid locations with their observation missing, but the decomposition (4.1) remains simple. In the case of the coal ash data r = 23 and c = 16, although as is shown by Figure 2, there are many empty locations. The resistant way to effect (4.1) is not by means, but by medians. This leads to the EDA technique known as median polishing (Tukey, 1977). Put quite simply, the algorithm successively sweeps medians out of rows and columns and rows and columns etc., accumulating them in "row", "column" and "all" areas, and leaving behind the table of residuals. Crucially, at each step of the algorithm, relation (4.1) is preserved. Table 2 shows the result of median polish on the coal ash data.

A plot of the row values and of the column values would yield a picture almost identical with Figure 4 (where the raw medians of rows and of columns were considered), showing a definite trend in the E-W direction, and much less non-stationarity in the N-S direction. But most importantly, the table of residuals (and their spatial locations) is available, and can be analysed in its own right. Furthermore this table is the result of a *resistant* analysis. Had the data been analysed by mean polish, we would have seen the unusualness of the outliers at their spatial location diluted, causing it to (confusingly) appear elsewhere in the table.

One final comment is that this technique for removing trend, is very orientation dependent. What is there special about the grid directions chosen for drilling, which will ensure that

																ROW
									-0.20	0.00	3.29	0.00	0.51			0.00
						1.35	1.15	-1.32	-0.00	0.00	-1.05	0.64				0.10
				0.10	0.01	0.06	-0.20	0.56	0.00	-1.32	-0.85	-0.00	-1.00			0.12
			0.01	-0.40	0.89	1.74	0.03	0.00	0.11	-0.41	-0.25	-0.00	0.00			-0.71
			-0.00	2.38	-0.74	-1.06	-1.34	-1.22	0.00	-1.97	0.76	0.53	0.23	1.53		0.25
			0.90	-0.10	0.00	-1.80	0.34	0.26	-0.03	0.07	0.25	-1.69	-0.49	0.00		-0.18
		0.27	-0.46	0.00	0.10	-0.00	0.68	-0.64	-0.78	0.58	-0.41	0.56		-0.17	0.00	-0.33
0.53	-0.67	-0.14	-0.26	0.00	0.53	0.42	-0.88	0.16	0.51	-0.43						0.04
-0.54	0.19	1.77	-1.39	-0.15	-1.48	0.15	-0.18	0.72	0.41							-0.14
0.00	0.35	-0.18	-0.75	-0.00	0.25	0.00	-0.36	-0.37	3.03							-0.39
		2.74	-0.59	0.92	-0.13	-0.50	0.09	1.28	1.40	-0.88	-0.09					-0.27
	-0.09	1.78	0.38	2.05	0.65	-0.96	0.00	-1.15	-0.07	-1.24	0.09					-0.76
	0.93	-0.23	-0.73	1.43	-0.20	0.92	0.00	-0.31	1.73	0.00	0.60					-0.39
	0.00	-0.50	-0.70	0.00	0.71	-1.39	0.00	0.83	-1.07	1.92						0.38
		0.59	1.06	-0.59	0.28	-0.58	0.33	-1.03	-0.84	0.00						0.20
	-0.96	-2.00	0.35	-0.09	2.35	0.09	1.72	-0.72	-0.45	1.67						0.20
		-0.20	0.06	0.23	-0.25	-0.64	-1.53	0.63	0.04	0.00						0.56
		-0.39	-0.45	6.66	-0.42		2.62	0.66	0.39	-0.63						0.78
		0.14	1.01	1.03	-1.25	-0.29	-0.04	0.04	-0.19							-0.40
		0.89	0.44	-0.65	0.08	0.08	-0.08	-0.17	-0.99							0.01
			-0.39	-0.19	0.00	2.93	0.43									-0.46
			0.32	0.10	-0.04	-0.38	0.00									-1.52
			0.00	0.16	-1.30											0.10
0.78	0.95	0.21	0.67	0.35	0.70	0.36	-0.16	0.16	-1.03	-0.82	-1.25	-0.91	-0.34	-1.69	-0.42	9.82
COLUMN																ALL

Table 2: Median polish of coal ash data.

"all + row + column", captures all the non-stationarity (in the mean)? To check for possible interaction between row and column, one can allow for an extra (interaction) parameter k:

data = all + row + column + k(row × column/all) + residual.

This "plus one fit" (as it is called by Tukey, 1977) for the coal ash data, showed k to be zero, so the purely additive decomposition of (4.1) is reasonable.

It has long been thought that working with residuals, introduces bias into variogram estimation. The next subsection discusses this interesting and complex issue.

4.2 Analysis of Residuals

It is our intention in the last section, to indicate briefly what can be done with an analysis of (resistant) median polish residuals. But first we must address the very thorny question of estimation bias that results when the trend is estimated.

In what is to follow, we will compare data configurations and estimation techniques to show that the residuals from (4.1) can have small bias problems. We will work with the covariance function $C(\cdot)$ and will use the usual estimator

$$C^*(h) = \frac{1}{N_h} \cdot \sum_{i=1}^{N_h} (Z_{t_i} - Z^*)(Z_{t_i+h} - Z^*). \tag{4.2}$$

The main reason is convenience, since if we worked with $\gamma(\cdot)$, we would have to postulate a *linear* drift in order to investigate bias effects.

The usual way of illustrating bias in the estimator is to consider the transect configuration:

$$\times \quad \times \quad \times \quad \times \quad \ldots \quad \times \overset{1}{\frown} \times \tag{4.3}$$

of n data points, assume unknown mean and second order stationarity, and evaluate (for some fixed h):

$$E\left\{\frac{1}{n-h} \sum_{t=1}^{n-h} (Z_t - \bar{Z})(Z_{t+h} - \bar{Z})\right\}$$

$$= C(h) - \left\{2 \sum_{k=1}^{\ell} C(k) + C(0)\right\}/n + O(1/n^2), \tag{4.4}$$

where $\bar{Z} = \frac{1}{n} \Sigma Z_t$, and ℓ is a fixed integer such that $C(g) \simeq 0$, $g > \ell$. Notice that provided the C's are positive (as they are for all well known geostatistical models), the bias is negative, constant, and $O(1/n)$ for *all lags*. Unfortunately this unhappy fact has discouraged researchers away from residuals (the residuals here are of course $\{Z_t - \bar{Z}\}$). But there are two ways in which our problem is *not* the one described above.

First, we have a spatial data configuration:

$$h\left\{\begin{array}{ccccc} \times & \times & \times & \overset{1}{\times} & \times \\ \times & \times & \times & \times & \times \end{array}\right. , \tag{4.5}$$

not a transect as described in (4.3). Suppose there are n-h such N-S differences possible. Then it is a simple matter to evaluate (for some fixed h):

$$E\left\{\frac{1}{n-h}\sum_{t=1}^{n-h}(Z_{1,t} - \bar{Z}_1)(Z_{h+1,t} - \bar{Z}_2)\right\}$$

$$= C(h) - \left\{2\sum_{k=1}^{\ell} C(\sqrt{h^2+k^2}) + C(h)\right\}/n + O(1/n^2). \tag{4.6}$$

So the bias is still negative and $O(1/n)$, but it is not constant. Provided the C's are positive and decrease to zero (as they do for all well known geostatistical models)

$$|O(1/n) \text{ bias of } (4.4)| \geq |O(1/n) \text{ bias of } (4.6)|.$$

Consider the spherical variogram model,

$$C(h) = \begin{cases} 1 - [3/2(h/a) - 1/2(h/a)^3]; & h \leq a \\ 0 & ; h > a. \end{cases} \tag{4.7}$$

Table 3 compares the $O(1/n)$ term in (4.4) and (4.6) for various choices of a. Clearly configuration (4.5), from which *spatial* relationships are estimated, shows much less of a bias problem. It is our contention therefore that bias problems, particularly at higher lags, have been over-dramatized due to a non-realistic choice of data configuration (see Matheron, 1971, P.196).

a = 1	h	1	2	3	4	5
	(4.4)	1.0	1.0	1.0	1.0	1.0
	(4.6)	0.0	0.0	0.0	0.0	0.0

a = 5	h	1	2	3	4	5	6	7
	(4.4)	3.8	3.8	3.8	3.8	3.8	3.8	3.8
	(4.6)	3.1	1.9	0.8	0.2	0.0	0.0	0.0

a = 10	h	1	2	3	4	5	6	7	8	9	10
	(4.4)	7.5	7.5	7.5	7.5	7.5	7.5	7.5	7.5	7.5	7.5
	(4.6)	7.1	6.1	5.0	3.8	2.7	1.7	0.9	0.4	0.1	0.0

Table 3: The leading |bias term| in (4.4) and (4.6), compared for the spherical variogram (4.7).

The second way in which our problem is not the one described by (4.3) and (4.4), is that we are using *resistant* methods to estimate the drift. Intuitively this should again ameliorate the bias problem, since resistant methods do not give rise to linear constraints on residuals (e.g. $\Sigma(Z_t - \bar{Z}) = 0$).

For ease of exposition, we will do our comparisons on data configuration (4.3). Let the estimate of $C(\cdot)$ based on the residuals $\{Z_t - \tilde{Z}\}$, where $\tilde{Z} \equiv$ median of Z_t's, be defined as:

$$C^{\dagger}(h) = \frac{1}{n-h} \sum_{t=1}^{n-h} (Z_t - \text{med}(Z_t))(Z_{t+h} - \text{med}(Z_t)). \tag{4.8}$$

Just as in Section 2, replace $\tilde{Z}$ by the formal asymptotically equivalent representation

$$\tilde{Z} = \mu + \frac{1}{n} \sum_{t=1}^{n} \frac{\text{sgn}(Z_t - \mu)}{2f(\mu)} . \tag{4.9}$$

Now for $\{Z_t\}$ jointly Gaussian, second order stationary, mean μ, and covariance function $C(\cdot)$, it is straightforward to show that

$$\begin{aligned} &E[(Z_t - \mu)(Z_{t+h} - \mu)] = C(h) \\ &E[(Z_t - \mu)\ \text{sgn}(Z_{t+h} - \mu)] = \frac{2C(h)}{\sqrt{2\pi C(0)}} \\ &E[\text{sgn}(Z_t - \mu)\ \text{sgn}(Z_{t+h} - \mu)] = \frac{2}{\pi} \sin^{-1}\left(\frac{C(h)}{C(0)}\right). \end{aligned} \tag{4.10}$$

The latter relation of (4.10) relies on a result by Sheppard (1899) on median dichotomy, viz.,

$$\frac{1}{2\pi|R|^{\frac{1}{2}}}\int_0^{\infty}\int_0^{\infty} \exp\left([x_1 x_2]R^{-1}\begin{bmatrix}x_1\\x_2\end{bmatrix}\right) dx_1 dx_2 = \frac{1}{4} + (2\pi)^{-1}\sin^{-1}\rho,$$

where $R = \begin{bmatrix}\sigma_1^2 & \sigma_1\sigma_2\rho \\ \sigma_1\sigma_2\rho & \sigma_2^2\end{bmatrix}$. Finally one is able to obtain,

$$E(C^{\dagger}(h)) = C(h) - \left\{\frac{(4-\pi)}{2} C(0) + 4\sum_{k=1}^{\ell} C(k)\right.$$

$$\left. - 2C(0)\sum_{k=1}^{\ell} \sin^{-1}\left(\frac{C(k)}{C(0)}\right)\right\}/n + O(1/n^2). \qquad (4.11)$$

Comparing (4.11) with (4.4) is very interesting. Provided the C's are positive and decrease to zero (as they do for all well known geostátistical models), and because

$$x \leq \sin^{-1}x \leq \frac{\pi}{2}x,$$

$$|O(1/n) \text{ bias of } (4.11)| \leq 2\sum_{k=1}^{\ell} C(k) + \left(\frac{4-\pi}{2}\right)C(0)$$

$$\leq |O(1/n) \text{ bias of } (4.4)|.$$

Consider the spherical model of (4.7). The table below compares the $O(1/n)$ term in (4.4) and (4.11) for various choices of a in that model.

	a = 1	a = 5	a = 10
|O(1/n) in (4.4)|	1.0	3.8	7.5
|O(1/n) in (4.11)|	0.4	3.0	6.4

Clearly the estimator $C^{\dagger}(\cdot)$, based on resistant estimation of the drift, shows much less of a bias problem. Using medians not only buys one resistance to outliers, but the median also possesses superior bias properties.

In actual fact, our case is a hybrid of the two departures summarized in the two tables above, leading to an improvement in bias from both sources. It is at the early lags where it is most important to have a good estimate. But it is precisely at these lags where the most number of pairs exist. Provided a spatial configuration of paired differences is appropriate, and resistant methods are used to estimate drift, the bias can be controlled and the variogram of the residuals is a sensible quantity to estimate.

5. VARIOGRAM OF RESIDUALS FROM MEDIAN POLISH

Proceeding with the residuals from median polish (given in Table 2) as if they were a fresh data set, we then go through the exploratory data analytic steps outlined in the previous sections, and compute estimates of the variogram. Cressie and Hawkins (1980) developed a robust estimator of the variogram designed to deal with atypical points *automatically*, rather than deleting them in some ad hoc way. The estimator $2\bar{\gamma}$ given by (3.5) is clearly seen to be doing what it should when we look at the associated square root differences plot of Figure 8, and see there the rather minor influence of the outlier. Nevertheless, we will compute both the classical variogram estimator $2\hat{\gamma}$ (given by (3.4)) and the robust estimator $2\bar{\gamma}$ (given by (3.5)), on both the original data and on the residuals. We have chosen to do this in the E-W direction because we know there is trend in the E-W originals (see Figure 4).

Figure 10 shows the robust and classical estimators for the E-W originals. The effect of the drift is plainly obvious, leading to a steadily increasing variogram. However when the resistant residuals (from median polish) are analysed, most of the apparent correlation in Figure 10 seems to have been due to the drift; see Figure 11 (cf Starks and Fang, 1982).

We have not discussed in this paper, the theoretical variograms we would eventually fit to these graphs, nor the kriging equations we would eventually use to predict block values. There are at least two ways to skin this cat. Because of the non-stationarity, we might think about universal kriging (Matheron, 1971, Chapter 4), or we might quite simply work with the *stationary* residuals, perform simple kriging (or robust kriging; see Hawkins and Cressie, 1984), and add in the drift afterwards. We have in fact done *both*, not only on this data set but also on an iron ore body in Australia; this work will be published elsewhere. Most encouragingly, the results show essentially no difference, leaving me no choice but to endorse the resistance movement in geostatistics.

ACKNOWLEDGEMENTS

The author would like to acknowledge the assistance given by Gary Glonek, towards the preparation of this paper. Research was partially supported by an ARGS grant, and Flinders University.

AUTHOR'S PRESENT ADDRESS

Department of Statistics,
Iowa State University,
Ames, Iowa 50011.

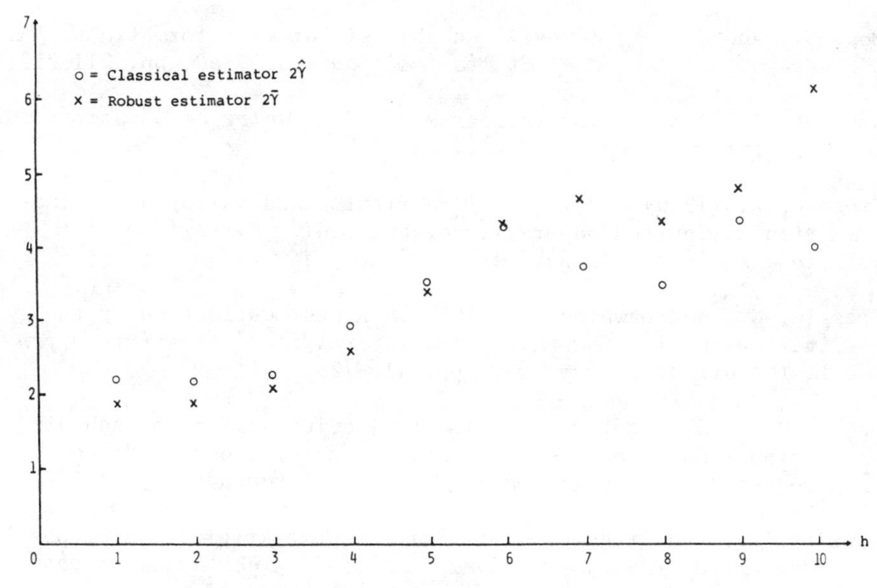

Figure 10: E-W variogram of coal ash data.

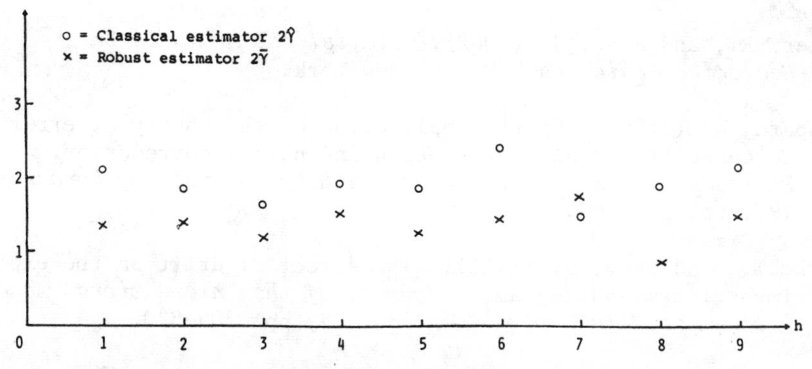

Figure 11: E-W variogram of coal ash residuals (from median polish).

REFERENCES

Box, G., and Cox, D. (1964). An analysis of transformations. *Journal of the Royal Statistical Society B, 26,* pp. 211-252.

Chauvet, P. (1982). The variogram cloud. *Centre de Géostatistique internal report, N-725.*

Cressie, N. (1979). Straight line fitting and variogram estimation. (Invited paper with discussion). *Bulletin of the International Statistical Institute, 48.*

Cressie, N., and Hawkins, D. (1980). Robust estimation of the variogram, I. *Journal of the International Association for Mathematical Geology, 12,* pp. 115-125.

Gomez, M., and Hazen, K. (1970). Evaluating sulfur and ash distribution in coal seams by statistical response surface regression analysis. *U.S. Bureau of Mines Report, RI 7377.*

Hawkins, D., and Cressie, N. (1984). Robust kriging - a proposal. *Journal of the International Association for Mathematical Geology, 16,* to appear.

Journel, A., and Huijbregts, C. (1978). *Mining Geostatistics.* Academic Press: London.

Matheron, G. (1963). Principles of geostatistics. *Economic Geology, 58,* pp. 1246-1266.

Matheron, G. (1971). *The Theory of Regionalized Variables and its Applications.* Cahiers du Centre de Morphologie Mathématique, No. 5: Fontainebleau.

Miller, R., and Kahn, J. (1962). *Statistical Analysis in the Geological Sciences.* Wiley: New York.

Sheppard, W. (1899). On the application of the theory of error to cases of normal distribution and normal correlation. *Philosophical Transactions of the Royal Society of London A, 192,* pp. 101-167.

Starks,T., and Fang, J. (1982). The effect of drift on the experimental semi-variogram. *Journal of the International Association for Mathematical Geology, 14,* pp. 309-320.

Tukey, J. (1977). *Exploratory Data Analysis.* Addison-Wesley: Reading, Mass.

STATISTICAL INFERENCE OF THE SEMIVARIOGRAM AND THE QUADRATIC MODEL

MARCO ALFARO

Mining Dpt. University of Chile. Casilla 2777.
Santiago Chile

Abstract: Since the semi-variogram is the basic tool on Geostatistics, the following question appear inmediately: is it possible to find an accurate estimation of a semi-variogram from a single spatial outcome of a random function?

The results indicate us that in the non-Gaussian case, special care must be taken in the control of Geostatistical simulation of random functions.

Finally, we introduce a new model of semi-variogram in the space of 3 dimensions.

I. Definitions.

Let Z(x) be an intrinsic random function defined on a set $S \subset \mathbb{R}^n$. Then, the random variable:

$$\gamma_R(h) = \frac{1}{2S'} \int_{S'} \overline{Z(x+h) - Z(x)}^{\,2} dx \qquad (1)$$

where S' is the intersection of S with its translation over a distance h, is called regional or local semi-variogram.

Let $\gamma(h) = \frac{1}{2} E(\overline{Z(x+h) - Z(x)}^{\,2})$ be the theoretical semi-variogram of Z(x). An inmediate result is that $E(\gamma_R(h)) = \gamma(h)$

The variance of the random variable $\gamma_R(h)$:

$$\sigma_F^2 = D^2(\gamma_R(h)) = E(\overline{\gamma_R(h) - \gamma(h)}^{\,2}) \qquad (2)$$

is called the fluctuation variance of the regional or local

G. Verly et al. (eds.), Geostatistics for Natural Resources Characterization, Part 1, 45–53.

semi-variogram. σ_F^2 characterizes $\gamma_R(h)-\gamma(h)$ fluctuations and plays an important role in the control of simulations of the random function Z(x) . Actually, if σ_F^2 is large, then the differences $\gamma_R(h)-\gamma(h)$, in average, will also be large.

The behavior pattern of semi-variograms for small values of distance h is of great importance for Geostatistics. Our purpose is to study the relative fluctuation variance, i.e. $\sigma_F^2/\overline{\gamma(h)}^2$ for small distances: $|h|\to 0$. Consider the following definition:

Definition: Let Z(x), $x \in \mathbb{R}^n$ be an intrinsic random function defined on a set $S \subset \mathbb{R}^n$. Its semi-variogram is microergodic near the origin if:

$$\lim_{|h|\to 0} \frac{\sigma_F^2}{\overline{\gamma(h)}^2} = 0 \tag{3}$$

If the semi-variogram of Z(x) is microergodic, then for a small h it is possible to determine the theoretical semi-variogram γ(h) quite accurately by taking the local semi-variogram from a single spatial outcome of the random function.

Calculations of the fluctuation variance of the local semi-variogram demands that the 4th moments of the increments of the random function Z(x) be known. Actually, it follows from (1) and (2) that:

$$\sigma_F^2 = E(\overline{\gamma_R(h)}^2) - \overline{\gamma(h)}^2$$

$$\sigma_F^2 = \frac{1}{4S'^2}\iint_{S'\,S'} E(\overline{Z_{x+h}-Z_x}^2 \cdot \overline{Z_{y+h}-Z_y}^2)\,dx\,dy - \overline{\gamma(h)}^2 \tag{4}$$

I.2. Study of the Microergodicity in the Gaussian Case.

Let Z(x) be an intrinsic random function with gaussian increments. In this case, we can show that (cf. G. Matheron, 1965) if $\gamma(h)=a|h|^\lambda+.$ near the origin, then the relative fluctuation variance for a small |h| is:

$$\sigma_F^2/\overline{\gamma(h)}^2 = A|h|^{4-2\lambda} + B|h|^n \tag{5}$$

where n = dimension of the space. There are two possible cases:
i) For $\lambda = 2$, the random function is derivable in quadratic mean i.e. the regionalized variable changes regularly in the space and its semi-variogram is not microergodic since the relative fluctuation variance approaches a constant value A when $|h|\to 0$.

ii) For $0<\lambda<2$, the random function is only continuous in quadratic mean, the regionalized variable is much more irregular and its semi-variogram is microergodic near the origin.

Let us discuss intuitively this result:

If $\lambda = 2$, the relative fluctuation variance is small only if $S \to \infty$ When S is small, the realization may be assimilated to a parabolic segment (see fig.1) and the local semi-variograms results to be different for S_1 and S_2. There exists redundant information in spite of the infinite number of data.

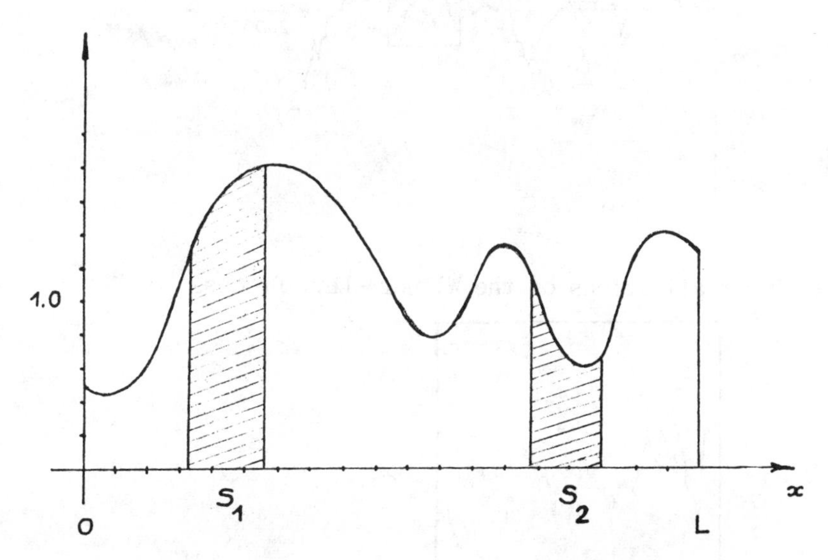

Fig. 1: Realization of a Gaussian Model on $\mathbb{R}^1$

If $0<\lambda<2$, then the information is far less redundant and the microergodicity occurs (see fig.2 and fig.3)

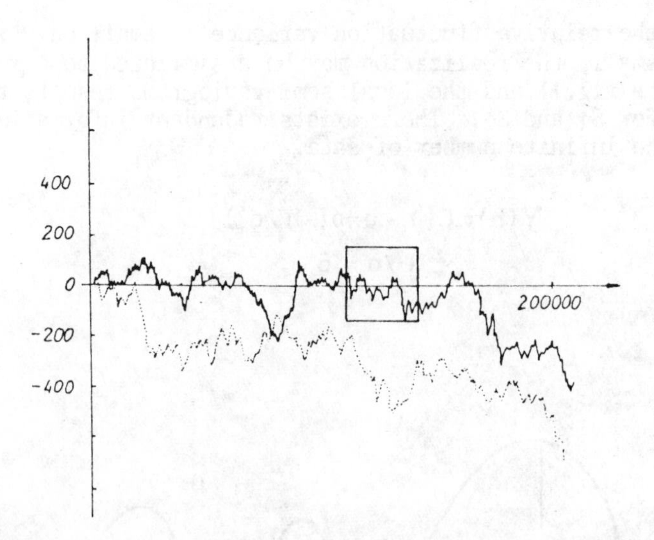

Fig.2: Two realizations of the Wiener -Levy Process

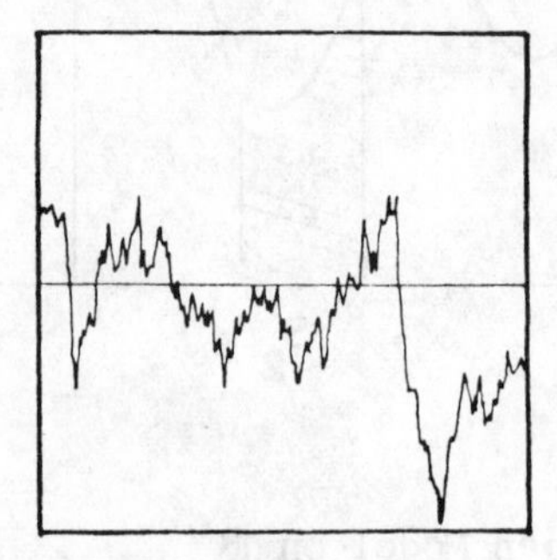

Fig.3: Enlarged portion of Fig.2
magnified 4 times.

The above results are classics (cf. G. Matheron, 1978), and lean on the Gaussian hipothesis. It is necessary to find more general relations but the problem is very difficult.

II. Study of the Microergodicity for the non-Gaussian Case.

In the geostatistical practice, simulation of a non-Gaussian random function are commonly abtained by graphical anamorphosis i.e transformation by a non-decreasing function.

Let $Y(x)$ be a stationary Gaussian random function such that:

$$E(Y(x))=0 \quad , \quad D^2(Y(x))=1 \quad , \quad \rho(h)=E(Y(x)Y(x+h))$$

then, if φ is any given function, the random function

$$Z(x) = \varphi(Y(x))$$

is non-Gaussian.

We are interested in finding out under what conditions the semi-variogram of the random function $Z(x)$ is microergodic. According to (4) we must find the 4th moment:

$$E(\overline{Z_{x+h} - Z_x}^2 \cdot \overline{Z_{y+h} - Z_y}^2) = E(\overline{\varphi(Y_{x+h}) - \varphi(Y_x)}^2 \cdot \overline{\varphi(Y_{y+h}) - \varphi(Y_y)}^2) \quad (6)$$

The above expression may be developed for some cases of function φ; the characteristic function of the k-dimensional Gaussian law must be used.

Now, we will examine two examples (for calculations details see M. Alfaro, 1979).

Example 1:

Let $Z(x) = \overline{Y(x)^2}$; (6) depends on the 8th moments of the Gaussian random function $Y(x)$.

Suppose that a particular outcome over an interval $[o,L]$ of $\mathbb{R}^1$ has been attained. Let $\rho(h)=\exp(-h/a)$. Then, it can be shown that the semi-variogram of $Z(x)$ is $\gamma_z(h) = 2(1 - \exp(-2h/a))$ and that:

$$\lim_{|h|\to 0} \sigma_F^2/\overline{\gamma(h)}^2 = 2\frac{a}{L} - \frac{a^2}{L^2}(1 - \exp(-2L/a)) \neq 0$$

Example 2:

Let $Z(x)=\exp(Y(x))$; expression (6) depends on the exponential moments of the random function $Y(x)$ which can be found by means of the characteristic function of the Gaussian law. The unidimensional law of $Z(x)$ is the lognormal law, which is very common in Geostatistics.

Suppose that there is a particular outcome on the interval $[0,L]$. Let $\rho(h)=1 - |h|/a + \ldots$; then it can be shown that the semi-variogram of $Z(x)$ is $\gamma_z(h) = e^2 |h|/a + \ldots$, and that

$$\lim_{|h|\to 0} \sigma_F^2/\overline{\gamma(h)}^2 = \frac{e^4 a}{2L} = 27.3\frac{a}{L}$$

In order to obtain a small fluctuation variance (say for example 0.01), it is necessary that L = 2730 a.

The lognormal case is very unfavorable as far as the fluctuations of the semi-variogram are concerned: there is a great disparity between functions $\gamma_R(h)$ and $\gamma(h)$ even for small distances.

On fig.4 there is a particular outcome of a stationary Gaussian random function $Y(x)$, whose semi-variogram is linear at the origin (spherical model, L/a = 2). If $Z(x)=\exp(Y(x))$, then different semi-variograms are obtained on fields S and S': the local semi-variogram depends on the local mean (proportional effect). See also fig.5.

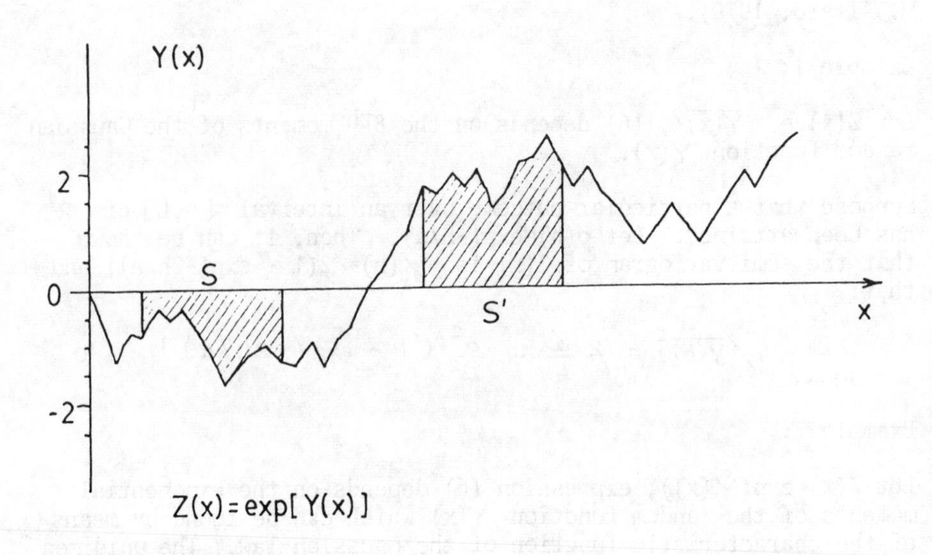

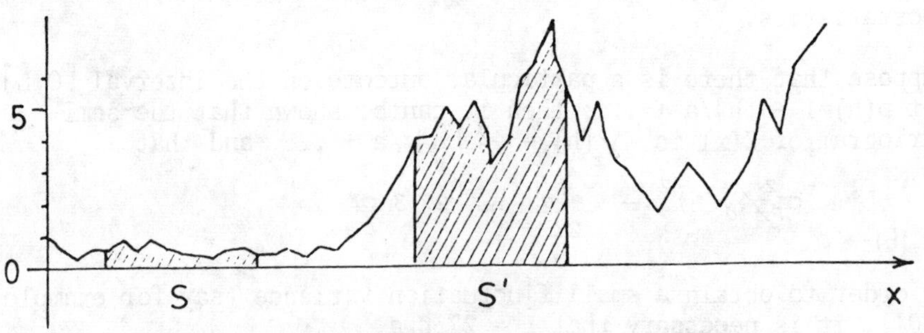

Fig. 4. Proportional Effect.

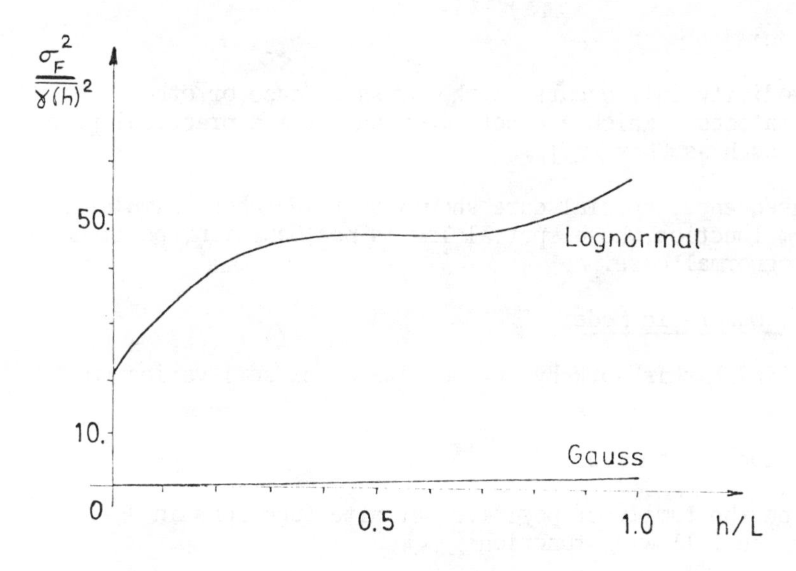

Fig. 5: Relative fluctuation variance for the model $\gamma(h) = C\,(1-\exp(-h/L))$, Lognormal and Gaussian Case.

The above results let us state the following theorem:

Theorem: Let Z(x) be $Z(x) = \varphi(Y(x))$, where Y(x) is a stationary Gaussian random function. Let us suppose that the semi-variogram of Y(x), near the origin is $\gamma(h) = a\,|h|^{\lambda} + \ldots$, then the semi-variogram of Z(x) is microergodic when $0 < \lambda < 2$ if the following relation is true:

$$|\varphi'(x)| = \text{constant}, \quad \forall\, x$$

Examine a non-rigurous demonstration of this theorem. Expression (4) is written as:

$$\sigma_F^2 = \frac{1}{4S'^2}\iint_{S'S'} E\left(\overline{\varphi(Y_{x+h})-\varphi(Y_x)}^2\ \overline{\varphi(Y_{y+h})-\varphi(Y_y)}^2\right)dx\,dy - \overline{\gamma_Z(h)}^2$$

If |h| is small, them since Y(x) is continuous in quadratic mean, it can be written:

$$\varphi(Y_{x+h}) - \varphi(Y_x) \cong \varphi'(Y_x)(Y_{x+h} - Y_x)$$

Then if |h| is small:

$$\sigma_F^2 = \frac{1}{4S'^2}\iint_{S'S'} E\left(\overline{\varphi'(Y_x)}^2\,\overline{\varphi'(Y_y)}^2\left(\overline{Y_{x+h}-Y_x}^2\cdot\overline{Y_{y+h}-Y_y}^2\right)\right)dx\,dy - \overline{\gamma_Z(h)}^2$$

Here we have the conditions prevailing in the Gaussian case when the derivative of φ is a constant.

Conclusion:

Microergodicity only occurs in the Gaussian case or other transformations which are not important from a practical point of view, such as $Z_x = |Y_x|$.

As a consequence, special care should be taken about simulations of random function whose spatial law is non-Gaussian, particularly in the lognormal case.

III. The Quadratic Model

We will finish this work by introducing a new semi-variogram model:

Construction of the model:

Let $\mathcal{H}^n$ be the family of positive definite functions in $\mathbb{R}^n$. Consider the following function:

$$C_1(h) = \begin{cases} 1 - 4|h| + 3|h|^2 & \text{if } |h| \leq 1 \\ 0 & \text{if } |h| > 1 \end{cases}$$

This function belongs to $\mathcal{H}^1$; actually, the Fourier transform of $C_1(h)$ is:

$$\mathcal{F}(C_1(h)) = \int_{-\infty}^{\infty} e^{iuh} C_1(h)\,dh = 2\int_0^1 \cos(uh)C_1(h)\,dh = \frac{4}{u^2}\left(2+\cos(u) - \frac{3\sin(u)}{u}\right) \geq 0$$

and according to Bochner theorem, $C_1(h) \in \mathcal{H}^1$

The rotating lines method of G. Matheron (cf. Journel Huijbregts, 1978) states that:

$$C_3(h) = \frac{1}{h}\int_0^h C_1(h)\,dh = (1-|h|)^2 \qquad (|h| \leq 1)$$

belong to $\mathcal{H}^3$. Therefore $C_3(h)$ can be used as a covariance model of a stationary random function in $\mathbb{R}^3$ space.

The corresponding semi-variogram is $\gamma(h) = C_3(0) - C_3(h) = 2|h| - |h|^2$ if $h \leq 1$. Then we may define the quadratic model of range "a" and sill "C" as Follows:

$$\gamma(h) = \begin{cases} C\left(2\frac{|h|}{a} - \frac{|h|^2}{a^2}\right) & \text{if } (|h| \leq a) \\ C & \text{if } (|h| > a) \end{cases}$$

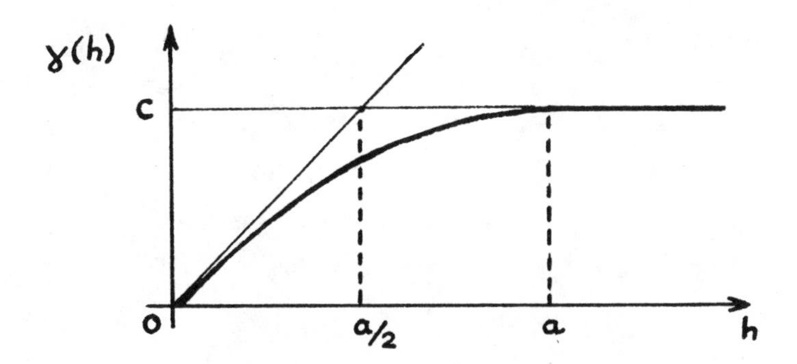

Fig. 6: Quadratic Model

This model is similar to the spherical model, but it is more simple. It has been successfully applied to the evaluation of ore blocks at the Chuquicamata deposit.

The model proposed may be very useful in geostatistical practice to fit nested models. The fact that its formulation contains only $|h|$ and $|h|^2$ terms makes the quadratic model a serious rival as compared to the spherical model (which contain $|h|$ and $|h|^3$ terms) as far as calculation time is concerned.

IV. References.

M. Alfaro, 1979 : Robustesse des simulations de fonctions alêatores. Thesis. E.N.S. Mines, París, France. 161 pp.

A. Journel
Ch. Huijbregts 1978 : Mining Geostatistics. Academic Press, N.Y. 600 pp.

G. Matheron, 1965 : Les variables regionalisées et leur estimation Masson, París, France. 305 pp.

G. Matheron, 1978 : Estimer et Choisir. E.N.S. Mines, París, France, 202 pp.

G. Matheron, 1970 : Recherche de Schemas polynomiaux E.N.S. Mines, París, France (N.177).

USE OF THE JACKKNIFE METHOD TO ESTIMATE AUTOCORRELATION FUNCTIONS (OR VARIOGRAMS)

C. F. Chung

Geological Survey of Canada

ABSTRACT

An application of the jackknife method to estimate autocorrelation functions(or variograms) of stationary random functions is discussed. Using the jackknife estimators and the corresponding jackknife variances, the models of the autocorrelation functions are fitted by the weighted least squares method. The method is particularly effective to study robustness of the estimators when the number of data points is small. The technique is applied to simulated data sets with known autocorrelation functions.

INTRODUCTION

To study spatial variation or "regionalized" variables(Matheron, 1962), the theory of stationary random functions has been applied. In application, an important but difficult problem is to postulate the underlying autocorrelation model and to estimate the unknown parameters in the model from the data. This paper does not discuss the problems of how to postulate the model but deals with estimation of its parameters.

Contrary to the importance of the autocorrelation function, only three studies, Agterberg(1970), Huijbregts(1971), and Cressie and Hawkins(1980) have been published on the estimation problems, after their application had been introduced in geosciences and mining engineering by Matheron(1962). However, extensive statistical research have been performed on the estimation of autocorrelation functions in time-series analysis where the observations are made at regular intervals and the

G. Verly et al. (eds.), Geostatistics for Natural Resources Characterization, Part 1, 55–69.

models are relatively simple. The latter two papers have discussed variograms instead of autocorrelation functions. The variogram(Jowett, 1955; Matheron, 1962) is defined as $E(Z(x)-Z(y))^2$ compared to $E((Z(x)-E(Z(x)))(Z(y)-E(Z(y))))$ for autocorrelation function. for a stationary random function,

$$\frac{1}{2}\text{variogram} = \text{variance} - \text{autocorrelation function}$$

Hence, the estimation problems for autocorrelation function and variogram are similar to each other.

Agterberg(1970) has discussed the problem in relation to the irregularly spaced data and data which contain a large scale systematic variation(trend or drift). Huijbregts(1971) has studied how to reconstruct the variogram when the observations are made in intervals (areas in the two dimensional space) rather than at points(i.e. the data are for $\int_A Z(x)\, dx$ where A represents the intervals, instead of for $Z(x)$). Cressie and Hawkins(1980) have discussed several robust methods of estimating variograms and have proposed to use M-estimators after applying the fourth root power transformation.

In this study, the jackknife method is proposed to compute correlograms followed by estimation of the unknown parameters in the autocorrelation models by the weighted least squares technique using jackknife estimators and the corresponding jackknife variances. The jackknife technique is a general method to estimate variances of the resulting estimators and it can be used in conjunction with any estimation procedure.

In order to illustrate the methodology discussed here, two realizations of standard normal processes in the one-dimensional space with known autocorrelation functions have been simulated. These simulated data sets were used for the experiments.

ESTIMATION OF AUTOCORRELATION

Let $Z(t)$ be a stationary random function in the one dimensional space with the autocorrelation function $B(s)$ defined as:

$$B(s) = E[Z(t) - m)(Z(t+s) - m) \quad / \sigma^2 \tag{1}$$

where $m = E(Z(t))$ and $\sigma^2 = var(Z(t)) = E(Z(t) - m)^2$. $B(s)$ can be interpreted as the population correlation coefficient of the pair of random variables $(Z(t), Z(t+s))$.

When $z(t)$ is an observed value of $Z(t)$, $z(t)$ is called a realization of $Z(t)$. In practice, only a few $z(t)$ at discrete points $t = t_1, t_2, \ldots, t_n$, for $Z(t)$, are usually available. In time-series analysis, the t_i 's are regularly spaced. i.e., $\delta = t_{i+1} - t_i$ is constant for all i. However, the t 's are usually irregularly spaced in the spatial analysis.

The problem to be discussed is how to estimate B(s) from $\{z(t) : t = t_1, \ldots, t_n\}$. As in time-series analysis, the correlogram based upon $\{\hat{B}(s_j) : j = 1, 2, \ldots, p\}$ is first computed from the data $\{z(t)\}$.

To estimate B(s) at $s = s_j$, let $s_j = d$ for simplicity of notation. Among n observed values $\{z(t_i) : i = 1, 2, \ldots, n\}$, suppose that there are n_d pairs $\{(z(t_k), z(t_{k'})) : k = 1, 2, \ldots, n\}$ of the observations with $t_{k'} - t_k = d$. n_d is normally much smaller than n/2 depending upon d , since the t_i's are, in general, irregularly spaced.

The most commonly employed method to estimate B(d) is:

$$\hat{B}(d) = \frac{1}{n_d \hat{\sigma}^2} \sum_{k=1}^{n_d} (z(t_k) - \bar{z})(z(t_{k'}) - \bar{z}) \tag{2}$$

where $\bar{z} = \frac{1}{n} \sum_{i=1}^{n} z(t_i)$ and $\hat{\sigma}^2 = \frac{1}{n} \sum_{i=1}^{n} (z(t_i) - \bar{z})^2$

However, no attempt was made to study the distribution of $\hat{B}(s)$ because of its complexity. Kendall (1973) has also used $\hat{B}(s)$ to compute correlogram in time-series analysis. Although the distributions of sample correlation coefficients and serial correlations of time-series data have been extensively studied (Chapter 32, Johnson and Kotz, 1970), the exact characteristics of the statistic $\hat{B}(d)$ in (2) are not known.

Suppose that m = 0 and $\sigma^2 = 1$ in (1), then B(d) in (1) becomes E(Z(t)Z(t+s)). In this situation, $\hat{B}(d)$ in (2) is reduced to

$$\hat{B}(d) = \frac{1}{n_d} \sum_{k=1}^{n_d} z(t_k)\, z(t_{k'}) \tag{3}$$

Assume that the n_d values $\{z(t_k) z(t_{k'}) : k = 1,2,\ldots,n$ and $t_{k'} - t_k = d\}$ are mutually independent observations of the random variable Z(t)Z(t+d), $\hat{B}(d)$ in (3) is the moment estimator of B(d) and thus, it is a consistent estimator. In addition, if Z(t) is assumed to represent a normal process, then

$$\tilde{B}(d) = \hat{B}(d) \left(1 + \frac{1 - \hat{B}(d)^2}{2(n_d - 4)}\right) \tag{4}$$

is the minimum variance unbiased estimator of B(d)(Johnson and

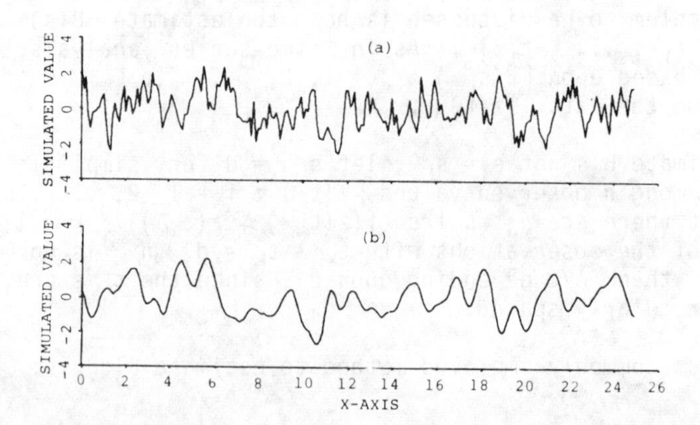

Figure 1. Two sequences of 250 regularly spaced realizations of normal processes of mean 0 and variance 1 with autocorrelation functions: (a) $B(s) = \exp(-2s)$, (b) $B(s) = \exp(-s^2)$.

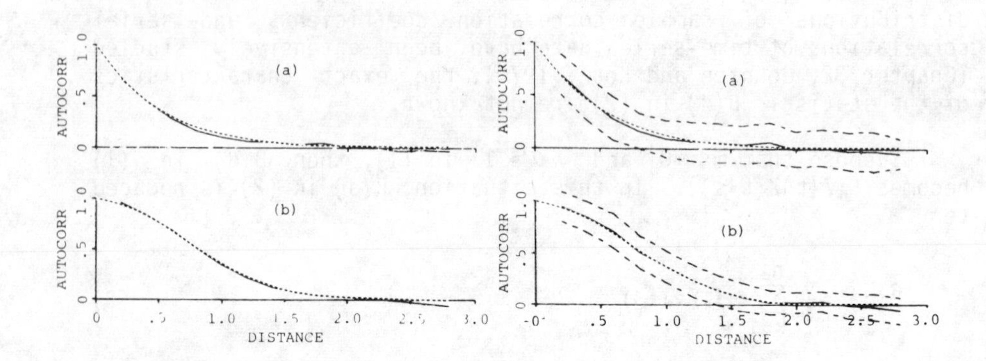

Figure 2. Correlograms (a) and (b) computed using eq.(4) based on the realizations shown in Figures 1.a and 1.b, respectively and shown as solid lines. The superimposed dotted lines are the models: (a) $B(s) = \exp(-2s)$, (b) $B(s) = \exp(-s^2)$.

Figure 3. Correlograms computed using jackknife method of eq.(4) based on Fig. 1. The jackknife bands (estimated values + 2 x jackknife standard deviations) are plotted as broken lines. The models are also superimposed as in Fig. 2.

Kotz, 1972). Hence $\hat{B}(d)$ can be a "reasonable" estimator of $B(d)$ when n_d is large.

However, the n_d pairs of observations $\{(z(t_k), z(t_{k'}))\}$ are not independent in autocorrelated data. Furthermore the underlying conditions of $Z(t)$ are not known in practice.

Therefore, it is very important to have a measurement of the variability of $\tilde{B}(d)$ in (4). A natural choice to obtain such a measurement is by using the jackknife method.

When the underlying distribution of $Z(t)$ is not known or $Z(t)$ contains random noises (outliers), it is desirable to have a robust estimator of $B(d)$ which performs reasonably well under a variety of underlying condition (Andrews et.al., 1972) as Cressie and Hawkins(1980) have proposed. As will be shown later, the Andrew's M-estimators did perform well on the simulated data with random noise.

After estimating $B(s_j)$ for $j = 1, 2, \ldots, p$, the correlogram is plotted and then a model with unknown parameters is postulated. Using the correlogram, the parameters are usually estimated by the least squares method(Draper and Smith, 1968).

However, when the jackknife method is employed to compute the correlogram not only the $B(d)$'s but also the corresponding jackknife variances are computed. The weighted least squares method is then used to estimate the unknown parameters.

A detailed description of the jackknife method and the M-estimators is given in the appendix.

SIMULATED TEST DATA

Two stationary normal processes with known autocorrelation functions are considered in this section. For each of the two models, a sequence of realizations of the normal process with zero mean and variance 1 is simulated at regularly spaced discrete points in the one-dimensional space. Each sequence consists of 250 numbers at every multiple of 0.2 in the interval [0 , 49.8] They are shown in Figure 1. The autocorrelation functions used are:

$$\begin{aligned} &\text{(i)} \quad B_1(s) = \exp(-2s) \\ &\text{(ii)} \quad B_2(s) = \exp(-s^2) \end{aligned} \tag{5}$$

From the realizations in Figure 1, the correlograms were computed by using equation (4). They are shown in Figure 2 together with the B (s) and B (s)(dotted lines). Figures 4 and 5 illustrate the correlograms computed by using Andrews' M-estimators directly and the M-estimators after the fourth root power transformation as Cressie and Hawkins(1980) have proposed, respectively.

As in Figure 2, the equation (4) reproduces the models in (5) very well. However, the robust method(Andrew's M-estimators) overestimates all the values except for the first lag of the second model. The Cressie-Hawkins' method works well. The jackknife method applied on (4) results almost identical to that of (4) and the results are shown in Figure 3 together with two broken lines representing the esimated values + (2 times the corresponding jackknife standard deviations). These "jackknife bands" may be used for the 95% confidence intervals of the esimators(Tukey,1958).

EXPERIMENTS

(I) Within each of the series shown in Figure 1, 25 out of 250 were randomly selected and replaced by the random values generated by normal variate of mean 0 and variance 9 . That is, 10% of the original series from N(0,1) with covariance functions in (5) are replaced by random noises of N(0,9). The modified series are shown in Figure 6. Unlikely to Figure 1.b, the smoothness describing strong correlation is not immediately apparent in Figure 6.b.

All three methods, the ordinary(equation (4)), the robust(Andrew's M-estimator) and the Cressie-Hawkins' estimators are applied to the modified series and are shown in Figures 7, 9 and 11, respectively. The 250 values were randomly divided into 25 groups of 10 values each and these grouping were used for applying the jackknife methods. Figures 8, 10 and 12 illustrate the results from the corresponding jackknife methods of the ordinary, robust and Cressie-Hawkins' estimators, respectively. By comparing the estimators and the corresponding jackknife estimators(Figures 7-12), it is obvious that the two matching correlograms are almost identical to one another.

Clearly the ordinary estimators(Figure 7) pathetically fails to capture the models. It implies that, when the data set contains over 10% random noise, the equation (4) is not reliable, even if the data set contains 225 "good" values. As expected, the jackknife bands in Figure 9 are much wider than those of Figure 3. The only comfort we have is that the models shown in dotted lines are within the bands.

The M-estimators depict the models very well as shown in Figure 9 opposed to the results shown in Figure 4. In

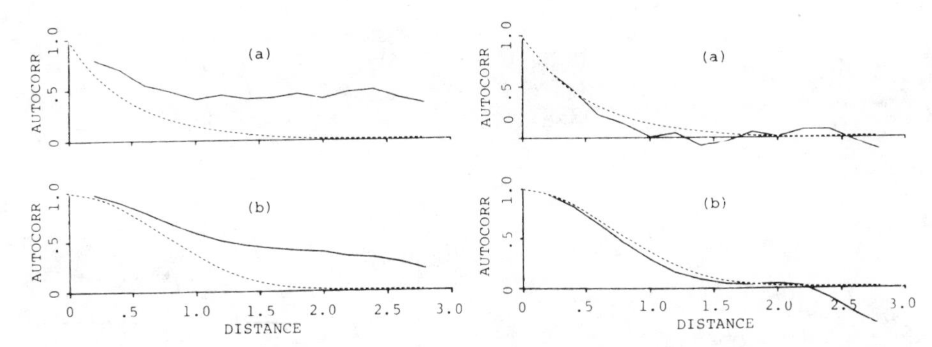

Figure 4. Correlograms computed using M-estimators (eq. (9)) based on Fig. 1.

Figure 5. Correlograms computed using the Cressie-Hawkins' estimators based on Fig. 1.

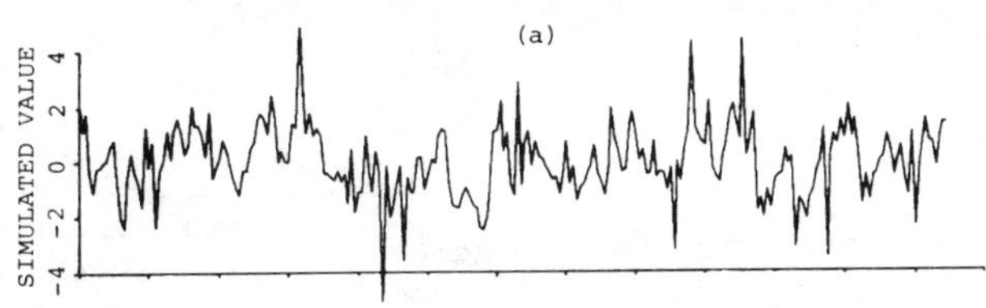

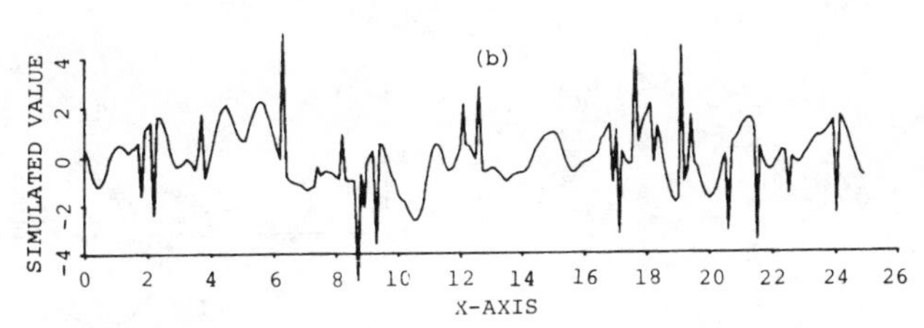

Figure 6. Two sequences with random noises. These two realizations are generated from the sequences in Fig. 1 by selecting 25 out of the 250 values randomly and replacing by random numbers from Normal (0,9).

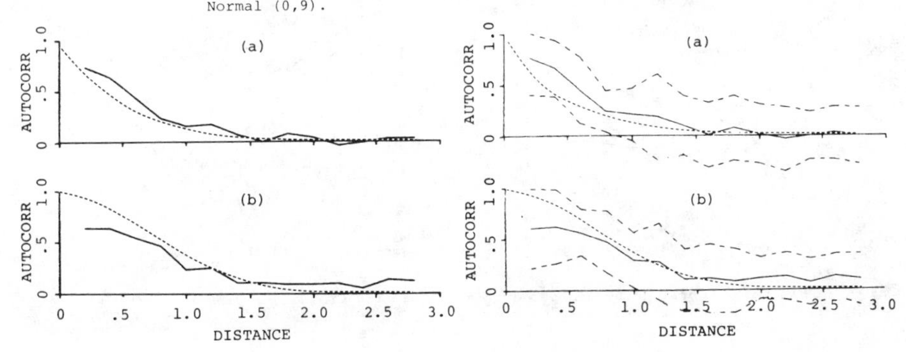

Figure 7. Correlograms computed using eq. (4) based on Fig. 6.

Figure 8. Correlograms computed using jackknife method of eq. (4) based on Fig. 6. The jackknife bands are superimposed as the broken lines.

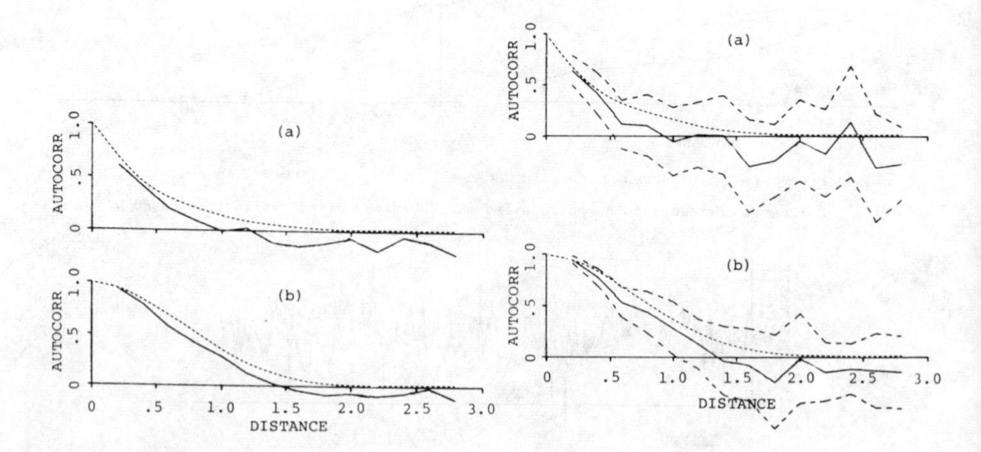

Figure 9. Correlograms computed using M-estimators based on Fig. 6.

Figure 10. Correlograms computed using jackknifed M-estimators based on Fig. 6.

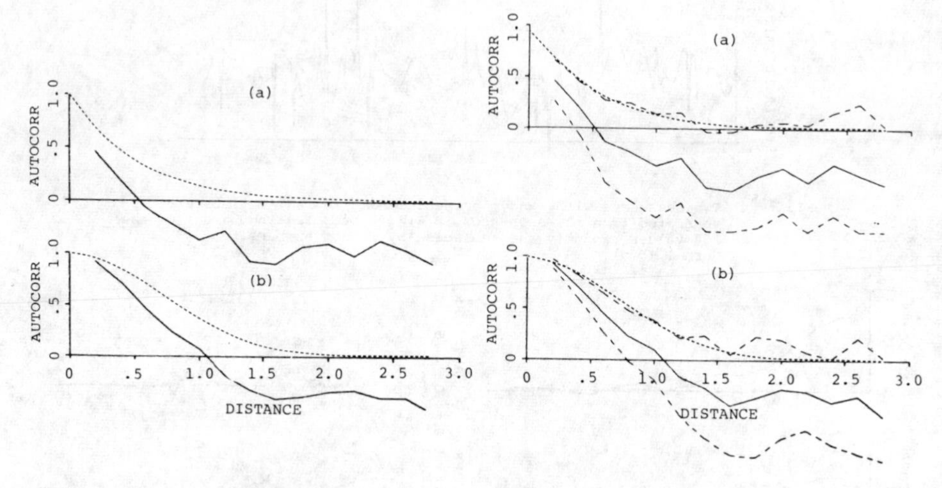

Figure 11. Correlograms computed using Cressie-Hawkins' estimators based on Fig. 6.

Figure 12. Correlograms computed using jackknifed Cressie-Hawkins' estimators based on Fig. 6.

particular, the jackknife band of second model in Figure 10.b is interesting and in fact it represents an important reason to use the jackknife methods. From Figure 10.b, the jackknife variance of the estimator of $B_2(0.2)$ is very small and the band is getting wider as it goes farther from the origin. The same property is observed in Figure 12.b. It implies that the estimators of the first few lags are more significant than those of the lags with long distance from the origin.

When the weighted least squares method is applied for estimating the unknown parameters of the model using the jackknife variances, the significant estimators with smaller variances will naturally get more weights than those with the larger jackknife variances. This will be shown in the next experiments.

The Cressie-Hawkins' estimators underestimate the models. Also as in the ordinary estimators, it is difficult to postulate models in (5) properly from the correlograms in Figure 11.

From these experiments, the correlograms by the M-estimators are certainly superior when there are random noises in the data sets.It is also reinforced by the following experiments.

(II) 100 points within each of the modified series used in (I) were selected at random. The series of the 100 points selected each are shown in Figure 13. The 100 points are divided into 10 groups of 10 points each. These groups are used for applying the jackknife methods. As in (I), the three methods and their corresponding jackknife methods are applied to the series and the correlograms computed are illustrated in Figures 14-19.

As expected, the ordinary method again completely fail to reconstruct the models. The method is not able to estimate even the first lag of the autocorrelation function in reasonable range. However when the parameters were estimated from the correlograms assuming the models are properly postulated, the estimator of the first model was good(see Table 1).

However the other robust methods depict the first three lags of the correlogram of the model B (h)(Figures 17.b and 19.b) very well. In particular, the first two lags of the correlogram are almost identical to those of the model: the jackknife bands are very narrow. It implies that those two estimators are statistically significant. On other hand, the jackknife bands in Figures 17.a and 19.a are so wide that it is not practically possible to reconstruct the autocorrelation function in (5). When the data set contains more than 10% of random noise and the sample points are relatively small(100 points), in general, it may not be reasonable to postulate the model properly from the

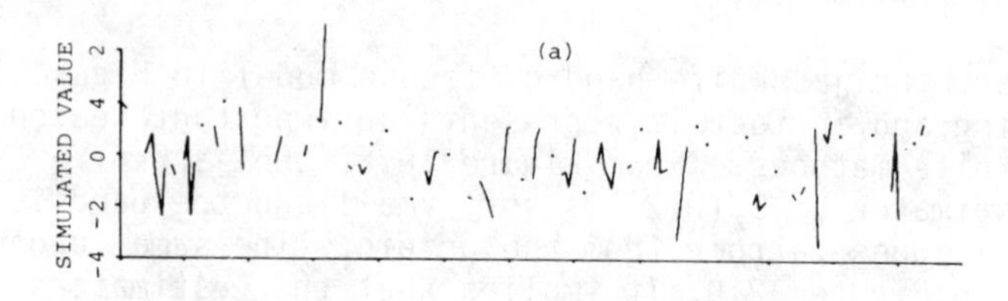

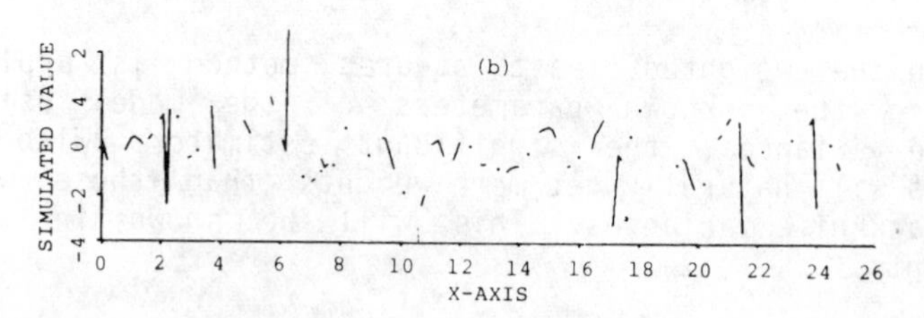

Figure 13. Two realizations of 100 randomly selected points from Fig. 6.

		Ordinary estimators eq. (4) Fig. (14)	Jackknifed ordinary estimators Fig. (15)	M-estimators eq. (9) Fig. (16)	Jackknifed M-estimators Fig. (17)	Cressie-Hawkins' estimators Fig. (18)	Jackknifed Cressie-Hawkins' Fig. (19)
$B_1(s)=\exp(qx)$ $q=-2.0$	estimator	-1.738	-2.105	-1.587	-1.917	-3.404	-3.512
	variance	0.310	0.374	0.218	0.242	0.240	0.131
	95% conf.int.	(-2.85,-0.63)	(-3.33,-0.88)	(-2.52,-0.65)	(-2.90,-0.93)	(-4.38,-2.42)	(-4.24,-2.79)
$B_2(s)=\exp(qx^2)$ $q=-1.0$	estimator	-55.086	-55.114	-1.100	-1.052	-1.460	-1.296
	variance	1967.7	2085.1	0.015	0.015	0.019	0.014
	95% conf.int.	(-144,34)	(-146,36)	(-1.35,-0.85)	(-1.29,-0.81)	(-1.74,-1.19)	(-1.54,-1.06)

Table 1.

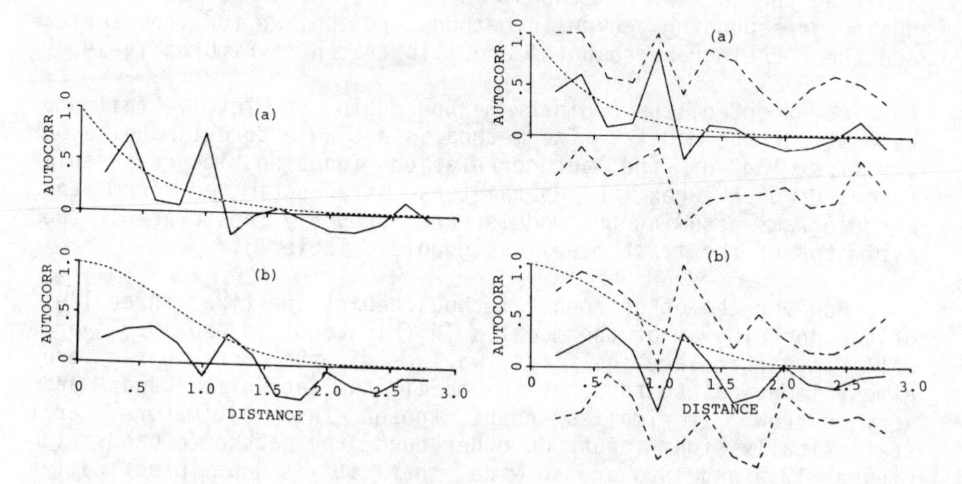

Figure 14. Correlograms computed using eq.(4) based on Fig. 13.

Figure 15. Correlograms computed using jackknife method of eq.(4) based on Fig. 13.

computed correlograms.

However, assume that the models of the $B\ (s)$ are known as:

$$B\ (s) = \exp(\ q\ s\ )$$
$$B\ (s) = \exp(\ qs^2\ ) \qquad (6)$$

for the correlograms in Figures 14, 16 and 18, respectively. Using the nonlinear least squares method, the q's in (6) were estimated (see Table 1). The computations were performed by using SPSS subprogram NONLINEAR(Robinson, 1979).

On the same data set, the jackknife method was employed to obtain the correlograms with the jackknife confidence intervals. These are shown in Figures 15, 17 and 19. The parameters in (6) were estimated by the weighted nonlinear least squares method using the jackknife variances as weighting components(see also Table 1).

From the Table 1, obviously the jackknifed M-estimators most accurately estimates the original parameters as expected from the Figures 14-19. The Cressie-Hawkins' estimators underestimate the parameters in both of the models. Although the estimators of the parameters by the ordinary estimators and its jackknife estimators in model 1 close to the true value $q = -2.0$, it may be a just coincident.

(III) Within each of the original 250 realizations shown in Figure 1, a set of 100 numbers were chosen at random for these experiments. These series of the 100 values chosen are shown in Figure 20. Using equation (4), the robust and the Cressie-Hawkin's methods the correlograms were computed and are shown in Figures 21,22 and 23, respectively. All three methods result very similar correlograms.

Since the selections were made at random, the numbers of pairs of observations used to compute the correlograms, are similar to each other and these 100 points are identical 100 points as in (II).

CONCLUDING REMARKS

Much work remains to be done to understand the exact or an approximate distribution of the $\widetilde{B}(s)$ in (4). In addition further study is required to improve upon the procedure of computing correlogram by equation (4).

Several assumptions which are unrealistic in practice have been made in this study. For instance, most spatial data in the

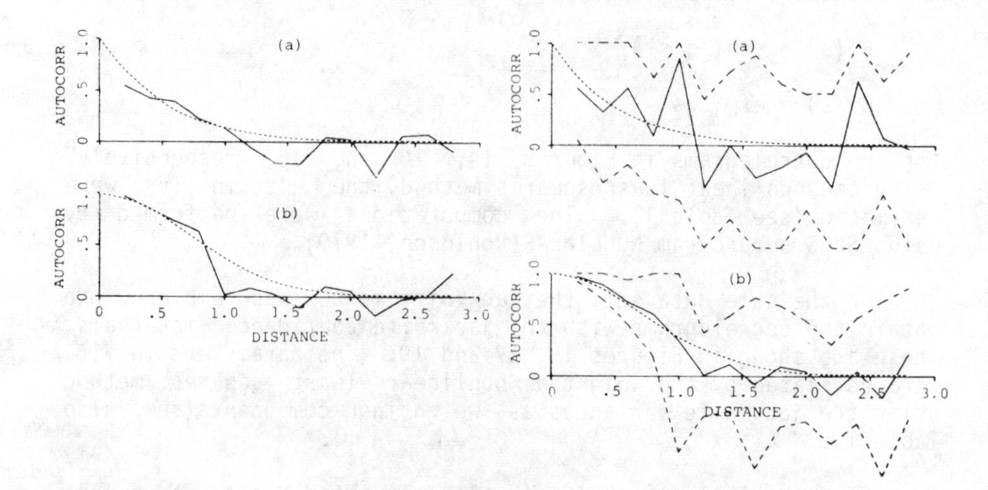

Figure 16. Correlograms computed using M-estimators based on Fig. 13.

Figure 17. Correlograms computed using jackknifed M-estimators based on Fig. 13.

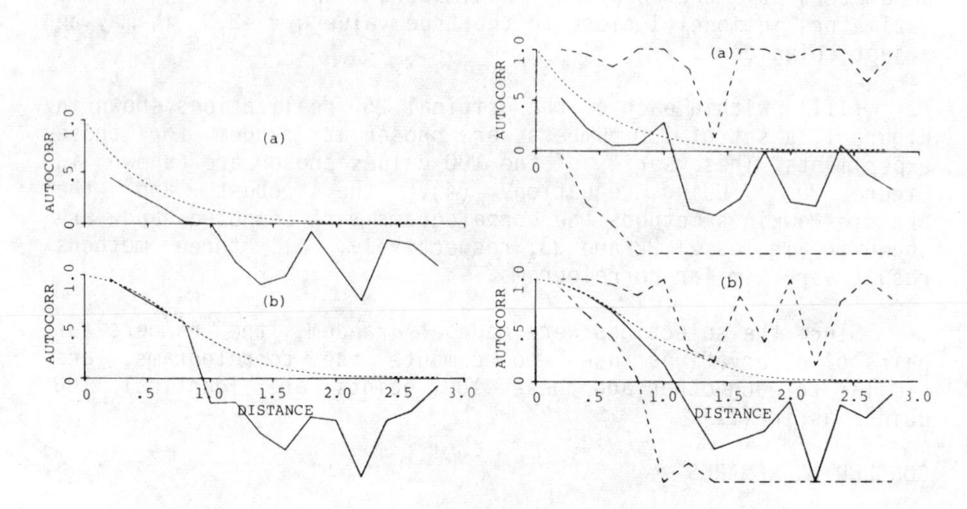

Figure 18. Correlograms computed using Cressie-Hawkins' estimators based on Fig. 13.

Figure 19. Correlograms computed using jackknifed Cressie-Hawkins' estimators based on Fig. 13.

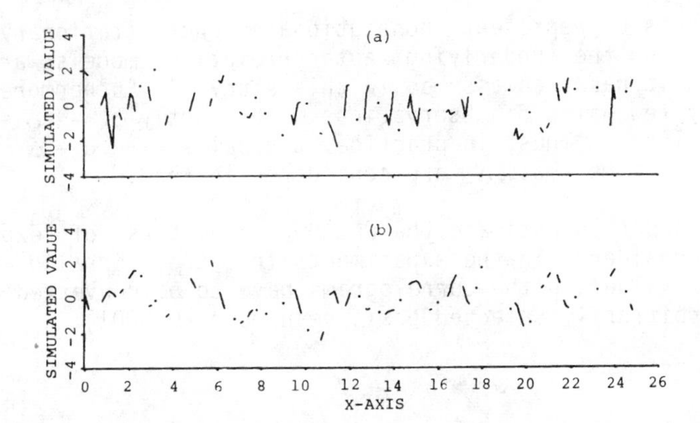

Figure 20. Two realizations of 100 randomly selected points from Fig. 1.

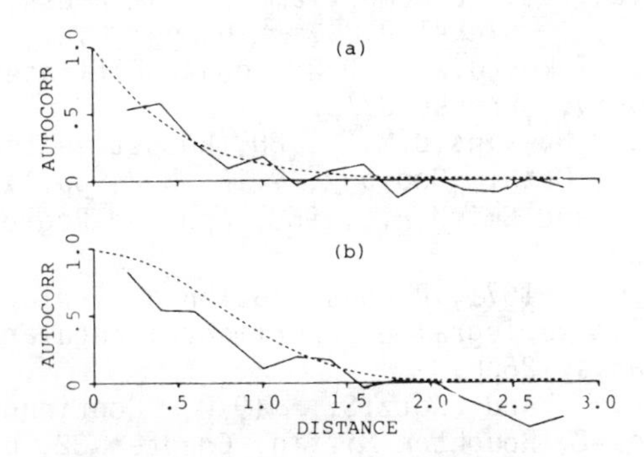

Figure 21. Correlograms computed using eq.(4) based on Fig. 20.

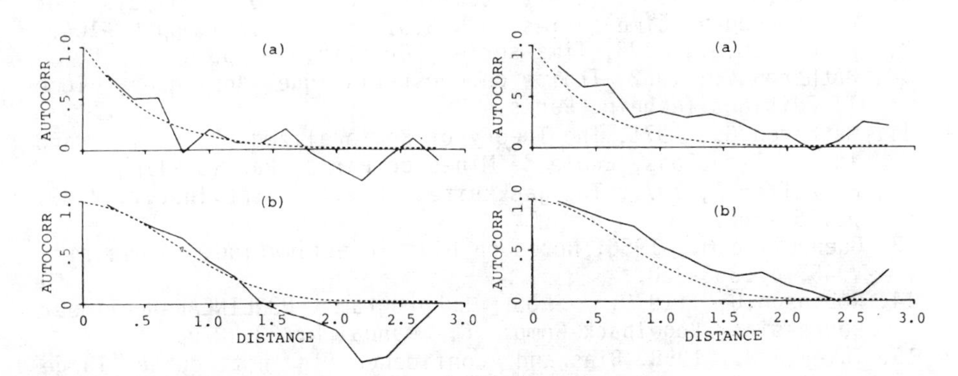

Figure 22. Correlograms computed using M-estimators based on Fig. 20.

Figure 23. Correlograms computed using Cressie-Hawkins' estimators based on Fig. 20.

geosciences represent nonstationary, not stationary random function and the underlying autocorrelation models are never known as it has been assumed in this study. Futhermore, there are very few pairs of observations with exactly $t_k - t_{k'} = d$ for any spacing d. Thus, in practice, the pairs with $d - \Delta < t_k - t_{k'} < d+\Delta$ for some small Δ are considered instead.

In order to estimate the unknown parameters of exponential models considered in the experiments from the correlograms, all negative values in the correlograms have to be converted into a small arbitrarily selected positive number(=0.0001).

REFERENCES

1. Agterberg,F.P., 1970, Autocorrelation functions in geology, Geostatistics, Ed. D.F.Merriam, Plenum Press, pp.113-141.
2. Andrews,D.F., Bickel,P.J., Hampel,F.R., Hubel,P.J., Rogers,W.H., Tukey,J.W., 1972, Robust Estimates of Location, Princeton Univ. Press, 371p.
3. Cressie,N and Hawkins,D.M., 1980, Robust estimation of the variogram: I, Math. Geology, V.12, N.2, pp.115-125.
4. Draper,N.R. and Smith,H., 1968, Applied Regression Analysis, Wiley, 407pp.
5. Huijbregts,Ch., 1971, Reconstitution du variogramme ponctuel a partir d'un variogramme experimental regularize, Ecole de Mines de Paris, 26pp.
6. Johnson,N.L. and Kotz,S., 1970, Continuous Univariate Distributins-2, Houghton Miffin, Chapter 32, p.220-252.
7. Johnson,N.L. and Kotz,S., 1972, Continuous Multivariate distributions, John Wiley, Chapter 35, p37-83.
8. Jowett,G.H., 1955, Least squares regression analysis for trend-reduced time-series, J.R.S.S., Ser.B, V.17, pp.91-104.
9. Kendall,M.G., 1973, Time-series, Griffin, 197pp.
10. Matheron G., 1962, Traite de Geostatistique Appliquee, Tome 1, Editions Technip, Paris.
11. Matheron,G., 1971, The Theory of Regionalized Variables and its applications, Ecole de Mines de Paris, Paris, 211pp.
12. Mosteller,F, 1971, The jackknife, Rev.Int.Stat.Inst., V.39, pp.363-368.
13. Quenouille,M., 1956, Notes on bias in estimation, Biometrika V.43, pp.353-360.
14. Robinson,B, 1979, SPSS Subprogram NONLINEAR-Nonlinear Regression, Vogelback Comp. C. Manual N.433, 27pp.
15. Tukey,J.W., 1958, Bias and confidence in not quite large samples, Ann. Math. Stat., V.29, pp.614.

APPENDIX

The jackknife method was first introduced by Quenouille(1956) for the purpose of reducing bias and later used by Tukey(1958) for obtaining approximate confidence interval. Mosteller(1971) has emphasized the use of the method to obtain the variance of the resulting estimator by sample reuse.

Suppose that p is an unknown parameter to be estimated from a sample of size n. Let $\hat{p}_a$ be an estimator of p based on all n data. When n = gh , let $\hat{p}_{-i}$ be the corresponding estimator of p based on a sample of size (g-1)h , where the i-th group of size h has been deleted. Define the i-th pseudo-estimator

$$\hat{p}_i = g\hat{p}_a - (g - 1)\,\hat{p}_{-1} \quad \text{for all } i = 1, 2, \ldots, g.$$

Then the estimator p:

$$\hat{p} = \frac{1}{g}\sum_{i=1}^{g} \hat{p}_i \tag{7}$$

is called the jackknife estimator of p. Since the $\hat{p}_i$'s are approximately independent(Tukey, 1958), the variance of p can be computed by:

$$s^2 = \frac{1}{g(g-1)}\sum_{i=1}^{g} (\hat{p}_i - \hat{p})^2 \tag{8}$$

where s^2 is called the jackknife variance of $\hat{p}$.

In this appendix, only one M-estimator, that of Andrews, is discussed because its use was proposed by Cressie and Hawkins(1980) to estimate the variogram.

In general, the M-estimator of the location parameter of $\{ y_k : k = 1, 2, \ldots ,n \}$ is a solution for T from an equation of the form:

$$\sum_{k=1}^{n} \Psi\, \frac{Y_k - T}{S} = 0 \tag{9}$$

In Andrew's M-estimator, the function:

$$\Psi(x) = \begin{cases} \sin(x/2.1) & |x| < 2.1\,\pi \\ 0 & \text{otherwise} \end{cases} \tag{10}$$

where s is the median of $\{ y_k \}$.

COMPUTING VARIOGRAMS IN FOLDED STRATA-CONTROLLED DEPOSITS

M. Dagbert(*), M. David(**), D. Crozel(**),
A. Desbarats(**)(1)

(*) Geostat Systems International Inc., Montreal, Canada
(**)Ecole Polytechnique, Montreal, Canada
(1) Now at Stanford University

ABSTRACT

In this paper we present two different approaches to compute variograms along folds in layered deposits. The common objective of the two methods is to measure the vector distance between samples in a "natural" coordinate system with axes parallel to strike, down-dip direction and true thickness of the layer. Because of the folded structure, the orientation of the reference system is changing at every point of the deposit.

Both methods necessitate the knowledge of the stratigraphic position of every sample. The first method is applicable to deposits with about cylindrical folds i.e. where the profile of the folds does not vary too much from one section to the next. In this case one of the coordinates is measured along these profiles which must be digitized. The other two coordinates are the distance along the axis of the folds and the stratigraphic position of the sample. In the second method, the true thickness vector at any point of a layer is determined by kriging the gradient of the stratigraphic position.

G. Verly et al. (eds.), Geostatistics for Natural Resources Characterization, Part 1, 71–89.

1 - INTRODUCTION

In the recent years, there has been a growing concern in the geostatistical community to have geostatistical models of orebodies which closely fit existing geological models for the same orebodies. Good illustrations of that concern are the paper by T. Barnes (1) where a paraboloidal reference system is used to compute variograms in a porphyry molybdenum deposit and the paper by J.M. Rendu et al. (2) where several other "natural" reference systems are proposed. The purpose of this paper is to present methods to define similar "natural" reference systems in strata controlled deposits. A typical example of deposits of this type are hematite banded iron ore deposits where one can observe a succession of layers of almost constant thickness and intrinsic mineralogical characteristics for the ore in each band (3). Very frequently these deposits are no longer isoclinal and in some cases they have been intensely folded and/or faulted. This means that it is not realistic to consider the usual enclidean distance that can be measured between two points at the present time as the distance which existed between the two points at the time of the ore deposition.

Then if a meaningful variography and/or kriging is to be done, it is necessary to try and reproduce these original distances. In this paper, we present two methods to solve this problem. The first method has a somewhat limited scope since it can only be used when folds are cylindrical and not overturned. We called it the "limited unfolding" technique. We have completely programmed it and used it successfully in the estimation of the ore reserves of a large uranium deposit in Australia. The original idea behind this method is to be credited to A. Debarats (4). The second method has a more general scope. It has been proposed by D. Crozel (5) but it has not yet been fully tested on a real case.

2 - THE LIMITED UNFOLDING TECHNIQUE

2.1 - Limitation of the method

This method is applicable to deposits where the different strata have been subject to cylindrical folding. Cylindrical folding is basically what you get when you squeeze a sheet of paper between two opposite sides(Figure 1).

With cylindrical folding, the axes, of the folds are straight lines parallel to the same direction. This privileged direction along which no more "stretching is possible would be used as the "strike axis" of our "natural" reference system. By projecting a point on that axis we directly obtain the "strike" coordinate of that point (Figure 2).

The other two axes of the "natural" reference system are within sections perpendicular to the strike axis. The "down dip" coordinate is measured along the profile of the fold and the "depth" coordinate is measured perpendicular to that profile (figure 3).

However, it is necessary to introduce a second limitation to the method so that the automatic computation of natural coordinates is feasible. This limitation is that there should not be any overturned fold with respect to an average dip plane. Given that average dip plane which is parallel to the strike axis, an overturned fold would be such that in a section perpendicular to the strike axis, there are points along the fold that project at the same location on the trace of the average dip plane. On figure 4, the fold is overturned if the trace of the dip plane is horizontal but it is no longer overturned if a 45° dipping axis is considered.

2.2 - Computation of natural coordinates

The method starts with the digitizing of the stratigraphic limits drawn by the geologists on sections perpendicular to the strike axis. Stratigraphic limits can be numbered starting from the top strata and increasing with depth. This numbering system defines a stratigraphic score (SS) which can be as a "natural" depth coordinate (figure 5).

In order to determine the SS of any point (sample or grid point in a block to be estimated) of the deposit, we use a simple linear interpolation method.

If the point is exactly on a section which has been digitized, we first have to find on each digitized contour the two points with the closest projection on the trace of the average dip plane to the projection of the point under study (figure 6).

This defines a "stack" of segments, with increasing SS value. In that stack there is a segment immediately above the point of interest with the next segment immediately below the same point. Once these two segments are found, the SS value of the point is determined by simple linear interpolation between the two segments (figure 7).

If the point falls between two sections which have been digitized the linear interpolation is extended to the interval between the two sections (figure 8).

The determination of the down-dip coordinate of a point is based on very similar linear interpolation techniques. Once a reference plane perpendicular to the average dip plane and parallel to the strike axis has been defined, the first step is to compute the down-dip coordinate of any point on the digitized strata limits (DDC) as a function of the coordinate of that point along the trace of the average dip plane (DC) like it is shown on figure 9.

Given a point M on a digitized section, we first have to find on the strata limit immediately above the two points with the closest DC values. The down-dip coordinate corresponding to point M is then determined from linear interpolation between the two points (figure 10).

A more general method would use both the hanging wall and the footwall limit of the stratum where point M is located (figure 10 again). If point M falls between two sections, the linear interpolation can also be extended to the interval between the two sections (figure 11).

It is possible to write a computer program which uses these simple interpolation methods to determine the strike and down-dip coordinates together with the stratigraphic score of any point given the regular 3D coordinates of that point and a set of digitized geological sections with limits of strata.

2.3 - Use of Natural Coordinates

The transformed coordinates of points according to the limited unfolding technique can be used for three basic purposes: first, to plot maps of the distribution of ore and waste samples within a stratum or a substratum ("slice maps"), second to compute meaningful variograms of grade in strata and third to perform the kriging of the average grade of the portion of a block or stope in a given stratum.

Slice Maps. Slice maps are obtained by specifying an upper and lower limit for the stratigraphic score. All the samples with a SS value between the two limits are plotted on a map where the two perpendicular axes of the reference system bear the strike and down dip coordinates respectively (Figure 12).
The corresponding map is equivalent to the map of an unfolded substratum.

Variogram Computation. On figure 13, we present an example of "slice variograms" computed in the natural reference system derived from the limited unfolding technique. In this case, the variable of interest is an ore indicator defined for each 5 m interval of almost vertical drill holes. The zone of interest is one particular stratum of the deposit. Variograms have been computed along the down dip coordinate axis and the E-W strike axis. In both cases, we use a 22.5^{o} tolerance angle around the direction being investigated in the surface defined by the strike and down-dip axis. Perpendicular to that plane, i.e. along the stratigraphic score axis, slices are defined by a maximum difference of 1/10 of a stratigraphic score unit (there is a difference of 1 stratigraphic score unit between the foot-wall and the hangingwall of the stratum).

With the same data, we have computed on figure 14 more traditional variograms along the principal direction of the average dip plane which is parallel to the E-W strike axis and dipping 45° to the south. Like before variograms are computed along the strike direction and the average dip direction with a 22.5^{o} tolerance angle in the plane of the two axes. Perpendicular to that plane, the maximum difference in coordinates is 5 m.

As expected the strike variogram is not too much affected by the change of the reference system. However the down dip variogram shows a better continuity when it is computed in the natural system rather than the conventional system. This is not really surprising since, in the latter case, because of the curvatures in the profile of the strata, one ends up comparing samples which are not in the same slice of substratum when longer and longer distances are considered. In that case, we would have missed an interesting anisotropy in the geometry of the mineralized lenses if we had used a conventional system parallel to the average dip plane of the strata.

Block Kriging. As illustrated on figure 15, it is possible to use the natural reference system of coordinates derived from the limited unfolding technique to do the estimation of block grades by the kriging method. In the procedure which has been implemented, the portion of each stratum in the block is considered independently of the other portions i.e. the grade of each portion is estimated with only the neighbor samples in the same stratum and the variograms characteristic of that stratum. The portion of the block in a given stratum is discretized by points on a small grid. Natural coordinates (SC, DDC and SS) of each point are computed. When both discretization points and samples are referenced in the same natural system it is possible to define search zones with a shape parallel to the local profile of the stratum. Such a search zone is simply defined as an ellipsoid in the natural reference system of coordinates. Once neighbor samples have been collected, kriging is performed entirely in this new reference system. Estimates and estimation variances of the different block portions are then combined to get the average grade and the associated estimation variance for the entire block.

3 - A MORE GENERAL METHOD

The procedure described in this section has been developed by D. Crozel (5). It has been programmed and tested in a fairly limited case where it seems to provide good results although some additional work is necessary to check that it is applicable to more complex structural geometries than those which can be processed by the limited unfolding technique.

The basic idea behind this method is illustrated on figure 16. If we have two samples at points A and B and then again at points C and D, the first pair is comparable to the second one in terms of grade similarity even if vector AB has not the same orientation as vector CD. This is because the components of the two vectors on a "natural" reference system with a "stratigraphic" axis (s) and a "down-dip" axis (d) are about the same.

Thus the idea before computing any variogram using the usual coordinate reference system is to estimate the orientation of the "natural" reference system at any point in the deposit. One vector of this system is on the stratigraphic axis. To determine that vector, we can go back to the concept of stratigraphic score (SS) which has already been used in the limited unfolding technique. An SS value can be assigned to any point in the deposit with coordinates x, y, and z. This defines a regionalized variable SS (x,y,z). The stratigraphic axis at point M is parallel to the gradient vector of the SS function at point M:

$$\vec{s} = \frac{\text{grad SS}}{\|\text{grad SS}\|} \quad \text{with grad SS} = \left(\frac{\partial SS}{\partial x}, \frac{\partial SS}{\partial y}, \frac{\partial SS}{\partial z}\right)$$

We can estimate the gradient vector or the partial derivative of the SS function at any point M by a linear interpolation method like kriging. The data points of this kriging system are on the digitized profiles of the strata SS(x,y,z) SS_i on several section R(x,y,z) R_k across the deposit. Since the SS function is expected to show a strong drift along the direction perpendicular to the average dip plane of the strata, universal kriging has to be favored. As for a spatial covariance or variogram model, it is difficult to infer one from the digitized profiles since these data points are not well conditioned for experimental covariance or variogram computation. However, we expect a fairly good continuity of the stratigraphic score function specially at short distance since this is actually a subjective variable which is defined by the geologist to reflect a continuous geological model. Hence we will tend to arbitrary select a linear generalized covariance $C(h) = -h$ with a linear drift to estimate both the stratigraphic score SS and the components of its gradient

$$\left(\frac{\partial SS}{\partial x}, \frac{\partial SS}{\partial y}, \frac{\partial SS}{\partial z}\right)$$

at any point of the deposit and in particular at any sample location.

Once the stratigraphic axis (vector $\vec{n}_z$) of the natural reference system has been estimated from the digitized profiles of strata on sections, it is necessary to specify the position of the last two vectors of the system in the plane perpendicular to the stratigraphic axis. At this point, some arbitrary decision has to be taken as to whether the stratigraphic axis has been derived from the vertical axis through a rotation around x and z or from a rotation around y and z. In the first case, the new x-axis of the natural reference system is defined by the unit vector n_x such that:

$$n_x = \frac{\vec{n}_z \wedge \vec{z}}{\|\vec{n}_z \wedge \vec{z}\|}$$

where z is the unit vector of the original reference system and the last axis is defined by the unit vector $\vec{n}_y$.

$$\vec{n}_y = \vec{n}_z \wedge \vec{n}_x$$

No study has been done yet on the implication of choosing one rotation instead of the other.

Once the natural reference system (n_x, n_y, n_z) is defined at any sample location, we must find a way to use these systems in variogram computation. The logical step is to measure the vector distance between two samples with respect to the natural reference system at one of the extremity of this vector. Then another arbitrary decision will have to be made as to what extremity to choose. To go around this problem, we can think of using an "average"natural reference system for the two samples which is simply defined by the following relationship:

$$\vec{N}_z = \frac{\vec{n}_z^1 + \vec{n}_z^2}{\|\vec{n}_z^1 + \vec{n}_z^2\|}$$

$$\vec{N}_z = \frac{\vec{N}_z \wedge \vec{z}}{\|\vec{N}_z \wedge \vec{n}_z\|}$$

$$\vec{N}_y = \vec{N}_z \wedge \vec{N}_x$$

A disadvantage of this average system is that it does not provide a robust estimate of the distance between two samples along the stratigraphic axis, i.e. their stratigraphic difference. In fact, two samples at the same stratigraphic level may have a fairly long vector distance component along the $\vec{N}_z$ vector as shown by figure 17.

To overcome this problem, Crozel (1982) has proposed to use as the component of the vector distance along the stratigraphic axis a standardized SS difference of the type.

$$\Delta_z \frac{2|(SS(2) - SS(1))|}{\|Grad_1(SS)\| + \|Grad_2(SS)\|}$$

Problems associated with the direct use of natural reference systems to measure vector distances in variogram computation seem to increase when the samples being compared start to be far apart, something which did not happen with the limited unfolding technique which really makes use of distances measured along surfaces and not rectilinear vectors.

4 - CONCLUSIONS

This paper has presented two solutions to the problem of computing realistic variograms in strata-controlled folded deposits and both solutions have their own limitations and/or dark areas. The first method may look more attractive to the geologist because it involves only lengthy but fairly simple computations. At least in one real case it has proved to be workable and giving good results. More research work is definitely needed with the second method which looks more general in scope and certainly more elegant in its derivation of the orientation of the natural reference system at any point.

Acknowledgements

The help of our fellow co-workers at GSII, Gilbert Sergerie and Jean-Maurice Brodeur, in programming the limited unfolding technique is gratefully acknowledged.

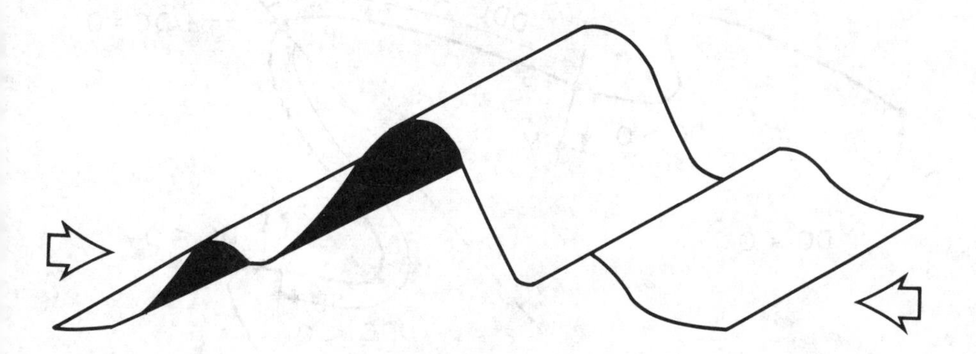

Figure 1 - A simple illustration of cylindrical folds.

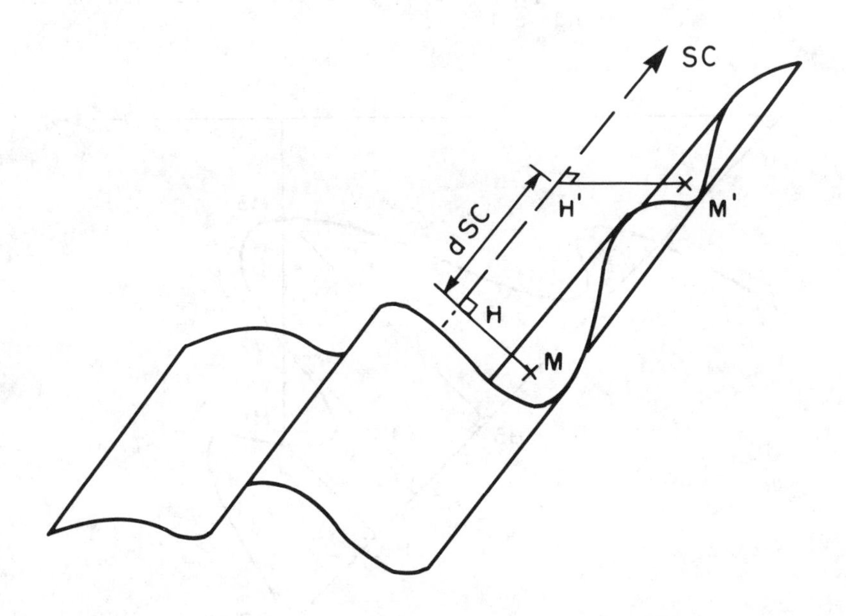

Figure 2 - Definition of the strike coordinate (SC) of points in cylindrical folds. dSC is the difference in strike coordinates for points M' and M.

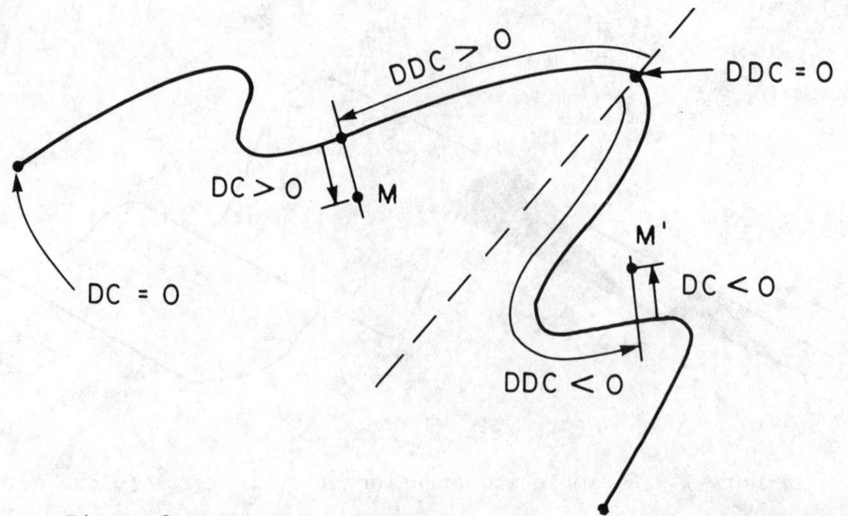

Figure 3 - General definition of the down-dip coordinate (DDC) and the depth coordinate (DC) in section planes perpendicular to the strike axis of cylindrical folds.

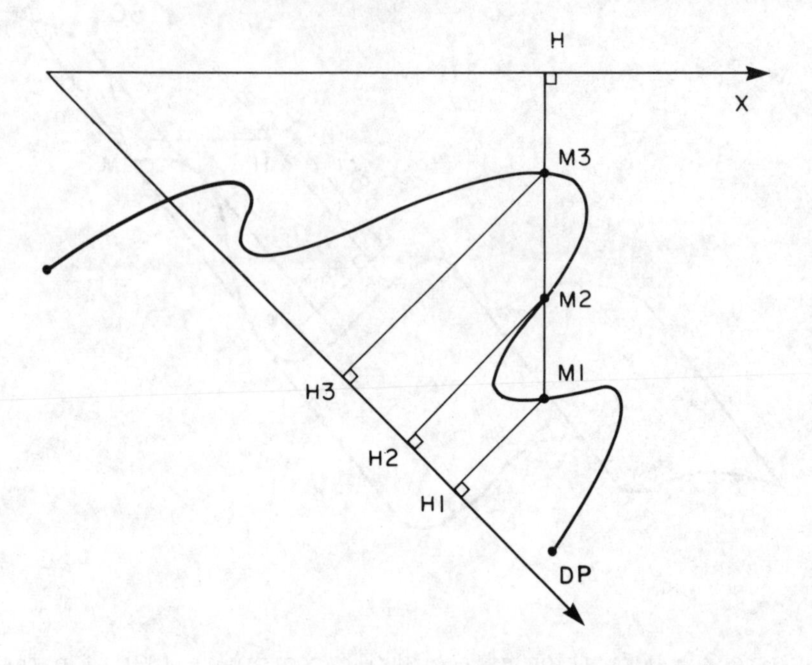

Figure 4 - Example of folds which are overturned with respect to the horizontal axis (X) but no longer overturned when the trace of the average down dip plane (DP) is considered.

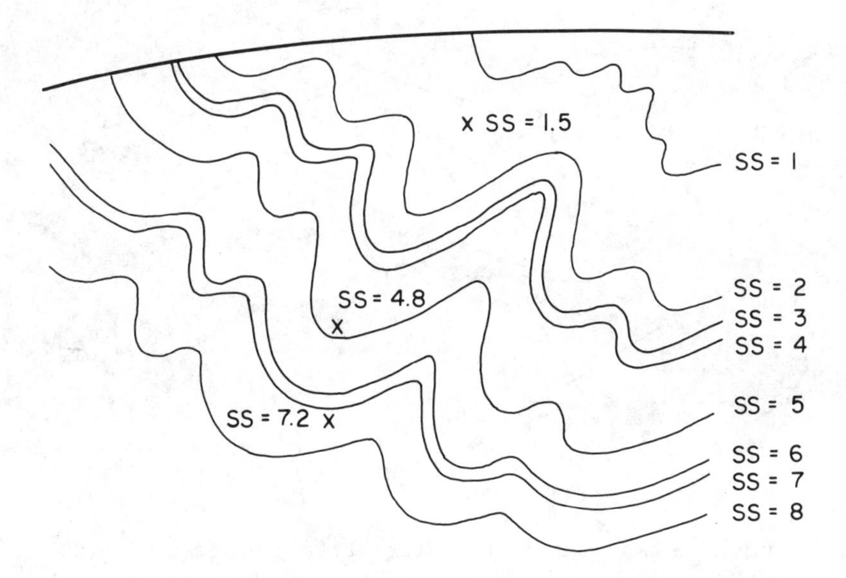

Figure 5 - Numbering of strata and definition of the stratigraphic score (SS) of any point.

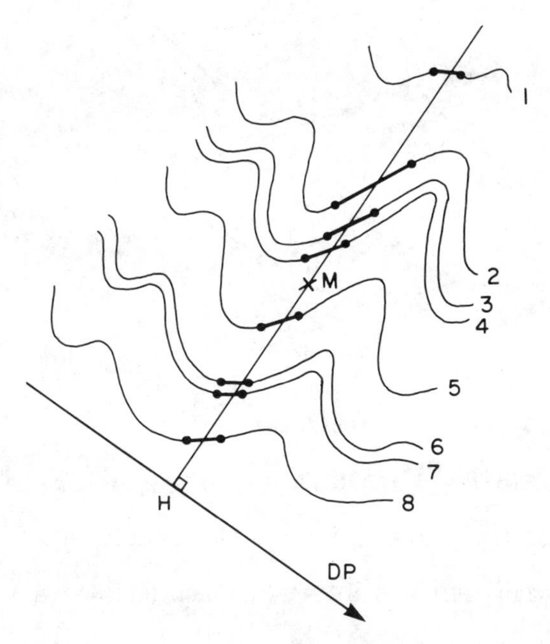

Figure 6 - Definition of the portions of the digitized strata limits which have to be considered in order to determine the stratigraphic score (SS) of point M.

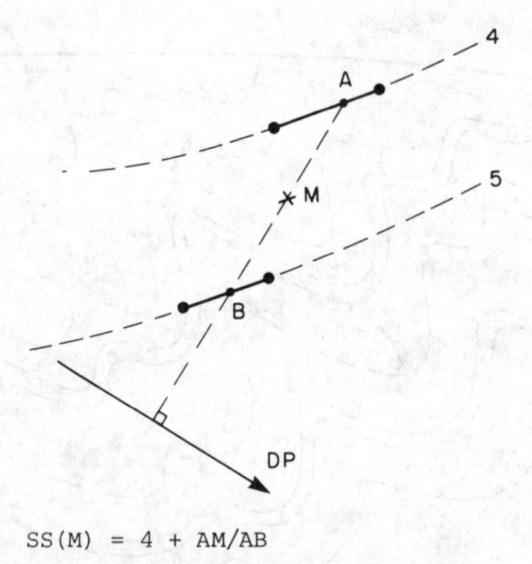

SS(M) = 4 + AM/AB

Figure 7 - Determination of the stratigraphic score (SS) of a point M between two strata limits.

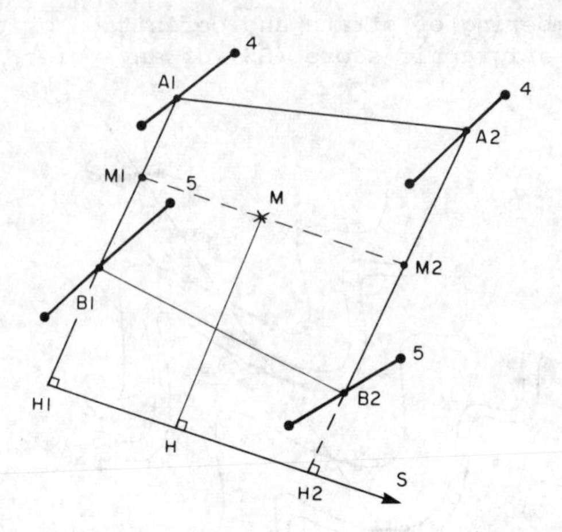

SS(M1) = 4 + A1M1/A1B1

SS(M2) = 4 + A2M2/A2B2

SS(M) = (HH1.SS(M2) + HH2.SS(M1))/H1H2

WITH: HH1 = S(M) - S(M1) AND HH2 = S(M2) - S(M)

Figure 8 - Determination of the stratigraphic score (SS) of a point M between two sections.

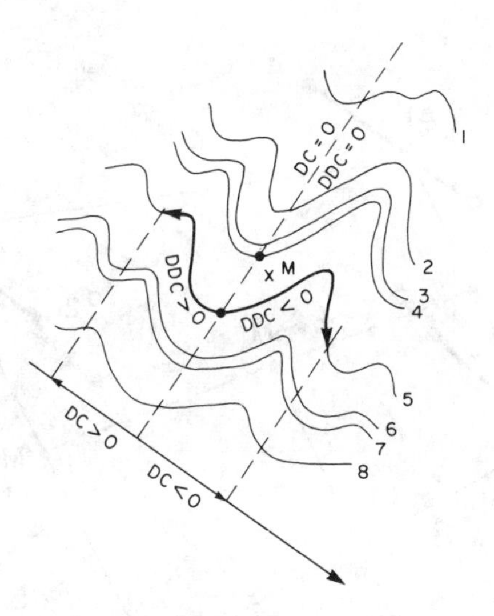

DDC = f(DC) AND IF DC1 > DC2, DDC1 > DDC2

Figure 9 - Determination of the down-dip coordinate of the points along the digitized limits (DDC) as a function of the coordinate of the same points along the trace of the average dip plane (DC).

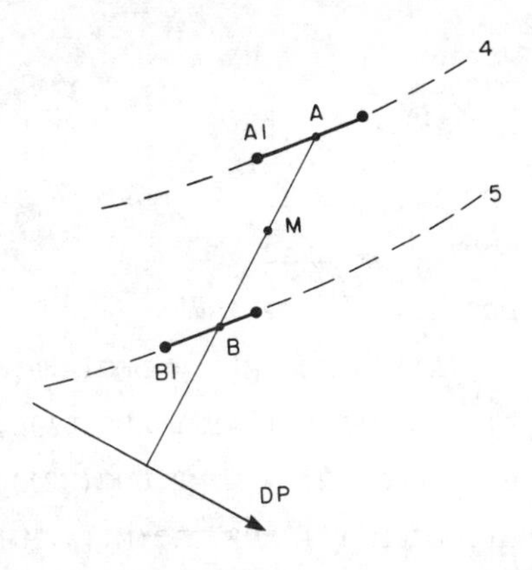

DDC(A) = DDC(A1) + A1A
DDC(B) = DDC(B1) + B1B
DDC(M) = DDC(A)
OR DDC(M) = (BM.DDC(A) + AM.DDC(B))/AB

Figure 10 - Determination of the down-dip coordinate (DDC) of a point M on a section using down-dip coordinates of points along the hanging wall only or along both the hangingwall and the footwall.

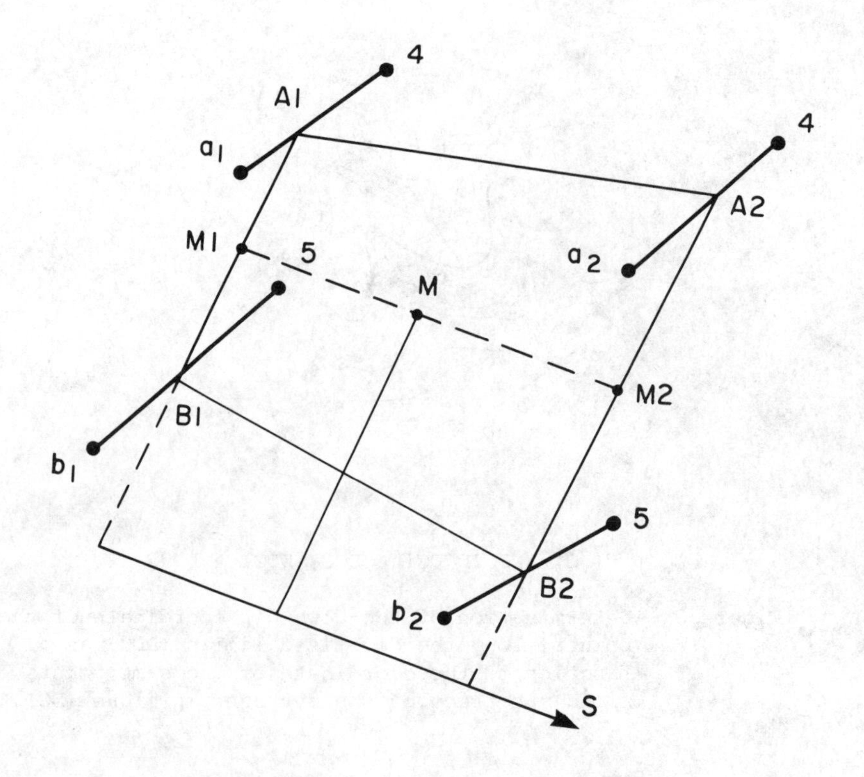


```
       DDC(A1) = DDC(a1) + a1A1
       DDC(B1) = DDC(b1) + b1B1
       DDC(A2) = DDC(a2) + a2A2
       DDC(B2) = DDC(b2) + b2B2
       DDC(M1) = DDC(A1)    DDC(M2) = DDC(A2)
OR     DDC(M1) = B1M1.DDC(A1) + A1M1.DDC(B1))/A1B1
       DDC(M2) = B2M2.DDC(A2) + A2M2.DDC(B2))/A2B2
AND    DDC(M) = MM1.DDC(M2) + MM2.DDC(M1))/M1M2
```

Figure 11 - Determination of the down dip coordinate (DDC) of a point M between two sections.

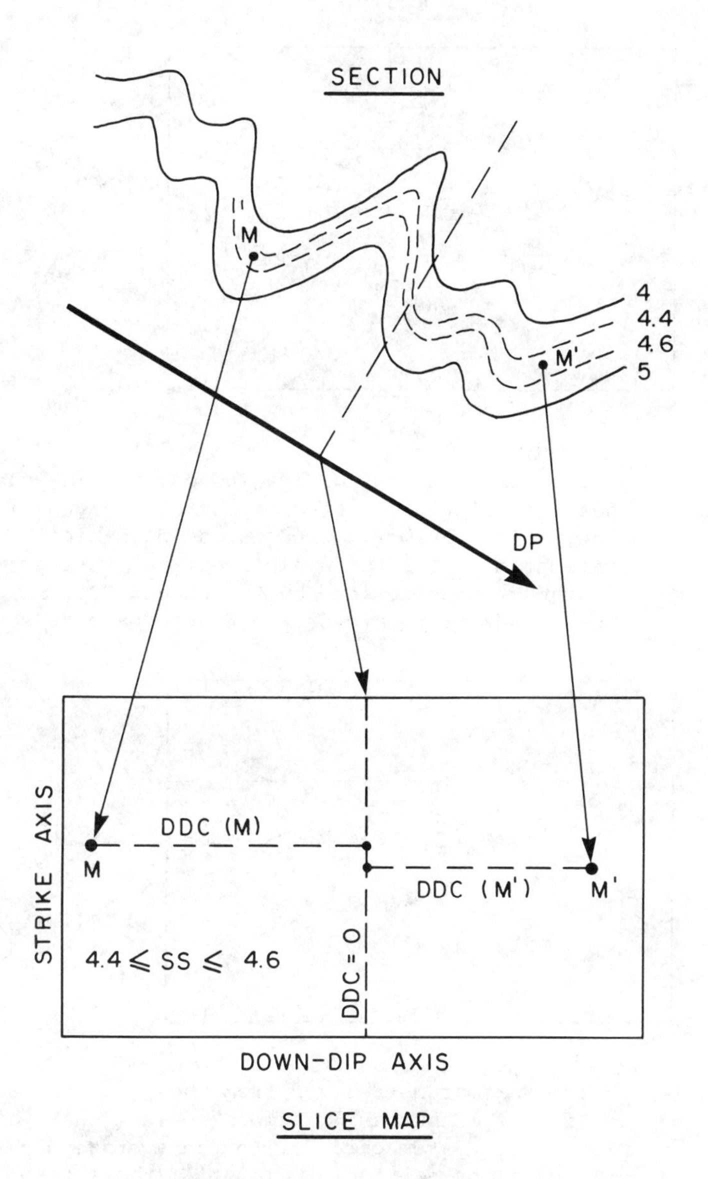

Figure 12 - Illustration of the method to produce slice maps once strike and down dip coordinates as well as stratigraphic scores can be defined for every point in the deposit.

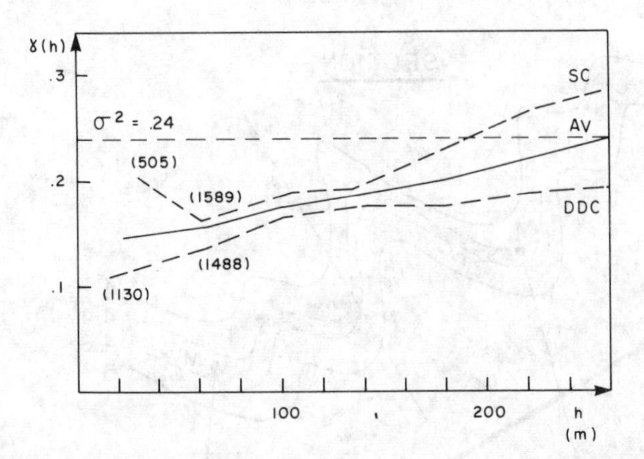

Figure 13 - Slice variograms of an ore indicator using the natural reference system of coordinates. SC is the strike variogram, DDC is the down-dip variogram. AV is the omnidirectional variogram in the surface defined by the SC and DDC axis. Numbers of pairs are shown between parenthesis.

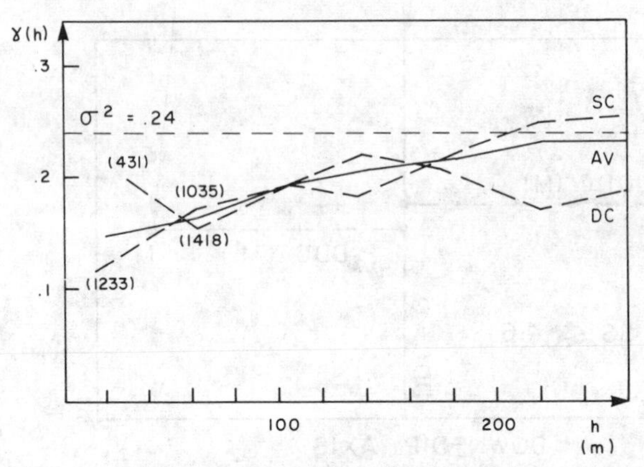

Figure 14 - Slice variograms derived from the same data as in figure 13 but using a more conventional reference system parallel to the average dip plane of the stratum. SC is still the strike variogram, DC is the variogram computed along a direction dipping 45° to the south. AV is the omnidirectional variogram in the surface defined by the SC and DC axis. Numbers of pairs are shown between parenthesis.

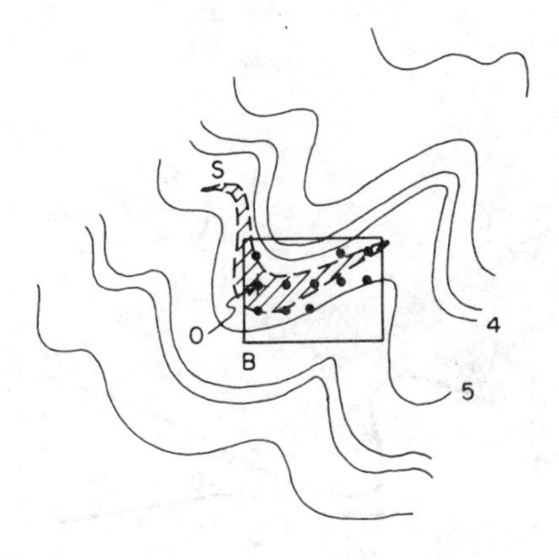

Figure 15 - Illustration of the kriging of the grade of a block B using the natural reference system of coordinates. The portion of the block in the stratum between SS = 4 and SS = 5 is discretized (point 0). S is the search zone around point 0.

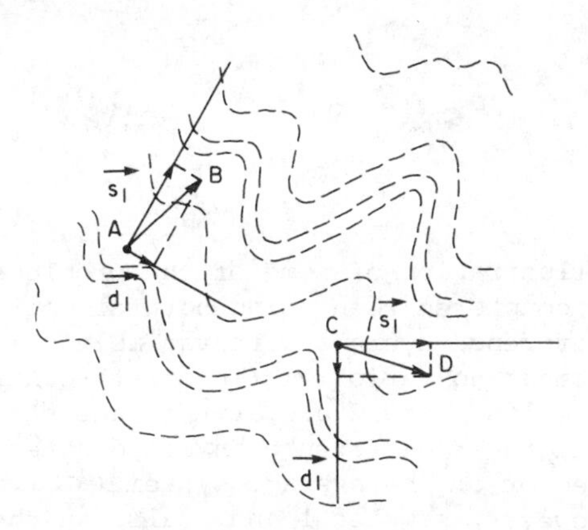

Figure 16 - Illustration of the basic idea behind the general method of defining the natural reference system at any point of a strata controlled deposit.

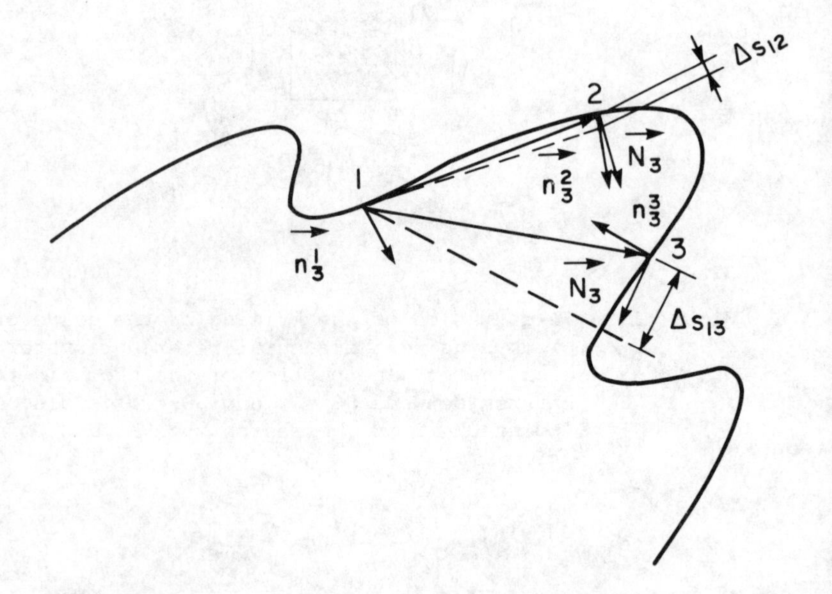

Figure 17 - Illustration of some of the problems that may be encountered when using natural coordinate reference systems with variable orientation when computing variograms in strata-controlled deposits. Samples 1 and 2 are at the same stratigraphic level and their difference in coordinates along the $\vec{N}_z$ vector of the average system $(\vec{n}_z^1 + \vec{n}_z^2)$ is small. However, samples 1 and 3 also on the same stratigraphic level have a large difference of coordinates, Δs_{13}, along that $\vec{N}_z$ vector.

References

(1) Barnes, T.E., "Orebody modeling. The transformation of coordinate systems to model continuity at Mount Emmons", 1982, 17th APCOMP Symposium, AIME, New York, pp. 765-770.

(2) Rendu, J.M., and Readdy, L., "Geology and the semivariogram. A critical relationship", 1982, 17th APCOMP Symposium, AIME, New York, pp. 771-783.

(3) O'Leary, J., "Application of geology and geostatistics at the Scully Mine Orebody, Wabush, Labrador", 1973, Ph. D. Thesis, Royal School of Mines, 196 pp.

(4) Desbarats, A., "Estimation géostatistique de couches plissées", 1982, Internal report, Ecole Polytechnique, Montreal, 19 pp. (in french - available from the author).

(5) Crozel, D., "Procédure de calcul de variogrammes stratigraphiques", 1982, Internal report, Ecole Polytechnique, Montreal, 15 pp. (in french, available from the author).

THE VARIOGRAM AND KRIGING: ROBUST AND RESISTANT ESTIMATORS

P.A. Dowd

Department of Mining and Mineral Engineering,
The University of Leeds, U.K.

Abstract : Previously published techniques for robust and resistant variogram estimation are reviewed and several alternative methods are introduced. The performance of various estimators is assessed on two typical practical examples.

Two suggestions for resistant methods of kriging are discussed and illustrated.

I - ROBUSTNESS AND RESISTANCE

As there is no commonly accepted definition of robustness it is as well to begin this paper by setting out exactly what we expect to achieve by developing and using "robust" estimators.

First a distinction must be made between robustness and resistance. An estimator is said to be resistant when a change in a few data will not substantially change the value of the estimate. In geostatistical applications we are usually content with resistance to outliers, i.e. *values which, within their physical context,* appear to be grossly high or low. Note that an outlier cannot be defined simply as a value above or below a certain limit - it depends on the values which surround it.

Robustness can mean many things. In general it refers to a lack of susceptibility to the effects of incorrect assumptions (usually the effects of non-normality). In this paper robustness will be taken explicitly to mean *distributional* robustness and implicitly to include robustness to inherently incorrect assumptions about the independence of data. (Very little is known about the effects of the latter). A distributionally robust estimator is one which has a high efficiency for some *idealised* or assumed distribution and performs well for a wide variety of deviations

G. Verly et al. (eds.), Geostatistics for Natural Resources Characterization, Part 1, 91–106.

from this distribution.

Clearly, for variogram estimation and kriging, robustness will be related to departures from normality whereas resistance may or may not depend on the type of distribution: in geostatistical applications an outlier may be as likely to arise from a normal distribution as from a highly skewed distribution.

In this paper two specific ways - location and scale estimation - of obtaining robust estimates of variograms are examined. In addition, a resistant method of kriging is discussed.

Note that estimating a scale parameter for a variable is equivalent to estimating a location parameter for the logarithm of the variable (cf Huber 1972).

Good reviews of robust estimators of location and scale parameters can be found in Huber (1964 and 1972) and Hogg (1974).

II - TESTING AND DERIVING ESTIMATORS

Any *general*, objective comparison of estimators requires a model. However, the comparison applies *only* to that model : estimators using data from distributions which do not belong to the class of distributions defined by the model may be considerably biased and their performance may be significantly impaired.

A very common model used both to derive estimators and to assess their performance is the *contaminated normal* defined by the class of distribution functions:

$$F = (1-\varepsilon)G + \varepsilon H$$

where G is the standard normal distribution function
H is a contaminating distribution function (often normal)
and $0 \leq \varepsilon < 1$ is the degree of contamination.

By varying ε and H the performance of an estimator can be assessed in the light of departures from normality.

Contaminated normal models were combined with simplified, one-dimensional correlation models (see for example Cressie and Hawkins 1980) to provide data for comparing variogram estimators. However, in testing estimators significant differences began to emerge between their performances on the models and on some real "problem" data. The difference in performance appears to be related to:

(i) possible significant differences between the class of distributions in the model and those encountered in practice (e.g. the breakdown point for ε-contamination is exceeded)
(ii) differences in correlation structure
(iii) a *residual* or *localised* non-stationarity.

The first two are somewhat related and make it difficult to use geostatistical simulation to provide models unless the simulation is conditioned to a real data set (which defeats the purpose of the model). Outliers in the models do not appear to arise in the same way as they do in real data sets. The major problem in (i) is in determining *a priori* whether a real data set is compatible with the model. Non-stationarity, or drift can only be assessed from estimates based on discrete data and its assumed *overall* absence on the scale required does not imply its absence at all other scales. In practice, for any given value of h the mean difference $[z(x_i) - z(x_i+h)]$ will differ from zero even though, *on average*, over all values of h it is zero as illustrated in figure 1.

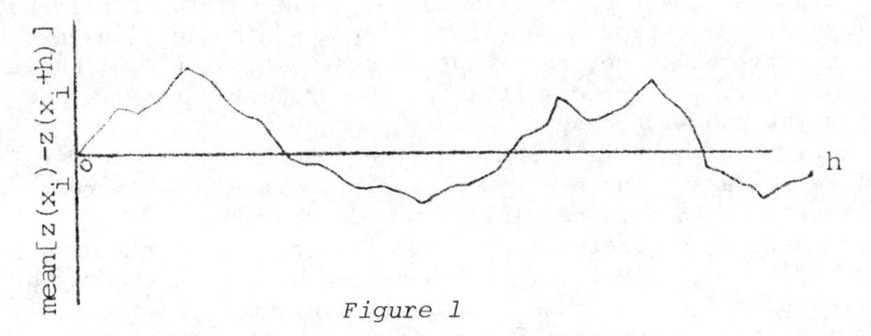

Figure 1

This type of *residual* drift is difficult to quantify and even more difficult to build into a model.

A practical method of testing performance in *particular* situations is to apply the estimators to data sets in which, often through painstaking and laborious geostatistical and geological analysis, outliers (in the form of individual data and geological groupings) have been identified and removed or in which, estimated variograms are significantly different from known optimal models. Two typical examples are given here. Optimal models are determined, and performances are compared, by cross validation with varying amounts of data removed.

The behaviour of most of the estimators described in this paper on contaminated normal models is already well documented in the literature of robust statistics.

III - THE VARIOGRAM

Consider a random variable $Z(x)$ defined at points x. The variogram is defined as the variance of the increment $[Z(x)-Z(y)]$:

$$2\gamma(x,y) = \mathrm{Var}[Z(x) - Z(y)]$$

Under the intrinsic hypothesis of geostatistics this variance reduces to:

$$2\gamma(h) = E[Z(x) - Z(x+h)]^2 \tag{1}$$

where h is the vector separating the points x and x+h (=y).

In this paper discussion of the variogram will be restricted to the intrinsic case.

The traditional estimator of the variogram from n data values $\{z(x_i), i=1,n\}$ is the arithmetic mean of the squared differences $[z(x_i) - z(x_i+h)]$:

$$2\gamma^*(h) = \sum_{i=1}^{n(h)} \frac{[z(x_i)-z(x_i+h)]^2}{n(h)} \tag{2}$$

where n(h) is the number of data pairs used in the calculation.

If the Z(x) and Z(x+h) have a joint normal distribution this estimator is a robust estimator of scale for the *differences* but it is not resistant to outliers. In the non-gaussian case it is neither robust nor resistant. As an estimator of location for the *squared differences* it is neither robust nor resistant even for the gaussian case.

Note that there is a limit to the *robustness* of experimental variograms. Any experimental variogram is only a reflection of the *true underlying* variogram and even with the "best" data sets enormous fluctuations about the true variogram should be expected especially for large values of h. i.e. there may be significant differences between experimental variograms calculated from different subsets of data (see Matheron (1965) for a theoretical analysis). For this reason only the first few points (in practice never more than half the total length in a given direction) can be considered significant.

III-1 - A review of some proposals

Armstrong and Delfiner (as reported in Armstrong, 1980) examine a resistant estimator of location for the squared differences and a robust, resistant estimator of scale for the differences.

Their resistant estimate of location is

$$2\gamma_q^*(h) = Q_q[z(x_i)-z(x_i+h)]^2 \tag{3}$$

where Q_q is the qth quantile of the squared differences for lag h. A value of $q=0.5$ gives the median, the simplest resistant estimate of location. Armstrong reports limited success with these estimators in "cleaning up" variograms affected by outliers.

These quantile variograms are not robust estimates of location for such strongly asymmetric distributions as those of the squared differences and they provide biased estimates of $\gamma(h)$. In addition kriging requires an estimate of the variance in (1) and this cannot be derived from the quantile variograms without making the (usually) false assumption that the Z(x) are multivariate gaussian and consequently:

$$[Z(x) - Z(x+h)]^2 \sim 2\gamma(h)\chi_1^2 \tag{4}$$

Note that under this assumption the bias correction for (3) for $q = 0.5$, *for large numbers* of samples, is straightforward and is obtained by dividing γ^*_q by 0.455.

Their scale estimate for the differences is adapted from Huber's scale estimator (Huber, 1964):

The variogram estimate is determined iteratively from:

$$\frac{1}{n} \sum_{i=1}^{n} \psi_c^2 \left\{ \frac{y_i}{\sqrt{2\gamma_A^*(h)}} \right\} = \beta_c$$

where $y_i = z(x_i) - z(x_i+h)$

$\psi_c(x)$ is defined by:

$$\psi_c(x) = \begin{cases} -c & x < -c \\ x & -c \leq x \leq c \\ c & x > c \end{cases} \tag{5}$$

and $\beta_c = E\{\psi_c^2 (y_i / \sqrt{2\gamma_A^*(h)})\}$

is calculated on the assumption that:

$$\frac{y_i}{\sqrt{2\gamma_A^*(h)}} \frown N(0,1)$$

Huber has shown that this is an optimal estimator for contaminated normal distributions, including non-symmetrical contamination.

Cressie and Hawkins (1980) attempt to convert the problem into one of finding a robust, resistant estimator of location of a symmetrical distribution. They make the assumption given in (4) by Armstrong and Delfiner, define a new variable:

$$y_i(h) = |z(x_i) - z(x_i+h)|^{\frac{1}{2}} \tag{6}$$

and estimate the centre of symmetry, y^*_h, for each distance interval h.

They examine ten estimators of location by testing their performance on one actual (lognormal) and six simulated (one normal, one Laplace and four symmetrically contaminated normal) data sets from which they conclude that maximum likelihood estimators (M-estimates) are the most robust and resistant. However, as the arithmetic mean of the $y_i(h)$ also performed well they advise using this estimator in practice for large data sets.

Having calculated the y_h^* they determine a corrective factor to obtain the variogram estimate:

$$2\gamma^*_{ch}(h) = y^{*4}_h / (0.457 + 0.494n^{-1} + 0.045n^{-2})$$

which, for large n reduces to (7)

$$2\gamma_{ch}^*(h) = \frac{y_h^{*4}}{0.457}$$

Note that taking the median of the y^*_h, (7) can be rewritten as:

$$\frac{1}{.457}\{\text{median}[(z(x_i)-z(x_i+h))^2]^{\frac{1}{4}}\}^4$$

which is asymptotically equivalent to γ^*_q in (3) for $q = 0.5$, provided the bias correction is made.

III.2 - Some further proposals

A simple scale estimator of the variogram is:

$$2\gamma_D^*(h) = \{\frac{1}{n}\sum_{i=1}^{n} | z(x_i) - z(x_i+h)|\}^2 \quad (8)$$

although this is a resistant estimator of scale it is only 88% as efficient as the traditional estimator in (2) for a normal distribution. However, even for very slight contamination of the normal the estimator rapidly becomes more efficient than (2).

Another simple scale estimator can be derived from the median deviation using a correction to ensure consistency at the normal distribution:

$$2\gamma_H^*(h) = 2.198 \times \{\text{median}|z(x_i)-z(x_i+h)|\}^2 \quad (9a)$$

or, by putting $y_i(h) = z(x_i) - z(x_i+h)$:

$$2\gamma_H^*(h) = 2.198 \times \{\text{median}\,|y_i(h) - \tilde{y}(h)|\}^2 \quad (9b)$$

Where $\tilde{y}$ is the median of the $y_i(h)$.

This estimate of scale is very resistant but has a low asymptotic efficiency of around 40% with strictly normal data. However in most practical geostatistical applications it proves to be a very effective scale estimate.

A statistic proposed by Lax (cf Lax 1975) and derived from the asymptotic variance of a bisquare M-estimator can be adapted to give an estimator of scale for the differences:

$$\text{Let } y_i(h) = z(x_i) - z(x_i+h)$$

$$\text{and } w_i = \frac{y_i(h) - \tilde{y}(h)}{K(\text{MAD})}$$

where:

$$\tilde{y}(h) = 0$$

and MAD is the median absolute difference: (10a)

$$\text{median } |z(x_i) - z(x_i+h)|$$

or
$\tilde{y}(h)$ is the median of the $y_i(h)$ and MAD is the median absolute deviation from the median: (10b)

$$\text{median } |y_i(h) - \tilde{y}(h)|$$

Then

$$2\gamma_L^*(h) = \frac{n_i(h) \sum\limits_{i\varepsilon I} [y_i(h) - \tilde{y}(h)]^2 (1-w_i^2)^4}{\left[\sum\limits_{i\varepsilon I} (1-w_i^2)(1-5w_i^2) \right]^2} \qquad \text{(10a and b)}$$

In practice values of k from 6 to 9 give good results depending on the data.

This estimator is asymptotically unbiased, consistent at the normal and has an asymptotic efficiency of almost twice that of the estimators in (9a) and (9b). It is asymptotically equivalent to the Cressie-Hawkins estimator using the Tukey biweight and appears closely related to (5).

III-3 Testing estimators

In general, quantile variograms have not given good results on models or real data and are not included in these examples. For a wide range of experimental variograms (including these examples) and many models the Huber estimator, γ_A^* in (5), has been found to give results which are almost indistinguishable from those given by the Lax estimator, γ_L^* in (10). As the former requires considerably more calculation than the latter it is not included. The simple form (arithmetic mean) of the Cressie-Hawkins estimator is used.

Example 1 : Athabasca tar sands

The full structural analysis is given in Dowd and Royle (1977). Data were available from holes over a 50 sq. mile area. Vertical variograms for % oil showed "text book" spherical models with ranges of 36 feet. Horizontal variograms as shown in figure 2 appeared random. The histogram of oil content is shown in figure 3; there was no apparent drift. Little was known of

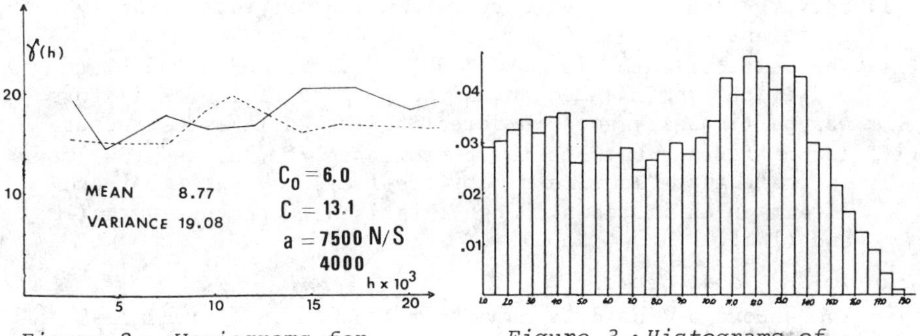

Figure 2 : Variograms for saturated wt. % oil

Figure 3 : Histograms of saturated wt. % oil

the *local* geology but from *regional* geology long range structures with good continuity were expected. After a laborious division of the deposit into various combinations of horizontal "slices" and calculation of associated variograms the cause of the random effects was identified as several relatively thin layers of supposedly weathered material. The analysis is summarised in figure 4. Variograms in all but the "weathered" zones showed exceptionally good spherical structures and a "global" spherical model variogram was ultimately fitted with parameters as shown in figure 2.

A summary of the performance of the estimators is given in table 1.

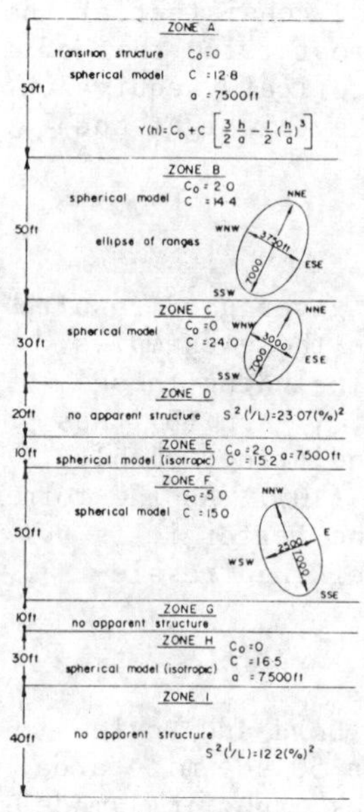

Figure 4

DIRECTION	North/South				East/West			
LAG	1	2	3	4	1	2	3	4
model	12.1	16.3	19.0	19.1	16.1	19.1	19.1	19.1
γ^* (2)	15.7	15.4	15.3	18.4	19.6	14.6	18.0	16.6
γ^*_{ch} (7)	14.5	15.0	15.0	18.4	18.6	13.4	17.3	16.1
γ^*_H (9a)	24.1	18.0	18.2	23.3	24.9	15.1	19.3	16.7
γ^*_H (9b)	9.6	16.2	13.3	19.8	15.6	13.6	17.5	21.6
γ^*_L (10a) k=6	14.5	13.9	15.5	15.2	18.6	14.0	15.8	17.4
γ^*_L (10a) k=9	13.0	12.7	14.0	13.8	16.8	12.9	14.2	14.3
γ^*_L (10b) k=6	9.8	12.7	11.4	15.1	14.1	12.0	14.4	16.0
γ^*_L (10b) k=9	9.3	16.7	10.6	12.8	13.5	11.4	13.0	13.4

Table 1

Note:

(i) the similarity in results for $\gamma_{ch}*$ and γ_L*

(ii) only γ_H* and γ_L* produce results which are significantly different from $\gamma*$

(iii) The best estimator is γ_H* in 9b, i.e. the median absolute deviation from the median, although none of the estimators was particularly good in detecting the East West structure

(iv) the difference between the performance of (9a) and (9b) may indicate that *"localised"* drift effects have an important effect on estimates although this is not apparent with (10a) and (10b).

Example 2: A coal deposit

The second set of data came from a southern African coal deposit in the Karroo Series. The variable studied was % volatile matter × intersection thickness in metres. The histogram of the data is shown in figure 5 and the experimental variogram is shown in figure 6. There is no geometric anisotropy and no significant drift could be detected.

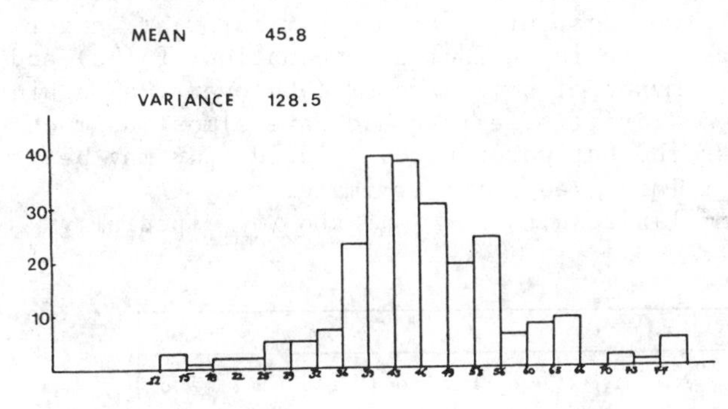

Figure 5: Histogram of accumulations of volatile matter

An optimal variogram model was determined by rigorous cross-validation by kriging each value in the data set with varying amounts of data removed. Variogram parameters were adjusted until all the criteria for an optimal model were satisfied i.e. global and conditional unbiasedness and equality of experimental and theoretical estimation variances. The optimal model is significantly different from the experimental variogram and is shown in figure 6.

Each of the variogram estimators was used to estimate the variogram and the results are summarised in table 2 and figure 7.

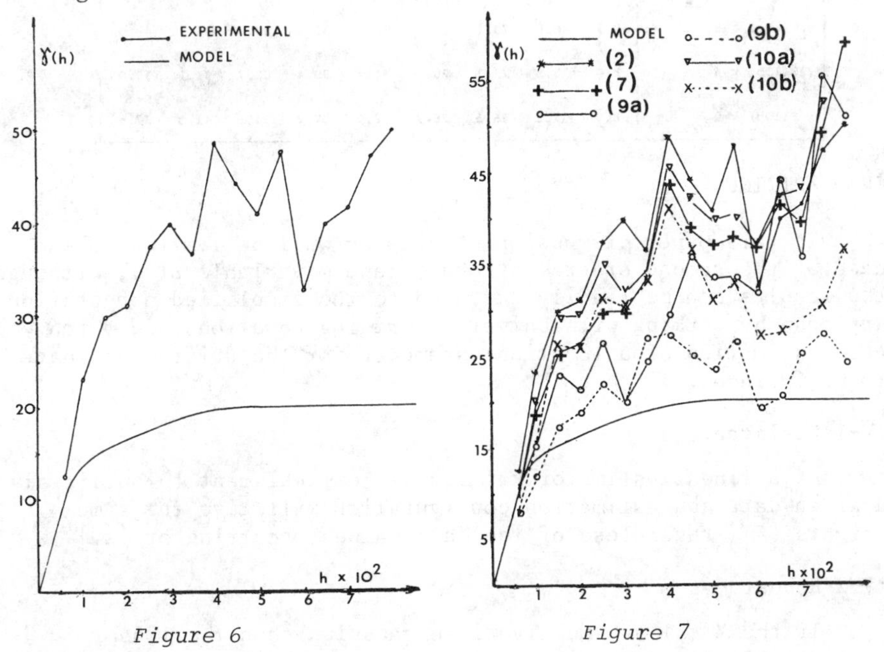

Figure 6 *Figure 7*

Taking the model as the criterion for performance the best *overall* estimator is (9b), i.e. the median absolute deviation from the median. Estimators (7) (Cressie and Hawkins), (9a) (median absolute difference) and (10a) (Lax's estimator without the location estimate) all do well for the first lag but increasingly fail to reflect the model as the lag increases.

The median absolute deviation from the median underestimates the model at the first two lags but is a good estimator at longer intervals. The best estimate is in fact a combination of (9a) and (9b). The better performance at some lags of (9b) over (9a) again suggests that *"localised"* drift effects could have almost as much effect on robustness as the estimator used. Indeed γ_{ch}^* may be improved by including a localised drift estimate.

Note again the similar results for γ_{ch}^* and γ^*_L especially (10a) with k=6.

LAG INTERVAL	1	2	3	4	5	6	7	8	9
Model	10.6	14.0	15.4	16.4	17.2	18.4	19.2	19.6	20.0
γ^* (2)	12.5	23.1	29.9	31.0	37.4	39.9	36.5	48.7	44.0
γ_{ch}^* (7)	9.9	18.7	25.4	26.3	30.8	29.9	33.5	43.9	38.5
γ_H^* (9a)	10.6	15.2	23.2	21.1	26.6	19.9	24.3	29.4	35.9
γ_H^* (9b)	7.6	11.9	17.2	18.8	22.1	20.2	26.9	27.4	25.3
γ_L^* (10a) k=6	9.9	19.7	29.5	29.2	34.9	31.6	33.9	45.7	41.8
γ_L^* (10a) k=9	10.3	20.6	29.1	29.3	35.4	33.7	33.3	45.4	41.0
γ_L^* (10b) k=6	8.0	15.5	26.1	25.7	31.4	30.4	32.7	41.1	36.6
γ_L^* (10b) k=9	8.9	17.5	26.5	26.9	32.0	32.2	31.8	41.8	37.6

IV - KRIGING

In this paper discussion of kriging will be limited to the case of the second order stationary random variable Z(x), although the results can be readily extended to the simplified non-stationary case by working with Universal Kriging equations, in which case estimation of a location parameter for the drift would have to be included.

IV-1 Resistance

As a linear estimator kriging is not resistant to outliers: a given data and estimation configuration will give the same weights $\{\lambda_i\}$ regardless of the data values occurring at $\{x_i\}$.

IV-2 Robustness

If the $Z(x_i)$ are multivariate gaussian then the linear kriging estimator is the conditional expectation of Z_V given $\{Z(x_i)\}$ and the estimates obtained from kriging will be conditionally as well as globally unbiased. In all other cases linear kriging provides the best linear estimate of the conditional expectation (Journel and Huijbregts 1978) but the estimates will be more or less (locally) conditionally biased although experience has shown that, in many cases, marked departure from normality (in the absence of outliers) does not significantly affect this property.

Notwithstanding this *practical* robustness kriging is not a robust estimator. Recent advances in *non-linear* kriging have concentrated on transforming the data to normality (see for example Matheron 1976). In the simplest case of data which have a joint lognormal distribution the optimal estimator is to be found by lognormal kriging (cf Rendu 1979 and Dowd 1982). The problem with all these methods is that they rest on dubious assumptions which in practice either do not hold or are very difficult to verify.

In the remainder of this paper attention will be focused on *resistant* methods of kriging which will only be *robust* for particular circumstances.

There is of course no theoretical reason for persisting with kriging type estimators or for that matter with the variogram. There are, however, very good practical reasons: least squares methods are well understood, are mathematically tractible and easy to compute and the variogram shape can be linked intuitively to many geological structures. As a first step then traditional kriging techniques will form the point of departure and as much as possible of their desirable properties will be maintained.

IV-3 - Some resistant kriging estimators

Variogram estimates obtained from absolute differences such as (8) and (9) suggest an estimator which

$$\text{minimises } E|Z_v - Z_v^*|$$

$$\text{subject to } E(Z_v - Z_v^*) = 0 \qquad (11)$$

Global, regression-type fits of such estimators are well documented and are usually obtained by linear programming techniques. As a geostatistical kriging type estimator, a solution may be found by an iterative solution to a minimisation of $E\{w[Z_v - \Sigma\lambda_i Z_i]^2\}$ where the weights w are dependent on λ_i, Z_i and some measure of absolute error at each step of the procedure. The author's initial enthusiasm for this approach has been somewhat dampened by later computational and theoretical difficulties and further work is required. As a regression technique this is a very useful *resistant, robust* method of checking for conditional unbiasedness (cf Dowd and Scott 1982).

One simple *approximation* to the solution is to take Z_v^* as a *weighted median* of the Z_i as suggested by Henley (cf Henley 1981). The weights are those obtained from the traditional kriging equations.

To obtain the weighted median first sort the z_i values into ascending order and then interpret the kriging weights as relative frequencies of occurrence. The weighted median is then the z_i value, or the interpolated value, with cumulative frequency 0.5.

In Henleys example:

ordered data values	1.0	1.1	3.6	4.0	5.2	6.1	7.7
weights	0.1	0.3	0.05	0.15	0.1	0.25	0.05
cumulative weights	0.10	0.40	0.45	0.60	0.70	0.95	1.00

Linear interpolation gives the weighted median as 3.7.

There is, however, a difficulty in this interpretation when negative kriging weights occur. A rough solution is to replace each weight λ_i by:

$$\frac{|\lambda_i|}{\Sigma|\lambda_i|}$$

Significant negative kriging weights will only occur in very continuous deposits (zero nugget variance) and so are unlikely in situations where resistant kriging estimators are required.

The kriging variance is calculated using the traditional expression which, of course, is only an approximation to the actual estimation variance of the weighted median.

Some examples are shown in figure 8. This simple procedure has proved very effective in practice provided it is properly limited to outlier prone cases. Used in conjunction with an "automatic" detection of outliers it would be even more useful. Such a method is described below.

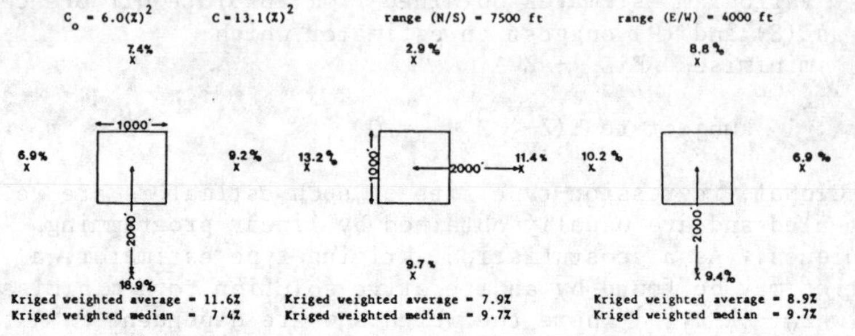

Consider first a method for kriging with unreliable data first suggested by Delhomme (cf Delhomme, 1974).

Suppose that some or all of the data $\{z(x_i), i=1,n\}$ have an error, or some degree of uncertainty $\{\varepsilon(x_i), i=1,n\}$

Assume that the errors are:

(i) unbiased : $E[\varepsilon(x_i)] = 0$

(ii) uncorrelated: $Cov[\varepsilon(x_i)\varepsilon(x_j)] = 0 \quad \forall\, i \neq j$

(iii) uncorrelated with the variable to be estimated:

$$\mathrm{Cov}[Z(x)\ \varepsilon(x_i)] = 0\ \forall\ i,x$$

The kriging estimator becomes:

$$Z_v^* = \sum_i \lambda_i[Z(x_i) + \varepsilon(x_i)]$$

and the estimation variance is:

$$E[Z_v - Z_v^*]^2 = E[Z_o - \sum_i \lambda_i Z(x_i)]^2 + \sum_i \lambda_i^2 s_i^2$$

where $\quad s_i^2 = \mathrm{var}\{\varepsilon(x_i)\} = E\{\varepsilon(x_i)\}^2$

Minimising the estimation variance subject to the non-bias condition (which remains unaltered) gives the following set of equations:

$$\sum_j \lambda_j \gamma_{ij} - \lambda_i s_i^2 + \mu = \bar{\gamma}_{iv} \quad (i=1,n)$$

$$\sum_j \lambda_j = 1$$

with kriging variance: (12)

$$\sigma_K^2 = \sum_j \lambda_j \bar{\gamma}_{iv} + \mu - \bar{\gamma}_{v,v}$$

where γ is the variogram of the $\{Z(x_i)\}$

A possible resistant kriging method can be defined as follows:

(i) Use a *resistant, robust* estimator to obtain an estimate, $\gamma^*(h)$, of $\gamma(h)$

(ii) Remove data value $z(x_i)$ from the data set and use standard kriging to krige it from all other data (in practice only those data in the neighbourhood of $z(x_i)$ which will contribute significantly to its estimation)

(iii) use the kriging weights from (ii) to obtain a *weighted median estimate* $z^*(x_i)$ and a kriging variance $\sigma_{K_i}^2$.

(iv) If $[z(x_i) - z^*(x_i)]^2 > k\sigma_{K_i}^2$ treat $z(x_i)$ as an outlier and calculate a value of s_i^2 to be used in the kriging equations in (12).

If $[z(x_i) - z^*(x_i)]^2 \leq k\sigma_{K_i}^2$, the value of $z(x_i)$ is not an outlier and $s_i^2 = 0$.

(v) Repeat steps (ii) to (iv) for all data values.

(vi) When kriging, use the equations in (12).

The value of k and the method of calculating s_i^2 may have to be determined from a study of the histogram and variograms of the data. In *practical* tests with known outliers values of k from 6 to 9 have been found to give consistent results when used with:

$$s_i^2 \begin{cases} = \dfrac{[z(x_i) - z^*(x_i)]^2}{c\,\sigma_{K_i}^2} & \text{if } [z(x_i) - z^*(x_i)]^2 > k\sigma_{K_i}^2 \\ = 0 & \text{otherwise} \end{cases}$$

Steps (ii) to (v) involve only marginally more computation than that entailed in the standard cross-validation kriging checks of variogram models. Solution time for the kriging equations in (12) is exactly the same as for the usual kriging system.

Note that a resistant estimator, such as the median, must be used in step (ii) otherwise all values in the neighbourhood of an outlier would themselves be classed as outliers.

As with traditional kriging this technique is an exact interpolator.

Note also that the three assumptions relating to the error are all automatically satisfied by kriging.

The value of k and c will depend on the data and the required amount of weighting of outliers. Values of k=6 to 9 and c = 0.5 to 1 have given good results in several practical tests.

As a simple example of the effect of s_i^2 on the weight, λ_i, attached to sample no i, consider the example shown in figure 9 where samples 1, 2 and 3 have fixed values and the value of sample 4 is allowed to vary.

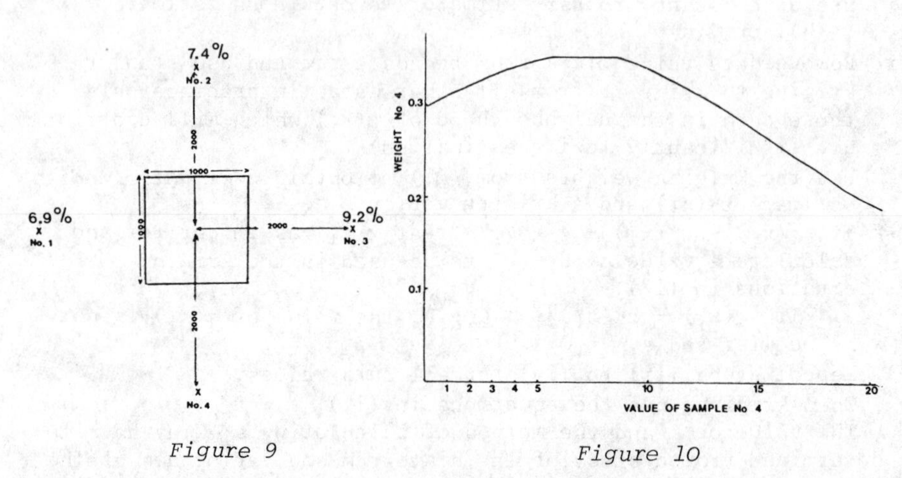

Figure 9 *Figure 10*

The variogram model assumed is the same as that given in figure 8; the value of k is 6 and c = 0.5. Figure 10 shows the weight attached to sample no. 4 as a function of its value. Note that the weight increases with the sample value until the latter

attains a range of values which are "likely" given the surrounding values. The "likely" range of values is a fonction of
(i) the variogram model (in particular, the sill value)
(ii) the positions of the data
(iii) the data values
(iv) the values of k and c.

Cressie (private communication) has suggested that a refinement of the weighted median would improve the technique; an example of a "one-step" refinement is given in Hawkins and Cressie (1984). Cressie also suggests that a more efficient, robust method is iteration to convergence, yielding Huber's robust M-smoother.

V - CONCLUSIONS

In a wide range of *practical* applications, two of which are described in this paper, the best variogram estimator has been found to be the median absolute deviation from the median, γ_H^* in (9b). All other variogram estimators mentioned in this paper, except the quantile, perform well in certain situations. The major problem in geostatistical applications is determining *a priori* whether those situations exist.

For a range of eleven experimental variograms, in which problems occurred, the MAD estimator (γ_H^* in 9b) was the most consistently good performer, although both γ_{ch}^* and γ_L^* performed well in some cases. It appears that *localised* drift effects have a significant effect on variogram estimates and these should be taken into account.

The weighted median provides a good, approximate, *resistant* kriging estimator in a range of practical problems, provided the nugget variance is greater than zero. The automatic method of weighting outliers during kriging has been found to be very effective the only practical problem being to determine *a priori* a scale of values for the s_i^2

ACKNOWLEDGEMENT

I am grateful do Dr. N. Cressie for helpful comments and suggestions during the preparation of the final version of this paper.

REFERENCES

Armstrong, M. (1981) "Application de la geostatistique aux problemes de l'estimation du charbon". Doctoral Thesis. Ecole Nationale Superieure des Mines de Paris, Fontainebleau France.

Cressie, N. and Hawkins, D. (1980) "Robust Estimation of the Variogram", *Journal of the International Association for Mathematical Geology*, Vol. 12 no. 2 pp115-126

Delhomme, J.P. (1974) "La Cartographie d'une Grandeur Physique a partir de donnees de differentes qualites" *Proceedings of the A.I.M. Confernece*. Montpellier 1974.

Dowd, P.A. (1982) "Lognormal Kriging - the general case" *Journal of the International Association for Mathematical Geology*, Vol. 14 no. 5 pp 475-499

Dowd, P.A. and Royle, A.G. (1977) "Geostatistical Applications in the Athabasca Tar Sands" *Proceedings of the 15th APCOM Symposium*. Australasian Inst. Mining and Metallurgy. pp235-242

Dowd, P.A. and Scott, I.R. (1982) "Geostatistical Applications in Stratigraphic Orebodies". Sixth Australian Statistical Conference, Melbourne, August 1982

Hawkins, D. and Cressie, N. (1984) "Robust Kriging a proposal" *Journal of the International Association for Mathematical Geology*. To appear.

Henley, S. (1981) "Non-parametric geostatistics" Applied Science Publishers, London and New Jersey 145 pp.

Hogg, R.V. (1974) "Adaptive Robust Procedures: a Review and some suggestions for further applications". *Journal of the American Statistical Association*, Vol. 69 pp. 909-922

Huber, P.J. (1964) "Robust estimation of a location parameter" *Annals of Mathematical Statistics*, Vol. 35 pp. 73-101

Huber, P.J. (1972) "Robust Statistics: a review" *Annals of Mathematical Statistics*, Vol. 43 no. 4, pp. 1041-1067

Journel, A.G. and Huijbregts, C. (1978) "Mining Geostatistics" Academic Press London, 600 pp.

Lax, D.A. (1975) "An interim report of a Monte Carlo Study of robust estimators of width" *Technical Report no. 93* (series 2) Dept. of Statistics, Princeton University

Matheron, G. (1965) "Les variables regionalisees et leur estimation" Masson et Cie, Paris 306 pp.

Matheron, G. (1978) "A simple substitute for conditional expection: Disjunctive Kriging" in *Proceedings of the First NATO A.S.I. on geostatistics*, Rome 1975, eds Guarascio et al. D. Reidel Publishing Co., Netherlands, pp. 221-236

Mosteller, F. and Tukey, J.W. (1977) "Data Analysis and Regression" Addison-Wesley Publishing Co. 588 pp.

THE VARIOGRAM AND ITS ESTIMATION

Henning Omre

Norwegian Computing Center, Oslo, Norway
Department of Applied Earth Sciences,
Stanford University, California

ABSTRACT

Robustness properties of various variogram estimators are discussed. A closer look at the variogram is made and conditions for the traditional non-parametric estimator to be optimal is presented. Frequently occurring deviations from these conditions are discussed. An alternative robust variogram estimator is defined. In an empirical test, this estimator is found to be more robust towards the deviations than other frequently used variogram estimators.

Key words: Geostatistics, variogram estimation, robustness.

I. INTRODUCTION

The variogram function is the backbone of geostatistical analysis. In most parts of geostatistical theory the variogram is assumed known or available through a sufficiently reliable estimate. In Matheron (1965) this assumption is justified with a thorough theoretical study of the variogram and its traditional non-parametric estimator. The study is based on regular sampling from a regionalized variable of a multinormal process. One conclusion of Matheron's study was that the reliability of the estimates of the variogram values are decreasing with increasing distance. They are fairly reliable at short distances, while for distances larger than half the extent of the area they are characterized as unacceptable. The main reason for this is the decreasing number of observation pairs available for estimation. These conclusions are used as rule of thumb in practical geostatistics.

G. Verly et al. (eds.), Geostatistics for Natural Resources Characterization, Part 1, 107–125.

The practical experience of geostatistics has shown that the traditional non-parametric estimator of the variogram values has several shortcomings. If the process is more heavy-tailed than the normal process the efficiency of the estimator is low. Too much weight is given to the extreme observations. In non-regular sampling this may cause large fluctuations in the estimates from one distance to another. At some distances the extreme observations are included in the estimator, at others not. Preferential sampling, which is not uncommon, may also have a large influence on the estimates. Problems also arise when measurement errors occur in the observations.

In literature, surprisingly few papers addresses the lack of robustness in the traditional non-parametric estimator of the variogram value. Cressie and Hawkins (1980) discusses an alternative set of robust estimators. The suggested estimators are heavily dependent on an underlying assumption of normality. Armstrong and Delfiner (1980) discusses two robust estimators. One is based on quantile estimation with an implicite distributional assumption while the other is associated with Huber's scale estimator. Several other authors mention the problem but do not suggest a solution.

The final part of this chapter is devoted to the definition of some useful terms and the general notation. In chapter two some interesting aspects of the variogram value and its traditional non-parametric estimator, are discussed. A classification of the expected deviations from the case where this estimator is optimal is also presented. An alternative robust estimator is suggested in chapter three. An empirical test on three different estimators is made in chapter four, and chapter five contains the conclusions. This paper is an abstract of parts of the author's PhD thesis, further discussion of the topic can be found in Omre (1984).

In the paper only the term "variogram" is used, even though "semi-variogram" in most cases would have been the correct term. This is not expected to cause confusion. The "variogram value" stands for the value of the variogram at a certain distance h. The distance h refers to a vector h, although the vector identifier is omitted. Similarly the reference location, x, is a vector. The regionalized variable considered is $\{z(x);\ x \in A\}$, and the underlying random function is $\{Z(x);\ x \in A\}$. The variable is assumed to be observed in the locations $x_i, i = 1,N$, which defines the set of observations s: $\{z(x_i);\ x_i \in A;\ i = 1,N\}$. The corresponding set of random variables is S: $\{Z(x_i);\ x_i \in A;\ i = 1,N\}$.

II. THE VARIOGRAM VALUE AND ITS ESTIMATION

Assume that $\{Z(x); x \in A\}$ is second-order stationary, and define:

$$E\{Z(x)\} = m; \text{ all } x \in A$$

$$\text{Var}\{Z(x)\} = \sigma^2; \text{ all } x \in A$$

$$\text{Cov}\{Z(x), Z(x+h)\} = C(h) \; ; \text{ all } x \in A$$

$$\text{Var}\{Z(x) - Z(x+h)\} = E\{(Z(x)-Z(x+h))^2\} = 2(\sigma^2 - C(h)) = 2 \cdot \gamma(h);$$

all $x \in A$

Consider an arbitrary $h = h_o$. In figure 1, for graphical convenience, the bivariate pdf of $(Z(x), Z(x+h_o))$, $f_{xh_o}(z_1,z_2)$, is presented. This pdf may vary with location x, but its two first moments are location invariant.

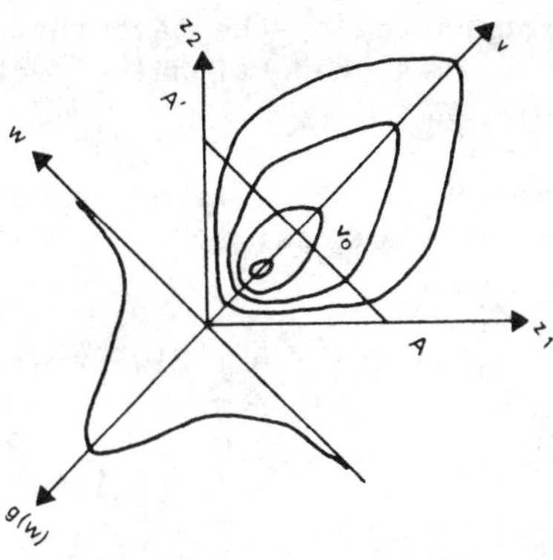

Figure 1.
Isograph representation of $f_{xh_o}(z_1,z_2)$.

Define the random variables:

$$V = \frac{1}{\sqrt{2}} \left(Z(x) + Z(x+h_o)\right)$$

$$W = \frac{1}{\sqrt{2}} \left(Z(x) - Z(x+h_o)\right)$$

The coefficients are included to ensure scale invariance. The corresponding pdf, $g(v,w)$, is graphically represented by rotating the axis in the (z_1,z_2)-system by $\pi/4$ in figure 1.

It is easily shown that:

$$\text{Var}\{W\} = E\{W^2\} = \tfrac{1}{2}E\{(Z(x) - Z(x+h_o)^2)\} = \gamma(h_o)$$

$$\text{Cov}\{V,W\} = \tfrac{1}{2}\left(\text{Var}\{Z(x)\} - \text{Var}\{Z(x+h_o)\}\right) = 0$$

hence the variance associated with the marginal pdf of $g(v,w)$ along the w-axis, $g(w)$, is identical to the variogram value.

In a similar manner it is possible to define a conditional variogram value as

$$\begin{aligned} \text{Var}\{W | V=v_o\} &= E\{W^2 | V=v_o\} \\ &= \tfrac{1}{2}E\{(Z(x) - Z(x+h_o))^2 | Z(x) + Z(x+h_o) = \sqrt{2}v_o\} \\ &= \gamma(h_o | Z(x) + Z(x+h_o) = \sqrt{2}v_o\} \end{aligned}$$

Graphically this is the variance associated with the conditional pdf along the profile A-A' in figure 1.

It is important to note that generally the conditional variogram-value is a function of v_o. The assumption of second-order stationarity has no influence on this. This dependence on v_o simply implies that the true variogram value is different in areas with high values than in areas with low. The only simple way to ensure invariance with regards to v_o is to have V and W independent, and not only uncorrelated. It is easy to show that if $(Z(x),Z(x+h_o))$ is binormally distributed, V and W are independent, and hence the conditional variogram value is invariant with regard to v_o. The above arguments are intuitively supported by the fact that a multi-lognormal random function shows a proportional effect, see Journel and Huijbregts (1978).

The variogram value is expressed through the conditional variogram value as:

$$\gamma(h_o) = E\{W^2\} = \int_{-\infty}^{\infty} E\{W^2 | V=u\} \cdot \text{Prob}\{V=u\}\, du$$

$$= \int_{-\infty}^{\infty} \gamma\left(h_o | Z(x) + Z(x+h_o) = \sqrt{2}u\right) \cdot \text{Prob}\{Z(x) + Z(x+h_o) = \sqrt{2}u\}\, du$$

In reality the variogram value has to be estimated by limited knowledge of the regionalized variable. A non-parametric and intuitively attractive estimator for the variogram value is:

$$\hat{\gamma}(h_o) = \frac{1}{2N_{h_o}} \cdot \sum_{(i,j)\varepsilon D_{h_o}} \left(Z(x_i) - Z(x_j)\right)^2$$

where

D_{h_o} : $\{(i,j) | Z(x_i), Z(x_j) \varepsilon S;\ x_i - x_j = h_o\}$

N_{h_o} = no. of elements in D_{h_o}

This corresponds to the estimator suggested in Matheron (1965), and is the one most commonly used in geostatistics.

From inference theory in statistics it is known that this estimator is optimal when $(Z(x),Z(x+h_o))$ is binormally distributed and the observations $\{(Z(x_i),Z(x_j));(i,j) \varepsilon D_{h_o}\}$ are all uncorrelated. This very particular situation is referred to as the Ideal Case for variogram value estimation. Experience from statistics tells that even small deviations from the Ideal Case may cause large deviations from optimality in the estimator, see Mosteller and Tukey (1977) and Devlin, Gnanadesikan and Kettenring (1975).

In reality deviations from the Ideal Case will occur. Three main deviation types can be identified. They are in declining order of importance:

- Distributional Deviations.
  The $(Z(x),Z(x+h_o))$ cannot be properly represented by a binormal distribution. The deviations involve characteristics of the phenomena itself, like Z(x) is non-negative for all x or the univariate distribution of Z(x) is skewed and heavy-tailed. From probability theory it is known that univariate normality does not imply bivariate normality. Hence a univariate transformation of the regionalized variable to univariate normality does not necessarily remove distributional deviations. Hence in most cases the conditional variogram values will be functions of the variable values.

  Distributional deviations are associated with Innovative Outliers as defined in Time-Series Analysis, see Fox (1972) and Kleiner, Martin and Thomson (1979). Their results indicates that the traditional estimator will remain consistent but that its efficiency will decrease drastically with increasing deviations from binormality.

- Sampling Deviations.
  The observations $\{Z(x_i);i=1,N\}$ are not sampled in a randomized way. Biased sampling has taken place, and the locations of the samples tend to be clustered in high or low value areas. These deviations do not include the characteristics of the phenomena, but only the sampling procedure. In practice, clustered sampling can be seen as an indicator of biased sampling. This, in turn, will make the traditional estimator for the mean biased. The influence on the estimates of the variogram values is not so obvious. In biased sampling the observation pairs used in the estimation are not representative of the values of the regionalized variable. If the conditional variogram value is a function of v_o, as generally is the case, then the traditional estimator based on biased sampling will probably be biased. The degree of biasedness in the traditional estimator depends on how biased the sampling is, and on the functional dependence on v_o.

- Outlier Deviations.
  The regionalized variable cannot be expected to be precisely observed always. Usually the observations are correct, but once in a while an erroneous observation is made. These deviations only include the sampling procedure. In practice,

outlier deviations may occur as human errors or misoperating sampling devices. If possible to detect, the erroneous observations should be deleted. A warning should, however, be given towards using this deviation model without justifying from the sampling procedure that outliers actually occur. Deleting extreme observations wrongly may cause serious underevaluation of the potential of an area.

Outlier deviations are associated with Additive Outliers as defined in Time-Series Analysis, see Fox (1972) and Kleiner, Martin and Thomson (1979). Their results indicate that the traditional variogram estimator is very sensitive to this kind of deviations and that the estimates will be upward biased.

III. A ROBUST VARIOGRAM ESTIMATOR

Assume that $\{Z(x);\ x \in A\}$ is strictly stationary to the bivariate level. This implies that the bivariate cdf of $[Z(x),Z(x+h)]$, $F_h(z_1,z_2)$, is independent of location x. The corresponding marginal cdf is F(z). The variogram value at distance h_o, $\gamma(h_o)$, is associated with the second order moment of $F_{h_o}(z_1,z_2)$, as

$$\gamma(h_o) = \tfrac{1}{2}E\{[Z(x) - Z(x+h_o)]^2\} = \tfrac{1}{2}\int_{-\infty}^{\infty}\int_{-\infty}^{\infty}(z_1-z_2)^2 d^2F_{h_o}(z_1,z_2)$$

Therefore, a robust estimator for $\gamma(h_o)$ may be obtained through a robust estimator for $F_{h_o}(z_1,z_2)$. This approach will be applied here. The latter estimator will make use of the fact that more information is available for estimating the marginal cdf, F(z), than for estimating the bivariate cdf, $F_{h_o}(z_1,z_2)$.

Assume that an estimator for the distribution of values over A, represented by F(z), is available. This estimator may be based on all elements in the set S. Let the estimator be:

$$[\hat{F}(z)]_S = \sum_{i=1}^{N} \beta_i \cdot I(x_i;z) \quad ; \quad \text{all } z$$

where

the subscript outside the paranthesis refers to the set on which the estimator is based.

$$I(x;z) = \begin{cases} 1 \text{ if } Z(x) \leqslant z \\ 0 \text{ else} \end{cases}$$

β_i;i=1,N are known weights assigned to each element in S.

The weights should be determined by the relative spatial location of the observations. Observations from sparsely sampled areas should receive more weight than observations which are located in clusters. Switzer (1977) gives some clues on how to optimally determine such weights. Simpler, intuitive opproaches may also be used.

Based on the set of observations, S, the following set of index-pairs can be defined:

$$D_{h_o}: \{(i,j) | Z(x_i), Z(x_j) \in S;\ x_i - x_j = h_o \text{ or } x_i - x_j = -h_o\}$$

$$N_{h_o} = \text{no. of elements in } D_{h_o}$$

In the following discussion $Z(x_i)$ and $Z(x_j)$ are interchangeable, hence the symmetric representation of pairs in D_{h_o}. The corresponding set of observation-pairs is:

$$S_{h_o}: \{(Z(x_i), Z(x_j)) | (i,j) \in D_{h_o}\}$$

Assume that the elements in S_{h_o} are ranked increasingly on the first coordinate of the elements. This should also be reflected the sequence of the elements of D_{h_o}.

Assume that the realization of the set S is known, which implies that the realization of S_{h_o} is known as well. Then an empirical representation of the bivariate cdf $F_{h_o}(z_1,z_2)$, called a h_o-scattergram, can be obtained. An example of a h_o-scattergram is shown in figure 2.

Note that $f_{xh_o}(z_1,z_2)$ is the pdf of $F_{h_o}(z_1,z_2)$, hence there is a close connection between figure 1 and figure 2. The bivariate observations in the h_o-scattergram is obtained from S_{h_o} while the marginal cdf's are defined by the estimate $[\hat{F}(z)]_s$. It is worth noting that the marginal cdf can be more precisely estimated than the bivariate characteristics since the elements in S_{h_o} is constructed from a subset of S.

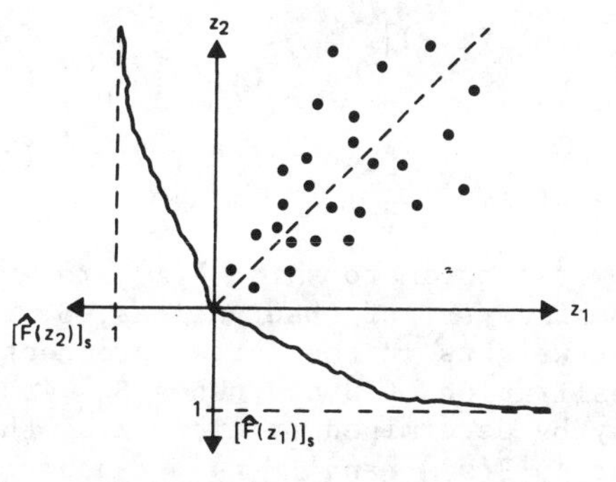

Figure 2.
An example of a h_o-scattergram which represent $F_{h_o}(z_1,z_2)$.

Define an estimator of $F_{h_o}(z_1,z_2)$ from the elements in S_{h_o} as:

$$[\hat{F}_{h_o}(z_1,z_2)]_{S_{h_o}} = \sum_{(i,j)\varepsilon D_{h_o}} \alpha_{ij}\cdot II(x_i,x_j;z_1,z_2) \quad ; \text{ all } z_1,z_2$$

where

$$II(x,y;z_1,z_2) = \begin{cases} 1 & \text{if } Z(x) \leqslant z_1 \text{ and } Z(y) \leqslant z_2 \\ 0 & \text{else} \end{cases}$$

α_{ij}; $(i,j) \varepsilon D_{h_o}$ are unknown weights to be determined.

Traditionally the bivariate observations are considered as the only information available, and these are all assigned equal weight. In the above case, however, additional information about the marginal cdf is present, hence equal weighting need not be optimal.

For the estimator of $F_{h_o}(z_1,z_2)$ to be unbiased and a valid cdf, it is necessary and sufficient that

$$\sum_{(i,j)\varepsilon D_{h_o}} \alpha_{ij} = 1$$

$$\alpha_{ij} \geqslant 0 \text{ ; all } (i,j) \varepsilon D_{h_o}$$

This estimator corresponds to assigning individual weights to all the elements in S_{h_o} and to estimate $F_{h_o}(z_1,z_2)$ from this set with its associated weights.

It is known, because of the symmetry in $F_{h_o}(z_1,z_2)$, that $F(z) = F_{h_o}(z,\infty) = F_{h_o}(\infty,z)$. From this an estimator for $F(z)$ based on the set S_{h_o} can be obtained:

$$[\hat{F}_{h_o}(z,\infty)]_{S_{h_o}} = \sum_{(i,j)\varepsilon D_{h_o}} \alpha_{ij}\cdot II(x_i,x_j;z,\infty)$$

$$= \sum_{(i,j)\varepsilon D_{h_o}} \alpha_{ij}\cdot I(x_i;z) = [\hat{F}(z)]_{S_{h_o}} \quad ; \text{ all } z$$

Now two estimators for $F(z)$ are available. One based on the complete set of observations, S, and the other based on S_{h_o}. The weights in the latter are considered unknown. Given the realization of S and hence S_{h_o}, the weights α_{ij}; $(i,j) \varepsilon D_{h_o}$ may be determined by minimizing the deviation between the estimates $[\hat{F}(z)]_S$ and $[\hat{F}(z)]_{S_{h_o}}$ under the constraints given on the weights.

In Omre (1984) an integrated weighted least square minimizing criterium is used. An approximate solution, which is fast to compute, is obtained. The weights can be expressed through a set with increasingly ordered elements, S'_{h_o}, which is identical to the first-coordinates of the elements in S_{h_o}. Given the realization of S and S_{h_o}, the realization of S'_{h_o} is

$$s'_{h_o}: \{z(x_i) | (i,j) \varepsilon D_{h_o}\} = \{z_{(i)}; i=1, N_{h_o}\}$$

To each element in S'_{h_o} there is an associated weight, $\alpha_{(i)}$; i=1, N_{h_o}. The approximate solution gives the following values to these weights:

$$\tilde{\alpha}_{(1)} = \int_{z_{(1)}}^{z_{(2)}} [\hat{F}(u)]_s \, du \Big/ \left(z_{(2)} - z_{(1)}\right)$$

$$\tilde{\alpha}_{(i)} = \int_{z_{(i)}}^{z_{(i+1)}} [\hat{F}(u)]_s \, du \Big/ \left(z_{(i+1)} - z_{(i)}\right) - \sum_{j=1}^{i-1} \tilde{\alpha}_{(j)} \quad ; \; i=2, N_{h_o}-1$$

$$\tilde{\alpha}_{(N_{h_o})} = 1 - \sum_{j=1}^{N_{h_o}-1} \tilde{\alpha}_{(j)}$$

The weights, $\tilde{\alpha}_{(i)}$; i=1; N, are obtained by letting the cdf step-function defined by the elements in s'_{h_o} and their associated weights deviate as little as possible from the estimate $[\hat{F}(z)]_s$. There is a one to one correspondance between the elements in S'_{h_o} and S_{h_o} and also between their respective weights. Hence a solution for α_{ij}; $(i,j)\varepsilon D_{h_o}$ is defined through $\tilde{\alpha}_{(i)}$; i=1, N_{h_o}. Denote the solution $\tilde{\alpha}_{ij}$; $(i,j) \varepsilon D_{h_o}$. It is obvious that this solution meets the required constraints.

This implies that an estimator for $F_{h_o}(z_1,z_2)$, dependent on the actual realization of S, is obtained:

$$[\hat{F}_{h_o}(z_1,z_2)]_{S_{h_o}} = \sum_{(i,j)\varepsilon D_{h_o}} \tilde{\alpha}_{ij} \cdot II(x_i,x_j;z_1,z_2) \quad ; \text{ all } z_1,z_2$$

From this estimator and the definition of the variogram value, it is reasonable to define the estimator of the variogram value as:

$$\hat{\gamma}(h_o) = \tfrac{1}{2} \int_{-\infty}^{\infty}\int_{-\infty}^{\infty} (z_1-z_2)^2 d^2[\hat{F}_{h_o}(z_1,z_2)]_{S_{h_o}} = \tfrac{1}{2} \sum_{(i,j)\varepsilon D_{h_o}} \tilde{\alpha}_{ij} \cdot \left(Z(x_i)-Z(x_j)\right)^2$$

Since $\sum_{(i,j)\varepsilon D_{h_o}} \tilde{\alpha}_{ij}=1$, this estimator is unbiased.

Consider the particular case where the sampling is in a regular grid. Intuitively, the weights in the estimator $[\hat{F}(z)]_S$ would be $\beta_i=1/N$; $i=1,N$. If the border effects are ignored, all elements in S will occur exactly the same number of times in S_{h_o}. Hence each element in S will contribute with exactly the same number of terms in $[\hat{F}(z)]_{S_{h_o}}$. In this case the deviation between $[\hat{F}(z)]_S$ and $[\hat{F}(z)]_{S_{h_o}}$ can be made equal to nought for all z by setting $\alpha_{ij}=1/N_{h_o}$, $(i,j)\ \varepsilon\ D_{h_o}$. This has to be the optimal solution to the above minimization. This implies that the estimator of the variogram value will be:

$$\hat{\gamma}(h_o) = \frac{1}{2N_{h_o}} \sum_{(i,j)\varepsilon D_{h_o}} \left(Z(x_i)-Z(x_j)\right)^2$$

which is identical to the traditional non-parametric estimator. In cases with irregular sampling the two estimators will not coincide.

It is important to note that during the discussion of the robust variogram estimator no distribution assumptions are made, neither are any additional parameters introduced. Hence the estimator can be classified as non-parametric. The robustness in the variogram estimator is introduced by the fact that the pairs of observations with interdistance h_o are weighted according to the two observations representativness of the distribution of values in the area. In the traditional estimator they would all receive equal weight.

IV. AN EMPIRICAL COMPARISON OF THREE VARIOGRAM ESTIMATORS

The traditional non-parametric estimator, the Cressie and Hawkins estimator and the estimator suggested in the previous chapter are evaluated. The comparison is based on sampling from simulated deposits. Deposits and sampling procedures reflecting the Ideal Case for variogram estimation, distributional deviations and sampling deviations are simulated.

The simulated deposits.
Two deposits are used, one with a multinormal flavor and the other showing distributional deviations. Both deposits are represented by a grid of dimension 120 x 120.

Deposit I is simulated by the following procedure:

- assign independent random numbers from a standard normal distribution to all the gridnodes

- impose spatial correlation by smoothing with a uniform circular filter with radius fifteen.

The resulting grid values will have a strong multinormal flavor, and hence meet the requirements of the Ideal Case. Border effects are avoided by actually simulating a larger area than the one used.

Deposit II is simulated by the following procedure:

- assign independent random numbers from a uniform distribution to all the gridnodes
- impose spatial correlation by smoothing with a uniform circular filter with radius fifteen. The resulting gridvalues will have a multinormal flavor due to the central limit theorem
- remove the multinormality by assigning to each node a value found by random drawing among the ten highest nodevalues in a neighborhood of two-and-a-half around the actual node
- obtain a plausible distribution of values by performing a univariate transformation of the nodevalues.

The resulting gridvalues will have the required univariate distribution, but not a multi-ϕ^{-1}-normal flavor. This simulation shows clear distributional deviations. Border effects are avoided by actually simulating a larger area than the one used.

The statistical characteristics of the simulated deposits are presented in figure 3A and B. The variogram is computed by the traditional non-parametric estimator based on all the 14 400 grid values.

The sampling procedure.
A two-step procedure is applied:

- Step I - random stratified sampling plan. The stratum size is 15 x 15 units, hence (8 x 8) = 64 unbiased observations are obtained
- Step II - a random stratified sampling plan around the three largest observations from step I. The stratum size is 5 x 5 units and eight strata are located around each observation. If an observation from this step falls outside the deposit it is ignored. Hence maximum (3 x 8) = 24 biased observations are obtained.

The observations from step I are unbiased, and hence no sampling deviations occur. The pool of observations from both step I and step II does simulate sampling deviations.

The variogram estimators.
Two of the most familiar estimators for the variogram value are compared to the alternative robust estimator.

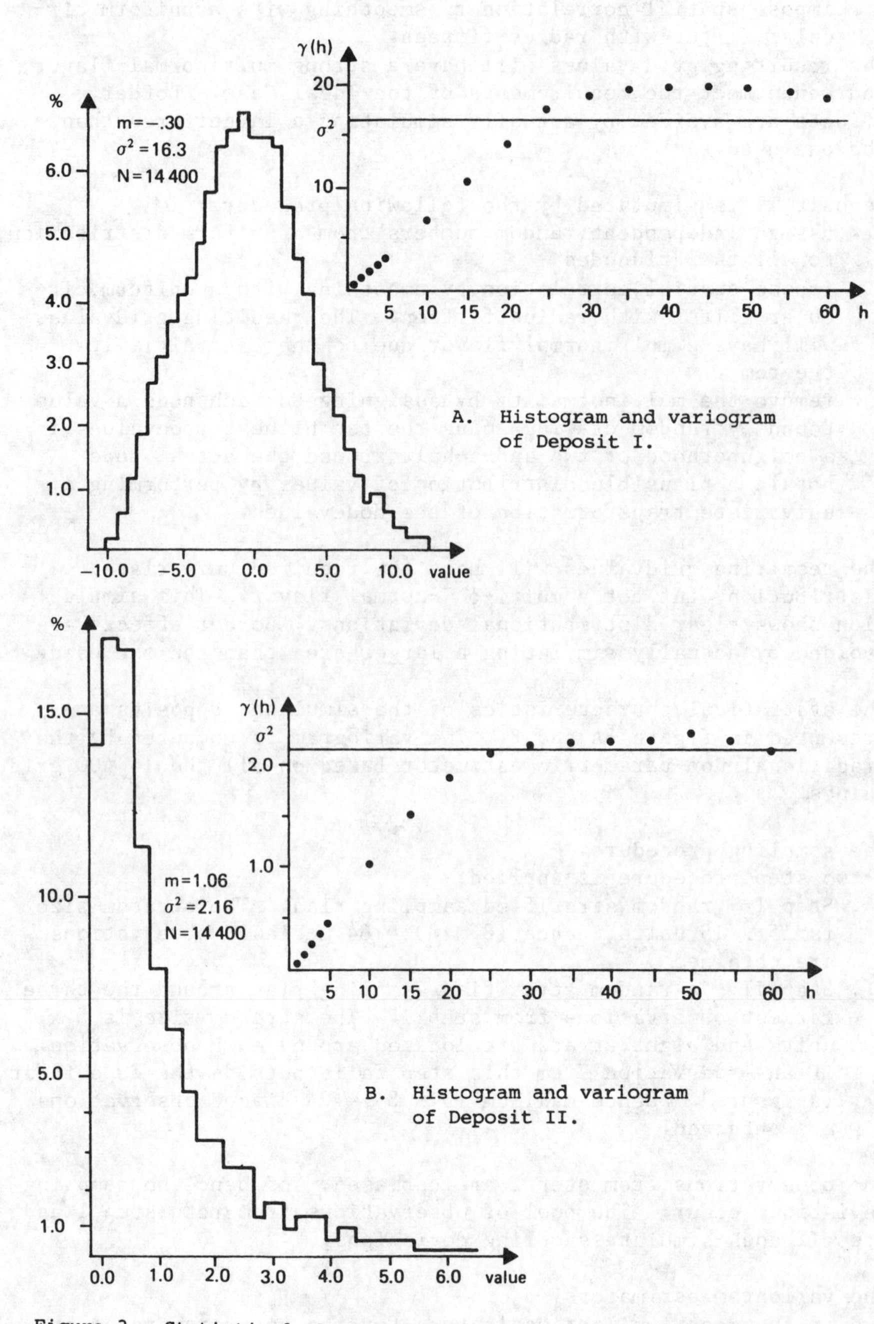

Figure 3. Statistical characteristics of simulated deposits.

- The traditional estimator.
  The estimator was suggested in Matheron (1965), its expression is:

$$\hat{\gamma}(h) = \frac{1}{2N_{h_o}} \sum_{(i,j)\varepsilon D_{h_o}} \left(Z(x_i)-Z(x_j)\right)^2$$

where

$$D_{h_o}:\{(i,j)\,|\,Z(x_i),Z(x_j)\ \varepsilon\ S\ ;\ x_i-x_j = h_o\}$$

N_{h_o} = no. of elements in D_{h_o}

This estimator is by far the one most frequently used.

- The Cressie & Hawkins estimator.
  The estimator was discussed among others in Cressie and Hawkins (1980). Cressie and Hawkins found it to be favorable in their test. The expression is:

$$\tilde{\gamma}(h) = \tfrac{1}{2}\cdot\left(.457 + \frac{.494}{N_{h_o}}\right)^{-1}\cdot\left[\frac{1}{N_{h_o}} \sum_{(i,j)\varepsilon D_{h_o}} \left(|Z(x_i)-Z(x_j)|\right)^{\frac{1}{2}}\right]^4$$

where

$$D_{h_o}\ :\ \{(i,j)\,|\,Z(x_i),Z(x_j)\ \varepsilon\ S\ ;\ x_i-x_j = h_o\}$$

N_{h_o} = no. of elements in D_{h_o}

- The robust estimator.
  This is the estimator suggested in chapter three. The estimation procedure is also specified in that chapter.

The numerical results.
The estimators are evaluated on three cases; the Ideal Case, in presence of distributional deviations and in presence of both distributional and sampling deviations.

The Ideal Case is simulated by deposit I and with sampling from step I only. The sampling procedure is performed independently one hundred times on the deposit. Each simulation is standardized to unit variance. For each of the hundred simulations the variogram values at distances multiples of five, with tolerance 2.5, are computed from the three variogram estimators. For the robust estimator all observations in a set of samples are given equal weight in the estimator of F(z). Figure 4.A presents further characteristics of the estimates from the simulations. From the results, all three estimators look unbiased in the Ideal Case. This is known from theory also. The variability in the traditional and the robust estimator are approximately equal. According to theory the traditional estimator should have been superior in the Ideal Case.

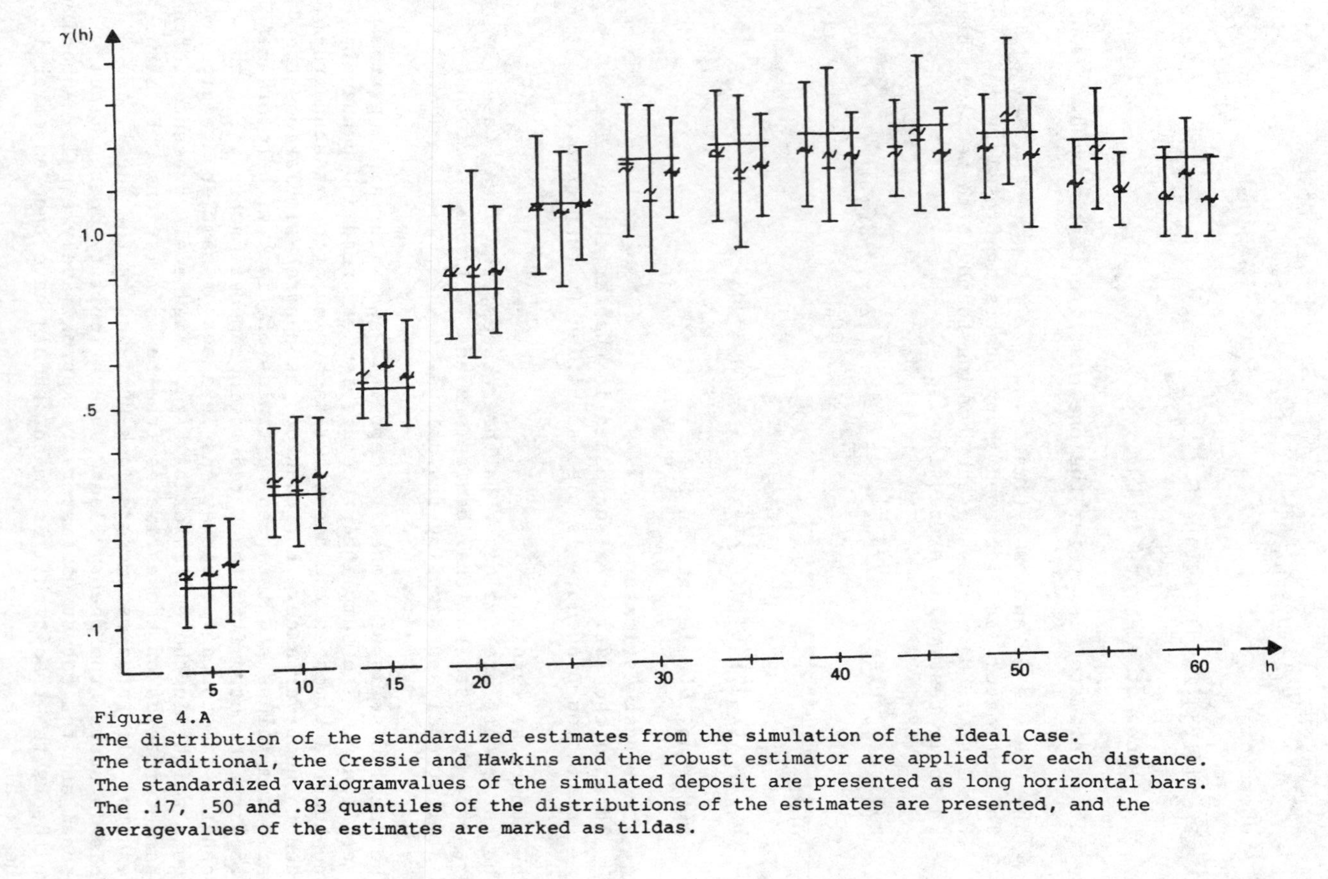

Figure 4.A
The distribution of the standardized estimates from the simulation of the Ideal Case.
The traditional, the Cressie and Hawkins and the robust estimator are applied for each distance.
The standardized variogramvalues of the simulated deposit are presented as long horizontal bars.
The .17, .50 and .83 quantiles of the distributions of the estimates are presented, and the averagevalues of the estimates are marked as tildas.

The reason for it not being so probably is that the simulation only approximates the Ideal Case and that the traditional estimator is only slightly superior to the robust one. The Cressie and Hawkins estimator shows larger variability than the two other estimators.

The distributional deviations are simulated by deposit II and unbiased sampling simulated by step I is performed. The sampling procedure is performed independently one hundred times on the deposit. Each simulation is standardized to unit variance. For each of the hundred simulations the variogram values at distances multiples of five, with tolerance 2.5, are computed from the three variogram estimators. For the robust estimator all observations in a set of samples are given equal weight in the estimator of F(z). Figure 4.B presents some further characteristics of the estimates from the simulations. The traditional and the robust estimator seem to be unbiased also in this case. The Cressie and Hawkins estimator does seriously underestimate the variogram values. The reason for this is that the former two are based on a non-parametric approach while the latter is based on normal distribution assumptions. The variability in the robust estimator is considerably less than the variability in the traditional estimator. This is so because the robust estimator uses an estimate of the distribution of values to stabilize the estimates of the variogram values from one distance to another.

The mixed distributional and sampling deviations are simulated by deposit II and by using observations from both step I and step II in the sampling procedure. The sampling procedure is performed independently one hundred times on the deposit. No standardization of the variance is done because the variance cannot be unbiasedly estimated from the observations. For each of the hundred simulations the variogram values at distances multiples of five, with tolerance 2.5, are computed from the three estimators. For the robust estimator, two groups of observations from each set of samples are distinguished in the estimator of F(z). The observation from sampling step I are given nine times the weight of the observations coming from step II. This corresponds approximately to the weights that would have been assigned to each observation in an estimator of the global average based on the polygon method. Figure 4.C presents some characteristics of the estimates from the simulations. In the presence of both distributional and sampling deviations, the estimators behave quite differently. The traditional estimator does seriously overestimate the variogram values. This is caused by the fact that the variability in high-value areas of the deposit is larger than in the low-value areas, and that these high-value areas are sampled more extensively. This effect is particularly strong at short distances and hence the shape of the

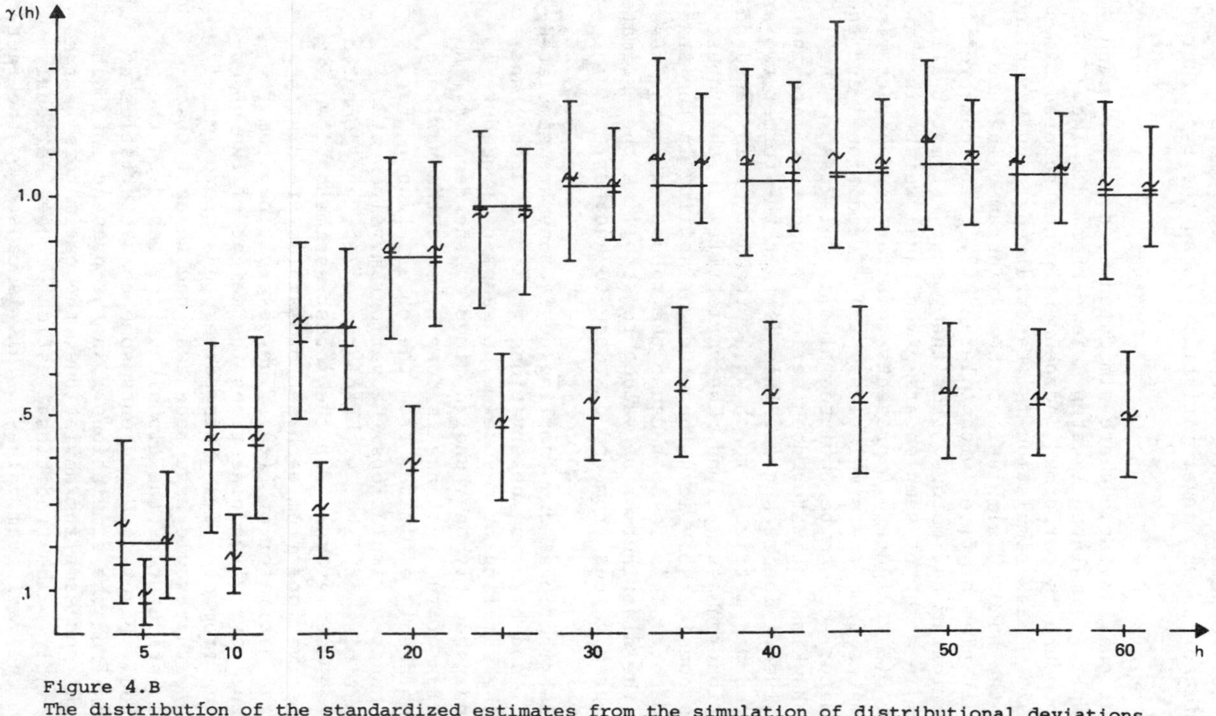

Figure 4.B
The distribution of the standardized estimates from the simulation of distributional deviations. The traditional, the Cressie and Hawkins and the robust estimator are applied for each distance. The standardized variogramvalues of the simulated deposit are presented as long horizontal bars. The .17, .50 and .83 quantiles of the distribution of the estimates are presented, and the averagevalue of the estimates are marked as tildas.

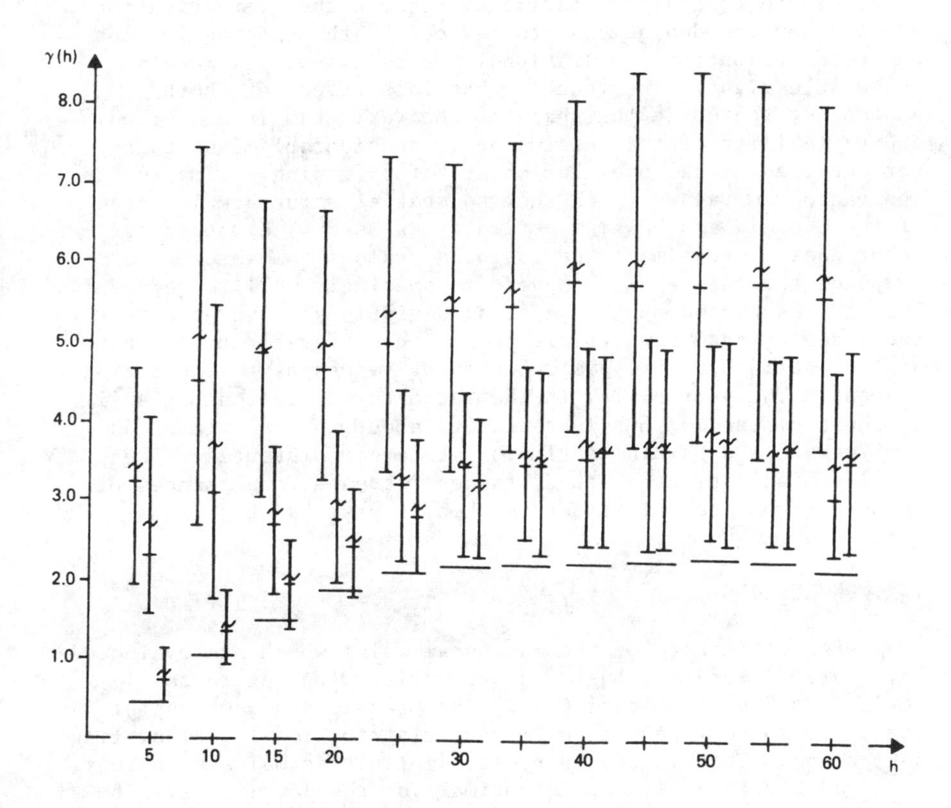

Figure 4.C
The distribution of the estimates from the simulation of distributional and sampling deviations.
The traditional, the Cressie and Hawkins and the robust estimator are applied for each distance.
The variogramvalues of the simulated deposit are presented as long horizontal bars.
The .17, .50 and .83 quantiles of the distribution of the estimates are presented, and the averagevalue of the estimates are marked as tildas.

variogram is distorted. The variability of the estimator is also large. For many samplesets in the simulation no spatial correlation could be recognized from the estimates from the traditional estimator. The Cressie and Hawkins estimator does also overestimate the variogram values. The bias is smaller than for the traditional estimator because the overestimation due to sampling deviations are reduced by the underestimation due to distributional deviations. Hence two wrongs make a reasonable right. The reducing has less effect at short distances, therefore the shape of the variogram is distorted. The variability of this estimator is considerably less than for the traditional one. The chances for finding estimates for the variogram values which show no spatial structure is large if the Cressie and Hawkins estimator is used. The robust estimator does overestimate the variogram values somewhat. The shape of the variogram, however, is amazingly well reproduced. The bias is caused by the sampling deviations. The effect of these deviations, however, is reduced considerably by the use of the estimate of the distribution of values as a stabilizing element. The variability in the estimator is reasonably small at short distances, but increases to about the same level as the Cressie and Hawkins estimator at larger distances. The small variability at short distances increases the chances of recognizing a spatial structure when it exsists.

V. CONCLUSIONS

The three estimators of the variogram value which are evaluated have very different robustness properties with regard to deviations from the Ideal Case. The Cressie and Hawkins estimator is relatively sensitive to both distributional and sampling deviations. This is caused by its implicit normal assumptions. The traditional estimator is optimal for the Ideal Case. It performs reasonably well as long as randomized sampling is done. If distributional and sampling deviations occur simultaneously, however, the traditional estimator is highly unstable. The suggested robust estimator is slightly less stable than the traditional estimator in the Ideal Case. If distributional or sampling deviations are present, the robust estimator is more reliable than the two others.

The Final conclusion seems to be that always using the robust estimator will cause only minor losses of efficiency in the Ideal Case while if distributional or sampling deviations occur, there will be a major gain in reliability.

VI. ACKNOWLEDGEMENTS

The research on robust variogram estimation is financially supported by The Royal Norwegian Council for Scientific and Industrial Research, The Geostatistical Programme, Stanford University and the Norwegian Computing Center.

VII. REFERENCES

Armstrong; M. and P. Delfiner, 1980: Towards a more robust variogram: A Case Study on Coal. Paper from École Nationale Superieure des Mines de Paris.

Cressie; N. and D.M. Hawkins, 1980: Robust Estimation of the Variogram: I. Mathematical Geology, Vol 12, No 2.

Devlin; S.J., R. Gnanadesikan and J.R. Kettenring, 1975: Robust estimation and outlier detection with correlation coefficients. Biometrica, Vol 62, No 3, p.531.

Fox; A.J., 1972: Outliers in Timeseries. Journal of the Royal Statistical Society Series B, Vol 34, No 3, pp 350-363.

Journel; A.G. and C.J. Huijbregts, 1978: Mining Geostatistics. Academic Press, 600 p.

Kleiner; B., R.D. Martin and D.J. Thomson, 1979: Robust Estimation of Power Spectra. Journal of Royal Statistical Soc., Series B, Vol 41, No 3, pp 313-351.

Matheron; G., 1965: Les Variables Regionalisées et leur Estimation. Masson & Cie, Paris.

Mosteller; F. and J.W. Tukey, 1977: Data Analysis and Regression. Addison-Wesley Publ. Co., 588 p.

Omre; H., 1984: Robust Variogram Estimation in Geostatistics. PhD-thesis of Dept. of AES, Stanford University - to appear Spring 1984.

Switzer; P., 1977: Estimation of Spatial Distributions from Point Sources with Application to Air Pollution Measurements. Bull. Int. Stat. Inst., XLVII(2), pp 123-137.

INFERENCE FOR SPATIAL AUTOCORRELATION FUNCTIONS

Paul Switzer

Stanford University

Abstract. Some general methods are proposed for generating confidence interval estimates for parameters of a variogram model. Particular attention is paid to scale and shape parameters and to joint estimation procedures for the scale parameter and the replication variance (nugget effect). The connection is made to formal testing of the hypothesis of the absence of spatial autocorrelation at a specified distance scale.

1. INTRODUCTION

This paper describes general and elementary approaches to statistical inference for parameters of a spatial autocorrelation model via the variogram. The first part concerns estimation of a scale parameter primarily intended to describe the linear part of the variogram for short interpoint distances. The second part concerns inferences for a shape parameter, basically describing the variogram for medium and long interpoint distances, and the associated problem of testing for the absence of spatial autocorrelation at specified distance scales. The third part concerns inference for the replication variability at zero distance (nugget effect) and simultaneous inference for the important problem of joint estimation of the nugget effect and the scale parameter of the variogram.

The basic ideas involve linear transformations of the data to uncorrelated quantities of constant variance and the comparisons of certain rank orderings. Fourth moment considerations play no role. Related work on inference for variograms may be found in Matheron [1] and in the time series literature such as Anderson

G. Verly et al. (eds.), Geostatistics for Natural Resources Characterization, Part 1, 127–140.

[2]. However, the time series literature depends either on having equally spaced data or on a constructive model for the data or both.

The main application of inference on variograms is to provide confidence intervals for estimation variances which are functionals of the variogram; thereby one obtains more realistic less optimistic interval estimates for interpolated values, block averages and other quantities which are calculated as linear functions of the data.

2. SCALE PARAMETER ESTIMATION

We consider first the case where the variogram model has been specified completely up to a scaling constant. This would be the case, for example, for the family of linear variograms. For a general stationary spatial process $Z(x)$ let

$$\gamma(h) = \tfrac{1}{2}E[Z(x) - Z(x+h)]^2 = \tau^2 \cdot \tilde{\gamma}(h)$$

where $\tilde{\gamma}(h)$ is completely specified and, for definiteness, $\tilde{\gamma}(h)$ may be taken to have unit slope at the origin. The linear variograms have $\tilde{\gamma}(h) = |h|$. The objective is to make inferences about τ, e.g. a confidence interval of possible τ values which are consistent with the data.

The general plan is to transform the observed random variables $Z(x_i)$, $i=1, \ldots, n$, to a set of uncorrelated random variables with zero mean and a constant variance depending on τ. Then standard procedures are used to obtain confidence intervals for this variance. It is straightforward to first obtain zero-mean linear combinations of the observations which are called "contrasts." For example, in the stationary case every difference $Z(x) = Z(x')$ is a contrast. Suppose we define m linearly independent contrasts $D_1, D_2, \ldots, D_m$. (The discussion of the choice of m and the choice of the D_i is deferred.) The vector of these contrasts may be represented as

$$D = A \cdot Z$$

where A is an $m \times n$ matrix with row sums equal to zero, and Z is the vector of observable data $Z(x_1), \ldots, Z(x_n)$.

The covariance matrix of the vector of contrasts, D, may be expressed as

$$\mathrm{cov}(A\cdot Z) = -(\tau^2)\ A\Gamma A' \equiv \tau^2\cdot\Sigma$$

where

$$\Gamma_{ij} = \tilde{\gamma}(x_i - x_j)\ .$$

Then Σ is a completely determined $m\times m$ covariance matrix. There now exist known linear transformations of the vector D which will produce a new vector of m uncorrelated contrasts Δ with constant variance, and Δ like D will have zero-mean vector, i.e.

$$\Delta = T\cdot D\ ;$$

$$\mathrm{cov}(\Delta) = \tau^2\cdot I_{m\times m}\ ;$$

$$E(\Delta_i) = 0,\quad E(\Delta_i^2) = \tau^2,\ \text{for}\quad i=1,\ \ldots,\ m\ .$$

The $m\times m$ matrix T is not unique. A reasonable choice for T could be the matrix of eigenvectors of Σ, rescaled to give unit variances.

From the observable data vector Z one now computes the observed vector $\Delta = (T\cdot A)\cdot Z$ and the squares values Δ_i^2, i=1, ..., m, whose expectations are all equal to the unknown τ^2. If we act as though the uncorrelated Δ_i were statistically independent then we may construct approximate confidence intervals based on the central limit theorem. Specifically, let

$$\hat{\tau}^2 = \Sigma\Delta_i^2/m,\qquad S^2(\hat{\tau}^2) = \Sigma(\Delta_i^2 - \hat{\tau}^2)^2/m^2\ .$$

If z_α is the upper α point of a standard Gaussian distribution, then an approximate confidence interval for the scaling constant τ^2 of a family of variograms is given by

$$\hat{\tau}^2 - z_{\alpha/2}\cdot S(\hat{\tau}^2) \leq \tau^2 \leq \hat{\tau}^2 + z_{\alpha/2}\ S(\hat{\tau}^2)$$

with coverage probability $1-\alpha$, approximately.

As an illustration suppose we are interested in estimating the scaling constant of the variogram for short interpoint distances, and are willing to act as though this variogram were linear for short distances and flat at long distances. We may operate on disjoint contrasts of the form $D_i = Z(x_i) - Z(x_i')$ for selected data pairs. For simplicity of the illustration suppose that the separation between pairs is sufficiently large that the contrasts are uncorrelated with each other. The transformation matrix T can be taken to be the diagonal matrix with

$$T_{ii} = [2|x_i - x_i'|]^{-\frac{1}{2}} .$$

Then $\Delta_i^2 = \frac{1}{2}[Z(x_i) - Z(x_i')]^2/|x_i - x_i'|$.

3. SHAPE PARAMETER ESTIMATION

We now consider a richer parametric family of variogram models of the form

$$\gamma(h) = \tfrac{1}{2}\cdot E[Z(x) - Z(x+h)]^2 = \tau^2 \cdot \tilde{\gamma}(h;\theta) ,$$

where θ is a shape parameter, $\tilde{\gamma}$ is a fixed variogram for each θ, and, as before, τ^2 is a pure scaling constant. It is convenient, but not necessary, to parametrize the variogram family so that θ is proportional to the slope of the variogram at $h=0$, i.e.

$$\left.\frac{\partial}{\partial h}\tilde{\gamma}(h;\theta)\right|_{h=0} = k\theta, \quad k \text{ fixed and known.}$$

Common examples are the isotropic spherical and exponential families given respectively by

$$\text{Spherical:} \quad \tilde{\gamma}(h;\theta) = \frac{3}{2}(\theta h) - \frac{1}{2}(\theta h)^3 \quad \text{for} \quad 0 \leq (\theta h) \leq 1$$

$$= 1 \quad \text{for} \quad (\theta h) \geq 1$$

$$\text{Exponential:} \quad \tilde{\gamma}(h,\theta) = 1-e^{(-\theta h)} \quad \text{for all} \quad h \geq 0;\ \theta > 0 .$$

Here, the increment h is treated as a scalar.

The correlation between two linear contrasts, as defined earlier, will still be free of the scaling τ^2 but will depend on θ. For a given vector of m linear contrasts $A \cdot Z = D = (D_1, \ldots, D_m)$, their $m \times m$ covariance matrix is

$$\text{cov}(D) = -\tau^2 \cdot A\Gamma(\theta)A' \equiv \tau^2\Sigma(\theta)$$

where

$$\Gamma_{ij}(\theta) = \tilde{\gamma}(x_i - x_j;\ \theta) \ .$$

The transformation of D to a vector of m uncorrelated equal variance linear contrasts, Δ, now depends on the correct choice for θ. Let $T(\theta_0)$ be such an $m \times m$ transformation matrix for a particular $\theta = \theta_0$, i.e.

$$\Delta(\theta_0) = T(\theta_0) \cdot D \ ,$$

$$\text{cov}_{\theta_0}\{\Delta(\theta_0)\} \propto I_{m \times m} \ ,$$

and, in particular,

$$E_{\theta_0}\{\Delta_i^2(\theta_0)\} = \text{constant for all } i=1, \ldots, m \ .$$

However, if the correct value of the shape parameter is $\theta \neq \theta_0$, then the covariance matrix of $\Delta(\theta_0)$ will not be proportional to the identity matrix. In particular, the transformed linear contrasts which are the components of $\Delta(\theta_0)$ will typically not have equal variances, although means of these contrasts must necessarily remain zero, i.e.

$$E_\theta\{\Delta_i^2(\theta_0)\} \neq E_\theta\{\Delta_j^2(\theta_0)\} \quad \text{for} \quad \theta \neq \theta_0 \quad \text{and} \quad i \neq j \ .$$

Let $r_i(\theta_0;\theta)$ be the rank order (fixed) of $E_\theta\{\Delta_i^2(\theta_0)\}$ among the set of m such values for $i=1, \ldots, m$. Similarly, let $R_i(\theta_0)$ be the rank order (stochastic) of the observed value of $\Delta_i^2(\theta)$ among the set of m such values for $i=1, \ldots, m$. Then both rank vectors $r(\theta_0,\theta) = [r_1(\theta_0,\theta), \ldots, r_m(\theta_0,\theta)]$ and

$R(\theta_0) = [R_1(\theta_0), \ldots, R_m(\theta_0)]$, are permutations of the integers 1, ..., m. The two permutations will appear to be correlated if the correct parameter value is $\theta \neq \theta_0$.

However, if θ_0 is the correct value then the random variables $\Delta_i(\theta_0)$ have constant variance for all i and are uncorrelated. If we strengthen the assumptions by regarding the $\Delta_i(\theta_0)$ as exchangeable random variables when θ_0 is the correct shape parameter then the stochastic rank vector $R(\theta_0)$ will be a completely random permutation of $1, \ldots, m$ and therefore uncorrelated with the fixed rank vector $r(\theta_0, \theta)$. Hence, a test of θ_0 versus θ can be constructed using a measure of the correlation between the two sets of ranks, and all such tests will be distribution-free.

The inconvenient aspect of the above described test of the shape parameter is that it depends both on the parameter value being tested, θ_0, and an arbitrary "alternative" value, θ, through the fixed rank vector $r(\theta_0;\theta)$. To overcome, in part, the lack of uniformity of the test over θ, we consider only the "local" alternatives near the tested value θ_0, i.e. $\theta = \theta_0 + \varepsilon$ for sufficiently small $|\varepsilon|$. Then we are led to define the fixed rank vector $r(\theta_0)$ as follows:

$r_i(\theta_0)$ is the rank order of $\left.\frac{\partial}{\partial\theta} E_\theta\{\Delta_i^2(\theta_0)\}\right|_{\theta=\theta_0}$

among the set of m such values for $i=1, \ldots, m$.

The test for θ_0 is then based on the correlation of the fixed rank vector $r(\theta_0)$, just defined, with the random rank vector $R(\theta_0)$ defined above.

A common test of the independence of two rank vectors of length m is based on the rank statistic (Spearman's coefficient) as described in Kendall [3]:

$$K(\theta_0) = 1 - 6 \sum_1^m (R_i - r_i)^2 / (m^3 - m) .$$

If the rank vectors are independent then the upper $\alpha/2$ point of the random sampling distribution of K is given approximately by

$$K_{\alpha/2} = z_{\alpha/2} \cdot \sqrt{m} .$$

Therefore, a value of the shape parameter θ_0 would be rejected at level α if $|K(\theta_0)| > K_{\alpha/2}$. The set of θ_0 values which are not rejected comprise a confidence set for θ at level $1-\alpha$.

As an example, suppose our linear contrasts are chosen to be differences of data pairs, i.e.,

$$D_i = Z(x_i) - Z(x_i') \quad \text{for} \quad i=1, \ldots, m .$$

Further, suppose that the data pairs are disjoint and well separated from one another so that, for any reasonable value of θ, the data pairs are uncorrelated. For testing that $\theta=\theta_0$, a transformation to uncorrelated equal variance linear contrasts is given by

$$\Delta_i(\theta_0) = [Z(x_i) - Z(x_i')]/[\tilde{\gamma}(x_i - x_i'; \theta_0)]^{\frac{1}{2}} \text{ for } i=1, \ldots, m .$$

The rank vector, $R(\theta_0)$, is given by the ranks of the $\Delta_i^2(\theta_0)$ computed using the observed Z data. The fixed rank vector, $r(\theta_0)$, does not depend on the data and is given by the ranks of

$$\frac{\partial}{\partial\theta} E_\theta\{\Delta_i^2(\theta_0)\}\bigg|_{\theta=\theta_0} = \frac{\left.\frac{\partial}{\partial\theta}\tilde{\gamma}(x_i - x_i'; \theta)\right|_{\theta=\theta_0}}{\tilde{\gamma}(x_i - x_i'; \theta_0)}$$

In the case of the exponential and spherical variograms and many other parametric models a substantial further simplification is possible. For every possible value of the shape parameter θ_0 it can be shown that the above ranking criterion is a monotone function of the interpoint distances $|x_i - x_i'|$ provided all such distances are less than the range of the variogram. Hence, the fixed rank vector $r(\theta_0)$ would be the same for every θ_0 and corresponds to the ranking of the interpoint distances of the m data pairs.

4. TESTING RANDOMNESS

Consider the problem of testing for the absence of spatial autocorrelation (randomness) at interpoint distances exceeding some specified distance h_0. We may take a subset of the data points whose interpoint distances all exceed h_0. The hypothesis of randomness may be embedded in a parametric family of variogram functions. Since the convention is to parametrize variograms so that the parameter θ is proportional to the slope of the variogram at the origin it follows that the value $\theta=0$ will correspond to randomness.

Hence we may follow the prescription of Section 3 for testing the hypothesis that $\theta=0$ in some parametric family, i.e., if the value $\theta=0$ is not included in the level $1-\alpha$ confidence set for θ then the hypothesis of randomness is rejected at level α. In general, the test of randomness will depend on the choice of parametric family of variogram.

As in Section 3, the test of $\theta=0$ involves the correlation between a stochastic rank vector $R(0)$ and a fixed rank vector $r(0)$. The vector $R(0)$ does not depend on the particular family of variograms insofar as all calculations involve null correlations. However, the vector $r(0)$ will, in general, depend on the particular variogram family so there is clearly no unique test of randomness in this context.

In the particularly convenient case where the contrasts are differences of disjoing data pairs, as seen in Section 3, the rank vector $r(0)$ typically corresponds to the ranking of interpoint distances within data pairs. In this case also the stochastic rank vector corresponds to the ranking of the squared differences of the data values within pairs. Therefore, the test for randomness here is based on the rank correlation between $|Z(x_i) - Z(x'_k)|$ and $|x_i - x'_i|$ for $i=1, \ldots, m$.

Once again we note that the cutoff K_α is derived under the assumption that the data differences $Z(x_i) - Z(x'_i)$ are exchangeable random variables which is not quite the same as saying that they are uncorrelated with the same means and variances.

5. REPLICATION VARIANCE ESTIMATION

In many applications there is implicit replication variability which is not negligible relative to spatial variability and which is part of every observation. This variability, called the nugget effect in mining applications, is commonly modelled as additive white noise lacking spatial correlation. It is best estimated directly from replicated observations in standard ways. However, sometimes replicated observations are not available or are very few in number. It is then necessary to estimate replication variance by extrapolation of the spatial variogram to the origin, i.e. to very short interpoint distances. The presence of non-negligible replication variance will also affect the method for estimating other parameters of the variogram model.

The generalized variogram model is of the form

$$\gamma(h) = \tfrac{1}{2}E\{Z(x) - Z(x+h)\}^2 = \tau^2[\eta + \tilde{\gamma}(h;\theta)]$$

where $\tau^2\eta$ is the replication variance and, as before, τ^2 is a scale parameter, θ is a shape parameter, and $\tilde{\gamma}(0;\theta) = 0$ for all θ. In the following development the shape parameter θ is regarded as fixed and the notational dependence on θ is dropped.

Again we consider a vector of m zero-mean linear contrasts $D = A \cdot Z$, where

$$\text{cov}(D) = \tau^2 \cdot \Sigma(\eta)$$

where

$$\Sigma(\eta) = A \cdot [\eta \cdot I - \Gamma] \cdot A'$$

and

$$\Gamma_{ij} = \tilde{\gamma}(x_i - x_j) \ .$$

Now, for a specified η, the linear contrasts, D, may be transformed to another set of linear contrasts, Δ, using the rescaled eigenvectors of $\Sigma(\eta)$, i.e.,

$$\Delta(\eta) = T(\eta) \cdot D$$

with

$$\text{cov}_\eta(\Delta(\eta)) = \tau^2 \cdot I_{m \times m} \ .$$

When $\tau^2\eta$, the replication variance, cannot be estimated directly from replicated data one should estimate this variance *jointly* with the scale parameter τ^2 of the variogram using the most closely spaced data points. The estimation proceeds in two stages. First, a confidence interval for η is generated (which does not depend on τ^2) at level $1 - \frac{1}{2}\alpha$. Second, a confidence interval for τ^2 is generated for *each* η, at level $1 - \frac{1}{2}\alpha$. The resulting joint confidence set for τ^2 and η will have coverage probability greater than $1-\alpha$ by the Bonferonni inequality. This joint confidence set may then be transformed to a joint confidence set for τ^2 and $\tau^2\eta$, the actual replication variance.

The estimation of the replication parameter, η, proceeds as though it were a shape parameter of the variogram. For each putative value η_0, a rank vector $R(\eta_0)$ is obtained from the calculated values of $\Delta_i^2(\eta_0)$. This rank vector is matched with the fixed rank vector, $r(\eta_0)$, based on the values of

$$\frac{\partial}{\partial\eta} E_\eta\{\Delta_i^2(\eta_0)\}\bigg|_{\eta=\eta_0} .$$

If the two rank vectors are uncorrelated at level $\alpha/2$ then η_0 is part of the confidence set for η. It is important to note that the scale parameter τ^2 plays no role so far.

Now, if η_0 belongs to the confidence interval for η then we construct a confidence interval for the scale parameter τ^2, according to the method of Section 2, using the m calculated values $\Delta_1^2(\eta_0), \ldots, \Delta_m^2(\eta_0)$. In this way the joint confidence region for η and τ^2 is constructed. The illustration of the next section demonstrates these procedures.

The same paradigm for constructing joint confidence sets for τ^2 and η may also be used to construct joint confidence sets for τ^2 and θ, the scale and shape parameters of the variogram. Sometimes it may also be useful to have joint confidence sets for all three parameters but this will require extension of the methods described here and will be reported later.

6. ILLUSTRATION

When estimating the variogram for small arguments it is reasonable to use contrasts of closely spaced data points such as close spaced differences of data pairs. If the data pairs are sufficiently well separated from one another to be regarded as uncorrelated then it may be sensible to treat the variogram as linear for short distances and flat for long distances. Then the shape parameter θ is explicitly absent in the construction of the joint confidence region for τ^2 and θ, the scale and replication variance parameters.

For example, consider contrasts which are differences of such data pairs. Then the covariance matrix $\Sigma(\eta)$ is the diagonal matrix with

$$\Sigma_{ii}(\eta) = 2\eta + 2|x_i - x_i'| \quad \text{for} \quad i=1, \ldots, m .$$

Then

$$\Delta_i^2(\eta) = \tfrac{1}{2}[Z(x_i) - Z(x_i')]^2/[\eta + |x_i - x_i'|] ,$$

and the ranking criterion for the fixed rank vector $r(\eta)$ is a monotone function of the interpoint distances within pairs.

The table below gives hypothetical values of $[Z(x_i) - Z(x_i')]^2/2$ for 25 such data pairs at indicated distances. These "data" were used to calculate $\Delta_i^2(\eta)$, i=1, ..., 25, for values of η at intervals of 0.1. For each η, a ranking of the $\Delta_i^2(\eta)$ is correlated with the ranks of the interpoint distances using the statistic K of Section 3. It was found that this rank correlation was too large negative for $\eta < 0.3$ and too large positive for $\eta > 2.8$ at the $\alpha = 10\%$ significance level. Hence the 90% confidence interval for η is approximately

$$0.3 \leq \eta \leq 2.8 .$$

Hypothetical Numbers Used to Generate a Joint
Confidence Region for Parameters of a Variogram

(1,1)	(2,1)	(3,2)	(4,2)	(5,2)
(1,1)	(2,2)	(3,2)	(4,3)	(5,4)
(1,1)	(2,2)	(3,3)	(4,3)	(5,4)
(1,2)	(2,2)	(3,3)	(4,3)	(5,4)
(1,2)	(2,3)	(3,4)	(4,4)	(5,5)

First coordinate: Interpoint distance $|x_i - x_i'|$

Second coordinate: $\frac{1}{2}[Z(x_i) - Z(x_i')]^2$

For each of the accepted η values above, a 90% confidence interval for τ^2 was calculated from the already computed values of $\Delta_i^2(\eta)$ by the standard method described in Section 2. The collection of these τ^2 confidence intervals comprises the joint confidence region for η and τ^2 with coverage probability 80% (by the Bonferroni inequality). Examples of these confidence intervals are:

$$0.75 \leq \tau^2 \leq 0.95 \quad \text{for } \eta = 0.3$$

$$0.60 \leq \tau^2 \leq 0.75 \quad \text{for } \eta = 1.0$$

$$0.40 \leq \tau^2 \leq 0.50 \quad \text{for } \eta = 2.8 .$$

It is straightforward to convert the joint confidence region for η and τ^2 to a region for $\tau^2\eta$ and τ^2, i.e., the replication variance and scale parameter. This 80% confidence region is shown in Figure 1. The requirement of a higher confidence probability would have given a larger region.

If the data at the five fixed interpoint distances have been averaged, the resulting five points would appear to lie very close to a straight line, perhaps giving the impression that the variogram has been precisely estimated. In fact, we may derive confidence intervals for the variogram function itself at fixed values of its argument from the joint confidence region for its parameters This is done by projection of the region onto the straight lines $\tau^2\eta + \tau^2 h$ for selected values of h. This is shown in Figure 2. The coverage probability of 80% applies simultaneously at all arguments h so that, at a fixed single argument, the confidence interval is conservative. For example, a conservative 80% confidence interval for $\gamma(1) = \tau^2(\eta+1)$ is given by $1.2 \leq \gamma(1) \leq 2.0$).

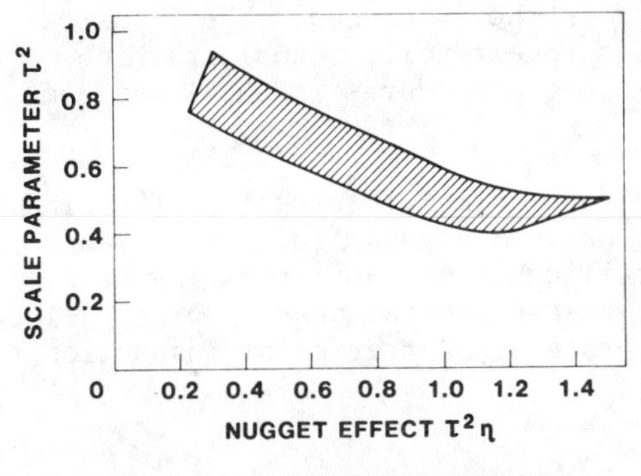

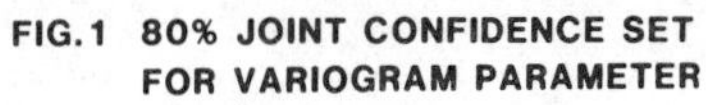
FIG.1 80% JOINT CONFIDENCE SET FOR VARIOGRAM PARAMETER

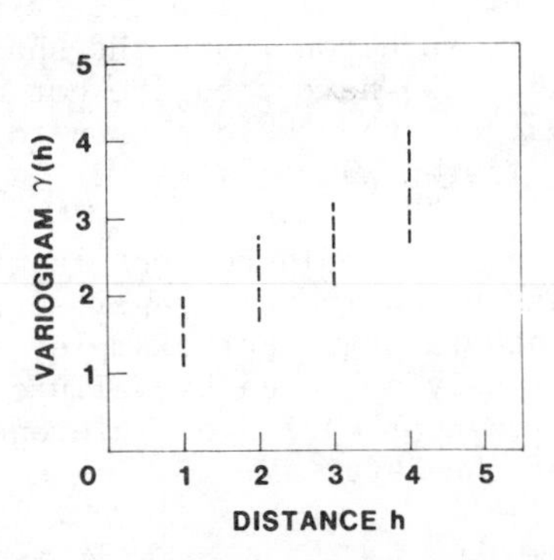

FIG. 2 PROJECTED SIMULTANEOUS CONFIDENCE INTERVALS FOR THE VARIOGRAM AT SELECTED DISTANCES

It must be kept in mind that such confidence intervals are tied to the parametric form adopted for the variogram. If one had a sufficient number of uncorrelated data pairs at a fixed distance h then a confidence interval for $\gamma(h)$ could be computed directly for that h without reference to any parametric model. It would then be interesting to compare it with the conservative intervals derived under parametric modelling.

The method of projecting joint confidence regions for parameters may also be used to obtain simultaneous confidence intervals for variances of interpolation errors and related quantities, since such variances may be expressed as functions of the parameters of the variogram. Indeed, this may be the most important use of the confidence regions.

7. CONCLUSION

The procedures we have described for inference on the parameters of variogram models are incomplete in many respects and do not lead to any automatic analysis of the second-order properties of spatial data. It will always be important to view the data carefully if only for the purpose of segregating stationary domains and selecting distance scales for modelling variograms. Among the issues not addressed here is the choice of the parametric family of variogram models and the goodness-of-fit issue. In principle, by introducing an extra parameter for model selection, it may be possible to address these issues by the methods here described. Also, we have said nothing about optimization of inference procedures, in particular regarding the choice of linear contrasts which are the basis of the analysis. Presumably, the more such contrasts the sharper is the inference. The illustrations we have used are not meant to exemplify actual practice. Rather, they were chosen to exhibit procedures in the least complicated way.

It should be kept in mind that formal inference of the kind described in this paper serves only as a guide to calibration of uncertainty. For example, significance probabilities may be grossly affected by failure to remove nonstationarity or by selective application of inference procedures after prior inspection of the data.

REFERENCES

[1] Matheron, G. (1965), *Les Variables Regionalisees et leur Estimation*, Masson, Paris.

[2] Anderson, T. W. (1971), *The Statistical Analysis of Time Series*, Wiley, New York.

[3] Kendall, M. G. (1970), *Rank Correlation Methods*, Griffin, New York.

PRECISION OF ESTIMATION OF RECOVERABLE RESERVES : THE NOTION OF CONDITIONAL ESTIMATION VARIANCE

Roland Froidevaux

David S. Robertson & Associates
Toronto, Canada.

ABSTRACT

An empirical approach to the problem of approximating the conditional estimation variance of an ore grade estimate is presented.

Through repetitive sampling of simulated deposits, (representing various types of regionalization), the effects of increasing the cut-off grade, and/or changing the size of the selective mining units on the precision of the ore grade, are analyzed and an approximation formula for the conditional estimation variance is proposed. The performance of this approximation is shown and its theoretical implications, as well as its limitations, are discussed.

INTRODUCTION

One of the tasks carried out routinely by geostatistics is the evaluation of total (or global) mining reserves. This evaluation generally entails two steps:

i) Inferring the distribution of the average grade of the selective mining units (SMU).

ii) Determining the recovered ore grade and tonnage by applying a cut-off grade to this distribution.

Along with the recovered grade and tonnage figures, the geostatistician generally provides an assessment of the

G. Verly et al. (eds.), Geostatistics for Natural Resources Characterization, Part 1, 141–164.

'global estimation variance', which represents the precision of the overall average grade estimate. This 'global estimation variance' is of limited interest for evaluating the merits of a mining project; it can even be misleading to a reader unaware of geostatistical theory, since said reader is likely to consider it to be the precision of recovered reserves. What is required is the reliability of the estimate of recoverable reserves, given a certain degree of mining selectivity and an envisioned cut-off grade.

The problem of assessing, correctly, the precision of mineable reserves is a critical one when ore reserves have to be classified (FROIDEVAUX, 1982). Indeed, most securities exchange legislations now require that the accuracy limits of an estimate be stated when reporting resources and reserves.

The need for techniques allowing to approximate the conditional estimation variance, therefore, is obvious.

However, despite its importance, the problem of assessing the reliability of total recoverable reserves has been neglected, largely, in the past, and only recently has the subject started receiving its deserved attention (DAGBERT & MYERS, 1982; HARRISON, 1983).

This paper presents an empirical approach to the problem of approximating the ore grade estimation variance: through repetitive sampling of simulated deposits (representing various types of regionalization) the effects of changes in cut-off grade and selective mining unit size on the ore grade precision are analyzed and approximation formula relating the conditional estimation variance to known geostatistical parameters is developed.

POSITION OF THE PROBLEM

Let us consider a deposit D which is recognized by N samples, uniformly distributed over it, at an average spacing of L. Let us assume further that this deposit will be completely mined-out by blocks of size V, and that each of these blocks will be classified either as ore, if its average grade is greater than z_c, or otherwise as waste.

The problem at hand, when dealing with total mineable reserve evaluation, is to estimate the recovered ore grade (i.e., the average grade of ore blocks) corresponding to various cut-off grades z_c and to assess the precision of these estimates.

What is required, therefore, is to infer the distribution function $F_V(z)$ of the true mean grade $Z_V(x)$ of the selective mining units V within the deposit D from the empirical distribution of the available sample data Z_{α_i}, i = 1 to N.

Generally the distribution function $F_V(z)$ of the mean grades $Z_V(x)$ is deduced from the 'declusterized' histogram (JOURNEL, 1983) of the data, by making a permanence of distribution hypothesis (direct or indirect correction of variance, affine correction of variance, et cetera).

Then, truncating the estimated distribution function $F^*_V(z)$ at selected cut-offs z_c will provide the estimates of the recovered ore grades $m_V(z_c)$. Figure 1 illustrates the relationships between the various distributions.

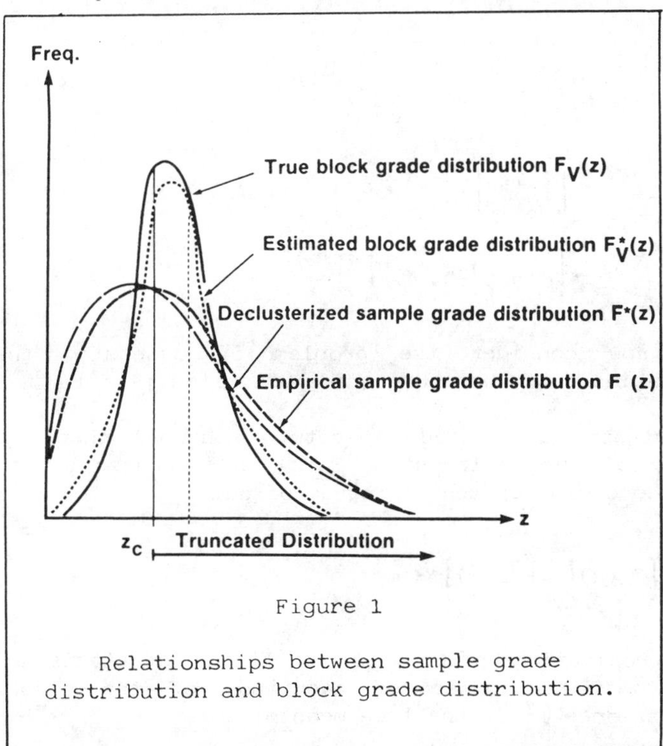

Figure 1

Relationships between sample grade distribution and block grade distribution.

For any given cut-off z_c, the truncated distribution function $F_V(z;z_c)$ of the mean grades can be characterized by:

i) the tonnage recovery factor

$$T_V(z_c) = \text{Prob}\left\{ Z_V(x) \geq z_c \right\}$$

ii) the conditional expectation (or true ore grade given a cut-off z_c)

$$m_V(z_c) = E\left\{Z_V(x) \mid Z_V(x) \geqslant z_c\right\}$$

iii) the conditional dispersion variance, i.e., the variance of ore block grades given a cut-off z_c

$$\sigma_V^2(z_c) = E\left\{\left[Z_V(x) - m_V(z_c)\right]^2 \mid Z_V(x) \geqslant z_c\right\}$$

Similarly, the truncated distribution function $F_V^*(z;z_c)$ can be described by:

$$\text{i) } T_V^*(z_c) = \text{Prob}^*\left\{Z_V(x) \geqslant z_c\right\}$$

$$\text{ii) } m_V^*(z_c) = E^*\left\{Z_V(x) \mid Z_V(x) \geqslant z_c\right\}$$

$$\text{iii) } \sigma_V^{2*}(z_c) = E^*\left\{\left[Z_V(x) - m_V(z_c)\right]^2 \mid Z_V(x) \geqslant z_c\right\}$$

Let us now consider the problem of assessing the estimation variance of these ore grade estimates $m_V^*(z_c)$.

If no cut-off is applied (in-situ resources) then the true mean m_V of the distribution $F_V(z)$ is estimated by the mean m^* of the declusterized histogram of data:

$$m_V = E\left\{Z_V(x)\right\} = E\left\{Z(x)\right\} = m^*$$

and the approximation principles of linear geostatistics provide us with an order of magnitude of the global estimation variance σ_E^2 of the true mean m:

$$\sigma_E^2 = E\left\{\left[m - m^*\right]^2\right\}$$

For example, if the N data are uniformly distributed over the deposit D at an average spacing of L (random stratified grid) the global estimation variance σ_E^2 can be approximated by a 'direct combining of elementary errors' (JOURNEL &

HUIJBREGTS, 1978 p. 410, sq.)

$$\sigma_E^2 \simeq \frac{F(L,L)}{N} \qquad (1)$$

where:

F(L,L) = value of auxiliary function F, i.e., the dispersion variance of a point with a square L x L.

In the more general case where a cut-off is applied (recoverable reserves), the true ore grade $m_V(z_c)$ is estimated by $m_V^*(z_c)$ which is deduced from the truncated block grade distribution $F_V^*(z)$.

The question which arises at this point is how to estimate the conditional estimation variance $\sigma_{EV}^2(z_c)$:

$$\sigma_{EV}^2(z_c) = E\left\{\left[m_V(z_c) - m_V^*(z_c)\right]^2\right\}$$

As a working hypothesis, it will be assumed that the conditional estimation variance (CEV) can be expressed as a function of the estimation variance of global in-situ resources, i.e.:

$$\sigma_{EV}^2(z_c) = K_V(z_c) \cdot \sigma_E^2 \qquad (2)$$

where $K_V(z_c)$ is a correction factor for support V of selective mining units and for cut-off z_c.

DERIVING EXPERIMENTAL CONDITIONAL ESTIMATION VARIANCES

The experimental values for the conditional estimation variance $\sigma_{EV}^2(z_c)$ are obtained by sampling n times consecutively the same simulated deposit using a random stratified grid. All n samplings provide the same number N of data, the same average distance between data, but the exact location of samples varies ramdomly from one sampling stage to another. Therefore, at each sampling stage i there is a different empirical sample distribution $F^{*(i)}(Z)$.

Assuming that the estimator $m_V(z_c)$ of the true ore grade is unbiased, the conditional estimation variance can be defined and estimated as:

$$\sigma^2_{EV}(z_c) = E\left\{\left[m_V(z_c)-m^*_V(z_c)\right]^2\right\} \simeq \frac{1}{n}\sum_{i=1}^{n}\left[m_V(z_c)-m^{*(i)}_V(z_c)\right]^2 \quad (3)$$

with

$m_V(z_c)$: true ore grade (known by using all the simulated information).

$m^{*(i)}_V(z_c)$: estimated ore grade corresponding to sampling stage i.

To develop a comprehensive understanding of the factors controlling the precision of total recoverable reserve estimates, it is necessary not only to consider the changes in the CEV at varying cut-offs and sizes of selective mining units, but also to see how these changes occur when dealing with different types of regionalization.

For the purpose of this study, six two-dimensional deposits, representing different types of regionalization, were considered. In each case the size of the deposit is 100u x 100u, and the grade of each elementary cell u x u is obtained through non-conditional simulation. These 10,000 units u x u (which constitute the true distribution function F(z) of elementary grades) are then regrouped into blocks of size 5u x 5u and 10u x 10u, thus providing the true distribution function $F_V(z)$ of block grades for V = 5u x 5u and V = 10u x 10u.

Table I summarizes the characteristics of these deposits. Full details on the simulation procedure used and on the statistical and structural characteristics of the six deposits can be found in FROIDEVAUX (1984).

Each deposit is then sampled 100 times consecutively, and at each sampling stage i the following operations take place:

1) 100 unit data u x u are collected (using a random stratified grid) and provide the distribution function $F^{*(i)}(z)$.

2) The distributions $F^{*(i)}_V$ of block grades for V = 5u x 5u and V = 10u x 10u are deduced from $F^{*(i)}$ by affine correction of variance.

3) The estimated ore grades $m_V^{*(i)}(z_c)$ are obtained by truncating the distribution $F_V^{*(i)}$ at the selected cut-off z_c.

TABLE I

Statistical and Structural Characteristics of the Simulated Deposits.

DEPOSIT	DISTRIBUTION	AVERAGE	VARIANCE	RANGE [1)]	N
1	Normal	0.0	1.0	3 u	10,000
2	Normal	0.0	1.0	11 u	10,000
3	Normal	0.0	1.0	18 u	10,000
4	Lognormal	0.50	.986	2.5u	10,000
5	Lognormal	0.50	1.013	8 u	10,000
6	Lognormal	0.51	0.233	22 u	10,000

1) All the regionalizations are isotropic and characterized by a spherical semi-variogram

Based on these 100 samplings, the experimental CEV (corresponding to the 3 SMU sizes V and to increasing cut-offs z_c) are then calculated using equation (3).

Figures 2 and 3 show the experimental results.

First, let us consider Figure 2 which shows the behaviour of $K_V(z_c)$ for the three deposits where the elementary grades are distributed normally. These experimental curves call for the following comments:

1) All the curves show a comparable general pattern: initial decrease of $K_V(z_c)$, as the cut-off z_c increases, with a minimum reached at a cut-off corresponding approximately to the median of the distribution $F_V(z)$, followed by a sharp increase.

2) In each case the value of $K_V(z_c)$ at its minimum is in the neighbourhood of the block to point variance ratio σ_V^2/σ^2. As a consequence, the larger the size of the SMU, the smaller the minimum value of $K_V(z_c)$.

3) The correction factor $K_V(z_c)$ is dependent on the range of influence of the regionalization. Indeed, for a fixed size of SMU, the ratio σ_V^2/σ^2 decreases with the range. This is a direct consequence of the classical geostatistical relationship:

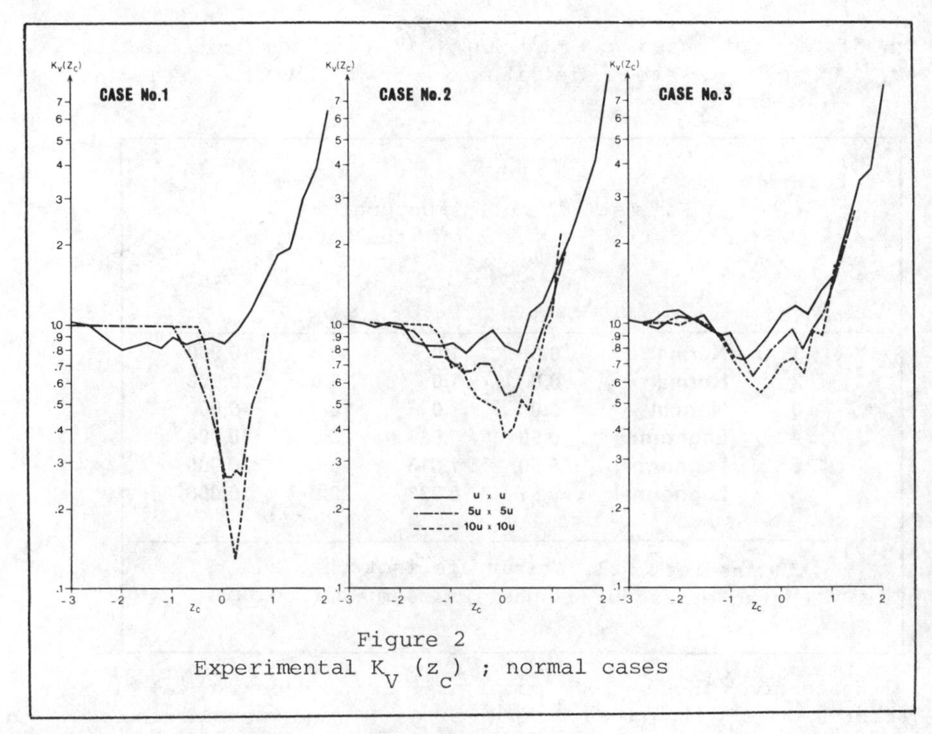

Figure 2
Experimental $K_V(z_c)$; normal cases

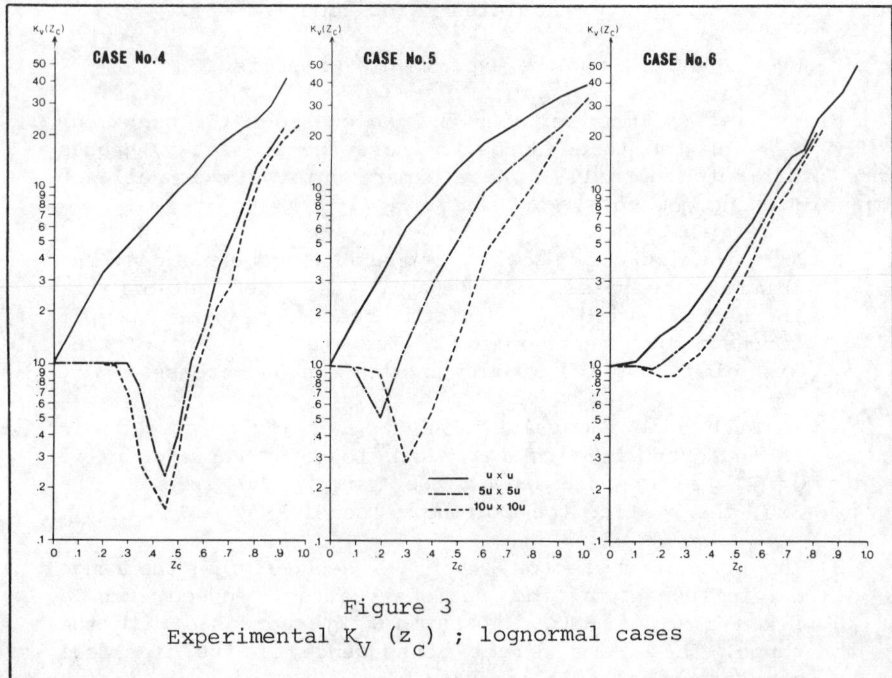

Figure 3
Experimental $K_V(z_c)$; lognormal cases

$$\sigma_V^2 = D^2(V/\infty)$$

and

$$D^2(V/\infty) = D^2(0/\infty) - \bar{\gamma}(V,V)$$

with $\bar{\gamma}(V,V)$ increasing as the range decreases.

Looking now at Figure 3 (which shows the results for the lognormal cases), it is seen that:

- When the cut-off z_c is applied to the sample grade distribution $F(z)$, $K(z_c)$ shows a monotonic increase as soon as z_c increases.
- In the general case where z_c is applied to $F_V(z)$ (with V = 5u x 5u and V = 10u x 10u) the behaviour of $K_V(z_c)$ is similar to what was observed in the normal cases: initial decrease, with a minimum reached at the median, followed by a steady increase.

Based on these experimental results, it is clear that the traditional 'global estimation variance' σ_E^2 is not good enough when one tries to quantify the reliability of an ore grade estimate.

The problem at this point is to find an acceptable approximation for the corrective factor $K_V(z_c)$.

APPROXIMATING THE CONDITIONAL ESTIMATION VARIANCE

In order to develop an approximation of the conditional estimation variance, firstly, it is necessary to identify the parameters that control its variation and, secondly, to develop a suitable function of these parameters.

Review of the Controlling Parameters

a) Tonnage Recovery Factor: This is the most obvious one. It is known from geostatistical theory that, when the total volume V to be characterized decreases, the corresponding estimation variance increases.

Applying a cut-off amounts to reducing the volume

recovered; this must cause the estimation variance of recovered reserves to increase.

b) Conditional Dispersion Variance: The global estimation variance (when no cut-off is applied) is, by definition, a function of the initial dispersion variance of sample data within the deposit.

Once a cut-off is applied, the dispersion variance of the remaining data (conditional dispersion variance $\sigma^2_V(z_c)$) has to be different from the initial one and, hence, the conditional estimation variance will be different from the global one.

Whether the conditional dispersion variance increases or decreases will depend upon the type of distribution:

- In a normal distribution, the low values of the left-hand tail of the distribution will be eliminated first as the cut-off is raised. In fact, half of the outliers which have a large contribution to the overall dispersion variance will be discarded as the cut-off is raised. Consequently, the conditional dispersion variance will decrease as the cut-off grade increases.

- For a lognormal distribution, the situation is quite the opposite. Raising the cut-off will result in the initial elimination of a large proportion of the sample data which are close one to another and which have a low contribution to the a-priori variance. As a consequence, the conditional dispersion variance will increase with the cut-off.

Figure 4 shows typical behaviour of conditional dispersion variances for the normal and lognormal cases.

c) The Conditional Semi-variogram: the characteristics of the underlying conditional semi-variogram, i.e., the semi-variogram calculated considering only the sample grades Z(x) which are greater than a cut-off z_c, and which is defined as:

$$\gamma(h,z_c) = \frac{1}{2} E\left\{\left[Z(x+h)-Z(x)\right]^2 \middle| Z(x) \geqslant z_c ; Z(x+h) \geqslant z_c\right\}$$

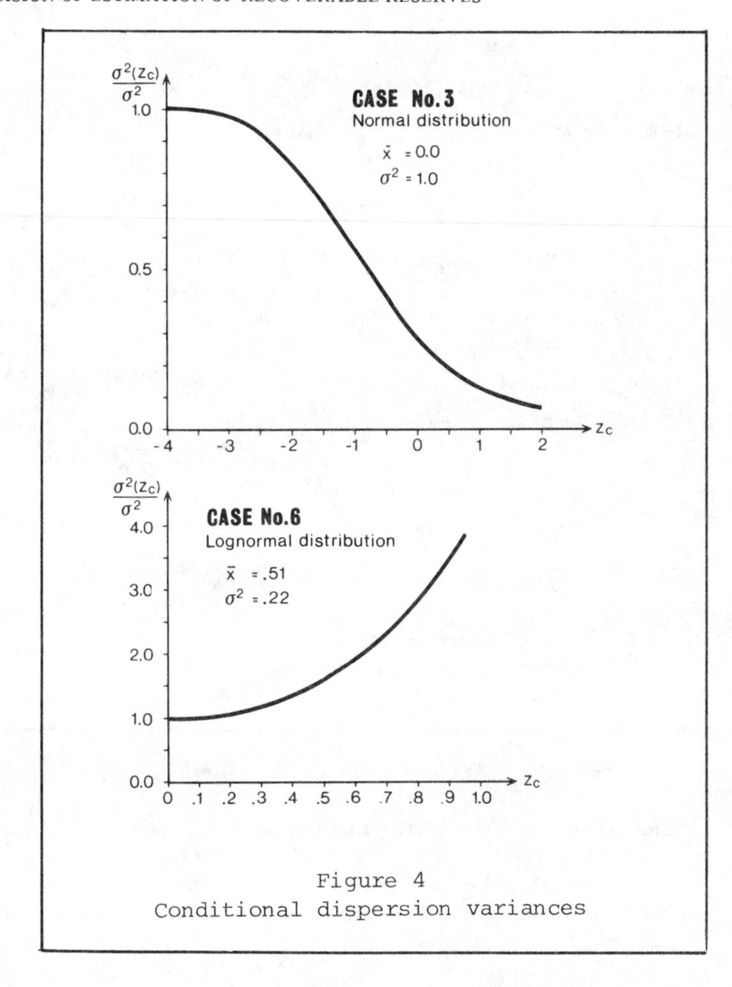

Figure 4
Conditional dispersion variances

affect the conditional estimation variance in two different ways:

i) Through the conditional range $a(z_c)$ which generally becomes shorter when the cut-off z_c increases. Such a decrease in the range of influence will cause a relative increase in the conditional estimation variance. Figure 5 shows the exhaustive conditional semi-variograms (calculated using all the simulated elementary cell grades) at various cut-offs for case 3 (normal distribution) and for case 6 (lognormal distribution).

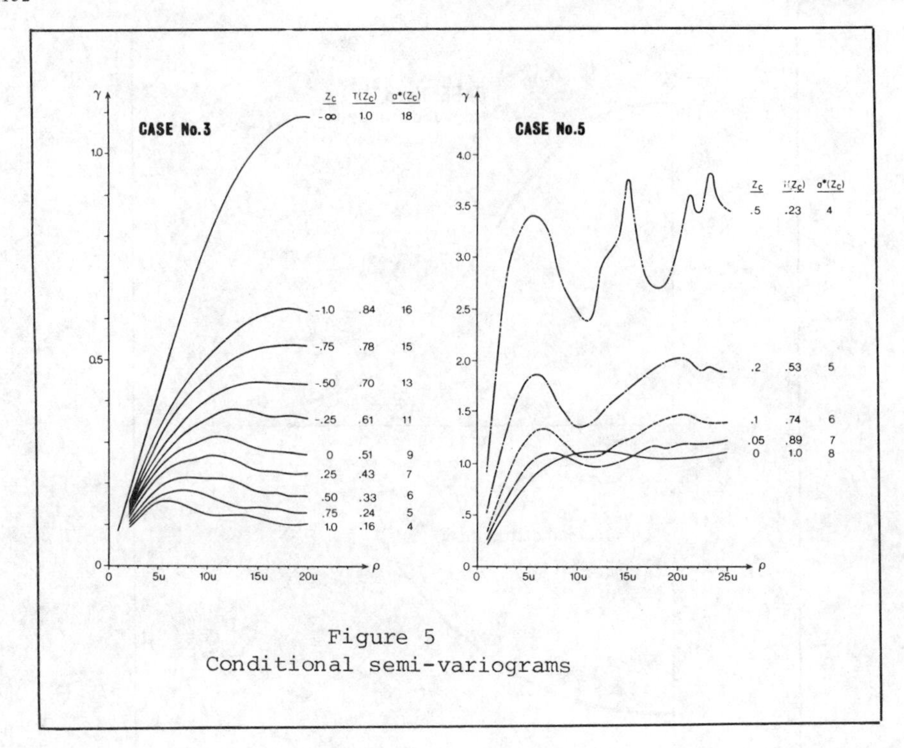

Figure 5
Conditional semi-variograms

For the six simulated deposits, it was found (FROIDEVAUX, 1984) that the conditional range $a(z_c)$ could be expressed as a function of the tonnage recovery factor $T(z_c)$ using one of the following equations:

For the normal cases:

$$a(z_c) = a \cdot T(z_c), \tag{4a}$$

and for the lognormal cases:

$$\frac{a(z_c)}{a} = \frac{e^{-4\left[1-T(z_c)\right]}+1}{2} \tag{4b}$$

It should be emphasized, however, that these relationships are not based on any kind of formal theory, and that, in different situations, totally different approximations might be required.

ii) Through the conditional sill $\sigma_V^2(z_c)$ which will influence the dispersion variance of the SMU average grades in the deposit by means of the $\bar{\gamma}(V,V)$ value:

$$D^2(V/D) = D^2(O/D) - \bar{\gamma}(V,V)$$

The resulting effects on the conditional estimation variance are immediate and have already been discussed in the previous paragraph.

It appears, therefore, that a good knowledge of the underlying conditional semi-variogram (point support) is very important for establishing the conditional estimation variance. Unfortunately, such a conditional variogram is not easy to estimate in practice, even less so if the non-conditional (global) variogram itself is difficult to establish.

Approximation for the Corrective Factor $K_V(z_c)$

In order to approximate the unknown theoretical $K_V(z_c)$, it is postulated, tentatively, that it is a simple linear function of the relative increase or decrease (from their initial stages when no cut-off is applied) in conditional dispersion variance, tonnage recovery and range:

$$K_V(z_c) = A_1 \cdot A_2 \cdot A_3 \cdot A_4 = \frac{\sigma_V^2(z_c)}{\sigma_V^2} \cdot \frac{F_1(L,L|z_c)}{F_1(L,L)} \cdot \frac{1}{T_V(z_c)} \cdot M_V(z_c) \quad (5)$$

where:

- $F_1(L,L|z_c)$ represents the value of the standardized auxiliary function F corresponding to a conditional range $a(z_c)$ and an average sample spacing of L.

- $A_1 = \sigma_V^2(z_c)/\sigma_V^2$ is the conditional variance factor and accounts for the dependence of $K_V(z_c)$ on changes in conditional dispersion variances.

- $A_2 = F_1(L,L|z_c)/F_1(L,L)$ is the conditional range factor and represents the relative increase of $K_V(z_c)$ due to the shortening of the range $a(z_c)$ of the conditional semi-variogram.

- $A_3 = 1/T_V(z_c)$ is the recovery factor and takes care of the variations in $K_V(z_c)$ introduced by changing the size of the support.

- $A_4 = M_V(z_c) = 1-2\ F_1(V,V)\ \min\left\{T_V(z_c), 1-T_V(z_c)\right\}$ is the median factor and reflects the dependence of $K_V(z_c)$ on changes in the size of SMU V, this dependence being, as described before, most pronounced when the cut-off z_c is close to the median of $F_V(z_c)$.

Although cumbersome and redundant in its present form, expression (5) was used to predict the 'theoretical' values of $K_V(z_c)$ for the six simulated deposits.

For $\sigma_V^2(z_c)$ and $T_V(z_c)$ the 'true' values were used, i.e., values obtained from the exhaustive distribution of the 10,000 simulated elementary grades, whereas $a(z_c)$ was estimated using equation (4a) or (4b).

Its performance is shown on Figures 6, 7 and 8 and calls for the following comments:

i) The approximation performs reasonably well for all the normal cases (cases 1, 2 and 3), no matter what the size of the SMU is.

ii) In the lognormal cases (4, 5 and 6), the equation 5 performs satisfactorily only when the cut-off is applied to point distributions. For larger sizes of SMU it fails to provide (except for case 4) an acceptable estimate of $K_V(z_c)$. In case 5 it grossly over estimates $K_V(z_c)$ at low cut-offs and in case 6, it over estimates constantly and systematically the experimental values.

Such a poor performance in estimating $K_V(z_c)$ for lognormal cases when V is larger than the sample support, finds it explanation in the conditional bias, i.e., the average difference between the true ore grade and the estimated ore grade.

Let us consider Table II which lists the conditional bias $CB_V(z_c)$:

$$CB_V(z_c) = E\left\{\left[m_V(z_c) - m_V^*(z_c)\right]\right\} / m_V(z_c)$$

and the average error on the conditional dispersion variance estimate $E_V(z_c)$:

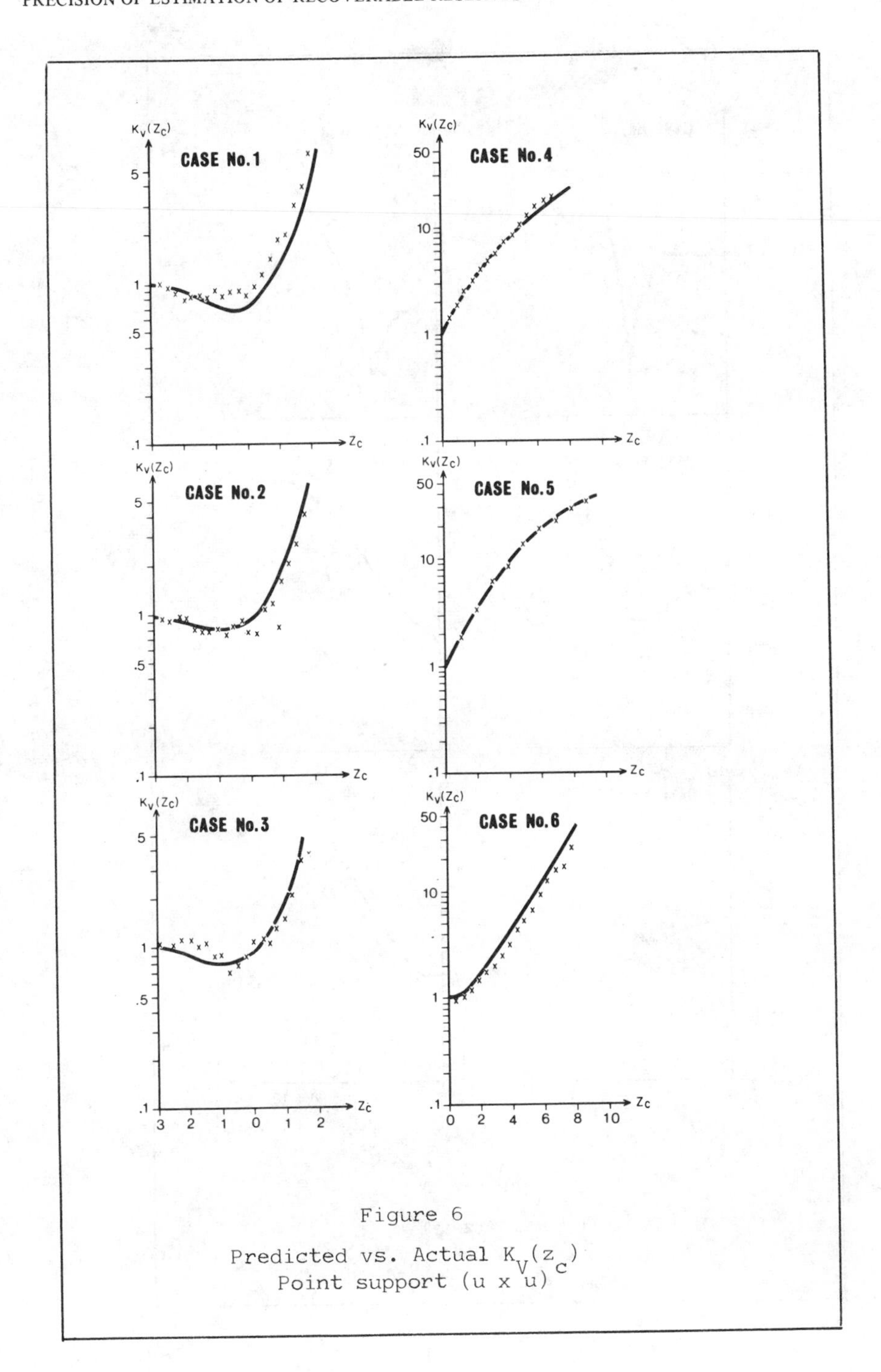

Figure 6

Predicted vs. Actual $K_V(z_c)$
Point support (u x u)

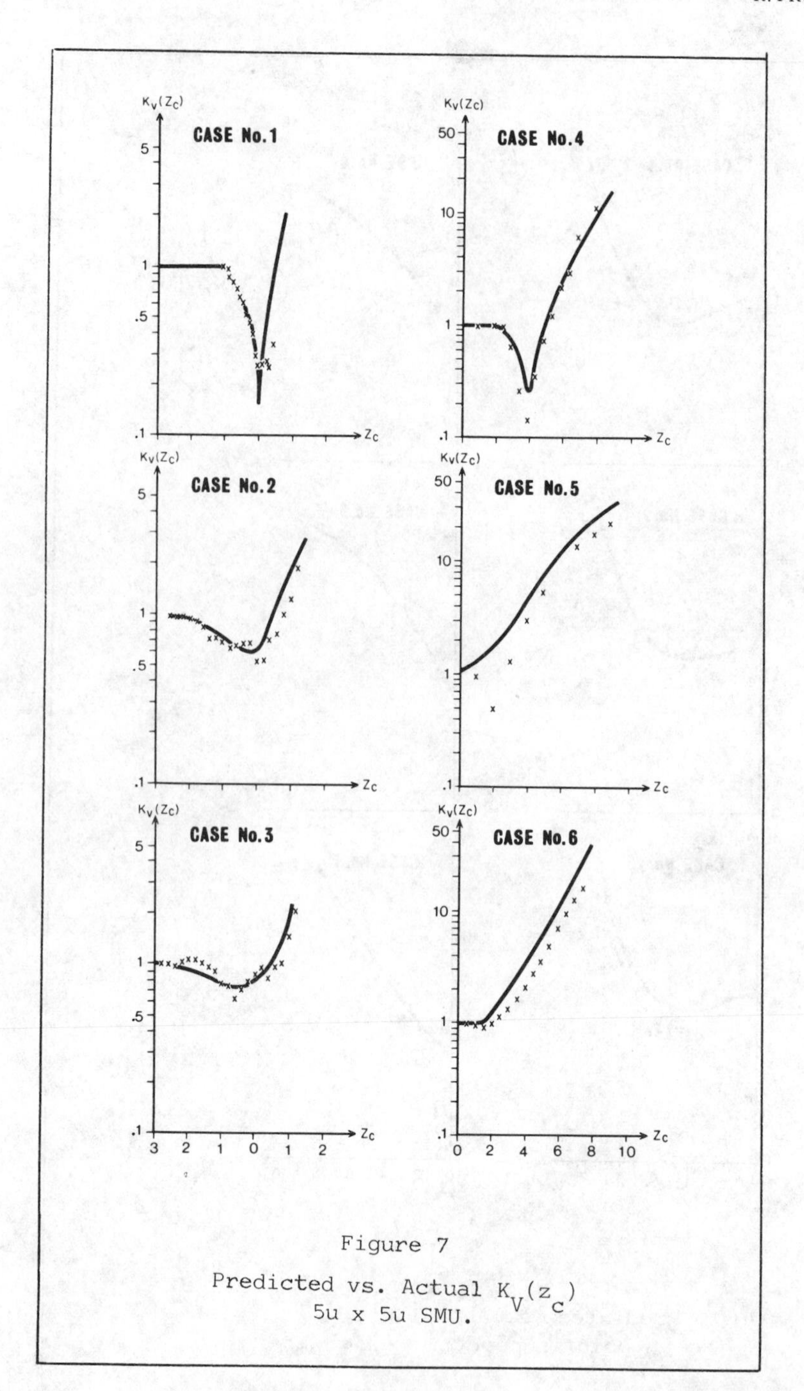

Figure 7

Predicted vs. Actual $K_V(z_c)$
5u x 5u SMU.

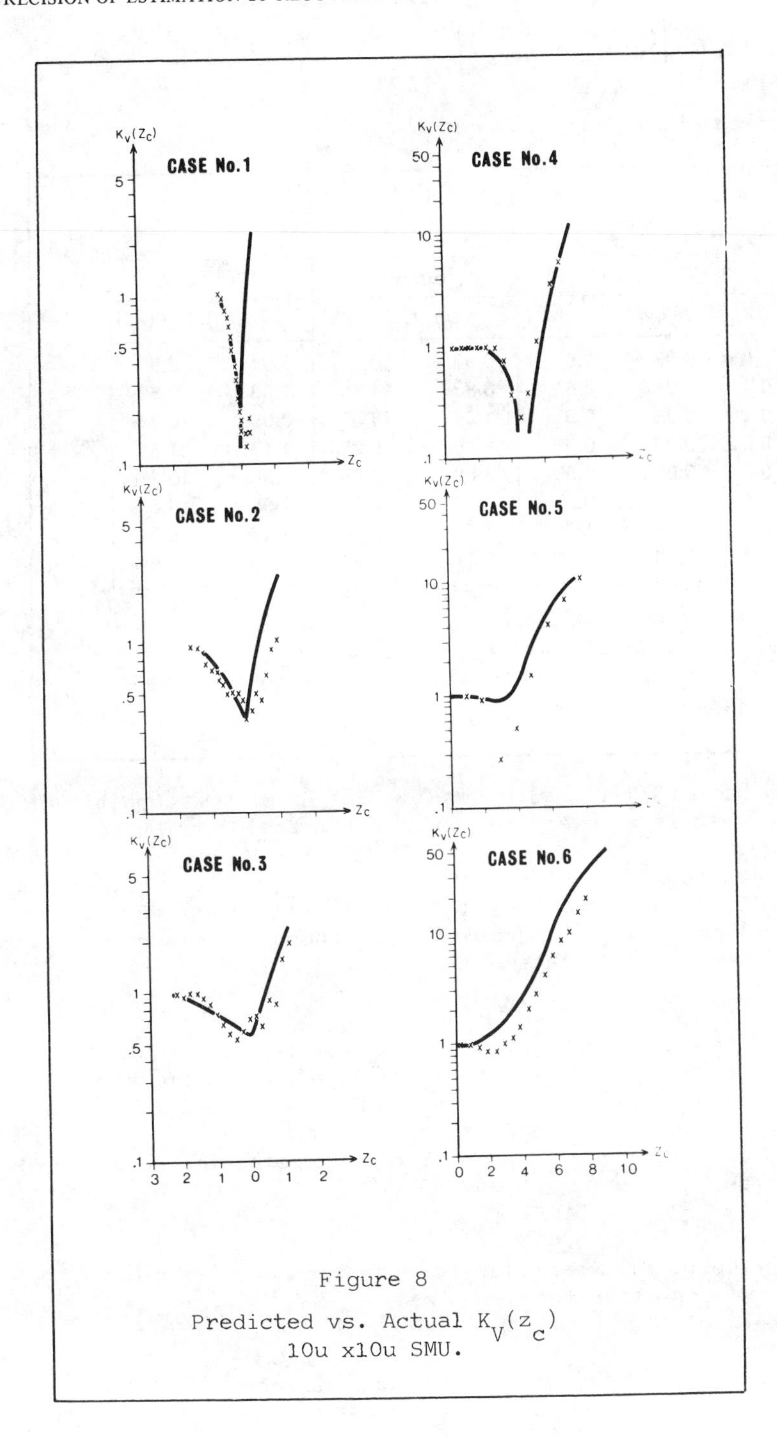

Figure 8

Predicted vs. Actual $K_V(z_c)$
10u x10u SMU.

$$E_V(z_c) = E\left\{\sigma_V^{2*}(z_c)/\sigma_V^2(z_c)\right\}$$

at increasing cut-offs for case 5.

Table II

	uxu		5ux5u		10ux10u	
z_c	$E_V(z_c)$	$CB_V(z_c)$	$E_V(z_c)$	$CB_V(z_c)$	$E_V(z_c)$	$CB_V(z_c)$
0	**0.95**	**0.9%**	**0.92**	**0.9%**	**1.04**	**0.9%**
0.1	**0.94**	**0.5%**	**0.83**	**14.3%**	**1.03**	**2.5%**
0.2	**0.94**	**0.3%**	**1.04**	**18.5%**	**0.95**	**20.2%**
0.3	**0.93**	**0.2%**	**1.00**	**9.2%**	**1.13**	**21.5%**
0.4	**0.93**	**0.5%**	**1.08**	**6.0%**	**1.58**	**10.1%**
0.5	**0.91**	**0.2%**	**1.11**	**6.8%**	**1.99**	**0.8%**
0.6	**0.89**	**1.2%**	**1.20**	**0.1%**	**2.26**	**-l.7%**
0.7	**0.89**	**1.0%**	**1.17**	**4.3%**	**2.63**	**2.5%**

Conditional bias ($CB_V(z_c)$) and error on estimate of conditional variance ($E_V(z_c)$) linked to change of support hypothesis.

From these figures, it is evident that the conditional unbiasedness is not verified for block grade distribution.

Remembering that the estimated distribution of block grades $F_V^*(z)$ was obtained through an affine correction of variance under the hypothesis of permanence of distribution, the implications are rather clear:

i) The permanence of distribution hypothesis does not hold true when the initial distribution is lognormal. This is a well known result of the geostatistical theory, but the practical consequences of making such a hypothesis are clearly shown here.

ii) The conditional bias is most pronounced for cut-offs z_c less than the median.

iii) The conditional dispersion variance is slightly underestimated for cut-offs lower than the median, and overestimated, sometimes severely, for higher cut-offs. This means that the left tail of the sample distribution has been too heavily corrected (in terms of variance) whereas the right tail has not been corrected enough.

Since an underlying assumption for estimating the conditional estimation variance is that the estimator of the true ore grade $m_V(z_c)$ is unbiased, the poor performance of equation (5) can be attributed to the conditional bias.

The formulation of equation (5), however, is awkward and some of its factors are redundant. The main reason for this redundancy is that all the factors of equation (5) were considered as acting independently one from the other. Accounting for the interdependences between them obviously will result in a much more simplified expression.

Let us consider the general case where the selection is performed on the block grade distribution $F_V(z)$. Recalling that the distribution $F_V(z)$ was deduced by an affine correction of variance from the empirical distribution F (z), the following relationship holds true (JOURNEL & HUIJBREGTS, 1978 p. 471):

$$F_V(z_c) = \text{Prob}\{Z_V(x) < z_c\} = \text{Prob}\{Z(x) < z'_c\} = F(z'_c)$$

where

$$z'_c = \frac{\sigma}{\sigma_V}(z_c - m) + m$$

Assuming that the N data are evenly distributed over the deposit, the tonnage recovery factor can be written as:

$$T_V(z_c) = T(z'_c) \simeq N(z'_c)/N$$

with $N(z'_c)$ being the number of samples above the 'equivalent cut-off' z'_c.

It can be demonstrated also (see Appendix) that:

$$\sigma^2_V(z_c) = \sigma^2(z'_c)\,\frac{\sigma_V^2}{\sigma^2}$$

It appears, therefore, that only the samples whose grades are greater than the equivalent cut-off z'_c are considered for estimating the recovered ore grade $m_V(z_c)$. Consequently, the conditional range factor:

$$\frac{F_1(L,L|z_c)}{F_1(L,L)}$$

of expression (5) should be re-written as:

$$\frac{F_1(L,L|z'_c)}{F_1(L,L)}$$

and

$$M_V(z_c) = 1 - 2F_1(V,V)\min\left\{T_V(z_c),\ 1-T_V(z_c)\right\} =$$

$$= 1 - 2F_1(V,V)\min\left\{T(z'_c),\ 1-T(z'_c)\right\} = M_V(z'_c)$$

By substituting in equation (5), the approximation for the corrective factor $K_V(z_c)$ becomes:

$$K_V(z_c) = \frac{F(L,L|z'_c)}{F(L,L)}\cdot\frac{N}{N(z'_c)}\cdot M_V(z'_c) \qquad (6)$$

Finally, by introducing expression (6) into expression (2), the following approximation for the conditional estimation variance is obtained:

$$\sigma^2_{EV}(z_c) = \frac{F(L,L|z'_c)}{N(z'_c)}\,M_V(z'_c) \qquad (7)$$

If we consider now the particular case where the selection is performed on data support, we have:

$$z'_c = z_c$$

and

$$M_V(z'_c) = 1$$

and the conditional estimation variance reduced to:

$$\sigma^2_E(z_c) = \frac{F(L,L|z_c)}{N(z_c)} \qquad (8)$$

The similarity between equation (7), (8) and (1) is obvious and suggests that the classical methods of approximating the global estimation variance can be customized to provide the conditional estimation variances.

The approximations proposed here, however, call for some important remarks:

1. The simplified approximation given in equation (7) applies only when an affine correction of variance is used for inferring the distribution $F_Y(z)$. If another type of variance correction is used, then the general 'descriptive' formula of equation (5) should be considered.

2. Equations 5, 6, 7 and 8 assume that both the tonnage recovery factor $T(z_c)$ and the conditional semi-variogram $\gamma(h,z_c)$ are known precisely. In practice, however, estimates only of these two parameters are available.

3. The approximations assume implicitly that there is conditional unbiasedness, but such a stringent non-bias condition is not ensured by classical methods of estimating total recoverable reserves, especially not with highly positively skewed distribution.

4. All the results presented here are based on simulated deposits which, implicitly, assume a multinormal distribution model. Whether the approximations

proposed here are model dependent or not, still remains to be seen.

CONCLUSION

The proper evaluation of the conditional estimation variance attached to an estimated ore grade, represents a definite challenge for the mining geostatistician, since this variance is an essential quantity when assessing the precision of recoverable reserves.

In the absence of any exact method of calculating the estimation variance of an ore grade, an empirical approach is attempted in order to 'fill the gap' and provide an empirical approximation for it.

Based on results obtained on simulated deposits, it appears that the conditional estimation variance of the recoverable ore grade depends upon:

a) The statistical and structural characteristics of the regionalization.

b) The size of the selective mining units and the cut-off grade considered.

c) The type of transformation used to deduce the block grade distribution $F_V(z)$ from the available data.

This analysis of the factors responsible for changes in the conditional estimation variance allowed for the development of an approximation formula which has been used successfully to predict the experimental ore grade estimation variances in the case of the six simulated deposits, and which is consistent with the geostatistical theory.

However, the a-priori assumptions and requirements of the proposed approximation cannot be overlooked:

- It requires the knowledge of the conditional semi-variogram $\gamma(h,z_c)$ for any cut-off z_c considered.

- It considers that the tonnage recovery factor is known perfectly.

- It assumes implicitly that there is conditional unbiasedness.

These three requirements are not so easy to fulfill as it

may seem (especially the first one and the third one) and, although the approximations proposed here are sufficient to obtain an order of magnitude of the conditional estimation variance, more research will be required to make them fully operational.

APPENDIX

Demonstration of the relation $\sigma_V^2(z_c) = \sigma^2(z'_c)\cdot\sigma_V/\sigma^2$

Let $F_V(z)$ be the distribution function of the true mean grade $Z_V(x)$ of selective mining units V characterized by its conditional moments:

$$m_V(z_c) = E\left\{Z_V(x)\,\middle|\,Z_V(x) \geqslant z_c\right\}$$

and

$$\sigma_V^2(z_c) = E\left\{\left[Z_V(x)-m_V(z_c)\right]^2 \,\middle|\, Z_V(x) \geqslant z_c\right\}$$

Since $F_V(z)$ is derived from the empirical distribution $F^*(z)$ by an affine correction of variance, we have:

$$Z_V(z) = m+\frac{\sigma_V}{\sigma}(Z-m)$$

and

$$F_V(z) = \text{Prob}\left\{Z_V(x) < z\right\} = \text{Prob}\left\{Z(x) < z'\right\} = F(z')$$

with

$$z' = m+(z-m)\frac{\sigma}{\sigma_V}$$

This entails that the two variables $Z_V(x)$ and $\sigma_V/\sigma(Z(x)-m)+m$ have the same distribution and the same conditional moments.

Hence the conditional expectation $m_V(z_c)$ can be re-written as:

$$m_V(z_c) = E\left\{Z_V(x) \mid Z_V(x) \geqslant z_c\right\} = E\left\{\frac{\sigma_V}{\sigma}(Z(x)-m)+m \mid Z(x) \geqslant z_c'\right\}$$

$$= m\left[1-\frac{\sigma_V}{\sigma}\right]+\frac{\sigma_V}{\sigma}\, m(z_c')$$

and the conditional dispersion variance becomes:

$$\sigma_V^2(z_c) = E\left\{\left[m+\frac{\sigma_V}{\sigma}(Z(x)-m)-m+\frac{\sigma_V}{\sigma}m-\frac{\sigma_V}{\sigma}m(z_c')\right]^2 \mid Z(x) \geqslant z_c'\right\} =$$

$$= \frac{\sigma_V^2}{\sigma^2}\, E\left\{\left[Z(x)-m(z_c')\right]^2 \mid Z(x) \geqslant z_c'\right\} = \frac{\sigma_V^2}{\sigma^2}\cdot\sigma^2(z_c')$$

ACKNOWLEDGMENT

Prof. André Journel, Stanford University, has provided invaluable help for this research by his constant availability for discussion, constructive comments and suggestions.

REFERENCES

DAGBERT, M. and MYERS, J., 1982, Precision of total reserve estimates : old formulae applied to new situations: Proc. 17th APCOM, Golden, Colorado p. 393-405

FROIDEVAUX, R., 1982, Geostatistics and ore reserve classification. CIM Bulletin, vol. 75, No. 843, p. 77-83

FROIDEVAUX, R., 1984, Conditional estimation variance: An empirical approach. Mathematical Geology. (In Press).

HARRISON, D., 1983, A schema for classifying and reporting mineral resources: rationalizing the needs for consistency and accuracy with the demands for disclosure. Proc. AIME Annual Meeting, Atlanta. 6 p.

JOURNEL, A.G. and HUIJBREGTS, Ch.J., 1978, Mining Geostatistics: Academic Press, London, 600 p.

JOURNEL, A.G., 1983, Non parametric estimation of spatial distributions: Mathematical Geology, V.15, No. 3, p. 445-468.

ESTIMATION VARIANCE OF GLOBAL RECOVERABLE RESERVE ESTIMATES

Bruce E. Buxton

Department of Applied Earth Sciences
Stanford University

ABSTRACT

New developments are presented which put the prediction of the variance of global recoverable reserve estimates into the framework of existing geostatistical theory. Predictive formulas are developed and some promising initial results are presented on the use of these new formulas for two simulated deposits.

INTRODUCTION

Critical management decisions about the viability of a proposed mining operation rely heavily upon reserve estimates made for the entire deposit (i.e. global reserve estimates). Such estimates, in fact any estimates, are made with full knowledge that they are, to some extent, in error. The analyst needs to know what the likely extent of the estimation errors is in order to make correct decisions. The global estimation variance (GEV) helps to quantify the likely extent of global estimation errors.

To date, not a great deal of work has been done in the area of assessing the precision of recoverable reserve estimates, although some important advances are beginning to be made. For example, see Dagbert and Myers, 1982; Harrison, 1983; and Froidevaux, 1984.

In this paper, prediction of the variance of global reserve estimates is done within the context of established geostatistical theory (see Journel, 1983). Regionalized variables are defined which correspond to desired recovery functions. Then

G. Verly et al. (eds.), Geostatistics for Natural Resources Characterization, Part 1, 165–183.

the GEV's for these regionalized variables are predicted by the method of compositing elementary extension variances. Global block recovery estimation variance is predicted through the additional use of the method of affine correction of variance.

COMPOSITING ELEMENTARY EXTENSION VARIANCES

Predicting global estimation variance by the method of compositing elementary extension variances is extensively discussed in Journel and Huijbregts, 1978, pp. 410-443. Only the highlights of this method will be given here as they apply to this study.

The average value of a regionalized variable $z(x)$ over domain D is to be estimated; for example, $z(x)$ is the ore grade at location x within a deposit D. The deposit is informed by N core-support data taken on a random stratified square grid with elementary grid cell of side dimension C. This global estimation can then be viewed as the compositing of the N elementary estimations where the average value of $z(x)$ within each elementary grid cell is estimated by its included datum.

Since each datum is used for the estimation of only one grid cell, the elementary estimation errors for the N cells can be considered to be approximately independent. The estimation variance of estimating the average value of $z(x)$ over a grid cell by a randomly located datum is simply the dispersion variance of a point (core support) within a square of dimension C, $D_z^2(o/C)$. And then, if the average value of $z(x)$ over the entire domain D is estimated by the average of the N elementary estimates (i.e. the average of the N data), then the global estimation variance is approximately

$$GEV \doteq D_z^2(o/C)/N \quad \text{or} \quad \bar{\gamma}_z(C,C)/N \tag{1}$$

since the dispersion variance can be obtained directly from the semi-variogram function for $z(x)$.

POINT RECOVERY FUNCTIONS

Suppose in this simplest case that selection will be made on core supports without error (i.e. no cores are misclassified during selection). Further, suppose that the regionalized variable, $z(x)$, is the ore grade for the core support located at point x in the deposit D, and that the random function, $Z(x)$, is stationary, is ergodic, and has finite variance within D. Define the following random functions:

indicator random function for cutoff grade z,

$$I(x,z) = 1 \text{ if } Z(x) \leqslant z \qquad (2)$$
$$= 0 \text{ otherwise,}$$

inverse-indicator random function,

$$I'(x,z) = 1 - I(x,z) \qquad (3)$$
$$= 1 \text{ if } Z(x) > z$$
$$= 0 \text{ otherwise,}$$

Z-inverse-indicator random function,

$$ZI'(x,z) = Z(x)I'(x,z) \qquad (4)$$
$$= Z(x) \text{ if } Z(x) > z$$
$$= 0 \text{ otherwise.}$$

Also, denote associated parameters as follows:

in situ average ore grade,

$$m = E(Z(x)) = \frac{1}{D}\int_D z(u)du, \qquad (5)$$

in situ point dispersion variance,

$$\sigma_o^2 = D_z^2(o/D) = Var(Z(x)), \qquad (6)$$

proportion of cores below cutoff,

$$F(z) = Prob(Z(x) \leqslant z) = E((I(x,z)) = \frac{1}{D}\int_D i(u,z)du. \qquad (7)$$

The point recovery functions can be defined in terms of the previous random functions, as follows:

recovered tonnage (ignoring density and specific gravity),

$$T(z) = E(I'(x,z)) = \frac{1}{D}\int_D i'(u,z)du, \qquad (8)$$

recovered quantity of metal,

$$Q(z) = E(ZI'(x,z)) = \frac{1}{D}\int_D z(u)i'(u,z)du, \qquad (9)$$

recovered ore grade,

$$m(z) = Q(z)/T(z). \qquad (10)$$

PREDICTED GLOBAL ESTIMATION VARIANCE FOR POINT RECOVERY

It can be seen in the previous expressions (8),(9) that the estimation of global recovered tonnage and quantity of metal is the estimation of the averages of two particular regionalized variables over the deposit D. Therefore, the GEV's for these recovery functions can be predicted as in expression (1) through the use of the semi-variograms for the inverse-indicator and Z-inverse-indicator functions, as follows:

predicted GEV for point recovered tonnage,

$$\sigma_{ET}^{2*}(z) \doteq \frac{\bar{\gamma}_{I'}(C,C,z)}{N} = \frac{\bar{\gamma}_{I}(C,C,z)}{N} \qquad (11)$$

$$\text{where } \gamma_I(h,z) = \frac{1}{2}E((I(x+h,z) - I(x,z))^2) = \gamma_{I'}(h,z)$$

predicted GEV for point recovered quantity of metal,

$$\sigma_{EQ}^{2*}(z) \doteq \frac{\bar{\gamma}_{ZI'}(C,C,z)}{N} \qquad (12)$$

$$\text{where } \gamma_{ZI'}(h,z) = \frac{1}{2}E((Z(x+h)I'(x+h,z) - Z(x)I'(x,z))^2).$$

Note that there are separate semi-variograms for the inverse-indicator and Z-inverse-indicator functions for each cutoff grade z. These semi-variograms can be estimated from the sample core-support data. Also note that the Z-inverse-indicator variogram is exactly the usual Z variogram for any cutoff grade which is less than zero (i.e. when no cutoff is applied).

The recovered ore grade is the quotient of the recovered quantity of metal and tonnage as shown in expression (10). Therefore, the GEV for recovered ore grade must be approximated by a first-order expansion utilizing the GEV's in expressions (11),(12) as follows (see Journel and Huijbregts, 1978, pp. 424-428):

predicted GEV for point recovered ore grade,

$$\frac{\sigma_{Em}^{2*}(z)}{m^2(z)} \doteq \frac{\sigma_{EQ}^{2*}(z)}{Q^2(z)} + \frac{\sigma_{ET}^{2*}(z)}{T^2(z)} - 2\rho_{ZI',I'}(z) . \frac{\sigma_{EQ}^{*}(z)}{Q(z)} . \frac{\sigma_{ET}^{*}(z)}{T(z)} \qquad (13)$$

where $\rho_{ZI',I'}(z)$ is the standard correlation coefficient between the regionalized variables zi'(x,z) and i'(x,z).

Use of the correlation coefficient $\rho_{ZI',I'}(z)$ assumes that the variables zi'(x,z) and i'(x,z) are intrinsically coregionalized and that recovered tonnage and quantity of metal are estimated from the same configuration of data. It can be shown that

$$\rho_{ZI',I'}(z) = s(z)m(z)/A(z) \tag{14}$$

where $s^2(z) = Var(I'(x,z))$
$m(z)$ = recovered ore grade
$A^2(z) = Var(ZI'(x,z))$.

In addition, note that the evaluation of expression (13) requires knowledge of the true recovery functions. In practice, the estimated recovery functions must be used, introducing a greater approximation to expression (13).

BLOCK RECOVERY FUNCTIONS

First, suppose that selection can be made without misclassification on the true block ore grades. Second, define the set of block random functions and parameters, analagous to expressions (2),(3),(8),(9),(10), by including the subscript V (to denote block support) in those expressions. Further define:

block ore grade,

$$z_V(x) = \frac{1}{V}\int_{V(x)} z(u)du, \tag{15}$$

in situ average ore grade (as before),

$$m = \frac{1}{D}\int_{D} z_V(u)du = E(Z(x)) \tag{16}$$

in situ block dispersion variance,

$$\sigma_V^2 = D_Z^2(V/D). \tag{17}$$

Also, it will be assumed that the statistical distribution of block grades can be obtained directly from the distribution of core grades by an affine correction of variance, as follows:

$$F_V(z') = F(z) \tag{18}$$

$$\text{where } \frac{z'-m}{\sigma_V} = \frac{z-m}{\sigma_o}.$$

It can be shown that expression (18) implies

$$\frac{m_V(z')-m}{\sigma_V} = \frac{m(z)-m}{\sigma_o} \quad (19)$$

where $m_V(z')$ is the recovered ore grade at cutoff grade z' from selection on block grades.

Given the affine correction hypothesis, block recovery functions can be obtained directly from point recovery functions as follows:

recovered tonnage,

$$T_V(z') = T(z), \quad (20)$$

recovered ore grade,

$$m_V(z') = m + \frac{\sigma_V}{\sigma_o}(m(z)-m), \quad (21)$$

recovered quantity of metal,

$$Q_V(z') = T_V(z')m_V(z')$$

$$= m\left[1-\frac{\sigma_V}{\sigma_o}\right]T(z) + \frac{\sigma_V}{\sigma_o}Q(z). \quad (22)$$

PREDICTED GLOBAL ESTIMATION VARIANCE FOR BLOCK RECOVERY

From expressions (20),(21),(22), and under the affine correction hypothesis, the GEV's for block recovery can be obtained directly from the GEV's for point recovery given in (11),(12),(13) as follows:

predicted GEV for block recovered tonnage,

$$\sigma^{2*}_{ETV}(z') \doteq \sigma^{2*}_{ET}(z), \quad (23)$$

predicted GEV for block recovered ore grade,

$$\sigma^{2*}_{EmV}(z') \doteq \frac{\sigma^2_V}{\sigma^2_o}.\sigma^{2*}_{Em}(z), \quad (24)$$

predicted GEV for block recovered quantity of metal,

$$\sigma^{2*}_{EQV}(z') \doteq m^2\left[1-\frac{\sigma_V}{\sigma_o}\right]^2.\sigma^{2*}_{ET}(z) + \frac{\sigma^2_V}{\sigma^2_o}.\sigma^{2*}_{EQ}(z) +$$

$$+ 2m\frac{\sigma_V}{\sigma_o}\left[1-\frac{\sigma_V}{\sigma_o}\right].\rho_{ZI',I'}(z).\sigma^{*}_{ET}(z).\sigma^{*}_{EQ}(z). \quad (25)$$

Use of these expressions assumes, as in the core-support case, that zi'(x,z) and i'(x,z) are intrinsically coregionalized, that recovered tonnage and quantity of metal are estimated from the same configuration of data, and that the true point recoveries are known. In addition, as presented, these expressions assume that the affine correction can be made with the true in situ average grade, core-support variance, and block-support variance. When these parameters are estimated, an additional error is introduced to the affine correction. This additional source of variance is not taken into account in this paper, but can be accounted for with some additional research (see Journel, 1983 and Froidevaux, 1984).

TRUE GLOBAL ESTIMATION VARIANCE

A preliminary study has been done to test the validity of expressions (11),(12),(13),(23),(24),(25) for predicting the variance of global recoverable reserve estimates. This study has been done by resampling from two simulated deposits.

As estimation variance is the average (over a large number of samplings) squared difference between estimated recoveries and the true recoveries that they estimate. True estimation variances can be calculated for a simulated deposit because the true recoveries are available and because the simulated deposit can be repeatedly sampled, leading to repeated estimated recoveries.

The two simulated deposits used for this study are called ST2B and ST2C. Both deposits are two-dimensional, 220 m in the north-south (NS) direction, and 110m in the east-west (EW) direction (see Figure 1). Core-support ore grades have been unconditionally simulated (see Journel and Huijbregts, 1978, pp.491-554) at 1m spacing for ST2B; so ST2B is exhaustively informed by 24200 core-support grades. ST2C was then created by simply rearranging the 24200 core-support grades in space. Therefore, ST2B and ST2C have exactly the same univariate core-support grade distribution, as shown in Figure 1, but different variograms and multivariate grade distributions.

Blocks (selective mining units) are identically defined in ST2B and ST2C to be 11m by 11m (see Figure 1). There are 200 such blocks in each deposit. A true block grade is then the average of the 121 core-support grades contained in the block. Since ST2B and ST2C have different multivariate grade distributions, they also have different block-grade distributions. ST2C has greater short-scale continuity for its point grades, and so it has greater in situ block grade variance since the in situ point grade variance is the same for both deposits. The in situ block grade variances are $\sigma^2_{VB} = .3052\%^2$ and $\sigma^2_{VC} = .4022\%^2$ for ST2B and ST2C respectively.

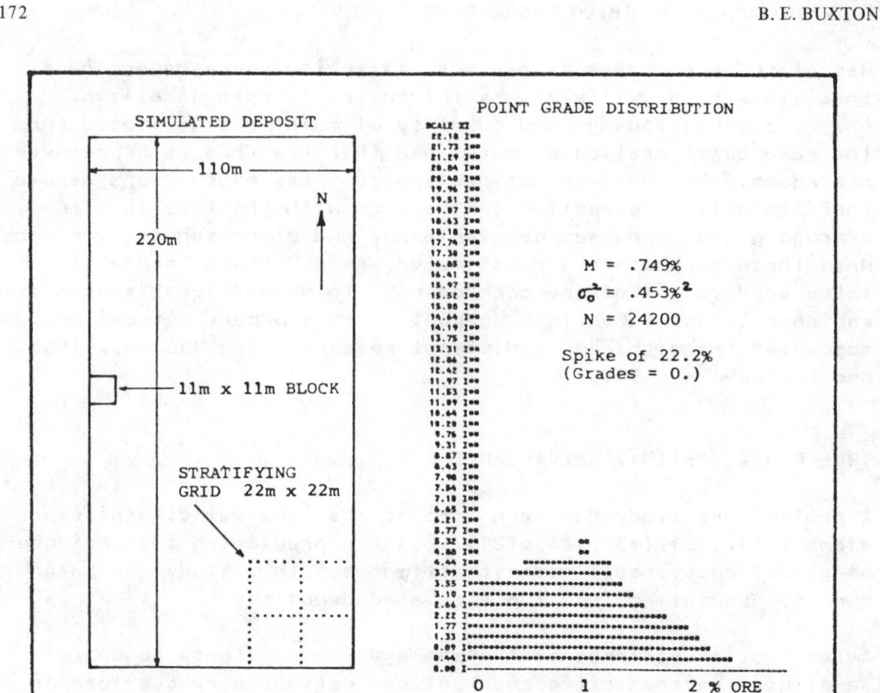

Figure 1. The ST2B and ST2C Simulated Deposits.

Because ST2B and ST2C are simulated deposits, all of the point grades, and hence, all of the block grades are known. Therefore, every point or block grade above cutoff is known for all cutoffs; and so, the true recovery functions can be calculated exactly. Eleven point-grade cutoffs have been selected, and the true point recovery functions for those cutoffs have been calculated and are presented in Table 1. (Note that the first cutoff corresponds to complete recovery.) Of course, true point recovery is the same for ST2B and ST2C. These eleven point-grade cutoffs correspond to eleven block-grade cutoffs for ST2B and for ST2C through the affine correction given by expression (18). Notice that the affine corrections for ST2B and ST2C are different since the in situ block variances are different. The true block recovery functions for the block-grade cutoffs have been calculated for ST2B and ST2C and are presented in Table 1. (Again the first cutoff corresponds to complete recovery.)

To obtain the true GEV's, the ST2B and ST2C deposits must be repeatedly sampled. As discussed earlier, the predictive formulas for GEV assume a random stratified sampling plan. Therefore, for this study, 100 repeated samplings have been taken from ST2B and ST2C, where each sampling selects 50 random stratified data as depicted in Figure 1. This configuration corresponds to a stratified grid cell which is 22m by 22m.

For each sampling, estimated point recoveries at each point-grade cutoff are calculated from the 50 data. Estimated recovered tonnage is the proportion of the 50 data above cutoff. Estimated recovered ore grade is the average of those data above cutoff. And estimated recovered quantity of metal is the product of estimated recovered tonnage and estimated recovered ore grade.

Also for each sampling, estimated block recoveries at each block-grade cutoff are calculated from the estimated point recoveries at each corresponding point-grade cutoff, using the affine correction relations given in expressions (20),(21),(22). When estimating block recovery the true in situ mean grade, point dispersion variance, and block dispersion variance are used in evaluating expressions (20),(21),(22). That is, for this study the influence of estimating these parameters in the affine correction has been eliminated.

At the end of the sampling process, 100 estimated point recovery functions for ST2B, 100 estimated block recovery functions for ST2B, 100 estimated point recovery functions for ST2C, and 100

POINT RECOVERY ST2B,ST2C				BLOCK RECOVERY ST2B				BLOCK RECOVERY ST2C			
z	$T(z)$	$Q(z)$	$M(z)$	z'_B	$T_{VB}(z'_B)$	$Q_{VB}(z'_B)$	$M_{VB}(z'_B)$	z'_C	$T_{VC}(z'_C)$	$Q_{VC}(z'_C)$	$M_{VC}(z'_C)$
-1.	1.00	.749	.749	-1.	1.00	.749	.749	-1.	1.00	.749	.749
0.	.778	.749	.963	.134	.840	.740	.881	.043	.810	.749	.925
.175	.714	.744	1.043	.278	.750	.722	.962	.208	.725	.739	1.020
.325	.655	.729	1.114	.401	.665	.692	1.040	.349	.670	.723	1.080
.510	.575	.696	1.211	.553	.590	.655	1.111	.524	.560	.674	1.204
.740	.466	.628	1.347	.742	.430	.552	1.285	.741	.440	.598	1.359
.905	.390	.565	1.450	.877	.370	.505	1.365	.896	.375	.544	1.451
1.03	.322	.500	1.551	.980	.310	.449	1.449	1.014	.325	.497	1.529
1.14	.271	.445	1.639	1.070	.255	.392	1.536	1.118	.305	.476	1.560
1.28	.215	.377	1.753	1.185	.230	.363	1.578	1.250	.235	.393	1.671
1.73	.093	.196	2.111	1.555	.100	.179	1.792	1.674	.100	.197	1.969

Table 1. True Recovery Functions.

estimated block recovery functions for ST2C are obtained. These estimated recovery functions estimate their corresponding true values given in Table 1. The true GEV at each cutoff is then the average squared difference, across the 100 samplings, between the true and estimated recovery at that cutoff. Also, when necessary, the true GEV was corrected for bias (i.e. when the average of the 100 estimated recoveries at a cutoff does not exactly equal the true recovery at that cutoff) since the GEV predictive formulas assume that the estimates are unbiased.

The true GEV's are presented in Figures 4 to 7. They are given in terms of relative estimation standard deviation at cutoff which is

$$SQRT(GEV)*100/TRUE\ RECOVERY. \qquad (26)$$

CHECKING THE AFFINE CORRECTION

The affine correction is only one of several possible block distribution hypotheses. Two of its primary advantages are that it is simple and easy to implement, as in expressions (20),(21), and (22), and that the estimation variance of block recovery estimates can be derived directly from the estimation variance of point recovery estimates as in expressions (23),(24),(25).

The predictive formulas for the GEV of block recovery given in expressions (23),(24),(25) assume that the statistical distribution of block grades can be obtained from the distribution of point grades by the affine correction of variance. If this assumption is not true, then the predictive formulas will provide a more extreme approximation. Therefore, the affine correction has been checked for ST2B and ST2C. (Of course in practice, this check is not possible since the true block distribution is not accessible.) The true block recovery functions have been plotted against the recovery functions obtained from the affine correction and true point recovery functions, as in expressions (20),(21), (22). The results of this check are given in Figure 2.

Figure 2 shows that for these two deposits the affine correction provides a very satisfactory approximation of the block recovery functions. The affine correction performs well in part due to the fact that the change in dispersion variance from point support to block support is not too large - 33% for ST2B and 11% for ST2C. However the affine correction also performs well in spite of the fact that the multivariate distributions of ST2B and ST2C are not simple. ST2B has multigaussian tendencies

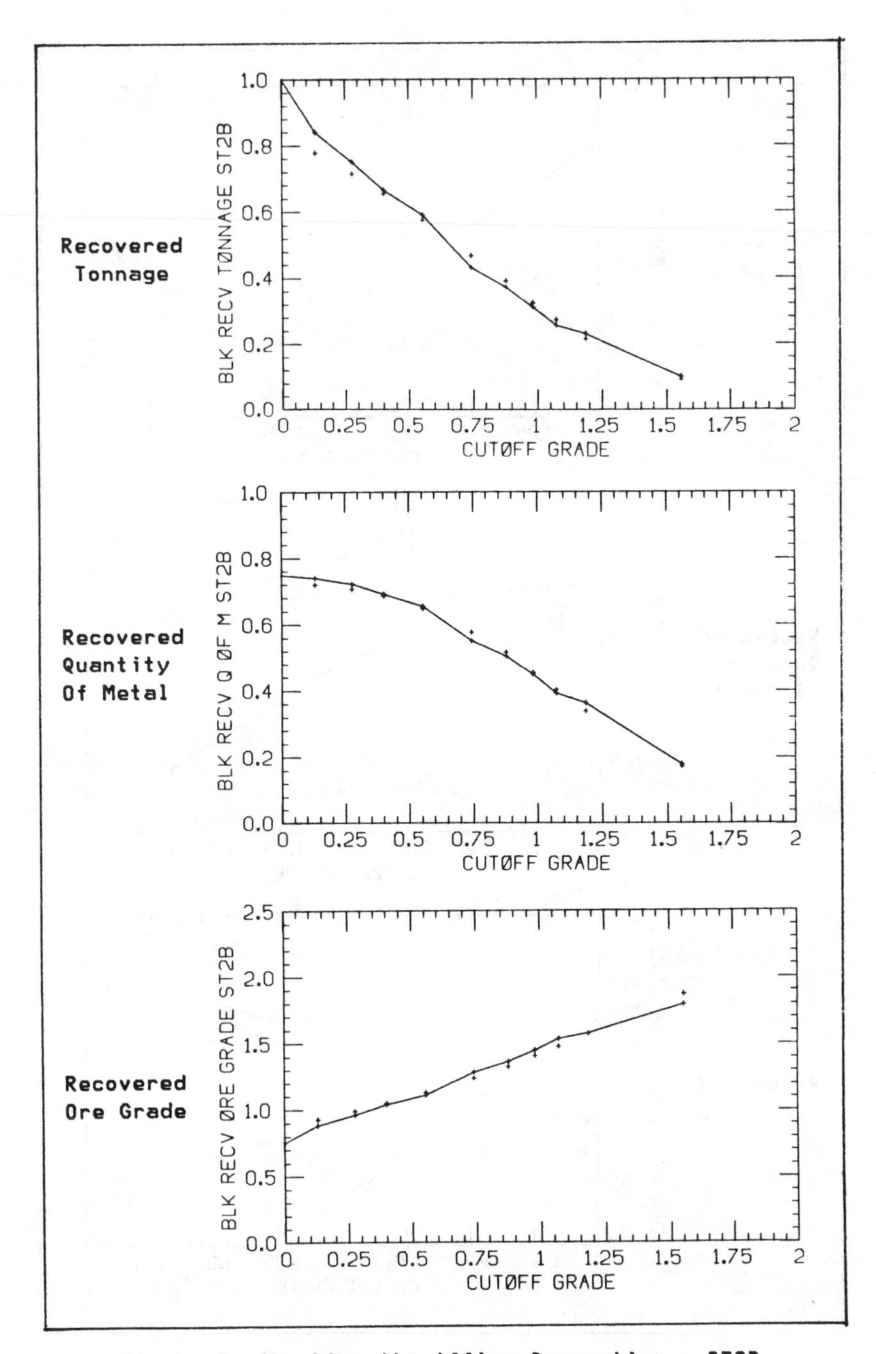

Figure 2. Checking the Affine Correction - ST2B
True Block Recovery (+——+)
Predicted Block Recovery From Affine Correction (+ +)

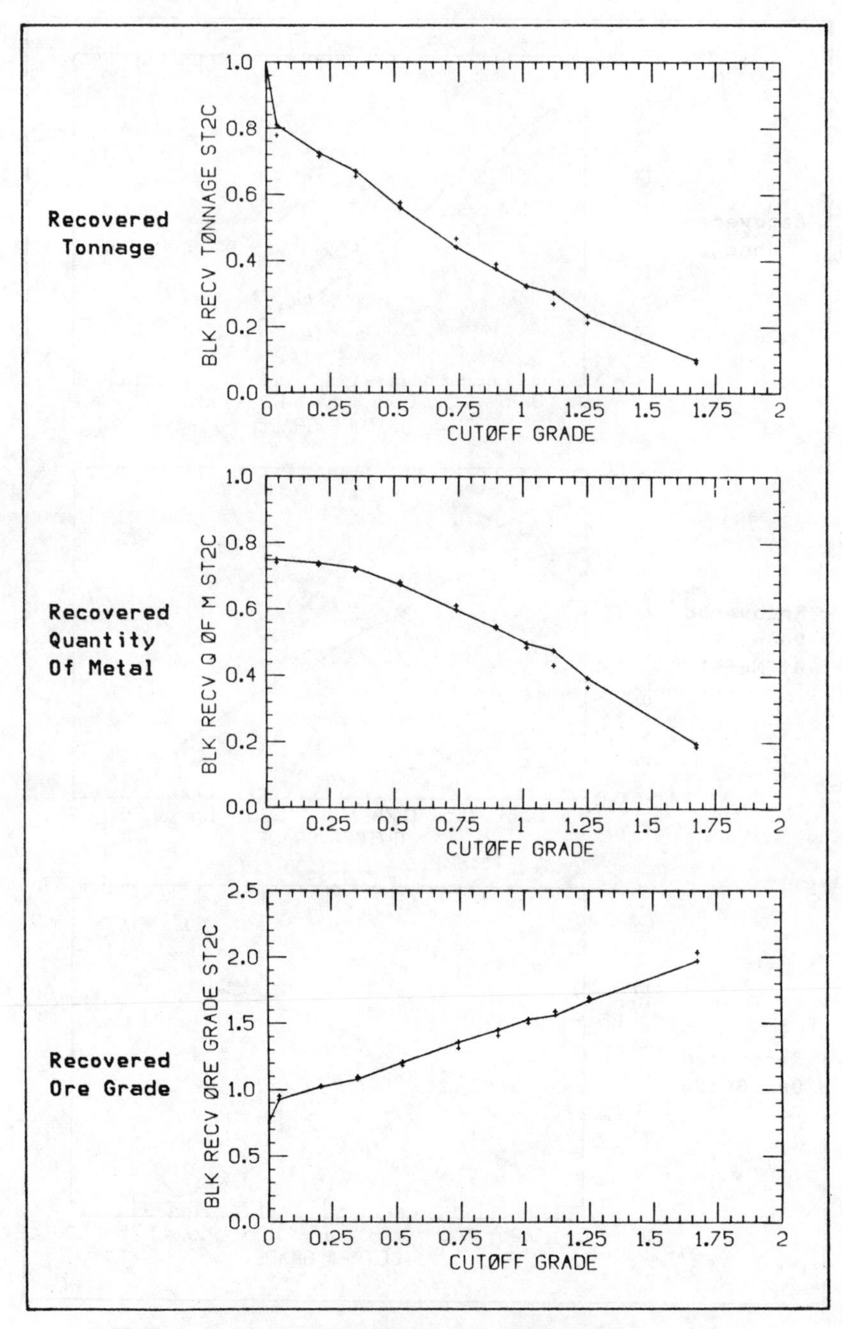

Figure 2. Checking the Affine Correction - ST2C
True Block Recovery (+——+)
Predicted Block Recovery From Affine Correction (+ +)

(imparted by the turning-bands simulation method, see Journel and Huijbregts, 1978) but with a large spike of zero grades as shown in Figure 1. ST2C has been constructed to avoid multi-gaussian tendencies, and it also has the large spike of zero grades.

COMPARISON OF PREDICTED AND TRUE GLOBAL ESTIMATION VARIANCES

Global estimation variance has been predicted for ST2B and ST2C with expressions (11),(12),(13),(23),(24),(25). In all cases the true recoveries, in situ mean ore grade, point dispersion variance, block dispersion variance, and correlation coefficients have been used. Also, the exhaustive semi-variograms (from all 24200 point grades) have been used to calculate the average semi-variogram values needed in expressions (11),(12). The exhaustive semi-variograms have been modeled by an isotropic linear structure. This type of model provides a satisfactory approximation, as shown in Figure 3, for separation distances up to 31 m (the maximum separation in the stratifying grid cell). (Only 2 of the 44 sets of semi-variograms are given in Figure 3 as an example.)

The results of this study for the ST2B and ST2C deposits are presented in Figures 4 to 7. The predicted GEV's are plotted against their corresponding true values. The GEV's are all given in terms of relative estimation standard deviation as defined in expression (26).

In general, the results presented in Figures 4 to 7 are very satisfactory, especially for recovered tonnage and quantity of metal. There is no consistent tendency toward over- or under-prediction except for the case of recovered ore grade. The consistent under-prediction of GEV for recovered ore grade is probably due to the first-order expansion used in expression (13). However, notice that this discrepancy for recovered ore grade is also exaggerated, relative to the discrepancies for tonnage and quantity of metal, by the vertical plotting scale.

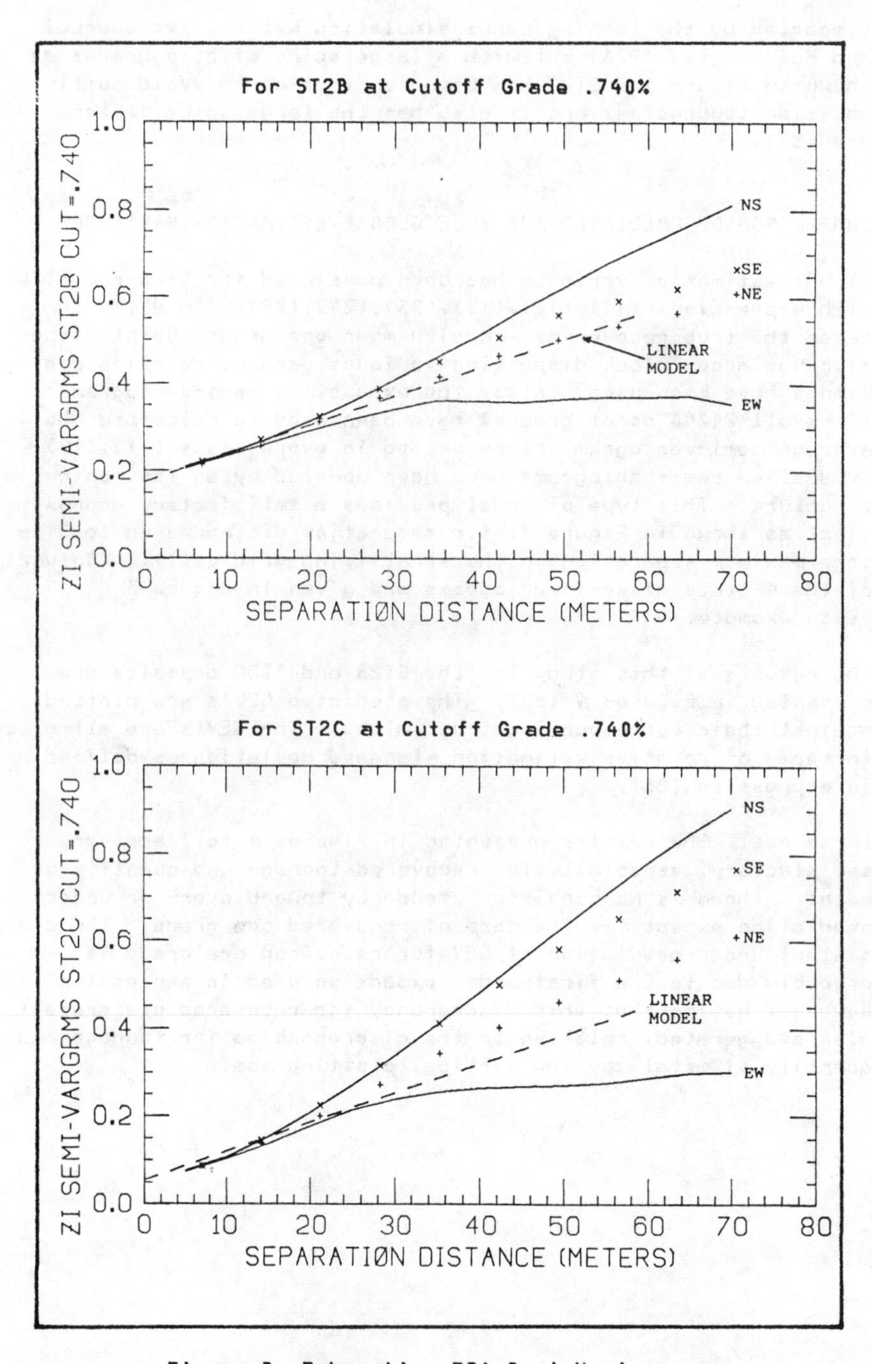

Figure 3. Exhaustive ZI' Semi-Variograms.

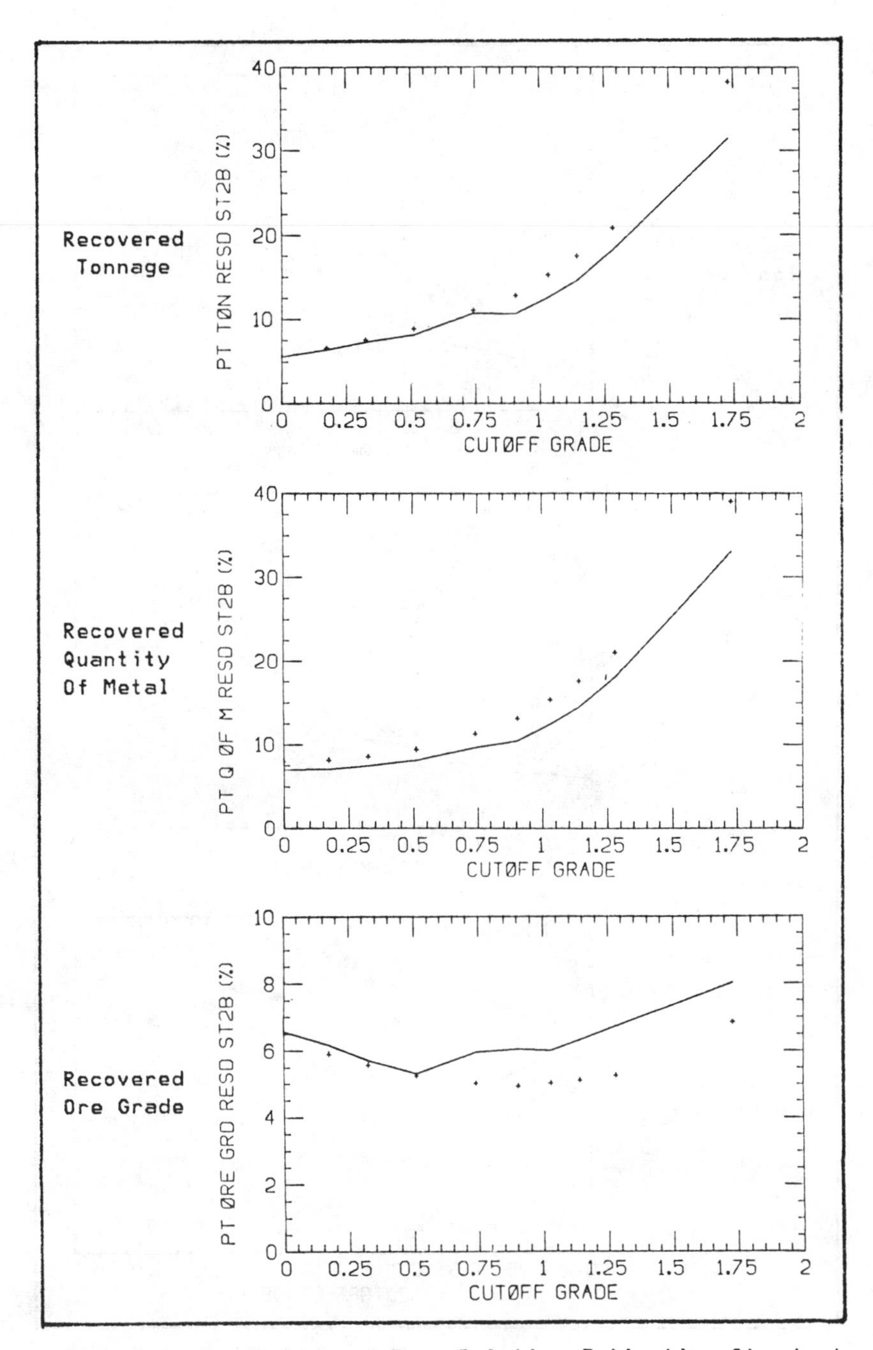

Figure 4. Predicted and True Relative Estimation Standard Deviation for Point Recovery From ST2B.
Predicted (+ +) True (+——+)

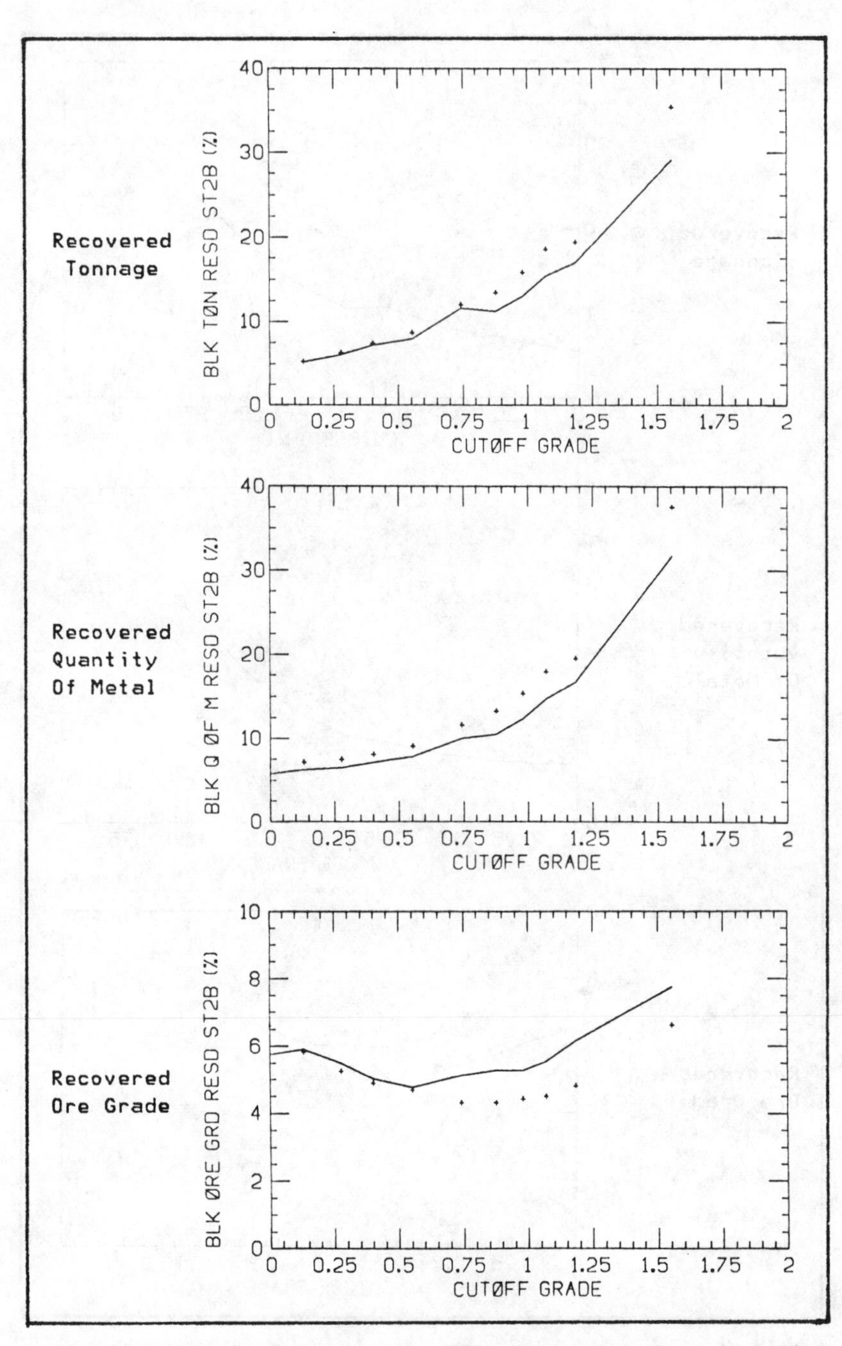

Figure 5. Predicted and True Relative Estimation Standard Deviation for Block Recovery From ST2B.
Predicted (+ +) True (+——+)

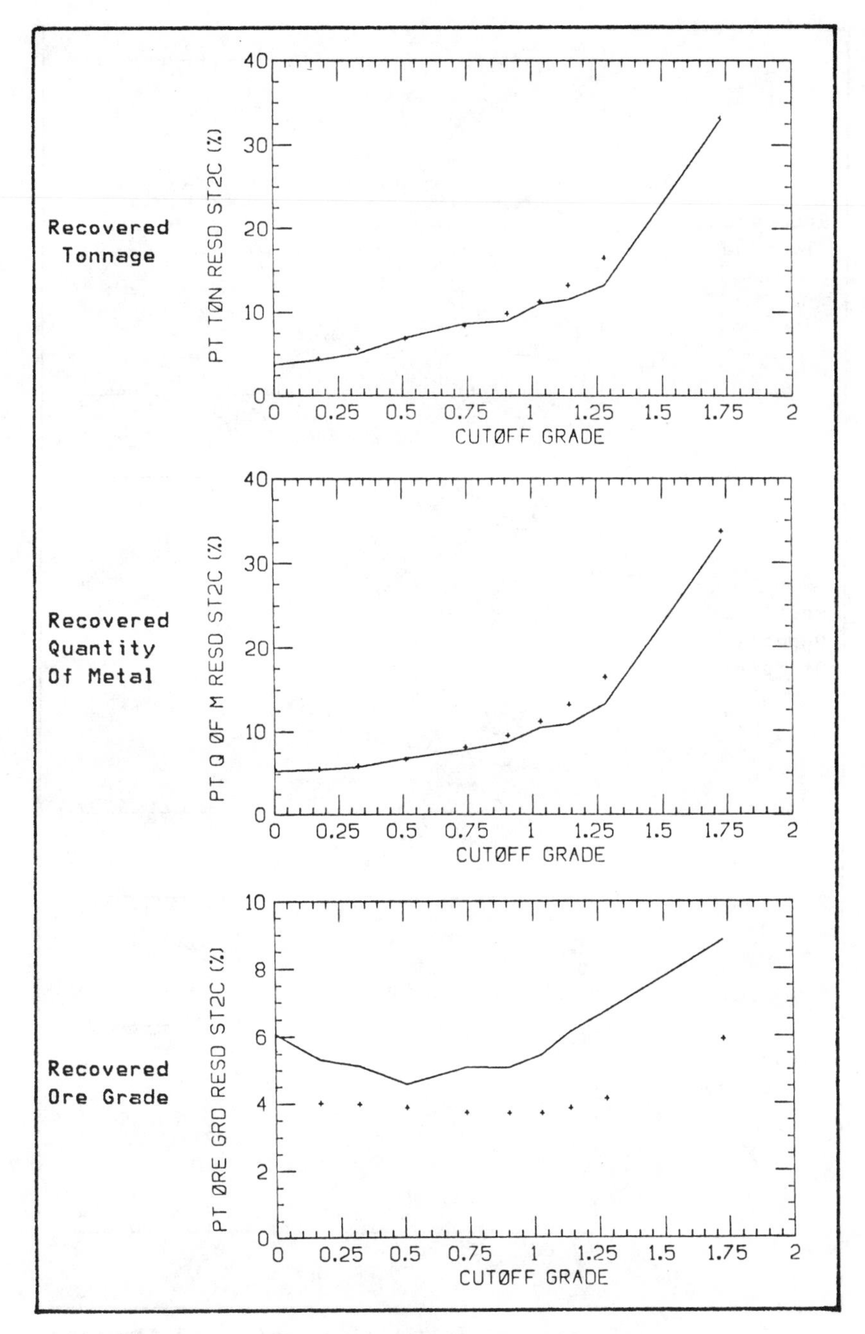

Figure 6. Predicted and True Relative Estimation Standard Deviation for Point Recovery From ST2C.
Predicted (+ +) True (+——+)

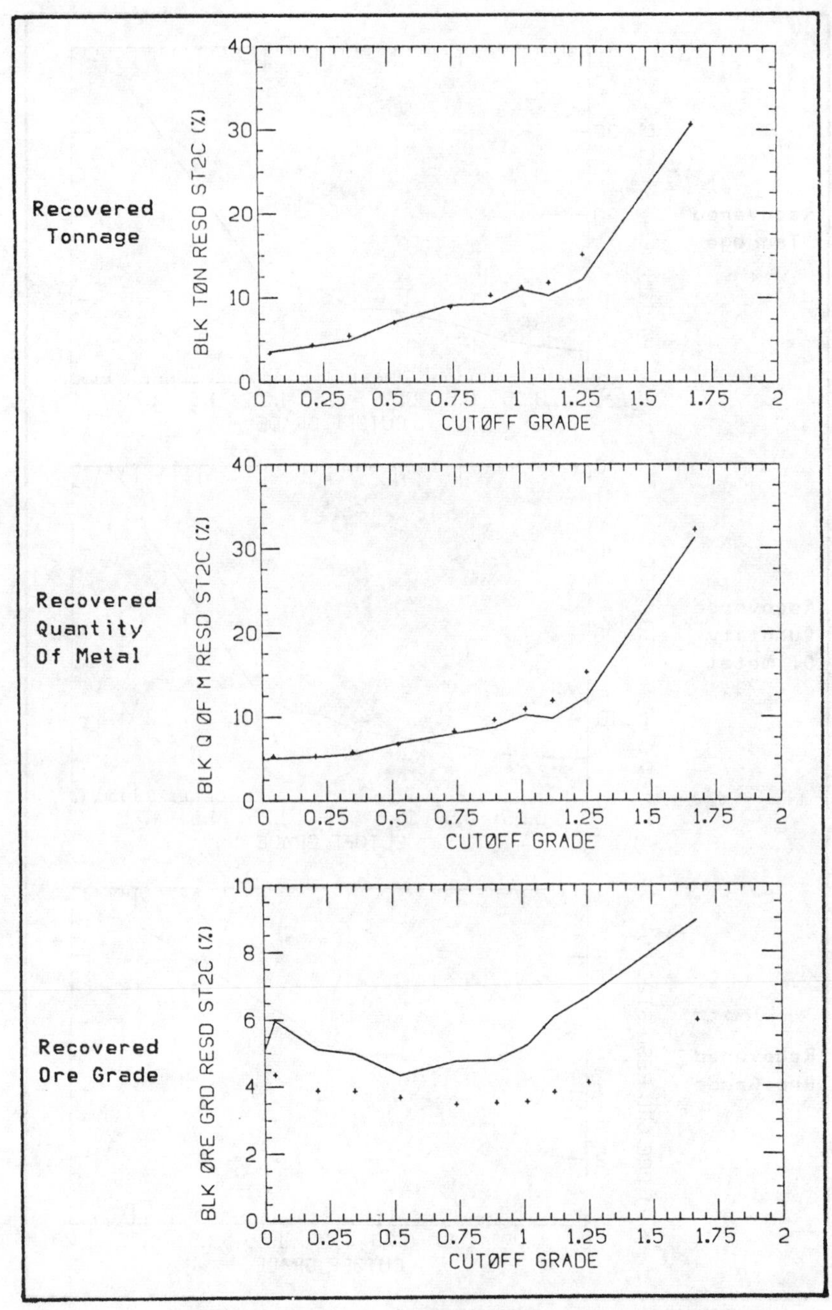

Figure 7. Predicted and True Relative Estimation Standard Deviation for Block Recovery From ST2C. Predicted (+ +) True (+——+)

CONCLUSION

In this paper, the prediction of the estimation variance of global recoverable reserve estimates has been placed in the framework of existing geostatistical theory. The results from the use of the predictive formulas derived in this study for two simulated deposits are quite encouraging. However, several problems must still be investigated before the method is complete. For example:

- in practice, the affine correction, used in expressions (20),(21),(22), is made with estimates of the in situ average ore grade, point dispersion variance, and block dispersion variance;
- in practice, expression (13), is evaluated with estimated recovered tonnage, quantity of metal, and ore grade;
- in practice, estimated semi-variograms are used to evaluate expressions (11),(12);
- real deposits can display a wide variety of grade distributions (e.g. highly skewed) which are different from the distributions for ST2B and ST2C;
- other block distribution hypotheses than the affine correction of variance can be used;
- and the first-order expansion in expression (13) may need to be refined.

REFERENCES

Dagbert,M. and Myers,J.,1982,"Precision of Total Reserve Estimates: Old Formulae Applied to New Situations", Proceedings of the 17th APCOM,Golden,Colorado.

Froidevaux,R.,1984,"Conditional Estimation Variances: An Empirical Approach",Mathematical Geology,in press.

Harrison,D.,1983,"A Schema for Classifying and Reporting Mineral Resources",Proceedings of the 112th Annual Meeting of the SME of the AIME,Atlanta,Georgia.

Journel,A.G., and Huijbregts,Ch.J.,1978,Mining Geostatistics, Academic Press,London.

Journel,A.G.,1983,"Estimation Variance of Global Recoverable Reserves",unpublished research notes.

COMBINING LOCAL KRIGING VARIANCES FOR SHORT-TERM MINE PLANNING

Young C. Kim and Ernest Y. Baafi

Department of Mining and Geological Engineering
The University of Arizona
Tucson, Arizona

ABSTRACT

Estimated grades of different working areas of a mine form the basis of any mine planning. A typical estimation method used in short-term mine planning is simple (linear) kriging. Very often, the mining engineer is not only interested in the individual reliability of the estimates from each working area, but also in the estimation variance of a combination of estimated grades from all working areas. This paper discusses a method for combining local kriging variances to obtain the precision of combined kriging estimator. The method relies on the fact that the local kriging weights and the Lagrange multipliers are linear in support, provided that the same samples are used during the local kriging processes. It is often not practical to use the same samples to krige all the working areas in the mine due to scattered nature of different working areas. Consequently, some adjustment is needed prior to taking advantage of the above property of kriging. This paper stresses the practical aspect of resolving the problem.

INTRODUCTION AND OBJECTIVE

The method of linear kriging in the estimation of mining blocks has become popular in the mining industry. The estimator is unbiased and has a minimum error of estimation. Figure 1 shows typical kriging results commonly used in either medium or short range mine planning. During mine planning, and especially for short term planning, not only are the estimates of individual

G. Verly et al. (eds.), Geostatistics for Natural Resources Characterization, Part 1, 185–199.

mining blocks required, but also a combined estimate of various mining blocks from different working areas. For example, in Figure 1 we may like to know the combined estimate of blocks A,B,C, D,E,F,G, H and I. The combined estimate of these nine blocks, i.e., A-I is usually obtained by weighting the local kriged estimates with their respective supports (i.e., volumes or weights).

In addition to the combined kriged value, the mining engineer may also like to know the accuracy (or estimation variance) of the combined estimate. Unfortunately a simple weighting rule does not apply in recombining local kriging variances. The objective of this paper is to discuss a method which combines the local kriging estimation errors to obtain a combined kriging estimation error. The method relies on the fact that both the kriging weights and the Lagrange parameters are linear in support provided that the <u>same</u> samples are used during the local kriging processes. By taking advantage of this property of linear kriging, the <u>combined</u> kriging weights and the Lagrange parameter can be computed from the local kriging weights and the associated Lagrange parameters. A standard relationship for kriging variance may then be used to obtain the combined kriging error of estimation.

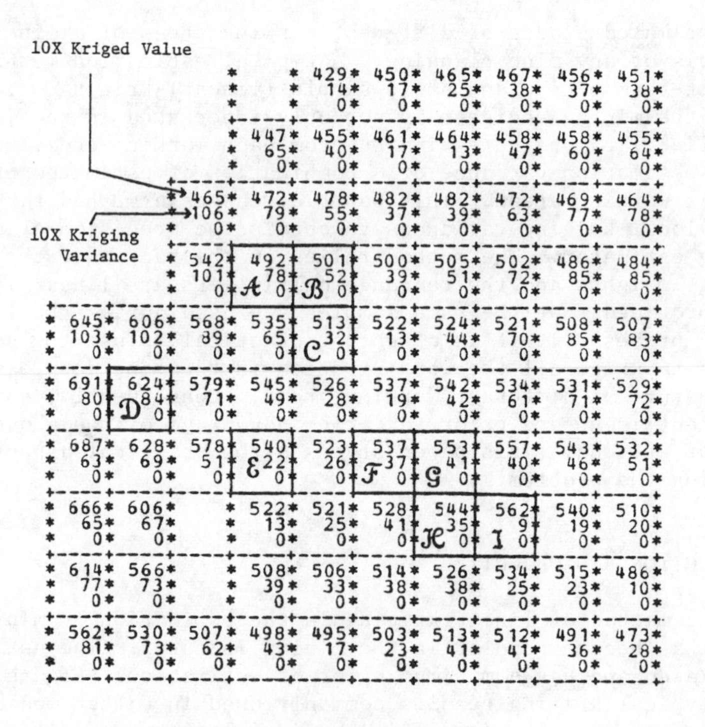

Figure 1. Map of Block Kriged Values

Often, it is not practical to use the same samples to krige all the working areas in the mine due to scattered nature of different working areas. Consequently, some adjustment is needed prior to taking advantage of the above property.

RECOMBINING KRIGING VARIANCES

Let $V_1, V_2, \ldots, V_n$ be n mining blocks of different volumes (supports), and let $Z^*_{V_i}$ $(i = 1,2, \ldots, n)$ be their respective kriged values. If the combined block of each V_i is D, then the kriging estimator of D, given as Z^*_D, reduces to equation (1).

$$Z^*_D = \frac{1}{D} \sum_{i=1}^{n} V_i \, Z^*_{V_i} \qquad (1)$$

where $$D = \sum_{i=1}^{n} V_i$$

The error of estimation of true grade Z_D would be given by equation (2).

$$Z_D - Z^*_D = \frac{1}{D} \sum_{i=1}^{n} V_i (Z_{V_i} - Z^*_{V_i}) \qquad (2)$$

The kriging estimator for Z_{V_1}, Z_{V_2}, ...k Z_{V_n} is given by equation (3).

$$Z^*_{V_i} = \sum_{k=1}^{m} \lambda_{ik} \, Z_{ik} \qquad i=1,2,\ldots,n \qquad (3)$$

where $Z_{i1}, Z_{i2}, \ldots, Z_{im}$ are sample values used to krige block V_i and λ_{ik} are the respective weights. In equation (3) above, it is assumed that the same number of points are used to krige each mining block. If it is difficult to satisfy the above assumption of using the same m samples to krige each block, it can still be satisfied by letting the weights of excluded assays to be zero, at least in theory.

Since each $Z^*_{V_i}$ is unbiased,

$$E\,(Z_{V_i} - Z^*_{V_i}) = E\,(Z_D - Z^*_D) = 0 \tag{4}$$

and

$$\begin{aligned} \mathrm{Var}\,(Z_D - Z^*_D) &= E[\{(Z_D - Z^*_D) - E(Z_D - Z^*_D)\}^2\,] \\ &= E\,[\,(Z_D - Z^*_D)^2\,] \\ &= E[\{\frac{1}{D}\sum_{i=1}^{n} V_i\,(Z_{V_i} - Z^*_{V_i})\}^2\,] \end{aligned} \tag{5}$$

Expanding equation (5) further, we obtain equation (6) below.

$$E[\{\frac{1}{D}\sum_{i=1}^{n} V_i\,(Z_{V_i} - Z^*_{V_i})\}^2]$$

$$= E[\sum_{i=1}^{n}\sum_{j=1}^{n} \frac{V_i}{D}(Z_{V_i} - Z_{V_i}^*)\ \frac{V_j}{D}(Z_{V_j} - Z_{V_j}^*)]$$

$$= \sum_{i=1}^{n}\sum_{j=1}^{n} \frac{V_i V_j}{D^2}\,E[(Z_{V_i} - Z_{V_i}^*)\,(Z_{V_j} - Z_{V_j}^*)]$$

$$= \frac{1}{D^2}\left\{\sum_{i=1}^{n} V_i^2\,\mathrm{Var}\,[\,Z_{V_i} - Z^*_{V_i}\,]\right.$$

$$\left. + \ 2\sum_{i=1}^{n}\sum_{j>i}^{n} V_i V_j\,\mathrm{Cov}[(Z_{V_i} - Z^*_{V_i}),(Z_{V_j} - Z^*_{V_j})]\right\}$$

$$= \frac{1}{D^2} \left\{ \sum_{i=1}^{n} V_i^2 \sigma_K^2 (V_i) \right.$$

$$\left. + \; 2 \sum_{i=1}^{n} \sum_{j>1}^{n} V_i V_j \operatorname{Cov}[(Z_{V_i} - Z_{V_i}^*),(Z_{V_j} - Z_{V_j}^*)] \right\} \qquad (6)$$

The quantity $\sigma_K^2 (V_i)$ in equation (6) represents the kriging variance for the block V_i simply because it is the variance of the kriging error for that block. The second term in equation (6) represents the error covariance between two errors of estimation, weighted by the respective block volumes V_i and V_j. The error covariance term in equation (6) must be considered because the same data points are often used to estimate each unit of the block and, therefore, the estimation errors are statistically dependent. If no common data points are used to krige different blocks, then the second term in equation (6) becomes zero, and the combined kriging variance reduces to equation (7).

$$\operatorname{Var} [(Z_D - Z_D^*)] = \frac{1}{D^2} \sum_{i=1}^{n} V_i^2 \sigma_K^2 (V_i) \qquad (7)$$

Even then, equation (7) is valid only as a first approximation, because there is a certain degree of correlation between Z_{V_i} and $Z_{V_i}^*$ for a regionalized variable (Journel and Huijbregts, 1978).

THE ERROR COVARIANCE TERM

If the variogram function $\gamma(h)$ for Z is known, the covariance term in equation (6) can be computed as shown below. First, expanding the covariance term in equation (6), we have

$$\begin{aligned} \mathrm{Cov}[(Z_{V_i} - Z^*_{V_i}),(Z_{V_j} - Z^*_{V_j})] = & \sum_{k=1}^{m} \frac{\lambda_{ik}}{V_i} \int_{V_i} \gamma(x - Z_{ik})\, dx \\ & + \sum_{k=1}^{m} \frac{\lambda_{jk}}{V_j} \int_{V_j} \gamma(y - Z_{jk})\, dy \\ & - \frac{1}{V_i V_j} \int_{V_i}\int_{V_j} \gamma(x - y)\, dx\, dy \\ & - \sum_{k=1}^{m} \sum_{r=1}^{m} \lambda_{ik}\lambda_{jr}\gamma(Z_{ik} - Z_{jr}) \end{aligned} \quad (8)$$

Equation (8) can then be substituted back into equation (6).

The complete mathematical derivation of equation (8) is given in the Appendix. In equation (8), the first two terms are the average variogram between the m sample points and blocks V_i and V_j, respectively. Also, the third term in equation (8) is the average variogram between the blocks V_i and V_j, while the fourth term is the average variogram between the m sample points used to krige blocks V_i and V_j.

THE KRIGING WEIGHTS

In equation (8), the kriging weights λ_{i1}, λ_{i2}, ..., λ_{im} are the weights assigned to sample points 1,2,..., m in the estimation of the grade of block V_i. If we are only interested in the estimation of the individual blocks V_i, these weights must minimize the respective local kriging variance, i.e., Var $(Z_{V_i} - Z^*_{V_i})$. Such weights are, in general, different if each block is kriged independently. However, since we are interested in minimizing the <u>combined</u> estimation variance σ^2_K (D) = Var $(Z_D - Z^*_D)$, and <u>not</u> the local estimation variances, the optimal kriging weights must minimize σ^2_K (D).

The weights which minimize σ^2_K (D) are usually obtained by differentiating a function of σ^2_K (D), i.e., equation (6) and then solving for the system of linear equations. Mathematically this approach is possible, but in practice its computation is time consuming when the number of blocks V_i to be combined is large. Consequently, equations (6) and (8) <u>will not</u>

be used in practice, unless the number of blocks is small. An easier approach must therefore be sought. Such an approach is discussed in the subsequent sections.

RELATIONSHIP BETWEEN SUPPORT AND SOLUTION TO KRIGING EQUATION

The kriging system of equations have a property that the kriging weights and the Lagrange parameters are linear in the support provided that the same samples are used in the estimations (Journel and Huijbregts, 1978). Therefore, if λ_{ik} is the solution of kriging block V_i using samples Z_{ik} for $k = 1, 2, \cdots, m$ and if the combined block D is also kriged with the same data points, the solution of the combined kriging system of equations can be obtained in terms of λ_{ik} using the following relationship.

$$\lambda_k(D) = \frac{1}{D} \sum_{i=1}^{n} V_i \lambda_{ik} \qquad k = 1,2,\ldots,m \tag{9}$$

where, λ_k (D) is the weight assigned to data point k which minimizes equation (6). Also, the Lagrange parameter $\mu(D)$ is given by equation (10).

$$\mu(D) = \frac{1}{D} \sum_{i=1}^{n} V_i \mu(V_i) \tag{10}$$

Equations (9) and (10) imply that if the same sample points are used to krige a group of blocks one at a time, it is possible to obtain the solution of combined kriging system of equations from the solutions of individual (local) systems of equations. Once the combined kriging weights and the Lagrange parameters are known, the combined kriging variance can be computed using the standard kriging variance relationship given in equation (11).

$$\sigma_K^2(D) = \bar{C}(D,D) - \sum_i \lambda_i(D)\,\bar{C}(D,x_i) + \mu(D) \tag{11}$$

where $\bar{C}$ is an average covariance term.

SOME PRACTICAL CONSIDERATIONS

If one is to take advantage of equations (9) and (10), the same sample points must be used to krige each mining block, even if some sample points may be too far away from a given block for any practical benefit. When all the workings in the mine are close together as in Figure 2, the only practical concern is using more than the necessary number of sample points that are used to krige each working area, thus consuming more of the computation time.

Often, it is not practical to use the same samples to krige all the working areas in the mine due to scattered nature of different working areas. Hence, some adjustment is needed. When working areas are too far apart, some working areas may conveniently be grouped together and then each group kriged with a set of sample points. In Figure 3, working areas which are enclosed by polygon A may be kriged using all the sample points inside polygon A only. Using equations (9), (10) and (11), their combined kriging variance $\sigma_K^2\ (V_A)$ will then be computed, where the support V_A is the combined supports of all working areas inside polygon A. Similarly, each working areas enclosed by polygon B will be kriged with all the sample points inside polygon B only. Their combined kriging variance $\sigma_K^2\ (V_B)$ will also be computed using equations (9) through (11). Since the estimate of V_A is statistically independent of the estimate of V_B, the combined kriging variance of V_A and V_B is obtained by direct weighting using equation (7).

Therefore, when mine workings are close together as in Figure (2), equations (9), (10) and (11) are directly applicable in the computation of the combined kriging variance. When the mine workings are scattered as in Figure 3, the mine workings are grouped in disjoint sets. The combined kriging variance of each group of mine workings is computed as usual using equations (9), (10) and (11). Next, the overall combined kriging variance of the disjoint groups is weighted using equation (7).

COMPUTER PROGRAM STPM6

A computer program STPM6 was written to take advantage of equations (9), (10) and (11). The program first kriges each mining block with the same sample points, and then utilizes the local kriging results to compute a combined kriging variance. The weights and Lagrange multiplier used in equation (11) are obtained from equations (9) and (10). Figure 4 summarizes the method of solution of program STPM6.

The program does <u>not</u> group working areas into disjoint sets which is needed when the working areas are scattered in the mine. Such a grouping must be done manually by the user. However, the program has an option to krige blocks with sample points inside a polygon whose coordinates are input by the user. This option ensures that the same sample points are used in the kriging process. The program additionally checks that identical sample points are used in the kriging of individual mining blocks. A run is terminated when different sample points are found to be used in the kriging of any block.

Since the same data configuration is used to krige each group of working areas, all the local kriging systems have the same left-hand matrix which reduces the computation time. Only the right-hand side vectors change between working areas.

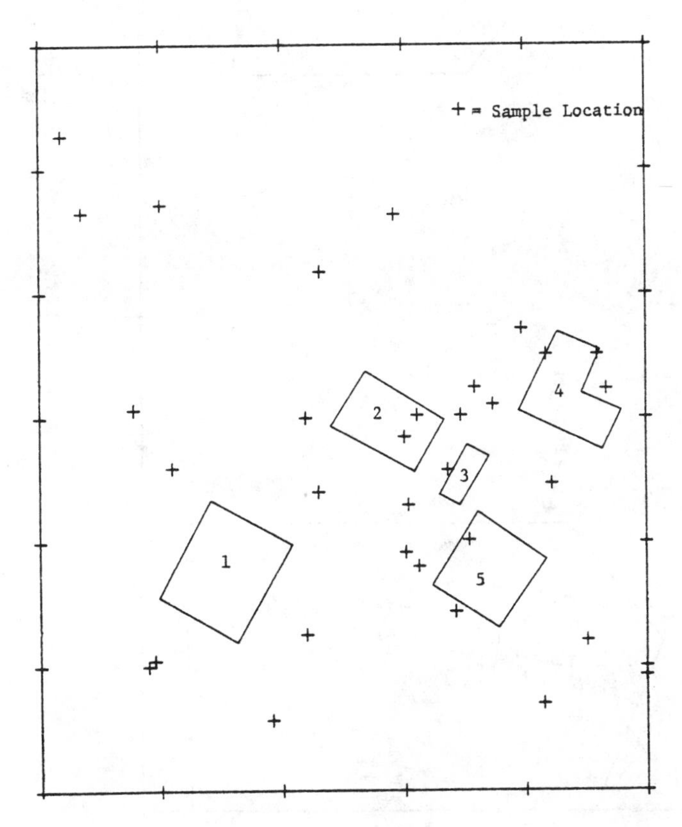

Figure 2. Combining Estimates of Five Working Areas

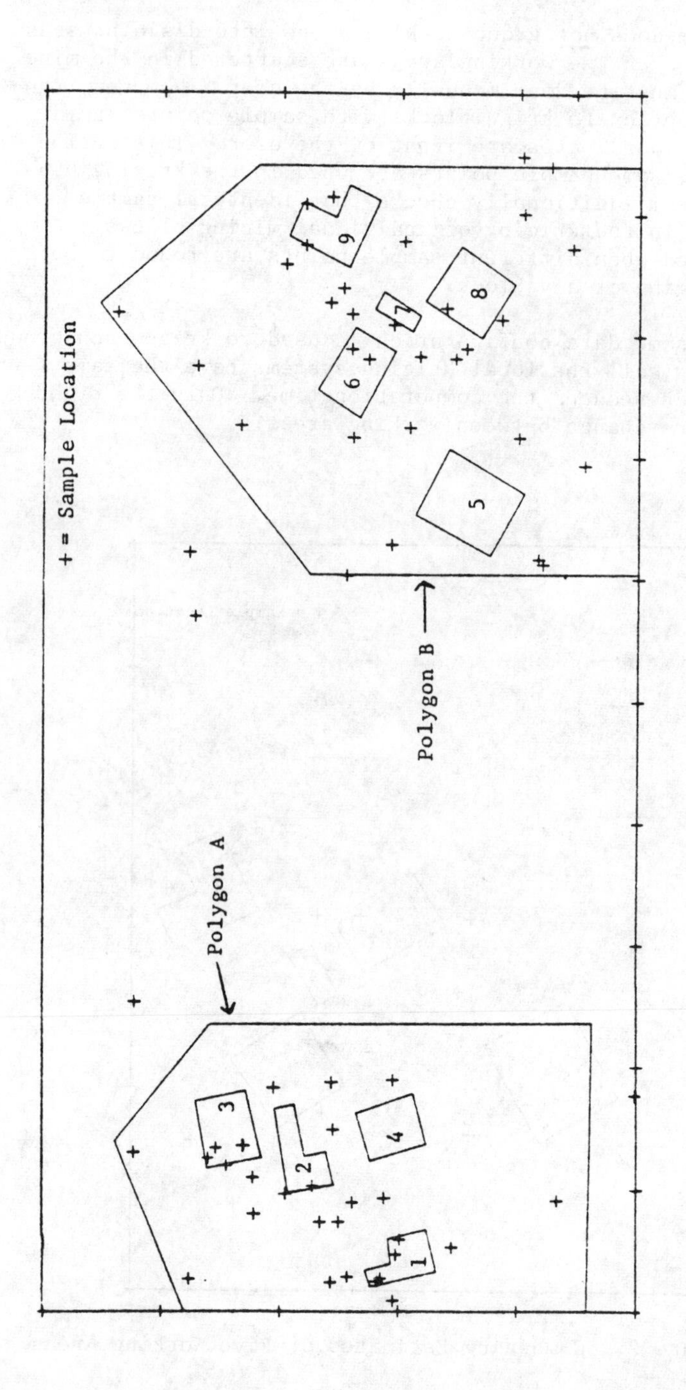

Figure 3. Nine Working Areas Enclosed by Two Polygons

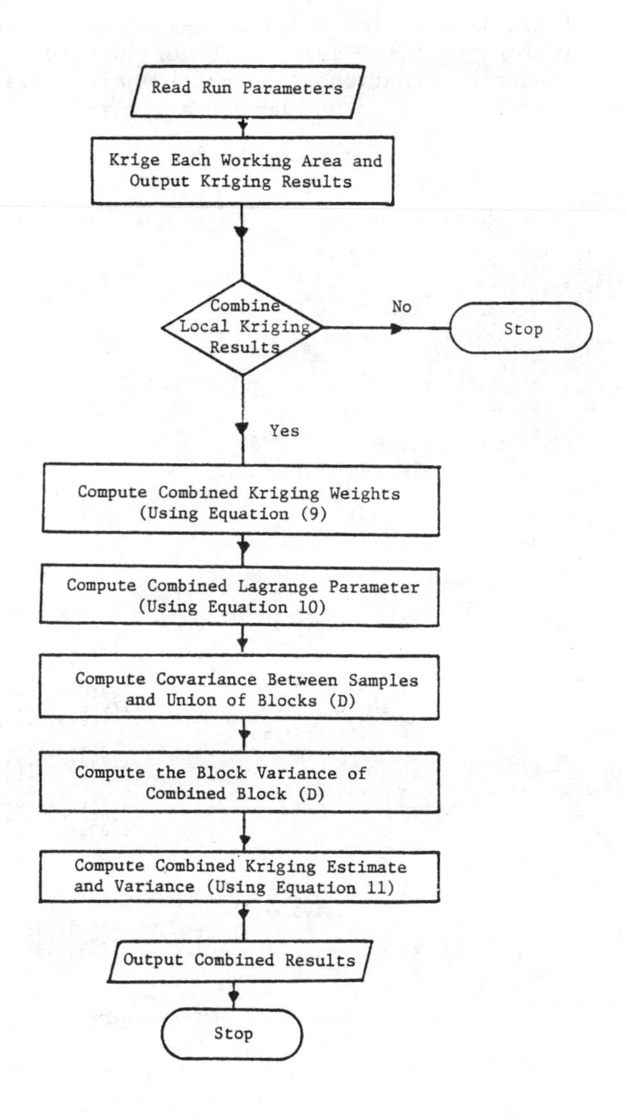

Figure 4. Summary of method of solution of Program STPM6.

Figure 5 is an example output of computer program STPM6. The output from program STPM6 includes parameter values used for a run, and the kriged values for each working area in addition to the combined kriging variance for a maximum of three variables.

To recapitulate the practical aspects of combining the kriging variances, one kriges each individual block one at a time as is commonly done, with exception that all the assays from a closed polygon is used for kriging. We do not, however, perform the overall kriging of the combined blocks within a given enclosed ploygon. Instead, equations (9) and (10) are utilized to obtain the necessary information on the overall combined blocks.

```
EXECUTION OF PROGRAM STPM6              TODAYS DATE  06/21/82
FILE NAMES
  MASTER FILE                  NEWMAS
  MASTER FILE DATE             06/18/82
  MINE BLOCK DATA FILE         AP82
  SAMPLE POLYGON FILE          HAREA1
NUMBER OF VARIABLES KRIGED              3
VARIABLES KRIGED  -            1    2   3
VARIOGRAM PARAMETERS
     ANGLE OF ANISOTROPY                0.00
     HORIZONTAL ANISO. FACT.            1.00
     VERTICAL   ANISO. FACT.            0.00
     SEARCH RADIUS                   4000.00
     MAXIMUM NO. OF SAMPLES            50
   VARIABLE               1
      SILL                            38.0000
      C(1)   8.5000 RANGE(1)              0.
      C(2)  14.0000 RANGE(2)           2500.
      C(3)  16.0000 RANGE(3)          10000.
   VARIABLE               2
      SILL                            18.0000
      C(1)   2.0000 RANGE(1)              0.
      C(2)   7.0000 RANGE(2)           3500.
      C(3)   9.0000 RANGE(3)          10000.
   VARIABLE               3
      SILL                             1.2000
      C(1)    .3700 RANGE(1)              0.
      C(2)    .9100 RANGE(2)          10000.
      BLOCK ID       NODES          AREA  VARIABLE      VALUE   KRIG VAR
AP82 3-EAST             4   .36208E+05
                                               1      .5350E+02 .7033E+01
                                               2      .1958E+02 .2482E+01
                                               3      .3111E+01 .1247E+00
AP82 2-EAST             6   .51470E+05
                                               1      .5237E+02 .4888E+01
                                               2      .1748E+02 .1463E+01
                                               3      .3446E+01 .1357E+00
AP82 10-EAST            4   .79920E+05
                                               1      .5531E+02 .4841E+01
                                               2      .1878E+02 .1457E+01
                                               3      .3409E+01 .1091E+00
AP82 SHORT WALL         4   .15172E+06
                                               1      .5477E+02 .6405E+01
                                               2      .2008E+02 .2131E+01
                                               3      .3509E+01 .1479E+00

      COMBINED KRIGED VALUES
      ----------------------
               SAMPLE SIZE          AREA  VARIABLE      VALUE   KRIG VAR
                              .31932E+06
                    20                         1      .5459E+02  .2076E+01
                    20                         2      .1927E+02  .8139E+00
                    20                         3      .3429E+01  .6137E-01
```

Figure 5. Sample Output of Program STPM6

CONCLUSION

The reliability of an estimate of a combination of kriged values is conveniently obtainable provided that the correct assumptions are made. If the same data points are used to krige a set of mining blocks, the combined kriging variance can be computed using the individual kriged results. When mine workings are scattered in a mine such that it is not possible to krige all the mine working with the same data points, some grouping of working areas may be needed prior to the use of the local kriging results.

APPENDIX

Expanding the covariance term in equation (6), we have

$$\mathrm{Cov}\,[(Z_{V_i} - Z^*_{V_i}),\ (Z_{V_j} - Z^*_{V_j})]$$

$$= E[(Z_{V_i} - Z^*_{V_i})(Z_{V_j} - Z^*_{V_j})] - [E(Z_{V_i} - Z^*_{V_i})]\,[E(Z_{V_j} - Z^*_{V_j})] \tag{12}$$

But,

$$E(Z_{V_i} - Z^*_{V_i}) = E(Z_{V_j} - Z^*_{V_j}) = 0$$

Hence,

$$\mathrm{Cov}\,[(Z_{V_i} - Z^*_{V_i}),\ (Z_{V_j} - Z^*_{V_j})]$$

$$= E[(Z_{V_i} - Z^*_{V_i})(Z_{V_j} - Z^*_{V_j})] \tag{13}$$

Substituting equation (3) into equation (13), we have

$$\mathrm{Cov}\,[(Z_{V_i} - Z^*_{V_i}),\ (Z_{V_j} - Z^*_{V_j})]$$

$$= \sum_{k=1}^{m} \sum_{r=1}^{m} \lambda_{ik}\lambda_{jr}\ [E(Z_{V_i} - Z_{ik})\ (Z_{V_j} - Z_{jr})] \tag{14}$$

But,

$$(Z_{V_i} - Z_{ik})\ (Z_{V_j} - Z_{jr}) = \frac{1}{V_i V_j} \int_{V_i} \int_{V_j} [Z(x) - Z_{ik}]\ [Z(y) - Z_{jr}]\ dxdy$$

Hence,

$$E[(Z_{V_i} - Z_{ik}),\ (Z_{V_j} - Z_{j_r})] = \frac{1}{V_i V_j} \int_{V_i} \int_{V_j} E[Z(x) - Z_{ik}][Z(y) - Z_{jr}]\ dxdy$$

$$= \frac{1}{V_i V_j} \int_{V_i} \int_{V_j} E[Z(x)\,Z(y) - Z(x)\,Z_{jr} - Z(y)\,Z_{ik}$$

$$+ Z_{ik} Z_{jr}]\ dx\,dy$$

$$= \frac{1}{V_i V_j} \int_{V_i} \int_{V_j} K(x-y)\ dxdy - \frac{1}{V_I} \int_{V_i} K(x - Z_{jr})\ dx$$

$$- \frac{1}{V_j} \int_{V_j} K(y - Z_{ik})\ dy + K(Z_{ik} - Z_{jr}) \tag{15}$$

Under the second order stationarity assumption employed in kriging, the relationship between the covariogram and the variogram is given by

$$K(h) = K(0) - \gamma(h) \tag{16}$$

Substituting equations (15) and (16) into (14), we obtain

$$\begin{aligned} \mathrm{Cov}[(Z_{V_i} - Z^*_{V_i}),(Z_{V_j} - Z^*_{V_j})] = & \sum_{k=1}^{m} \frac{\lambda_{ik}}{V_i} \int_{V_i} \gamma(x - Z_{ik})\,dx \\ & + \sum_{r=1}^{m} \frac{\lambda_{jr}}{V_j} \int_{V_j} \gamma(y - Z_{jk})\,dy \\ & - \frac{1}{V_i V_j} \int_{V_i}\int_{V_j} \gamma(x - y)\,dx\,dy \\ & - \sum_{k=1}^{m}\sum_{r=1}^{m} \lambda_{ik}\lambda_{jr}\,\gamma(Z_{ik} - Z_{jr}) \end{aligned} \tag{17}$$

In the derivation of equation (17), the unbiasedness condition of kriging is utilized to simplify the equation; i.e.,

$$\sum_{k=1}^{m} \lambda_{ik} = \sum_{r=1}^{m} \lambda_{jr} = 1.$$

Equation (17) can then be substituted back into equation (6) to obtain the combined variance Var $(Z_D - Z_D^*)$.

ACKNOWLEDGEMENT

The authors would like to acknowledge the contribution of D.E. Myers of the University of Arizona for the detailed mathematical drivation of the problem.

REFERENCES

1. Dagbert, M. and Myers J., "Precision of Total Estimates: Old Formular Applied to New Situations", 17th APCOM, 1981, pp 393-405

2. Journel, A. G. and Huijbregts, Ch. J., Mining Geostatistics, Academic Press, New York, 1978, 600 p.

3. Kim, Y.C., Knudsen, H.P., and Baafi, E.Y., "Development of Emission Control Strategies Using Conditionally Simulated Coal Deposits", University of Arizona, 1981, Proprietary Report Prepared for the Homer City Owners, Homer City, PA, 15748.

4. Royle, A. G., "Global Estimates of Ore Reserves", Trans. Instn. Min. Metall. (Section A), 86, 1977, pp A9-A17.

APPLICATION OF A GEOSTATISTICAL METHOD TO QUANTITATIVELY DEFINE VARIOUS CATEGORIES OF RESOURCES

Raymond L. Sabourin

Mine Evaluation Group, Mining Research Laboratories, CANMET, Energy, Mines and Resources, Canada

Abstract. In a previous article the author proposed a geostatistical method providing technical criteria required to classify reasonably assured resources by levels of assurance of their existence. This method has been used to define the "measured" and "indicated" geological resources of the Short Creek lignite coal zone in Southern Saskatchewan. The technical criteria provided by this geostatistical method are compared with the more traditional criteria previously used to classify the same coal resources. Several properties of the method are discussed and recommendations are made concerning its application.

1.0 INTRODUCTION

In the last two years, several articles have been written on the use of geostatistics for resource classification. At the Tenth Geochautauqua held in Ottawa, in October 1981, the author explained a number of concepts related to the above subject and proposed a methodology to quantitatively define various categories of resources. The paper presented at this conference was recently published in the Mathematical Geology Journal (1). In 1977, Royle (2) suggested the use of block kriging variances (KV) for the classification of resources. Since its beginning, geostatistics provided the possibility to calculate the relative precision of an estimate and Carlier (3) in 1963 used this criterion to define several categories of resources. More recently, Deihl and David (4), as well as Froidevaux (5) each proposed a different geostatistical method for the classification of resources. It can be seen that there are almost as many proposed solutions as there are geostatisticians working on the subject.

G. Verly et al. (eds.), Geostatistics for Natural Resources Characterization, Part 1, 201–215.

Each of these methods has its own merits. However, it is not the intention here to make a detailed comparison between them. Instead, the utilization of the method proposed by the author will be described and some of its properties discussed.

In this paper, the method is applied to the Short Creek lignite coal zone of the Estevan region in Southern Saskatchewan and the geological resources are classified in several categories defined by their assurance of existence. The classified tonnages are compared to those classified by a traditional method as well as to the tonnages classified by a local relative precision method. Several block dimensions are used in order to check the accuracy of the method. Finally, some properties of the method are discussed and several recommendations are made concerning the application of the method presented in this article.

2.0 DESCRIPTION OF THE METHOD

The purpose of the method is to classify local tonnage quantities (or metal quantities for ore deposits) using KV limits (σ_k^2) determined from the distribution variance (σ_b^2) of the blocks. These KV limits are calculated using the following relation:

$$\sigma_k^2 = K \sigma_b^2 \qquad (1)$$

where K is a constant chosen arbitrarily to represent various categories of resources. The KV limits will be defined in this paper using K = 0.2, K = 0.4 and K = 0.8 for the measured, indicated and inferred resources respectively. As proposed, this method will work only when σ_k^2 varies proportionally with σ_b^2 for different sizes of blocks. It will be shown later that the two variances vary each by the same amount and not proportionally for small blocks. These ratios will nevertheless be used to show that it is necessary to use a corrective factor with this method. The ratios are chosen arbitrarily at first, but must then be kept the same for all deposits. The distribution variance is calculated using an infinite field which means that the variance of samples (coal intersections) in the deposit is equal to the sill of the variogram model when it exists.

The KV limits are not the average KV as was originally proposed, not because these could not be applied, but because a rather simplified procedure should be sufficiently convenient to demonstrate the properties of the method. The KV limits will thus classify individually each evaluated block.

The estimated variable is the average thickness of a block because the coal density is taken to be constant. In reality, the density is varying mainly in relation to ash content, porosity and moisture of coal. Dry tonnages are being evaluated. Block kriging will be performed with known mean using the twenty-two intersections.

3.0 GEOLOGICAL ENVIRONMENT

The coal deposits in Southern Saskatchewan consist of several more or less continuous seams which split and regroup themselves from one intersection to the next. A coal deposit will be referred as a coal zone which may include several coal seams. Five main coal zones have been identified (6) in the Ravenscrag sedimentary formation near Estevan in Southern Saskatchewan (see Figure 1). The name of these zones are, starting with the more geologically recent: Short Creek, Roche Percée, Souris, Estevan and Boundary. The Ravenscrag formation is a large synclinal sedimentary basin with its axis slightly inclined in the South-East direction. The Northern edge of the formation was eroded by glaciers and the five coal zones are outcropping successively beginning with the Boundary zone farther North. The Southern boundary of the evaluated territory is the Canada-United States

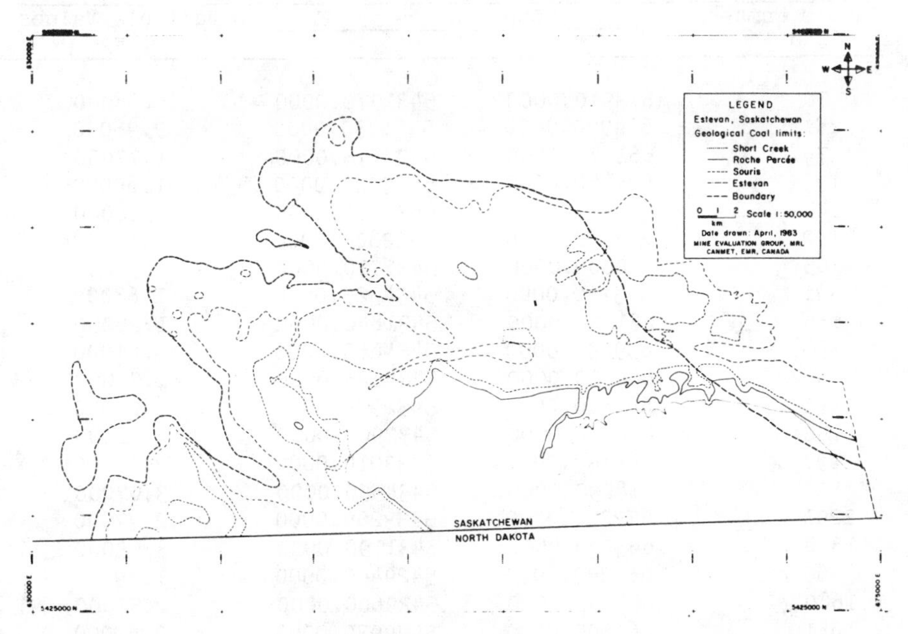

Figure 1. Geological limits of the five coal zones in Estevan, Southern Saskatchewan.

border line. This coal field covers an area of almost one thousand square kilometers. However, only a small fraction the total geological resources can be economically exploited by the actual mining methods. The classification methodology will be applied on geological resources of the Short Creek zone without taking into account economical or mining conditions which would define the reserves. However, the method is not limited to classify geological resources and can equally well be applied to the reserves.

4.0 STRUCTURAL ANALYSIS

Twenty-two coal intersections were used to calculate the average variogram of the geological thickness for the Short Creek zone (Figure 2). The intersection numbers, coordinates and coal thicknesses (meters) are given in Table 1. Figure 3 shows the distribution of the intersections within the Short Creek zone. The North, East and West boundaries are defined by the erosion edges of the zone. This geological limit has been schematically outlined for the purpose of this article, using one square kilometer blocks.

Sample Number	Coordinates		Variable Values
	X	Y	V2
70	673910.0000	5431370.0000	1.68000
281	664920.0000	5431230.0000	2.95000
74	661575.0000	5433575.0000	1.77000
139	669750.0000	5433020.0000	1.69000
141	664810.0000	5434460.0000	1.98000
142	663260.0000	5432800.0000	3.16000
185	660000.0000	5432710.0000	.31000
391	663160.0000	5434325.0000	1.68000
186	661630.0000	5431840.0000	1.28000
358	659890.0000	5434650.0000	.61000
357	666450.0000	5434625.0000	1.75000
195	664810.0000	5432930.0000	2.23000
196	666480.0000	5432900.0000	3.02000
197	668130.0000	5433010.0000	2.91000
198	668090.0000	5434550.0000	3.07000
2201	672260.0000	5433200.0000	2.77000
1668	663290.0000	5431190.0000	3.53000
1669	663292.0000	5429460.0000	1.89000
1680	664980.0000	5429600.0000	1.52000
1681	661690.0000	5429930.0000	2.61000
1682	658350.0000	5431810.0000	.55000
1683	659968.0000	5431140.0000	.64000

Table 1. List of basic data

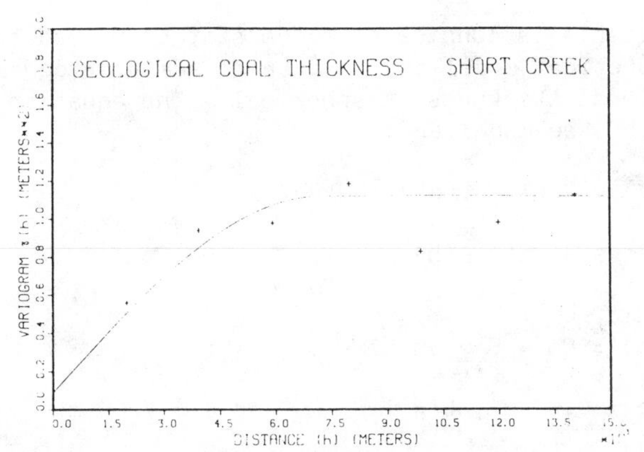

Figure 2. Average experimental variogram and its interpreted model.

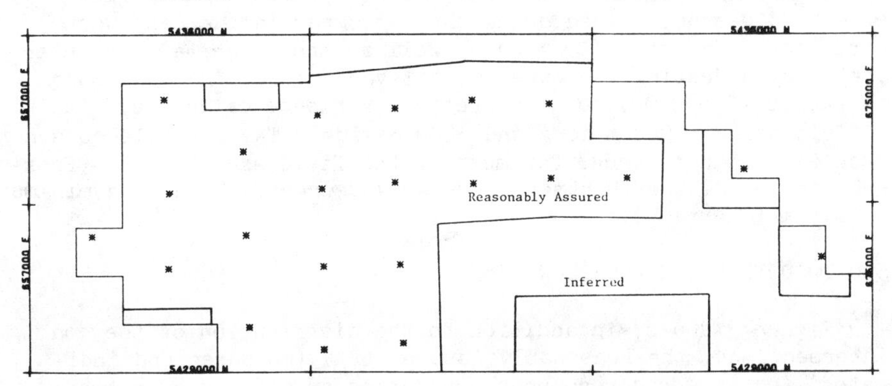

Figure 3. Distribution of coal intersections and resource boundaries from a traditional method in the Short Creek zone.

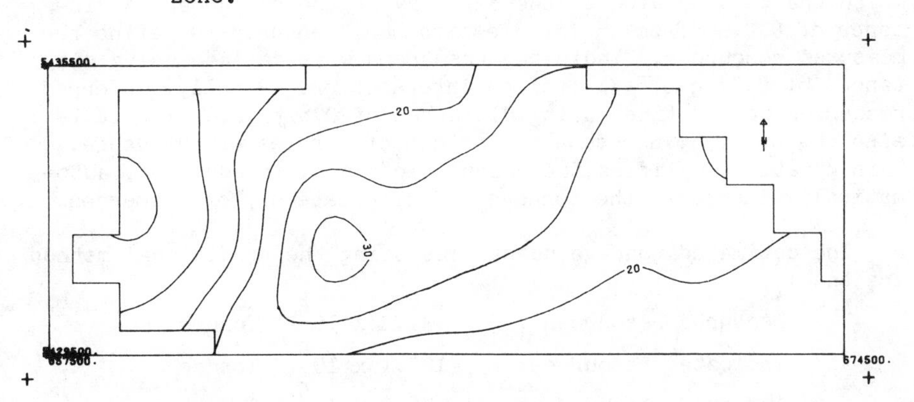

Figure 4. Geological coal thickness (meters x 10) isocurves using 1000 m x 1000 m blocks - Short Creek.

The variable is considered to be stationary and the interpreted theoretical model of the average semi-variogram for the geological coal thickness is spherical. The equation and the parameters of the model are:

$$1/2\ \gamma(h) = Co[1 - \delta(h)] + C(3h/2a - h^3/2a^3) \quad (2)$$

when $h \leq a$, with $\delta(h) = 1$ for $h = 0$
$= 0$ for $h \neq 0$

and $\gamma(h) = Co + C$ when $h > a$, (3)

with $Co = 0.09\ m^2$, $C = 1.03\ m^2$ and $a = 7200$ m.

The average semi-variogram has been used here because some directional variograms could not be interpreted. The nugget effect Co was determined using the variogram in the East-West direction. The sill (Co + C) as well as the range (a) was interpreted by adjusting the experimental variance of the coal zone thickness (0.882 m^2) to a theoretical variance calculated within a field being 15000 m long and 5000 m wide. The variable does not show a pronounced trend for most of the field except in the Eastern part where the thickness regularly decreases down to a minimum of 0.5 m (Figure 4).

5.0 TRADITIONAL EVALUATION

Figure 3 shows, in addition to the distribution of the coal intersections, the Reasonably Assured (RA) (measured and indicated) and inferred resources boundaries as defined by a traditional method for resource classification. The criteria used by this method are distances of influence from the intersections which change according to the category of the resource. A distance of 0.4 Km from an intersection has been used to define the measured resources. Indicated resources were defined using distances of 0.4 to 0.8 Km from an intersection. Finally, inferred resources were defined using distances of 0.8 to 2.4 Km. It is also customary to use squares instead of circles of influence. This greatly simplifies the calculations and, in addition, automatically increases the tonnage for each category of resources.

The estimated tonnage quantities using the traditional method are then:

Measured resources :	32.2×10^6	Tonnes
Indicated resources:	102.1×10^6	Tonnes
Inferred resources :	76.6×10^6	Tonnes
Total resources :	210.8×10^6	Tonnes

6.0 APPLICATION OF THE KV LIMITS METHOD

This method was proposed to limit the impact of the KV reduction for large blocks when resources are classified. The method should work well only when the KV (σ_k^2) and the distribution variance of the blocks (σ_b^2) vary proportionally in relation to the dimension of the blocks. It is certainly asking a lot that the ratio σ_k^2/σ_b^2 be the same for any size of blocks and, consequently, we will try to find out for which dimension of block the method can be applied. Let us remember that one of the purposes of the method is to use a criterion which permits to calculate more or less the same tonnage within a category of resource no matter what is the size of the block. The variation in the measured, indicated and inferred tonnages for several sizes of blocks must then be limited, for example, to within a range of values defined by the relative precision of these estimated tonnages.

The method will be applied as proposed and ratios 0.2, 0.4 and 0.8 between σ_k^2 and σ_b^2 will be considered to arbitrarily define three KV limits: $\sigma_{kM}^2 = 0.2\sigma_b^2$, $\sigma_{kI}^2 = 0.4\sigma_b^2$ and $\sigma_{kF}^2 = 0.8\ \sigma_b^2$ for the measured, indicated and inferred tonnage resources respectively. The values shown below on the contour maps represent twice the standard deviation of the block KV. Consequently, the above criteria should be transformed to: $2\sigma_{kM} = 2\sqrt{.2}\sigma_b$, $2\sigma_{kI} = 2\sqrt{.4}\sigma_b$ and $2\sigma_{kF} = 2\sqrt{.8}\sigma_b$. A new set of criteria will be calculated for a different size of block. The size of the chosen blocks are respectively: 1000 m x 1000 m, 500 m x 500 m and 250 m x 250 m. The values of the distribution variances are: $\sigma_b^2(1000) = 0.91957\ m^2$, $\sigma_b^2(500) = 0.97467$ m and $\sigma_b^2(250) = 1.00237$ m. Table 2 gives the criteria used to classify the resources of the Short Creek coal zone.

	Blocks (1000 m x 1000 m)	Blocks (500 m x 500 m)	Blocks (250 m x 250 m)
Measured (M)			
($2\sigma_{kM}$)	0.8577	0.8830	0.8955
(Tonnage)	105×10^6T	89×10^6T	81×10^6T
M + Indicated (I)			
($2\sigma_{kI}$)	1.2130	1.2488	1.2664
(Tonnage)	163×10^6T	154×10^6T	152×10^6T
M + I + Inferred (F)			
($2\sigma_{kF}$)	1.7154	1.7661	1.7910
(Tonnage)	213 x 106T	212 x 106T	211 x 106T

Table 2 - Classification criteria of the KV limits (meter) method and cummulative tonnage (tonnes) for the Short Creek zone.

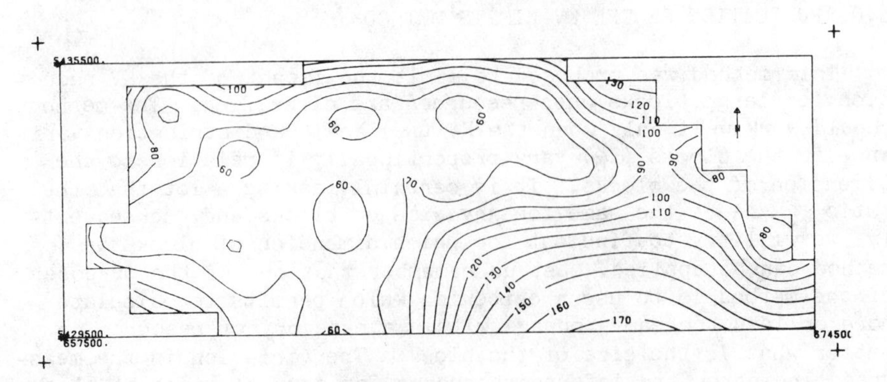

Figure 5. Isocurve map of twice the standard deviation (m x 100) of kriging variances for 1000 m x 1000 m blocks.

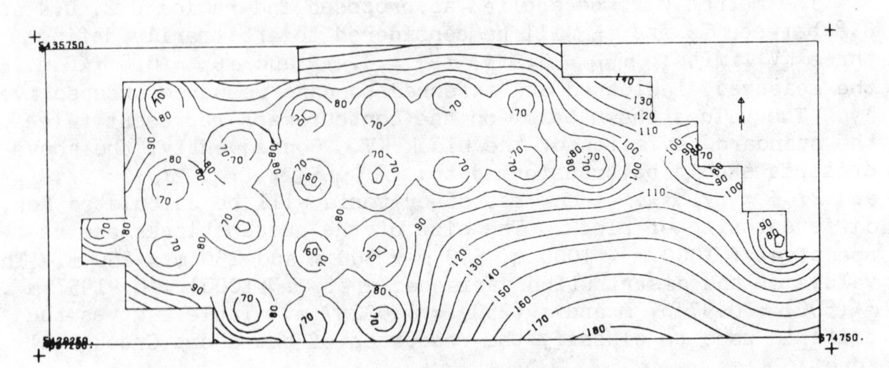

Figure 6. Isocurve map of twice the standard deviation (m x 100) of kriging variances for 500 m x 500 m blocks.

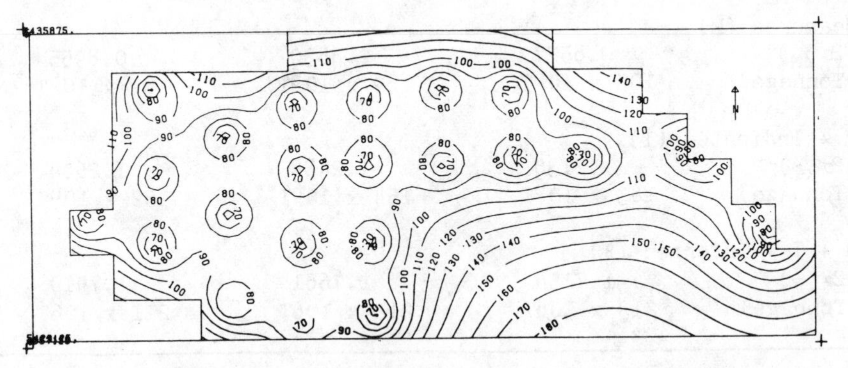

Figure 7. Isocurve map of twice the standard deviation (m x 100) of kriging variances for 250 m x 250 m blocks.

These criteria vary slowly with the size of the blocks. This can be explained by the large value of the range of the variogram compared to the dimensions of the blocks. Indeed σ_6^2 should vary more rapidly when the range of the variogram is small.

Figures 5, 6 and 7 give the contours for twice the standard deviation of the KV for each size of blocks. Figures 8, 9 and 10 show the local relative precision for the same size of blocks. Finally, Table 3 and Figure 11 give the cummulative estimated tonnage quantities for several classification criteria.

7.0 DISCUSSION

7.1 Traditional Method

As expected, kriging increases the zone of influence of clustered intersections compared to isolated ones. This phenomenon explains the large difference in tonnage between the measured resources obtained by the traditional and by the KV limits method. The traditional method does not take into account that a better estimation is provided when intersections occur at distances smaller than the range of the variogram. The classified measured tonnage by the KV limits method is almost three times as large as the measured resource tonnage of the traditional method. The indicated and inferred tonnages are also very different. Finally, by coincidence, the KV limits criteria give almost the same zone of influence for isolated intersections as the distances of influence used with the traditional method.

7.2 Relative Precision Method

The tonnage quantities classified by local relative precisions in Table 3, show that it is unwise, in this case, to use a local relative precision as small as 20 per cent for the measured resources. Which relative precisions should then be used for the classification of resources? Should these relative precisions be the same for all deposits?

The area covered by the same relative precision for smaller blocks is significantly reduced compared to the larger blocks. In addition, the relative precision vary with the fluctuations of the variable and this factor becomes critical especially when the variable decreases significantly in value. It can be seen, on Figures 8, 9 and 10, that the western part of the Short Creek zone will never be included in the RA resources, even though several intersections are present in this area.

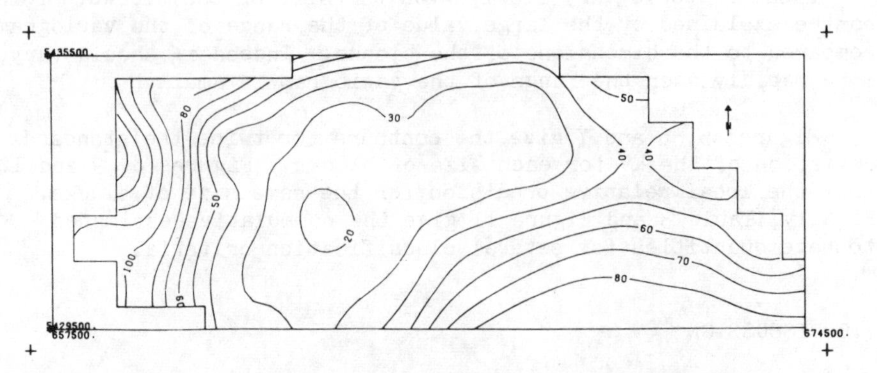

Figure 8. Isocurve map of twice the relative precision (x100) of kriging estimates for 1000 m x 1000 m blocks.

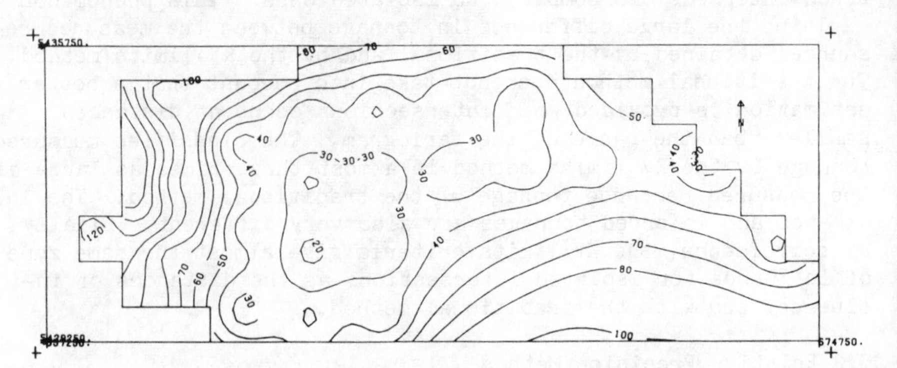

Figure 9. Isocurve map of twice the relative precision (x100) of kriging estimates for 500 m x 500 m blocks.

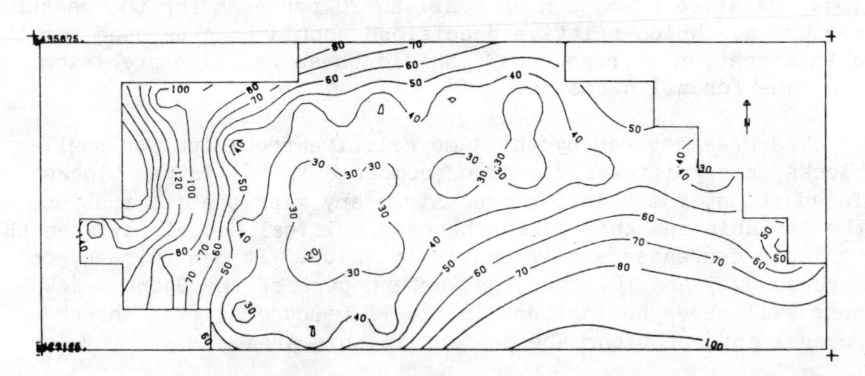

Figure 10. Isocurve map of twice the relative precision (x100) of kriging estimates for 250 m x 250 m blocks.

Use of a relative variogram could eliminate this drawback but this would imply the assumption that the variable is not stationary. Can we justify the use of a relative variogram when the absolute variogram is well defined? In any case, the use of a relative variogram would determine similar relative precision contour shapes as would the KV limits method with absolute KV.

7.3 KV Limits Method

Figure 11 shows that the chosen criteria for the same type of resources give different tonnages for different sizes of blocks. These tonnage differences are larger when the ratio σ_K^2/σ_b^2 is smaller.

Two reasons can explain these variations in tonnage. First, the error of estimation of a surface area increases with the dimension of the grid used to define that surface. The larger the size of the blocks, the more difficult it becomes to calculate with precision the surface area of, for example, the RA resource. This surface area will also be small when the KV limits are small. The size of the blocks could thus be changed with the type of resource to classify.

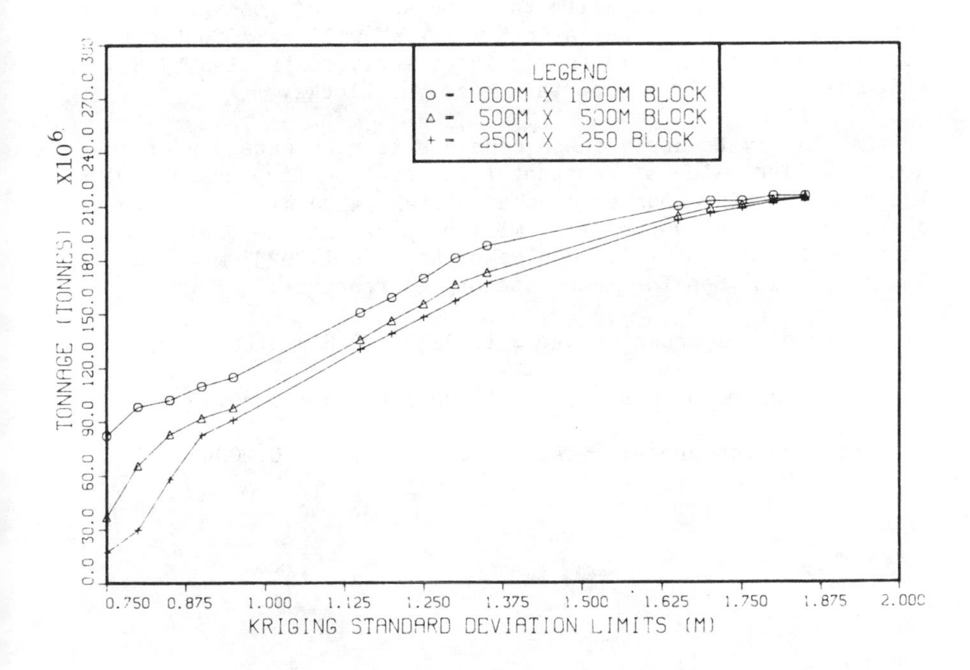

Figure 11. Cumulative tonnage for various KV limits and three sizes of blocks.

The second reason arises from the non-proportionality between the KV and the distribution variance which is critical for small ratios ($\sigma_k^2/\sigma_b^2 < 0.4$). Indeed, for small blocks, the KV is varying almost by the same amount as the distribution variance.

The KV is expressed theoretically by the following formula, when the mean value is known:

$$\sigma_k^2(Z) = D^2(Z) - \sum_{i=1}^{n} \lambda_i \sigma_{iZ} \qquad (4)$$

where $D^2(Z)$ is the distribution variance of the average (Z) thickness value of a block within the deposit, λ_i are the kriging weights for n intersections, and σ_{iZ} are the covariances between each intersection i and the block (Z). It can be verified that for small blocks the covariances σ_{iZ} will not vary too much and consequently the second term on the right of Eq. 4 is almost constant. The ratio σ_k^2/σ_b^2 can then be transformed to:

$$(\sigma_b^2 - A)/\sigma_b^2 \qquad (5)$$

which cannot vary proportionally.

If one wants to classify the same amount of tonnage for various sizes of blocks, the ratio $K = \sigma_k^2/\sigma_b^2$ will be affected by a corrective factor. This factor is directly controlled by Eq. 5 which is valid only for certain sizes of blocks.

Let us try to apply these concepts to each category of resources. The value of constant A in equation five changes with the category of resource and the initial value of σ_b^2. In this case, the initial value of σ_b^2 will be taken as the distribution variance for 250 m^2 x 250 m blocks, that is 1.00237 m^2. The value of A is then for each category of resource:

Measured resources: $A_M = 1.00237 \times 0.8 = 0.80190$

Indicated resources: $A_I = 1.00237 \times 0.6 = 0.60142$

Inferred resources: $A_F = 1.00237 \times 0.2 = 0.20047$

KV Limits	Blocks (1000 m x 1000 m)	Blocks (500 m x 500 m)	Blocks (250 m x 250 m)
0.75	82.2	36.9	17.1
0.80	98.4	65.6	29.8
0.85	102.2	83.0	58.1
0.90	109.8	91.9	82.6
0.95	114.9	97.9	91.0
1.15	150.6	135.7	130.4
1.20	159.1	146.1	138.9
1.25	169.7	155.4	148.1
1.30	181.0	166.2	157.0
1.35	187.8	173.0	166.8
1.65	209.7	204.2	201.7
1.70	212.6	208.5	205.6
1.75	212.6	210.3	208.8
1.80	215.3	213.1	212.0
1.85	215.3	214.5	213.9
Relative Precision Limits			
20%	7.5	1.9	0.5
40%	96.7	83.4	74.3
80%	193.3	190.0	188.2

Table 3. Cumulative tonnage (tonnes x 10^6) for various KV Limits and sizes of blocks

The corrected KV limits for the three sizes of block and for each category of resource are shown below on Table 4.

	Block (1000 m x 1000 m)	Block (500 m x 500 m)	Block (250 m x 250 m)
Measured (M)			
σ^2_{kM}/σ^2_b	0.1280	0.1773	0.2
$(2\sigma_{kM})$	0.6861	0.8313	0.8955
Tonnage	69 x 10^6T	76 x 10^6T	81 x 10^6T
M + Indicated (I)			
σ^2_{kI}/σ^2_b	0.3460	0.3830	0.4
$(2\sigma_{kI})$	1.1281	1.2219	1.2664
Tonnage	146 x 10^6T	150 x 10^6T	152 x 10^6T
M + I + Inferred (F)			
σ^2_{kF}/σ^2_b	0.7820	0.7943	0.8
$(2\sigma_{kF})$	1.6960	1.7598	1.7910
Tonnage	211 x 106T	211 x 106T	211 x 106T

Table 4. Corrected classification criteria of the KV limits method and cumulative tonnage for the Short Creek zone.

The results confirm the hypothesis that for relatively small sizes of blocks compared to the range of the spherical variogram, both σ_k^2 and σ_b^2 varies by almost the same amount. The theoretical demonstration will not be carried out in this paper.

From Table 4, it can be said that the maximum size of the block must be smaller for measured resources compared to M + indicated (I) or M + I + inferred (F) resources, if the difference in tonnage must stay within the confidence interval of the estimated total cumulative tonnage. The maximum sizes of the blocks for each category could be determined from a value of the distribution variance between constant A in Eq. 5, and the initial value σ_b^2.

8.0 CONCLUSION

This method of classification is at an early stage of development and further investigations are required. However, if one wants to use it as demonstrated in this article, the following recommendations should be considered:

1) The size of the blocks should be adjusted to each category of resources. The maximum size of blocks could be determined using a value of the distribution variance between constant A in equation five and the initial value σ_b^2 , for example: $A + 0.3\ (\sigma_b^2 - A)$. This distribution variance value would determine the maximum size of block for that resource category. The value 0.3 is chosen here as an example.

2) Corrective factors should be applied to the initial ratio σ_k^2/σ_b^2. These corrective factors are needed because the KV σ_k^2 and the distribution variance σ_b^2 vary approximately by the same amount for small blocks.

3) The initial ratios of 0.2, 0.4 and 0.8 for σ_k^2/σ_b^2 were arbitrarily chosen by the author to show some properties of the method. The tonnage (mainly in the measured and indicated resources categories) would be considerably reduced if the initial ratios would have been applied on points instead of blocks. In fact, no corrective factors could have been calculated if the nugget effect would be larger. To eliminate this problem, the initial ratio should never be applied to points because it involves using the nugget effect. Instead, the initial value of σ_b^2 should be equal to the sill value less the nugget effect, i.e., $\sigma_b^2 = C$. The above initial ratios could be legitimately used to define the three categories of resources.

References

(1) SABOURIN, R.L., Geostatistics as a tool to define various categories of resources: Mathematical Geology, Vol. 15, no. 1, pp. 131-143, 1983.
(2) ROYLE, A.G., How to use geostatistics in ore classification: World Mining, pp. 52-56, 1977.
(3) CARLIER, A., Contribution aux methodes d'estimation des gisements d'uranium: Commissariat à l'Energie Atomique, pp. 52-56, 1977.
(4) DIEHL, P., and DAVID, M., Classification of ore reserves/resources based on geostatistical methods: CIM Bulletin, pp. 127-136, February 1982.
(5) FROIDEVAUX, R., Geostatistics and ore reserve classification: CIM Bulletin, pp. 77-83, July 1982.
(6) IRVINE, J.A., WHITAKER, S.H. and BROUGHTON, P.L., Coal resources of southern Saskatchewan: A model for evaluation methodology; Part I-II: Ec. Geol. Report 30, Geological Survey, Energy, Mines and Resources Canada, 1978.

CONDITIONAL BIAS IN KRIGING AND A SUGGESTED CORRECTION

M. David, D. Marcotte, M. Soulié

Ecole Polytechnique de Montréal,
C.P. 6079, Succursale"A", Montréal, Québec, Canada.

Abstract : Kriging has long been said to be "usually" conditionaly unbiased. The relation between Z and Z* will be carefully considered and a solution to improve the conditional bias when the average is unknown or the distribution is not normal will be proposed. This method, called KSK, contains as two special cases simple kriging and ordinary kriging, which thus appear to be particular cases of the same method. Simulations of common types of mineral deposit have been made to verify that the bias can be corrected. In addition it is found that in the case of an arbitrary highly skewed distribution, relative variograms are satisfactory and lognormal kriging can be extremely biased.

I - INTRODUCTION

Kriging has now been in existence for twenty years (Matheron, 1963) and it is still widely known as a minimum variance estimator. From a mining point of view however, this is not its essential property. Ten years ago (David, 1972; David, 1973) discussions about conditional distribution and conditional bias started to appear. The fact that the cover of Journel and Huijbregts (1978) illustrates the bivariate distribution of the real and estimated values is probably not coincidental. Miners will consider that an estimation method is good if at the time of mining they find, - in average - what they were hoping for. More than often they don't. If Z_V is the real grade of block V and Z_V^* is the kriged grade of block V, Z_V^* is conditionaly unbiased if and only if the distribution of Z_V is normal and $E(Z_V)$ is known. A variety of kriging methods were investigated in Rendu (1979) and the bias was clearly illustrated. What this

G. Verly et al. (eds.), Geostatistics for Natural Resources Characterization, Part 1, 217–230.

paper will try to do is to cope with the bias and get rid of it as much as possible. Using simulated deposits, a number of situations will be reviewed, covering normal, lognormal and highly skewed distributions. Simple, ordinary and lognormal kriging will be considered using variograms, relative variograms or lognormal variograms. The impact of the bias on reserve estimation will be reviewed and a solution proposed.

The main remark used in this paper will be that ordinary kriging is equivalent to simple kriging performed with the optimally estimated mean (Matheron, 1971). As for computer reasons ordinary kriging is performed with a small number of points, it means that it is equivalent to having estimated the local mean with the same small number of points and then having done the simple kriging. We will see what improvement can be gained simply by getting a better estimate of the mean with a larger number of samples. It will be seen that this surprisingly simple tool does get rid of most of the bias. Other possible avenues to correct the bias will also be discussed at the end.

II - THE NORMAL CASE

II - 1 - Simple kriging with known average (KS)

Let Z^*_{vKS} be, the simple kriging estimator:

(1) $Z^*_{vKS} = m + \sum_i \gamma_i (Z_i - m) = (1 - \sum_i \gamma_i)\, m + \sum_i \gamma_i Z_i$

$m = E(Z)$ and the γ_i's are solution of

(2) $\sum_{j=1}^{n} \gamma_j \operatorname{cov}(Z_i, Z_j) = \operatorname{cov}(Z_i, Z_v) \quad i = 1,\ldots.n$

Z_v and Z^*_{vKS} have a binormal joint distribution.

Hence, using properties of the bivariate normal distribution, it follows that

$$E(Z_v | Z^*_{vKS}) = m + \frac{\operatorname{cov}(Z_v, Z^*_{vKS})}{D^2_{Z^*_{vKS}}} (Z^*_{vKS} - m)$$

where $D^2_{Z^*_{vKS}}$ is the dispersion variance of Z^*_{vKS} or considering (2)

(3) $E(Z_v | Z^*_{vKS}) = Z^*_{vKS}$

Z^*_{vKS} is conditionally unbiased.

The estimation variance σ^2_{KS} is then

(4) $\sigma^2_{KS} = D^2_{Z_v} - \sum_i \gamma_i \operatorname{cov}(Z_v, Z_i)$

II - 2 - Ordinary kriging with unknown average (KO)

The ordinary kriging estimator is

(5) $Z^*_{vKO} = \sum_i \gamma_i Z_i$

where γ_i 's are solution of

$$(6) \begin{cases} \sum_j \gamma_j \operatorname{cov}(Z_i, Z_j) - \mu = \operatorname{cov}(Z_i, Z_v) & i = 1, n \\ \sum_j \gamma_j = 1 \end{cases}$$

Again Z_v and Z^*_{vKO} have a binormal joint distribution. It follows, considering (5) and (6) that

$$(7) \quad E[Z_v | Z^*_{vKO}] = m + \left(\frac{D^2_{Z^*_{vKO}} - \mu}{D^2_{Z^*_{vKO}}}\right)(Z^*_{vKO} - m)$$

as $\mu \neq 0$, $E[Z_v | Z^*_{vKO}] \neq Z^*_{vKO}$

In average, as Rendu (1979) pointed out ordinary kriging overestimates high values and underestimates low values.

II - 3 - Estimating the unknown average (KSK)

If $\hat{m}$ is an estimator of unknown m, obtained by ordinary kriging, then

$$(8) \quad Z^*_{vKO} = (1 - \sum_{i=1}^{n} \lambda_i)\hat{m} + \sum_{i=1}^{n} \lambda_i Z_i$$

with $\hat{m} = \sum_{j=1}^{n} \alpha_j Z_j$ subject to $\sum \alpha_j = 1$

As usually, n is small, a better estimate Z^*_{vKSK} can be obtained with a better estimate of $\hat{m}$ using a larger number N of samples, or a set of better distributed samples, in which case this estimation of $\hat{m}$ is only done once. (i.e. the problem would not exist if we had unlimited computing capacity).

The estimation variance of this Z^*_{vKSK} is now (with $a = 1 - \sum_i^n \lambda_i$; app. 2)

$$(9) \quad E[(Z_v - Z^*_{vKSK})^2] = \sigma^2_{KS} + a^2\sigma^2_{\hat{m}} + 2a[\operatorname{cov}(\sum_i \lambda_i Z_i, \hat{m}) - \operatorname{Cov}(Z_v, \hat{m})]$$

The last term vanishes if the same data set is used to estimate the mean and the block value (case of KO).

If $\hat{m}$ is a good estimate (N large) the the last two terms are very small compared to σ^2_{KS} and we may write the relation: $\sigma^2_{KS} < \sigma^2_{KSK} < \sigma^2_{KO}$; using properties of the normal distribution (app.3)

$$(10) \quad E[Z_v | Z^*_{vKSK}] = m + \frac{\operatorname{cov}(Z_v, Z^*_{vKSK})}{D^2_{Z^*_{vKSK}}}(Z^*_{vKSK} - m)$$

where $\operatorname{cov}(Z_v, Z^*_{vKSK}) = D^2_{Z^*_{vKSK}} - a^2\sigma^2_{\hat{m}} - a[2\operatorname{cov}(\sum_i \lambda_i Z_i, \hat{m}) - \operatorname{Cov}(Z_v, \hat{m})]$

The last two terms being small relatively to $D^2_{Z^*_{vKSK}}$, the bias is small and decreases as the precision of the estimate of the mean increases with special cases being ordinary kriging and simple kriging. Comparing equations (7) and (10) leads to

(11) $\mu = a\sigma^2_{\hat{m}}$

This relation was derived differently by Matheron (1971). It expresses the Lagrange multiplier μ as a fonction of the imprecision on the estimation of the mean with again, as limiting cases, ordinary kriging and simple kriging.

III - SIMULATION AND COMPARISON OF KO, KS AND KSK

III - 1 - Normal case

For each distribution N(50,25) corresponding for instance to an iron ore deposit, investigated, 28 "deposits" of 10 000 points (100 x 100) regularly spaced on a 5 x 5 m grid have been simulated. A spherical model has been used with parameters $C_o = 5$, $C = 20$, $a = 100$. For each simulation one point has been selected at random in each block 50 x 50, providing 100 samples each time. This was repeated 5 times for each simulation. The "real" value of each block 50 x 50 is obtained as the arithmetic average of the 100 points it contains. The Z^*_{vKS}, Z^*_{vKO}, and Z^*_{vKSK} are obtained for each block in each of the 28 x (5 x 100) = 14000cases available.

The regressions of Z_v on each of these values are then calculated. Average results are presented in table 1, where ε is the estimation error, b_o is the intercept at the origin for $E[Z_v|Z^*_v]$ and b_1 is the slope for the same. As expected and well known simple kriging is better than ordinary kriging, but the interesting result is that simple kriging, using an estimated mean obtained once for the whole deposit, is almost equivalent to simple kriging with a really known average.

Table 1

	KO	KS	KSK
Error	0.0077	0.0040	0.0068
Absolute error	1.705	1.665	1.678
Squared error	4.659	4.437	4.502
b_0 global*	3.279	-0.2712	0.1285
b_1 global	0.9346	1.0055	0.9976
b_0 average*	2.983	-0.477	-0.16
b_1 average	0.941	1.010	1.003
r (correlation)	0.7905	0.7999	0.7965

* global means that 1 regression was computed on 14000 points.
* average means that the average of 28 regressions was computed.

Another way to analyse the results is presented in table 2 where the number of better performances (out of 28 simulations) is calculated for each method versus the other.

Table 2

Absolute error

	KO	KS	KSK
KO		1	3
KS	27		28
KSK	25	0	

Squared error

	KO	KS	KSK
KO		1	3
KS	27		28
KSK	25	0	

Slope

	KO	KS	KSK
KO		14	14
KS	14		11
KSK	14	17	

Table 2 confirms that KSK is superior to KO for the precision of the estimation, but there is a tie for the slope; we can see however the comparable performance of KS and KSK and table 2 confirms the strong advantage of these two over KO.

III - 2 - The lognormal case

The previous "good" results are always more or less expected in the case of a normal distribution; the lognormal case is by far much more common. This is where bigger differences are expected. Data sets mimicking gold or uranium values have been simulated. The average is 0.12 (0.12 oz/t or 0.12% U_3O_8) and the standard deviation is 0.24, or a coefficient of variation of 2. The variogram is again isotropic spherical with $C_o = 0.0198$, $C = 0.0378$, $a = 100$. Here 20 simulations have been made and each time, five random stratified sampling have been made on a grid 100 x 100, the original grid being again 10000 points on a 5 x 5 grid.

The same comparison as before have been made and are presented in table 3 and 4. This time the bias of ordinary kriging is severe, showing an average slope of only 0.857. Simple kriging is almost perfect (0.965) and our simple kriging with estimated mean emulates this result very well (0.964). It is worth remembering that the situation investigated is particularly difficult (coefficient of variation of 2) but frequent in the industry. Hence this is considered a very valuable result.

	KO	KS	KSK
ε	-0.002	-0.003	-0.002
ε^2	0.0106	0.0096	0.0098
$\|\varepsilon\|$	0.0587	0.0562	0.0571
b_0	0.018	0.005	0.006
b_1	0.857	0.965	0.964

Table 3: Global comparison on 100 data sets, of KO, KS and KSK

	KO	KS	KSK
KO		19	29
KS	79		48
KSK	71	49	

Absolute error

	KO	KS	KSK
KO		8	18
KS	86		64
KSK	72	8	

Squared error

	KO	KS	KSK
KO		32	32
KS	68		50
KSK	68	48	

Slope

Table 4: Paired comparison of KO, KS and KSK (best is read horizontally; 100 data sets)

In this case, the distribution being lognormal, the regression line does not give $E[Z_v|Z_v^*]$ but it is the least square best approximation. Again here, Z_{vKS}^* and of more practical importance Z_{vKSK}^* provide results which are much better than ordinary kriging.

III - 3 - Case of a highly skewed arbitrary distribution

We then proceeded to simulate a non lognormal highly skewed distribution which resembles those encountered in the high grade uranium deposits of Saskatchewan. The histogram can be seen on figure 1. The average is 0.82% and the standard deviation is 1.814%. This distribution has been normalized and a spherical variogram with a range of 100 m and a C_o/C ratio of 0.20 has been imposed to the normalized data. These data have been inversely transformed and 4 selections of 100 samples have been selected according to a random stratified grid. The actual variograms (absolute, relative, logarithmic) have been estimated and models have been adjusted. Figure 2 shows these variograms, they are highly variable from one selection to the other! In each case kriging has been done using these infered variograms rather than a theoretical one which we don't know any way. Now in addition to comparing ordinary kriging and our kriging with an estimated mean, we have the opportunity to compare the performance of the different variogram models. When the logarithmic variogram is used, logarithmic kriging is done. The equations for logarithmic kriging with an estimated mean (obtained by ordinary kriging) did not seem to be available and have been derived (Appendix 4). Kriging has been used to estimate large blocks (50 m x 50 m). The errors and regression parameters of Z_v on Z_v^* are shown in table 5.

- Looking at table 5, we can see as a by product that working with the relative variogram provides the smallest absolute error and sum of square. For the conditional bias, working with logarithms seems to give a better slope. Our main purpose, comparing KO and KSK, shows again that whatever the variogram, KSK is better than KO. Although the slope is not as well corrected as in the case of the well behaved distributions which we considered before. We will discuss later another possible way to correct the conditional bias further.

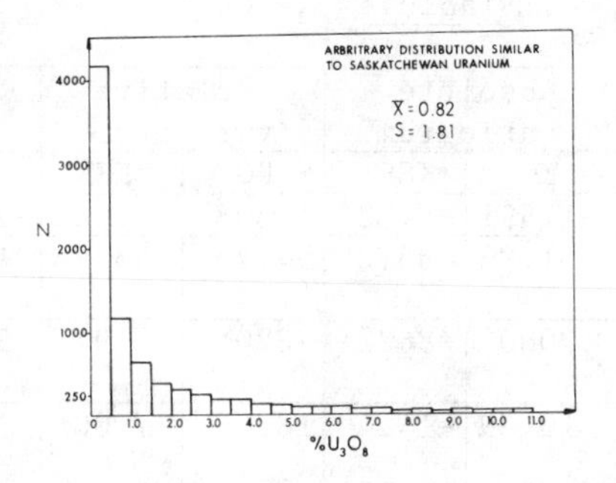

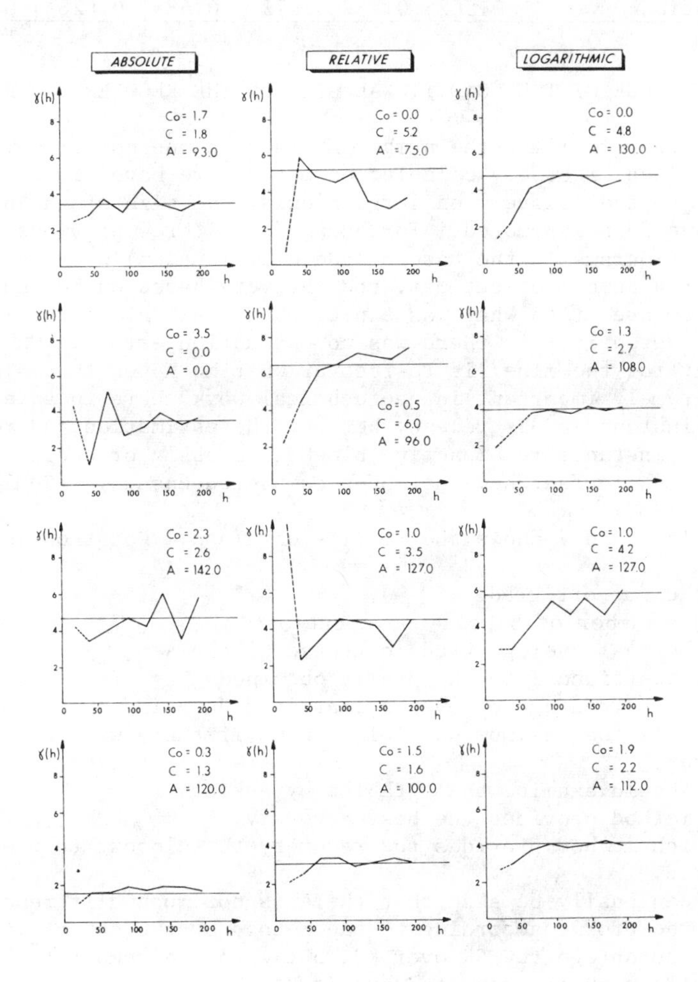

FIGURE 2: Variograms found for the arbitrary distribution

Table 5

Variogram	Absolute variogram		Relative variogram		Logarithmic variogram	
Method	KO	KSK	KO	KSK	KO	KSK
Error x 10^2	-1.591	-4.328	1.308	-1.936	-13.24	-22.40
Absolute error x 10^2	61.93	63.28	54.73	53.78	62.32	67.47
Sum of square x 10^4	9000	8622	7820	7191	14148	19095
Intercept b_0 x 10^2	10.39	-14.40	18.96	7.00	32.22	33.00
Slope b_1	0.856	1.085	0.806	0.914	0.965	1.021
Correlation	0.562	0.562	0.672	0.685	0.628	0.613

IV - INFLUENCE OF THE CONDITIONAL BIAS ON THE RECOVERY OF RESERVES

In order to limit the number of factors and not to have to choose a local recovery calculation method, we have limited our exercise to the recovery on large blocks, the size of the blocks which have been estimated. For each of the three previous cases (and 3 variograms in the case of the distribution) we have computed for a number of cut-off, the recovery based on KO and KSK and we compare it to what was expected, as well as to the maximum possible potential (if there was no estimation error). What is important for the miner is to recover what he hopes for. This is also extremely important in geotechnical work where in an earth dam for instance a discrepency between the estimation and reality means for instance recompacting blocks uselessly or accepting blocks which in fact have not been compacted enough. (Soulié and al., 1983).

Table 6 and 7 shows the results obtained. Notation is as follows:

t_c is the cut-off grade
T is the number of blocks above cut-off
t is the real average grade of these blocks
P is proportionnal to the profit obtained ($P = T(t - t_c)$)
% is the percentage of overestimation made using KO or KSK
Potential is the maximum possible profit if there was no estimation error.

We should examine these tables by asking:

- which method provides the best recovery
- and which method provides the recovery the closest to what was expected.

It can easily be seen that there is not much difference between the profit according to one method or the other with a marginal advantage to KSK over KO, but when one compares the profit made with the expected profit K.S.K is always highly superior, i.e. it gives less surprises. This is especially true when the distribution become worse and the cut-off is high. We can also see as a by product that the lognormal kriging applied to non lognormal data is a total disaster and that surprisingly the absolute variogram is not so bad, but from the point of view of the least discrepency between expected results and results achieved, the relative variogram proves the best if KO is used.

	KO				KSK				
t_c	T	t	P	%	T	t	P	%	Potential
1°- 28 simulations: Normal distribution (50,25) (14000 blocks of 50 m)									
45	13354	50.28	70482	0.3	13472	50.24	70553	-0.1	71763
47	11843	50.78	44624	1.5	12018	50.73	44779	0.0	47367
48	10497	51.17	33286	2.5	10635	51.14	33405	0.0	36669
49	8818	51.66	23474	4.0	8921	51.65	23596	0.0	27383
50	6956	52.23	15491	6.7	6982	52.24	15626	0.2	19621
2°- 20 simulations; Lognormal distribution m = 0.12, σ/m = 2.0 (10000 blocks)									
.02	9920	.117	962	2.6	9984	.116	962	2.6	965
.03	9524	.120	861	3.4	9909	.117	862	2.9	875
.05	7995	.134	673	5.8	8855	.126	676	3.2	720
.07	6149	.155	521	9.8	6802	.147	527	2.5	599
.10	4059	.188	356	18.3	4149	.187	362	4.5	461

Table 6: Recovery for the normal and lognormal case

	KO				KSK				
t_c	T	t	P	%	T	t	P	%	Potential
1°- Absolute variogram									
.10	363	.89	288*	3.0	399	.83	290	5.8	294
.20	310	.99	246*	7.0	386	.85	250	6.8	266
.40	231	1.13	169*	24.3	337	.94	183	6.4	225
.60	187	1.33	136	23.8	248	1.13	131	3.8	193
2°- Relative variogram									
.10	325	.97	283*	2.1	376	.87	289*	3.6	294
.20	275	1.09	244*	6.0	336	.94	249*	5.9	266
.40	192	1.32	176	20.8	231	1.20	186	11.2	225
.60	159	1.53	147	21.1	156	1.55	149	14.2	193
3°- Logarithmic variogram									
.10	337	.95	286*	20.7	344	.93	287*	32.8	294
.20	251	1.20	250	26.5	267	1.14	251	39.8	266
.40	190	1.45	200	36.7	191	1.44	199	53.6	225
.60	143	1.71	159	52.5	157	1.60	157	72.7	193

* The "no selection" option gives better profit than the above.

Table 7: Recovery for the skewed distribution

V - ANOTHER POSSIBLE METHOD OF CORRECTING THE BIAS

Two other methods had been envisaged to correct the bias and then dropped in view of the success of the KSK.

Each time a point or block is estimated by kriging the slope of the regression $E[Z_V|Z^*]$ can be computed as follows. It has been seen that in the normal case the slope is $(D^2(Z^*) - \mu)/D^2(Z^*)$. It is known that (David, 1972), $D^2(Z^*) = \sigma_Z^2 - \sigma_K^2 + 2\mu$, hence the slope is available for each point. Estimating the mid point through which this line goes can be done by solving the same kriging system with O'S on the right hand side, hence a correction is immediately available.

A second approximate method could be to use the cross validation technique; reestimating each known sample from its neighbour, plotting the resulting estimated versus the actual values and computing the regression line. Then kriging would be done point by point and each point would be corrected according to the previous line. Corrected points would then be grouped to obtain blocks of the desired size. Correction would be approximate only as all points are in different positions respective to the sample points, hence the bias is different, and the line obtained in the cross validation process would apply to points which are in the worst possible situation, very far from the sample points.

VI - CONCLUSION

It then seems that the conditional bias can be avoided in the case of a non normal distribution, and when the average is unknown, by using a two step kriging, estimating first a local average with a number of points larger than which will be used in the kriging, and then performing a simple kriging using this estimated mean. It thus appears that the conditional bias is linked to the number of points used to perform the kriging and it can be hoped that as computing power increases, this problem will be solved by itself. In the mean time, a method is suggested which reduces the discrepency between expected reserves and mineral reserves. We have also verified that a relative variogram performs well in the case of a highly skewed distribution where the use of lognormal kriging only leads to disasters. As a by product the equations of lognormal kriging with an estimated average have been obtained.

ACKNOWLEDGEMENTS

This research has been supported by grant 7035 of the National Research Council of Canada. Computer time has been supplied by University of Montreal and Geostat Systems International has supplied the necessary software.

APPENDIX

1 - Preambule

The simple kriging (K.S.) system is:

$$\sum_{i}^{n} \lambda_i \operatorname{Cov}(Z_i, Z_j) = \operatorname{Cov}(Z_v, Z_j) \qquad \sigma^2_{KS} = D^2_{Z_v} - \sum_{i}^{n} \lambda_i \operatorname{Cov}(Z_v, Z_i)$$

The system to estimate the average $\hat{m}$ by ordinary kriging is:

$$\sum_{i}^{N} \gamma_i \operatorname{Cov}(Z_i, Z_j) = \mu_o$$

$$\sum_{i}^{N} \gamma_i = 1$$

and we have

$$\sigma^2_K = \sigma^2_{\hat{m}} = \mu_o$$

The final estimation Z^*_{KSK} is:

$$Z^*_{KSK} = a\,\hat{m} + \sum_{i} \lambda_i Z_i \qquad \text{where } a = 1 - \sum_{i}^{n} \lambda_i$$

2 - Derivation of formula 9

$$\begin{aligned}
\sigma^2_{KSK} &= D^2_{Z_v} - 2 \operatorname{Cov}(Z_v, Z^*_{vKSK}) + D^2_{Z^*_{vKSK}} \\
&= D^2_{Z_v} - 2 \sum_{i} \lambda_i \operatorname{Cov}(Z_v, Z_i) + \sum_{i}\sum_{j} \lambda_i \lambda_j \operatorname{Cov}(Z_i, Z_j) \\
&\quad - 2\,a \operatorname{Cov}(Z_v, \hat{m}) + a^2 \sigma^2_{\hat{m}} + 2\,a \sum_{i} \lambda_i \operatorname{Cov}(Z_i, \hat{m}) \\
&= \sigma^2_{KS} + a^2 \sigma^2_{\hat{m}} + 2a\left(\operatorname{Cov}\left(\sum_{i} \lambda_i Z_i, \hat{m}\right) - \operatorname{Cov}(Z_v, \hat{m})\right)
\end{aligned}$$

3 - Derivation of formula 10

The slope of the regression line is:

$$\frac{\mathrm{Cov}(Z_v, Z^*_{vKSK})}{D^2_{Z^*_{vKSK}}}$$

But $D^2_{Z^*_{vKSK}} = \sum_i \sum_j \lambda_i \lambda_j \ \mathrm{Cov}(Z_i, Z_j) + a^2 \sigma^2_{\hat{m}} + 2a \ \mathrm{Cov}(\sum_i \lambda_i \ Z_i, \hat{m})$

$$= \sum_i \lambda_i \ \mathrm{Cov}(Z_v, Z_i) + a^2 \sigma^2_{\hat{m}} + 2a \ \mathrm{Cov}(\sum_i \lambda_i \ Z_i, \hat{m})$$

$$= \mathrm{Cov}(Z_v, Z^*_{vKSK}) + a^2 \sigma^2_{\hat{m}} + a[2\mathrm{Cov}(\sum_i \lambda_i Z_i, \hat{m}) - \mathrm{Cov}(Z_v, \hat{m})]$$

4 - Lognormal KSK

Let the estimator Y^*_{vKSK} be obtained from the logarithm of data $(Z_i, \ i = 1....n)$

$$Y^*_{vKSK} = (1 - \sum_i \lambda_i) \ \hat{m} + \sum_i \lambda_i \ Y_i + \frac{\sigma^2_Y}{2} - \frac{D^2_{Y_v}}{2}$$

where $\hat{m}$ = estimator of the mean of logarithm of point values.

i.e. $\hat{m} = \sum_i \gamma_i Y_i$ with $\sum_i \gamma_i = 1$

and the γ_i 's obtained are solution of

$$\sum_i \gamma_i \ \mathrm{Cov}(Y_i, Y_j) = \mu_o = \sigma^2_{\hat{m}} \qquad j = 1.....n.$$

The variance of this estimator is

$$D^2_{Y^*_{vKSK}} = a^2 \sigma^2_{\hat{m}} + 2a \ \mathrm{Cov}(\sum_i \lambda_i \ Y_i, \hat{m}) + \sum_i \sum_j \lambda_i \lambda_j \ \mathrm{Cov}(Y_i, Y_j)$$

$$= D^2_{Y_v} - \sigma^2_{KS} + a^2 \sigma^2_{\hat{m}} + 2a \ \mathrm{Cov}(\sum_i \lambda_i \ Y_i, \hat{m})$$

This estimator is a normal variable, hence,

$$E[\exp(Y^*_{vKSK})] = \exp(m + \frac{\sigma^2_Y}{2} - \frac{\sigma^2_{KS} + a^2 \sigma^2_{\hat{m}}}{2} + a \ \mathrm{Cov}(\sum_i \lambda_i \ Y_i, \hat{m}))$$

$$= M \exp(- \frac{\sigma^2_{KS} + a^2 \sigma^2_{\hat{m}}}{2} + a \ \mathrm{Cov}(\sum_i \lambda_i \ Y_i, \hat{m}))$$

where M is the mean of the lognormal variable hence

$$\exp[Y^*_{vKSK} + \frac{\sigma^2_{KS} - a^2\sigma^2_{\hat{m}}}{2} - a\,\mathrm{Cov}(\sum_i \lambda_i Y_i, \hat{m})] = Z^*_{vKSK}$$

is unbiased.

This result should be compared with the expressions for simple lognormal kriging and ordinary lognormal kriging given in Journel (1978). Again KSK can be viewed as the general expression for which KS and KO are special cases.

Now, the variance of the kriging error is

$$E[(Z_v - Z^*_{vKSK})^2] = D^2_{Z_v} - 2\,\mathrm{Cov}(Z_v, Z^*_{vKSK}) + D^2_{Z^*_{vKSK}}$$

with

$$D^2_{Z_v} = M^2\,[\exp(D^2_{Y_v}) - 1]$$

$$D^2_{Z^*_{vKSK}} \simeq M^2\,[\exp(a^2\sigma^2_{\hat{m}} + D^2_{Y_v} - \sigma^2_{KS} + 2a\,\mathrm{Cov}(\sum_i \lambda_i Y_i, \hat{m})) - 1]$$

and using Rendu's approximation (1979) as pointed out by Dowd (1983)

$$\mathrm{Cov}(Z_v, Z^*_{vKSK}) = M^2[\exp(a\,\mathrm{Cov}(Y_v, \hat{m}) + \sum_i \lambda_i \mathrm{Cov}(Y_v, Y_i)) - 1]$$

$$= M^2[\exp(a\,\mathrm{Cov}(Y_v, \hat{m}) - \sigma^2_{KS} + D^2_{Y_v}) - 1]$$

Thus the kriging error may be written

$$\sigma^2_{z^*_{vKSK}} = M^2\exp(D^2_{Y_v})\{1+\exp(-\sigma^2_{KS})[\exp(a^2\sigma^2_{\hat{m}}+2a\mathrm{Cov}(\sum_i \lambda_i Y_i, \hat{m}))$$

$$- 2\exp(a\mathrm{Cov}(Y_v, \hat{m}))]\}$$

Again, if we compare this expression with the expressions (35) and (45) presented in Dowd (1982), we see that KSK leads to a general expression for which KO and KS are particular cases.

This expression also shows that the better the estimate of the mean ($\sigma^2_{\hat{m}}$ small) the smaller the kriging error.

NOTE: (if we don't use Rendu's approximation, there is no simple expression either for $\sigma^2_{z^*_{vKO}}$ or $\sigma^2_{z^*_{vKS}}$ or $\sigma^2_{z^*_{vKSK}}$ and therefore comparison is impossible to make).

REFERENCES

David, M., 1972. Grade-tonnage curve: use and misuse in ore reserve estimation. I.M.M., vol. 81, July 1972, pp. 129-131.
David, M., 1973. Tools for planning, variances and conditional simulations. XIth International Symp. on Computer Application in the Mineral Industry, pp. D10-D23.
Dowd, P., 1982. Lognormal Kriging - General Case. I.A.M.G. Journal, Vol. 14, no. 5, pp. 475-500.
Journel, A. and Huijbregts, C., 1978. Mining Geostatistics. Academic Press, 600p.
Matheron, G., 1963. Principles of Geostatistics. Economic Geology, 58, 1246-1266.
Matheron, G., 1971. The theory of Regionalized Variables and its Applications. Les Cahiers du Centre de Morphologie Mathématique. Fasc. 5, CG Fontainebleau.
Rendu, J.M., 1979. Kriging, Logarithmic Kriging and Conditional Exceptation: Comparison of theory with Actual Results, 16th APCOM, AIME ed., pp. 199-212.
Soulié, M., Favre, M., et Konrad, J.M., 1983. Analyse géostatistique d'un noyau de Barrage tel que construit. To appear in Canadian Geotechnical Journal, Vol. 20, No. 3.

POSITIVE KRIGING

Randal J. Barnes and Dr. Thys B. Johnson

Mining Engineering Department
The Colorado School of Mines
Golden, Colorado, U.S.A., 80401

ABSTRACT

To eliminate the serious potential problem caused by negative estimation weights, a modified mathematical formulation of Kriging is presented. The necessary and sufficient conditions for a unique optimal solution, and an efficient and easily programmed computational algorithm are developed.

1. INTRODUCTION TO THE PROBLEM

A serious practical problem which has been categorically ignored by geostatistical theoreticians is that of negative Kriging weights. The Ordinary Kriging estimation procedure calculates the set of weights which minimizes the estimation variance subject to the constraint that the sum of the weights must equal one (an unbiased condition). Kriging does not place any restrictions on the sign of the resulting weights; subsequently, generated weights may often be negative.

The presence of negative weights yields several unsatisfactory results. Negative weights can produce negative estimated grades, an intolerable outcome. Negative weights can produce estimated grades greater than the highest sample value, a very dangerous outcome. Also, negative weights can cause highly erratic estimated grades: slight changes in locations can bring about substantial changes in estimates.

G. Verly et al. (eds.), Geostatistics for Natural Resources Characterization, Part 1, 231–244.

Schaap and St. George [16] addressed the problem of negative weights but did not tender a workable solution. Presented in this paper is a mathematical formulation for the elimination of negative weights, the necessary and sufficient conditions for an optimal solution, and an efficient and easily programmed computational algorithm. The problem formulated and its solution algorithm will be termed Positive Kriging.

2. LINEAR KRIGING AS A QUADRATIC PROGRAM

In the field of Operations Research, the problems which involve the minimization of a quadratic form subject to linear constraints are called Quadratic Programming problems. Historically, the first ventures into the theory of nonlinear programming were through quadratic programming. Subsequently, many methods for solving quadratic programming problems have been published. Kunzi and Krelle [9] discuss seven of them, and since 1961 much additional work has been carried out. In short, quadratic programming is a class of problems which has been heavily investigated, and is well understood within the field of Operations Research.

The problem of Ordinary Kriging (OK) can be stated as minimizing the estimation variance (a quadratic form), subject to the unbiased condition (a single linear constraint) [6]. Thus, from an Operations Research point of view, OK is nothing more than a simple quadratic programming problem. The OK problem can be written in a matrix equation form as

$$\text{Minimize } q(\underline{x}) = \sigma_v + \underline{x}'\underline{\Sigma x} - 2\underline{\sigma}'\underline{x} \qquad \{2.1a\}$$

$$\text{Subject to: } \underline{1}'\underline{x} = 1 \qquad \{2.1b\}$$

Where:

The underline indicates a matrix and the prime notation (') inidicates a matrix transpose.

$q(\underline{x})$ is the quadratic form representing the estimation variance.

$\underline{x}$ is the Nx1 dimensional matrix (vector) of unknown kriging weights.

σ_v is the predetermined block variance for the volume of material being estimated.

$\underline{\Sigma}$ is the symmetric, positive definite, NxN dimensional matrix of known sample to sample covariances.

$\underline{\sigma}$ is the Nx1 dimensional matrix (vector) of known sample to block covariances.

$\underline{1}$ is an Nx1 dimensional matrix (vector) of numeral ones.

As has been demonstrated in many geostatistical references (e.g. [3], [6], and [15]) the solution of the OK problem can be found using the classical method of Lagrange multipliers from the calculus of variation. Letting the variable "2λ" be the Lagrange multiplier associated with the single constraint, {2.1b} , necessary and sufficient conditions for a unique optimal set of weights are that the weights and the Lagrange multiplier must satisfy the OK system:

$$\underline{\Sigma x} + \underline{1}\lambda = \underline{\sigma} \qquad \{2.2a\}$$

$$\underline{1}'\underline{x} = 1 \qquad \{2.2b\}$$

These conditions may be written more conveniently in a partitioned matrix form [4] as

$$\begin{bmatrix} \underline{\Sigma} & \underline{1} \\ \underline{1}' & 0 \end{bmatrix} \begin{bmatrix} \underline{x} \\ \lambda \end{bmatrix} = \begin{bmatrix} \underline{\sigma} \\ 1 \end{bmatrix} \qquad \{2.3\}$$

Written in this form, it can be seen that the solution of the Ordinary Kriging system can be calculated using a simple matrix inversion approach (or the computationally more correct Gaussian elimination with back substitution). Since the known sample to sample covariance matrix, $\underline{\Sigma}$, is positive definitive, the inverse of $\underline{\Sigma}$ necessarily exists and thus,

$$\begin{bmatrix} \underline{\Sigma} & \underline{1} \\ \underline{1}' & 0 \end{bmatrix}^{-1} \qquad \{2.4\}$$

is guaranteed to exist.

The Positive Kriging problem can be stated as finding the set of <u>non-negative</u> weights which minimizes the estimation variance (a quadratic form), subject to the unbiased condition (a linear constraint). Thus, Positive Kriging is again nothing more than a quadratic programming problem (it is, however, more complicated than the OK problem). The Positive Kriging problem can be written in a matrix equation form as

$$\text{Minimize } q(\underline{x}) = \sigma_V + \underline{x}'\underline{\Sigma x} - 2\underline{\sigma}'\underline{x} \qquad \{2.5a\}$$

$$\text{Subject to: } \quad \underline{1}'\underline{x} = \underline{1} \qquad \{2.5b\}$$

$$\underline{x} \geqslant \underline{0} \qquad \{2.5c\}$$

The Positive Kriging problem can not be solved using only the classical Lagrange multiplier technique because the non-negativity constraints, {2.5c} , are inequality constraints. Fortunately, sufficient additional tools are available.

The Kuhn-Tucker theorem [8] is the central theorem for all constrained nonlinear optimization. It represents a generalization of the classical method of Lagrange multipliers for the determination of extrema under constraints, to include the case where these constraints not only contain equations but inequalities as well. More specifically, the Kuhn-Tucker theorem provides the necessary and sufficient conditions for a solution to the Positive Kriging problem. The development and proof of the Kuhn-Tucker theorem is beyond the scope of this paper; however, excellent discussion may be found in [1], [9], [12], and many other nonlinear programming texts.

Letting the variable "2λ" be the Lagrange multiplier associated with the unbiased condition {2.5b}, and letting the Nx1 dimensional matrix "$2\underline{\mu}$" be the vector of Lagrange multipliers associated with the non-negativity constraints {2.5c}, the Kuhn-Tucker necessary and sufficient conditions for a unique optimal set of Positive Kriging weights are that the weights and the Lagrange multipliers satisfy the Positive Kriging system:

$$\underline{\Sigma x} + \underline{1}\lambda - \underline{I\mu} = \underline{\sigma} \qquad \{2.6a\}$$

$$\underline{1}'\underline{x} = 1 \qquad \{2.6b\}$$

$$\underline{\mu}'\underline{x} = 0 \qquad \{2.6c\}$$

$$\underline{x} \geqslant \underline{0} \quad \text{and} \quad \underline{\mu} \geqslant \underline{0} \qquad \{2.6d\}$$

where

$\underline{I}$ is an NxN dimensional identity matrix.

$\underline{0}$ is an Nx1 dimensional matrix (vector) of zeros.

Conditions {2.6} may be more conveniently written in a partitioned matrix form as

$$\begin{bmatrix} \underline{\Sigma} & \underline{1} & -\underline{I} \\ \underline{1}' & 0 & \underline{0}' \end{bmatrix} \begin{bmatrix} \underline{x} \\ \lambda \\ \underline{\mu} \end{bmatrix} = \begin{bmatrix} \underline{\sigma} \\ 1 \end{bmatrix} \qquad \{2.7a\}$$

with side constraints

$$\underline{\mu}'\underline{x} = 0 \qquad \{2.7b\}$$

$$\underline{x} \geqslant \underline{0} \qquad \text{and} \qquad \underline{\mu} \geqslant \underline{0} \qquad \{2.7c\}$$

Since conditions {2.7c} require all of the x's and all of the μ's to be non-negative, condition {2.7b} requires

$x_i = 0$ or $\mu_i = 0$ for all i.

This is known as a complementary slackness condition, and variables

x_i and μ_i

are said to be a complementary pair.

Unfortunately, a simple matrix inversion procedure is not appropriate. The system of equations is no longer invertible (as it is no longer square), and the solution procedure must deal explicitly with the non-negativity constraints and the complementary slackness conditions.

3. THE POSITIVE KRIGING SYSTEM

The complementary slackness conditions have the following interpretation: if, at optimality, the Lagrange multiplier for the i'th non-negativity constraint is positive, then the i'th weight must be zero. Conversely, if the i'th weight is positive at optimality, then the Lagrange multiplier for the i'th non-negativity constraint is zero at optimality.

Using this interpretation the optimal weights may be subdivided into two categories: those which are positive at optimality and those which are zero at optimality. Borrowing some nomenclature from linear programming, for a given solution (not necessarily an optimal solution) the weights which are positive will be called the "basic" x's and the weights which are zero will be called the "non-basic" x's. In a parallel fashion, the set of μ's complementary with the basic x's will be called the "basic" μ's , and the set of μ's complementary with the non-basic x's will be called the "non-basic" μ's .

Thus, for any solution which satisfies {2.7}:

basic x's ≥ 0 non-basic x's $= 0$

basic μ's $= 0$ non-basic μ's ≥ 0

Without loss of generality, the rows and columns of the Positive Kriging system may be suitably rearranged to allow partitioning of the $\underline{x}$ vector and the $\underline{\mu}$ vector as

$$\underline{x} = \begin{bmatrix} \underline{x}_b \\ \underline{x}_n \end{bmatrix} \qquad \underline{\mu} = \begin{bmatrix} \underline{\mu}_b \\ \underline{\mu}_n \end{bmatrix}$$

where the subscripts "b" and "n" indicate basic and non-basic. Carrying out this rearrangement of rows (and symmetric columns), the Positive Kriging system may be written as

$$\begin{bmatrix} \underline{\Sigma}_{bb} & \underline{\Sigma}_{bn} \\ \underline{\Sigma}_{nb} & \underline{\Sigma}_{nn} \end{bmatrix} \begin{bmatrix} \underline{x}_b \\ \underline{x}_n \end{bmatrix} + \begin{bmatrix} \underline{1}_b \\ \underline{1}_n \end{bmatrix} \lambda - \begin{bmatrix} \underline{I}_{bb} & \underline{0}_{bn} \\ \underline{0}_{nb} & \underline{I}_{nn} \end{bmatrix} \begin{bmatrix} \underline{\mu}_b \\ \underline{\mu}_n \end{bmatrix} = \begin{bmatrix} \underline{\sigma}_b \\ \underline{\sigma}_n \end{bmatrix} \qquad \{3.1a\}$$

$$\begin{bmatrix} \underline{1}'_b & \underline{1}'_n \end{bmatrix} \begin{bmatrix} \underline{x}_b \\ \underline{x}_n \end{bmatrix} = 1 \qquad \{3.1b\}$$

$$\begin{bmatrix} \underline{\mu}'_b & \underline{\mu}'_n \end{bmatrix} \begin{bmatrix} \underline{x}_b \\ \underline{x}_n \end{bmatrix} = 0 \qquad \{3.1c\}$$

$$\underline{x} \geq \underline{0} \quad \text{and} \quad \underline{\mu} \geq \underline{0} \qquad \{3.1d\}$$

Where

$\underline{\Sigma}_{bb}$ is the square matrix of covariances between the samples associated with the basic x's.

$\underline{\Sigma}_{nn}$ is the square matrix of covariances between the samples associated with the non-basic x's.

$\underline{\Sigma}_{bn}$ is the rectangular matrix of covariances between the samples associated with basic x's and samples associated with non-basic x's.

$\underline{\Sigma}_{nb}$ is the rectangular matrix of covariances between the samples associated the the non-basic x's and samples associated with the basic x's; $\underline{\Sigma}_{bn}$ and $\underline{\Sigma}_{nb}$ are transposes of one another.

$\underline{\sigma}_b$ is the vector of covariances between samples associated with the basic x's and the volume to be estimated.

$\underline{\sigma}_n$ is the vector of covariances between samples associated with the non-basic x's and the volume to be estimated.

The identity matrices, null matrices, and vectors of numeral ones are defined in a suitably conforming manner.

Multiplying out the expanded system {3.1} and remembering that, by definition, the non-basic x's and the basic μ's are equal to zero, yields:

$$\underline{\Sigma}_{bb}\ \underline{x}_b + \underline{1}_b\ \lambda = \underline{\sigma}_b \qquad \{3.2a\}$$

$$\underline{\Sigma}_{nb}\ \underline{x}_b + \underline{1}_n\ \lambda - \underline{I}_{nn}\ \underline{\mu}_n = \underline{\sigma}_n \qquad \{3.2b\}$$

$$\underline{1}'_b\ \underline{x}_b = 1 \qquad \{3.2c\}$$

$$\underline{x} \geqslant \underline{0} \quad \text{and} \quad \underline{\mu} \geqslant \underline{0} \qquad \{3.2d\}$$

Condition {3.1c} , the complementary slackness condition, is implicitly satisfied at optimality by the definitions of basic and non-basic x's and μ's .

Rewriting conditions {3.2a} and {3.2c} in a partitioned matrix form yields

$$\begin{bmatrix} \underline{\Sigma}_{bb} & \underline{1}_b \\ \underline{1}'_b & 0 \end{bmatrix} \begin{bmatrix} \underline{x}_b \\ \lambda \end{bmatrix} = \begin{bmatrix} \underline{\sigma}_b \\ 1 \end{bmatrix} \qquad \{3.3\}$$

This system of equations has N+1 equations and N+1 unknowns, and may thus be solved directly. In fact, system {3.3} is identical to the Ordinary Kriging system {2.3} except it is solely in terms of the basic x's and the unbiased condition's Lagrange multiplier. System {3.3} will be called a Reduced Ordinary Kriging system in the basic variables $\underline{x}_b$.

PROPOSITION 1 - The optimal solution for a Positive Kriging problem includes some positive weights (the basic x's) and possibly some zero weights (the non-basic x's). Furthermore, the optimal values of the positive weights obtained by Positive Kriging are equal to the weights generated by the Ordinary Kriging procedure using only those samples associated with the optimal basic x's.

An immediate corollary to this proposition is as follows: if the OK solution includes only non-negative weights, then the

OK solution is also optimal for the Positive Kriging problem. Thus, if it were possible to know a priori which weights would be basic and which weights would be non-basic, a computer program for OK would be sufficient to solve Positive Kriging problems. However, in general, it is impossible to make an optimal a priori division of the samples between basic and non-basic weights. Note: the set of negative weights generated by an initial execution of the OK procedure is not necessarily equal to the optimal set of non-basic weights.

4. SIMPLEX-BASED ALGORITHMS

The Kuhn-Tucker conditions for the Positive Kriging problem, given in {2.7}, include (N+1) linear equations and (N) nonlinear equations in (N+1+N) variables. A solution to these (N+1+N) equations yields a stationary point to the original quadratic program. In the Positive Kriging problem, since $\underline{\Sigma}$ is positive definite, a stationary point is guaranteed to be the globally optimum solution; therefore, the problem consists of finding a non-negative feasible solution to {2.7a} which satisfies {2.7b} . This suggests the use of some variation of the Simplex Method of linear programming: non-negative feasible solutions to the (N+1) linear equations in (N+1+N) variables are easily obtained through simplex pivoting operations [2].

One technique for applying the simplex algorithm to quadratic programming problems is Wolfe's Method [18]. It consists of constructing an extended system of linear equations by introducing N additional artificial variables for which a simplex tableau and an associated basic feasible solution can be immediately given. The artificial variables are then forced to vanish by means of the Phase I Simplex Method [5]. The first basic feasible solution is chosen so that the complementary slackness condition is initially satisfied; then, the complementary slackness condition is maintained throughout the pivoting process by the application of an additional rule for the selection of the entering basic variables. The additional rule is the only change from the usual Simplex Algorithm of linear programming. A lucid discussion of Wolfe's Method, including a complete working FORTRAN code, can be found in [10].

Though Wolfe's Method is a straightforward application of the Simplex Algorithm, a more efficient, and simpler, method for directly solving the Kuhn-Tucker conditions {2.7} has been developed. This method, known as the Complementary Pivot Algorithm [11], was developed as a general procedure for solving a special class of problem: the linear complementary problem. An excellent presentation of the linear complementary algorithm can be found in [13] or [14].

The simplex-based algorithms have a major shortcoming when viewed from the Positive Kriging vantage point. Both Wolfe's Method and Lemke's Complementary Pivot Algorithm are general techniques created to solve a relatively wide variety of problems; they do not take the highly stylized structure and the specific requirements of the Positive Kriging system into account. On the average, the more efficient of the simplex-based algorithms (Lemke's Complementary Pivot Algorithm) requires four to six times the number of arithmetic operations needed to solve an equivalent sized Ordinary Kriging system using the standard Gaussian triangulation back substitution approach.

5. THE POSITIVE KRIGING ALGORITHM

Considering a usual block model estimation, negative weights are only a periodic problem occurring for a relatively small fraction of the blocks. Thus, in an average situation, a significant amount of expensive computer effort would be wasted if the Positive Kriging problem were explicitly solved when the Ordinary Kriging approach would have generated identical results. The obvious implication is that a Positive Kriging procedure should be invoked only when needed. A rational outline for the Positive Kriging of a block is to first carry out the Ordinary Kriging algorithm, followed by a check for negative weights, and then, only if needed, is the Positive Kriging code called.

Within the described rational Positive Kriging scenario, it would be highly advantageous to make full use of the computational effort expended in solving the initial OK. The Positive Kriging routine should be able to start with the output from an OK "SOLVE" routine, with the system in an upper-triangularized form. Neither of the simplex-based algorithms are conducive to such a modification; their direct application would require the reconstruction of the original $\underline{\Sigma}$ matrix and $\underline{\sigma}$ vector each time they were called, a relatively time-consuming task.

Additionally, the Positive Kriging routine should be designed to run most efficiently on the most common situations. In the vast majority of practical cases where negative weights do occur, few of the weights are ultimately non-basic. A typical example might be as follows: from 16 initially selected sample locations, there are 13 basic weights and 3 non-basic weights at optimality. Thus, it is intuitively appealing to consider an algorithm which starts with the OK solution using all 16 sample locations and proceeds by eliminating sample locations from the set of basic weights until the optimum set is found.

This section presents an algorithm for solving the Positive Kriging problem which is guaranteed to find a division between the basic and non-basic x's such that when the solution of the associated OK system, {3.3}, is solved only non-negative weights are generated. The algorithm is extremely fast and easy to program; however, it is theoretically possible, though highly unlikely, for the algorithm to reduce the set of basic x's too far; thus, it is a heuristic. To test if a generated solution is truly optimal condition {3.2b} can be used; however, experience has shown this to be an unwarranted expenditure of computational effort.

The Inverse Update Algorithm is one method of solving the Positive Kriging system which incorporates the desirable features previously mentioned. The first step in this algorithm is Ordinary Kriging; if the optimal Ordinary Kriging solution contains all non-negative weights the algorithm immediately terminates with the optimal weights. If, however, the Ordinary Kriging solution contains any negative weights then additional computations must be carried out to eliminate negative weights from the set of basics x's so that the resulting solution is non-negative.

To accomplish the necessary division of the set of weights the algorithm proceeds along a logical path, starting with the OK solution. At each iteration thereafter, the smallest (most negative) weight in the current solution is found; if the resulting smallest weight is non-negative the algorithm terminates using the current solution as optimal. If, on the other hand, the smallest weight is negative the corresponding x is eliminated from the current set of basic x's and the current solution (matrix inverse and subsequent weights) is updated to reflect the reduction in the number basic x's in the system. The algorithm then loops back up to the point where the smallest weight in the current solution is found.

To implement the Inverse Update Algorithm, three computational elements, described in the following section, are required: an efficient method for updating the system inverse from one iteration to the next; an efficient method for calculating the resulting updated solution; and, an efficient method for initializing the system inverse given the output from a standard OK solution routine.

6. COMPUTATIONAL DETAILS

The following proposition presents a useful identity for the inverse of a symmetric matrix partitioned on its rightmost column and bottommost row.

PROPOSITION 2 - Given that $\underline{A}$ is symmetric, $\underline{A}^{-1}$ exists, and

$$(1/\beta) = \underline{b}'\underline{A}^{-1}\underline{b} - \alpha \neq 0$$

then the following identity holds:

$$\begin{bmatrix} \underline{A} & \underline{b} \\ \underline{b}' & \alpha \end{bmatrix}^{-1} = \begin{bmatrix} \underline{A}^{-1} - \underline{A}^{-1}\underline{b}\beta\underline{b}'\underline{A}^{-1} & \underline{A}^{-1}\underline{b}\beta \\ \beta b'\underline{A}^{-1} & -\beta \end{bmatrix}$$

The identity of Proposition 2 can be directly used when it is necessary to extend a symmetric system by a row and a column; if the inverse of the $\underline{A}$ matrix is available, the inverse of the extended system can be efficiently calculated without involving a full matrix inversion routine. Furthermore, the identity of Proposition 2 can be algebraically manipulated to yield the required formula for eliminating a basic variable from the Positive Kriging system.

PROPOSITION 3 - Given that A is symmetric, A^{-1} exists, and

$$\begin{bmatrix} \underline{A} & \underline{b} \\ \underline{b}' & \alpha \end{bmatrix}^{-1} = \begin{bmatrix} \underline{C} & \underline{d} \\ \underline{d}' & -\beta \end{bmatrix}$$

then the following identity holds:

$$\underline{A}^{-1} = \underline{C} + (1/\beta)\underline{dd}'$$

Utilizing the result of Proposition 3, eliminating an x_i corresponding to a negative weight in the Positive Kriging system can be accomplished very efficiently: it is not necessary to reinvert the system matrix. As stated, Proposition 3 is only applicable when removing the symmetric last column and bottom row; however, the following proposition extends the result of Proposition 3 to the general case.

PROPOSITION 4 - Given that

$$\begin{bmatrix} a_{11}\cdots & b_{1j}\cdots & a_{1m} \\ \vdots & \vdots & \vdots \\ b_{j1}\cdots & b_{jj}\cdots & b_{jm} \\ \vdots & \vdots & \vdots \\ a_{m1}\cdots & b_{mj}\cdots & a_{mm} \end{bmatrix}^{-1} = \begin{bmatrix} c_{11}\cdots & d_{1j}\cdots & c_{1m} \\ \vdots & \vdots & \vdots \\ d_{j1}\cdots & d_{jj}\cdots & c_{jm} \\ \vdots & \vdots & \vdots \\ c_{m1}\cdots & d_{mj}\cdots & c_{mm} \end{bmatrix}$$

then the inverse of the matrix after row j and column j have been permuted to the far right and bottom of the matrix is

$$\begin{bmatrix} a_{11} \cdots a_{1m} & b_{ij} \\ \vdots \quad\;\; \vdots & \vdots \\ a_{m1} \cdots a_{mm} & b_{mj} \\ b_{j1} \cdots b_{jm} & b_{jj} \end{bmatrix}^{-1} = \begin{bmatrix} c_{11} \cdots c_{1m} & d_{1j} \\ \vdots \quad\;\; \vdots & \vdots \\ c_{m1} \cdots c_{mm} & d_{mj} \\ d_{j1} \cdots d_{jm} & d_{jj} \end{bmatrix}$$

Thus, consider an initial symmetric matrix, $\underline{K}$. Given the inverse of $\underline{K}$, the inverse of the matrix resulting from eliminating row j and symmetric column j from $\underline{K}$ can be efficiently calculated. First, as shown in Proposition 4, the j'th row and j'th column of the inverse matrix are implicitly permuted to the far right and the bottom of the matrix; then, the identity of Proposition 3 is applied.

Proposition 2, 3, and 4 supply the computational tools for updating the system inverse when eliminating any x_i from the set of basic weights; the following proposition supplies an equally effective method for calculating the solution (the resulting weights) of the updated system.

PROPOSITION 5 - Given that the conditions of Proposition 3 hold, and the solution to the system

$$\begin{bmatrix} \underline{A} & \underline{b} \\ \underline{b}' & \alpha \end{bmatrix} \begin{bmatrix} \underline{y} \\ z \end{bmatrix} = \begin{bmatrix} \underline{\sigma} \\ \Psi \end{bmatrix}$$

is given by

$$\begin{bmatrix} \underline{A} & \underline{b} \\ \underline{b}' & \alpha \end{bmatrix}^{-1} \begin{bmatrix} \underline{\sigma} \\ \Psi \end{bmatrix} = \begin{bmatrix} \underline{C} & \underline{d} \\ \underline{d}' & -\beta \end{bmatrix} \begin{bmatrix} \underline{\sigma} \\ \Psi \end{bmatrix} = \begin{bmatrix} \underline{y} \\ z \end{bmatrix}$$

then the solution to the reduced system $\underline{Ay} = \underline{\sigma}$ is given by

$$\underline{A}^{-1}\underline{\sigma} = \underline{y} + \underline{d}(z/\beta)$$

There are two important points to note in the result of Proposition 5. First, the solution of the reduced system can be calculated prior to updating the inverse of the reduced system; thus, the solution for the current set of basic variables can be calculated and inspected, and only if an additional iteration is required is it necessary to update the inverse ($\underline{A}^{-1}$). Second, since z and β are real numbers, all that is required to calculate the updated solution is 1 division, M multiplications, and M additions (assuming $\underline{d}$ is an Mx1 matrix).

Starting with the original inverse of the full OK system, the results of Propositions 2 through 5 can be iteratively applied to eliminate the most negative x_i from the set of basic weights until a set of all nonnegative weights is achieved. To initialize this process it is necessary to obtain the inverse of the original OK system matrix; however, the calculation of this inverse does not require the recomputation of the system matrix followed by a full inversion each time the Positive Kriging routine is called. If a modified Gaussian elimination algorithm is used to solve the OK system, as is most commonly done (eg. [6], [7]), the OK "SOLVE" routine expends most of the computational energy required to invert the initial system matrix. In fact, the standard modified Gaussian elimination procedure accomplishes a complete Doolittle LU decomposition which may be used to rapidly compute the requisite OK system matrix inverse (for details see [17]).

7. CONCLUSION

Negative weights pose a very real problem for many practicing geostatisticians. The Inverse Update Algorithm, presented in this paper, can be effectively applied to the Positive Kriging problem to eliminate the possibility of negative weights. This new algorithm solves the negative weights problem without adversely affecting the computational speed of an existing Ordinary Kriging program.

The discussion and results of this paper were presented within the format of Ordinary Linear Kriging. This format was chosen for clarity in presentation and does not limit the application of the results to more complex Kriging situations.

8. REFERENCES

[1] Bazaraa, M.S. and C.M. Shetty, Nonlinear Programming, Wiley & Sons, New York, NY, 1979.
[2] Dantzig, G.B., Linear Programming and Extensions, Princeton University Press, Princeton, NJ, 1963.
[3] David, M., Geostatistical Ore Reserve Estimation, Elsevier, New York, NY, 1977.
[4] Dhrymes, P.J., Mathematics for Econometrics, Springer-Verlag, New York, NY, 1978.
[5] Hadley, G., Linear Programming, Addison-Wesley, Reading, MA, 1962.
[6] Journel, A.G., and C.J. Huijbregts, Mining Geostatistics, Academic Press, New York, NY, 1978.

[7] Knudsen, H.P. and Y.C. Kim, A Short Course on Geostatistical Ore Reserve Estimation, University of Arizona, Tucson, AZ, 1977.
[8] Kuhn, H.W. and A.W. Tucker, "Nonlinear Programming", Proceedings 2nd Berkeley Symposium on Mathematical Statistics and Probability, J. Neyman (Ed.), University California Press, Berkeley, CA, 1951.
[9] Kunzi, H.P. and W. Krelle, Nonlinear Programming, Blaisdell Publishing Company, Waltham, MA, 1966.
[10] Kunzi, H.P., H.G. Tzschach and C.A. Zehnder, Numerical Methods of Mathematical Optimization, Academic Press, New York, NY, 1971.
[11] Lemke, C.E., "A Method of Solution for Quadratic Programs", Management Science, 8, pp. 442-445, 1962.
[12] Luenberger, D.G., Introduction to Linear and Nonlinear Programming, Wiley & Sons, New York, NY, 1976.
[13] Murty, K., Linear and Combinatorial Programming, Wiley & Sons, New York, NY, 1976.
[14] Phillips, D.T., A. Ravindran, and J.J. Solberg, Operations Research, Wiley & Sons, New York, NY, 1976.
[15] Rendu, J.M., An Introduction to Geostatistical Methods of Mineral Evaluation, S.A.I.M.M., Johannesburg, S.A., 1978.
[16] Schaap, W. and J.D. St. George, "On weights of linear Kriging estimator", I.M.M. Transactionsm 90, January 1981.
[17] Stewart, G.W., Introduction to Matrix Computations, Academic Press, New York, NY, 1973.
[18] Wolfe, P., "The simplex method for quadratic programming", Econometrica, 27, pp. 382-398, 1959.

CORRECTING CONDITIONAL BIAS

Kateri Guertin
Applied Earth Sciences Department
Stanford University

ABSTRACT

A non-linear correction function $K(Z^*)$ is proposed to transform any initial linear estimator Z^* into a conditionally unbiased estimator $Z^{**} = K(Z^*)$ with reduced conditional estimation variance. The correction is based upon an isofactorial representation of the bivariate distribution of the true grade Z and Z^*; it is designed to deal with the problems of change of support and clustered information. A detailed case-study is presented.

I. INTRODUCTION

In Earth Sciences, the study of the mineralization within a given area is often focused on the richest part of that mineralization: in a mining operation where a marginal cutoff grade z_0 is applied, only the profitable grades $z(x)$ greater than or equal to z_0 are selected to be mined from the population of all in-situ grades; hence, the study is not focused on the random function $Z(x)$ itself, but rather on the random function $Z(x)$ conditioned by the inequality $Z(x) \geq z_0$.

Commonly, the unknown realizations $z(x)$ are estimated by linear combinations of surrounding data, say $z(x)^* = \sum \lambda_\alpha z(x_\alpha)$. The selection decision must be based on this estimated value $z(x)^*$, the true grade $z(x)$ being unknown; but once the unit x is selected for mining, it is the true grade $z(x)$ which is recovered.

It follows that the random function of interest is

G. Verly et al. (eds.), Geostatistics for Natural Resources Characterization, Part 1, 245–260.

$Z(x)/Z(x)^* \geq z_0$, rather than $Z(x)$ or $Z(x)/Z(x) \geq z_0$; consequently, characteristics of estimation such as unbiasedness and estimation variance should be defined on $Z(x)/Z(x)^* \geq z_0$:

- conditional unbiasedness
  $E\{(Z(x)-Z(x)^*)/Z(x)^* \geq z_0\} = 0$ for all z_0,
- conditional estimation variance
  $E\{(Z(x)-Z(x)^*)^2/Z(x)^* \geq z_0\}$ to be minimized, for all z_0.

Linear estimators $Z(x)^* = \sum \lambda_\alpha Z(x_\alpha)$ are widely used in the mining industry. Unfortunately, they do not usually fulfill the essential criterion of conditional unbiasedness with possible severe consequences on the project at hand: an overestimation of the average grade actually recovered may jeopardize the whole mining project [1].

The present paper aims at defining a correction function

$$K(Z(x)^*) = Z(x)^{**},$$

to be applied to any initial linear estimator $Z(x)^*$ to turn it into a conditionally unbiased estimator $Z(x)^{**}$. Such a corrected estimator would insure accurate prediction of recoverable reserves from an ore deposit [2]: $Z(x)^{**}$ would be such that

$$E\{Z(x)/Z(x)^{**} \geq z_0\} = E\{Z(x)^{**}/Z(x)^{**} \geq z_0\}, \text{ for all } z_0.$$

This approach to conditional unbiasedness has several advantages. First, it allows correction of any linear estimator already implemented; indeed, it is often more convenient in engineering practice to apply a correction to the current estimator than adopt an entirely new type of estimator. Secondly, following this approach, accurate predictions of recoverable reserves can be readily obtained for any cutoff grade. Therefore, the solution proposed is a practical one that companies could easily implement as a correction factor into their current ore reserve estimation or grade control system.

In this paper, the probabilistic interpretation of the problem is first presented; then a theoretical correction function $K(Z(x)^*)$ is derived from the bivariate probability distribution of the true grade $Z(x)$ versus the initial estimator $Z(x)^*$. The practical determination of this correction function is approached through an isofactorial bivariate model of the probability distribution of $Z(x)$ and $Z(x)^*$. This model is based on the available parameters of the univariate distributions of $Z(x)$ and $Z(x)^*$ and is designed to deal with the problems of changes of support and clustered information. Results from experimental tests are finally presented and analysed.

II. PROBABILISTIC INTERPRETATION OF THE PROBLEM

Consider a deposit D as a set of N mining blocks of size v with true grades: $\{zv_i, i=1,N\}$. The information available from D is constituted by k core grades: $\{z(x_\alpha), x_\alpha \in D, \alpha=1,k\}$.

From that information, a set of N linear block grade estimates is obtained: $\{zv_i^*, i=1,N\}$.

Assuming stationarity of both random functions Zv(x) and Zv*(x) over D, the bivariate distribution of Zv and Zv* can be illustrated as a scattergram of Zv versus Zv* (figure 1).

The estimator Zv* is said to be conditionally unbiased if and only if, for all cutoff values z_0,

$$E\{Zv/Zv^* = z_0\} = z_0,$$

which entails that: $E\{Zv/Zv^* \geq z_0\} = E\{Zv^*/Zv^* \geq z_0\}$, i.e. the actual recovered grade is equal to the predicted one (in expectation).

Notice that the concept of 'non-conditional' unbiasedness for positive random variables is defined as the particular case when $z_0 = 0$: $E\{Zv/Zv^* \geq 0\} = E\{Zv\} = E\{Zv^*\} = E\{Zv^*/Zv^* \geq 0\}$. Conditional unbiasedness does entail unbiasedness; the reverse is not true. Hence, conditional unbiasedness is a much more severe criterion than non-conditional unbiasedness.

On the scattergram of figure 1, the conditional expectation $E\{Zv/Zv^* = z_0\} = Kv(z_0)$ or regression of Zv given $Zv^* = z_0$, clearly departs from the first diagonal: $Kv(z_0) \neq z_0$. This means that Zv* is conditionally biased. In this example, the overestimation of low Zv-grades balances the underestimation of high Zv-grades, providing an overall unbiasedness of Zv*. However, since mining will consider only those blocks such that $Zv^* \geq z_0$, the conditional bias of Zv* will affect the accuracy of the predicted recovery.

Provided the regression curve of Zv given Zv* is known, a new estimator Zv** can be defined:

$$Zv^{**} = E\{Zv/Zv^*\} = Kv(Zv^*).$$

On a scattergram of Zv versus Zv** (figure 2), the regression curve $E\{Zv/Zv^{**} = z_0\} = Lv(z_0) = z_0$, for all z_0, is now identical to the first diagonal, meaning that Zv** is conditionally unbiased.

Thus, the regression function 'Kv' of Zv given Zv* is in

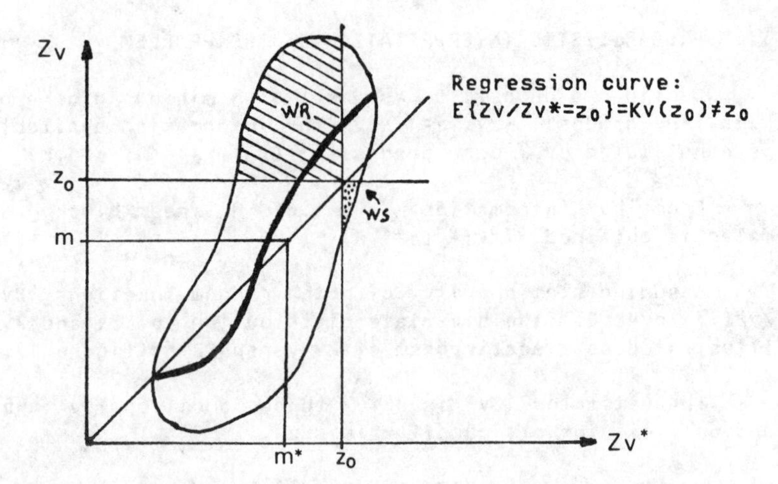

Figure 1. Scattergram of Zv vs Zv*, Zv* being the initial estimator. Notice the conditional bias: $Kv(z_0) \neq z_0$. At the cutoff grade z_0, the proportion of wrongly rejected units is much greater than the proportion of wrongly selected units.

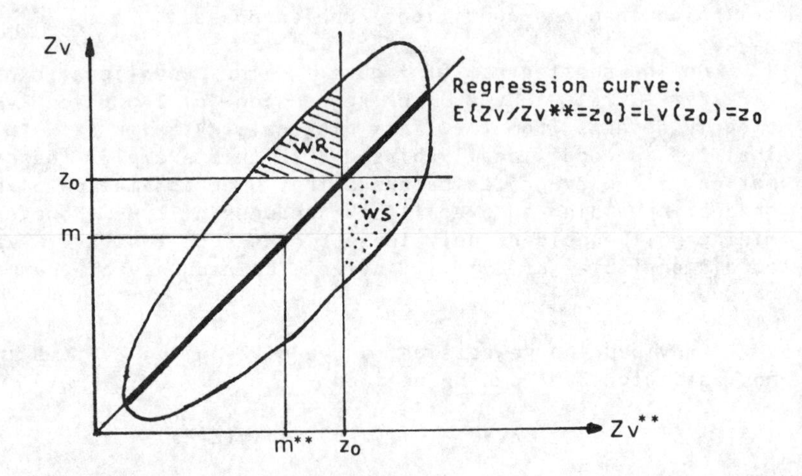

Figure 2. Scattergram of Zv vs Zv**, Zv** being the corrected estimator. Notice the conditional unbiasedness: $Lv(z_0)=z_0$. At the cutoff grade z_0, the proportion of wrongly rejected units roughly balances that of wrongly selected units.

fact the correction function that transforms Zv* into an improved estimator Zv** whose main properties are the following:

a) Given that the correction function is strictly increasing, Zv** is by definition conditionally unbiased and hence unbiased, whether the initial estimator Zv* is unbiased or not; indeed, for all cutoff grades z_0,

$$E\{Zv/Zv^{**} = z_0\} = E\{Zv/Zv^{*} = Kv^{-1}(z_0)\}$$
$$= Kv(Kv^{-1}(z_0)) = z_0 \ .$$

b) Given the initial estimator Zv*, the estimator Zv** = Kv(Zv*) is the best estimator of Zv, in terms of minimization of the conditional estimation variance; hence, Kv(Zv*) is the best correction function that can be applied to Zv*.

c) Consequently, in terms of non-conditional estimation variance, Zv** does better than Zv*:

$$E\{(Zv-Zv^{**})^2\} \leq E\{(Zv-Zv^{*})^2\}.$$

Practically, Zv** entails fewer large errors than Zv* does: the scattergram of Zv versus Zv** (figure 2) is more balanced with respect to the first diagonal than that of Zv versus Zv* (figure 1).

Notice also that selection given Zv** entails misclassification of units as shown on figure 2:
- the upper hatched area WR represents those units wrongly rejected, i.e. such that: $Zv \geq z_0$, but $Zv^{**} < z_0$:
- the lower dotted area WS represents those units wrongly selected, i.e. such that: $Zv < z_0$, but $Zv^{**} \geq z_0$.

Going from Zv* to Zv**, the misclassified units are redistributed such that the two areas WS and WR obtained by applying a cutoff grade to Zv**, are usually more balanced with each other than they are when the same cutoff is applied to Zv* (cf. figure 1, similar areas).

III. DETERMINATION OF THE CORRECTION FUNCTION Kv(Zv*)

Basically, the determination of the correction function Kv(Zv*) = E{Zv/Zv*} which defines the improved estimator Zv**, requires an estimate of the regression curve of the true block grade Zv given its initial estimator Zv*.

Unfortunately, data are defined on cores, which are supports much smaller than the usual block size v. No block grade Zv being currently available, the scattergram of Zv versus Zv* is not known and the regression curve Kv(Zv*) is not accessible.

One possible way around this problem consists in the definition of a bivariate distribution model for (Zv,Zv*) consistent with the available statistical parameters of the univariate probability distributions of Zv and Zv*, namely:

- the dispersion variance of Zv calculated from the variogram $\gamma_z(h)$ of the available core data $z(x_\alpha)$;
- the covariance core-block, also calculated from the variogram $\gamma_z(h)$;
- the histogram of Zv* deduced from the initial estimates zv_i*.

The derivation of the correction function Kv is based on an isofactorial representation of the bivariate probability distribution model of Zv and Zv*. Such a representation simplifies numerical calculations (due to the properties of orthonormal polynomials) and provides a polynomial development of the function Kv.

In this paper, the isofactorial binormal distribution model is used . The grade variables Z, Zv and Zv* are first transformed into univariate standard normal random variables, respectively Y, B and B*. Then, the isofactorial binormal distribution models of (Y,B) and (B,B*) are successively employed to express a block grade distribution model and the correction function Kv derived from it. The main steps of this modelization can be described as follows [3].

Under strict stationarity of Z(x) over D, the Hermite expansions of core and block grades are defined [2,p.574]:

$$Z(x) = \Phi(Y(x)) = \sum_{n=0}^{\infty} \Phi_n \, \eta_n(Y(x)) \tag{1}$$

where Y(x) is standard gaussian and Φ is the normal-score transform function of core grades, experimentally determined from the histogram of $z(x_\alpha)$;

$$Zv = \Phi_v(B) = \sum_{n=0}^{\infty} \Phi_{vn} \, \eta_n(B) \tag{2}$$

where B is standard gaussian and Φv is the normal-score transform function of block grades, not readily accessible.

For each core grade Z(x) located within a block v, assume that [4]

$$E\{Z(x)/Zv\} = Zv.$$

Consider also an isofactorial binormal distribution model for the corresponding couple (Y(x),B), with a coefficient of correlation 'r'. Then, a model of the distribution of the block grade Zv (based on the preceding hypothesis) can be expressed as

$$Zv = \sum_{n=0}^{\infty} \Phi_n \, r^n \, \eta_n(B). \tag{3}$$

The block-support correction factor 'r' is determined from the relation

$$D^2(v/D) = \sum_{n=1}^{\infty} \Phi_n^2 \, r^{2n}, \tag{4}$$

where the dispersion variance of the block grade within the deposit is estimated from the variogram $\gamma_z(h)$ of the core grades $z(x_\alpha)$.

Assuming strict stationarity over D of the initial linear estimator Zv*, its Hermite expansion is expressed as a function of the standard gaussian transform B*:

$$Zv^* = \Psi(B^*) = \sum_{n=0}^{\infty} \Psi_n \, \eta_n(B^*) \tag{5}$$

The normal-score transform function Ψ is experimentally determined from the histogram of the estimated block grades zv_i^*.

Supposing an isofactorial binormal distribution model for (B,B*), the correlation coefficient 'ρ' of (B,B*) is obtained from the following relations, where the average covariance between the block grade Zv and its estimator Zv* is known from the $\gamma_z(h)$:

$$\overline{\mathrm{Cov}}\{Zv,Zv^*\} = E\{(\sum_{n=0}^{\infty} \Phi_n \, r^n \, \eta_n(B))(\sum_{p=0}^{\infty} \Psi_p \, \eta_p(B^*))\} - \Phi_0\Psi_0 \tag{6}$$

$$= \sum_{n=1}^{\infty} \Phi_n \, \Psi_n \, (r\rho)^n. \tag{7}$$

Then, the conditionally unbiased estimator Zv** defined by the correction function Kv(Zv*) can be derived from the model of the block grade distribution (cf. relation 3) and expanded as follows:

$$Zv^{**} = Kv(Zv^*) = E\{Zv/Zv^*\} = \sum_{n=0}^{\infty} \Phi_n \, (r\rho)^n \, \eta_n(B^*) \tag{8}$$

where $B^* = \Psi^{-1}(Zv^*)$.

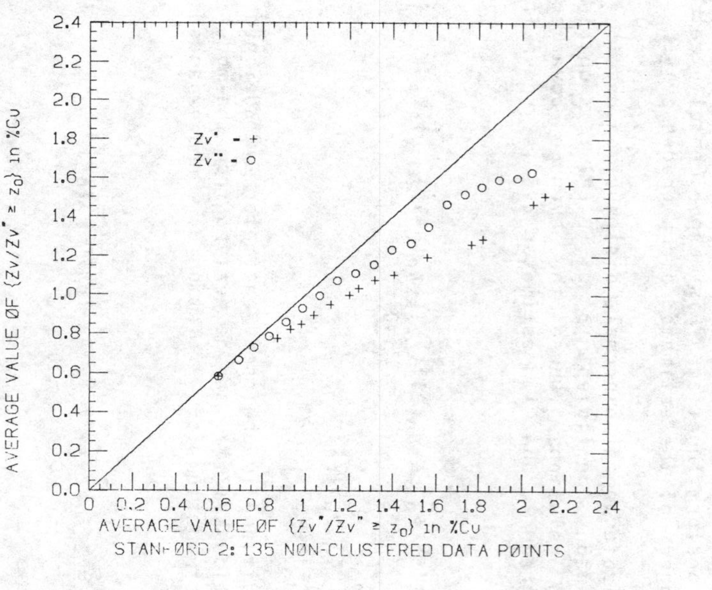

Fig.3 Cumulative conditional bias given a cutoff z_0; Zv″ holds for the initial estimator Zv* and its corrected estimator Zv**. Each symbol '+' or 'o' holds for a cutoff z_0, and $z_0 \in [0.0, 1.8]$, with increment of 0.1 %Cu.

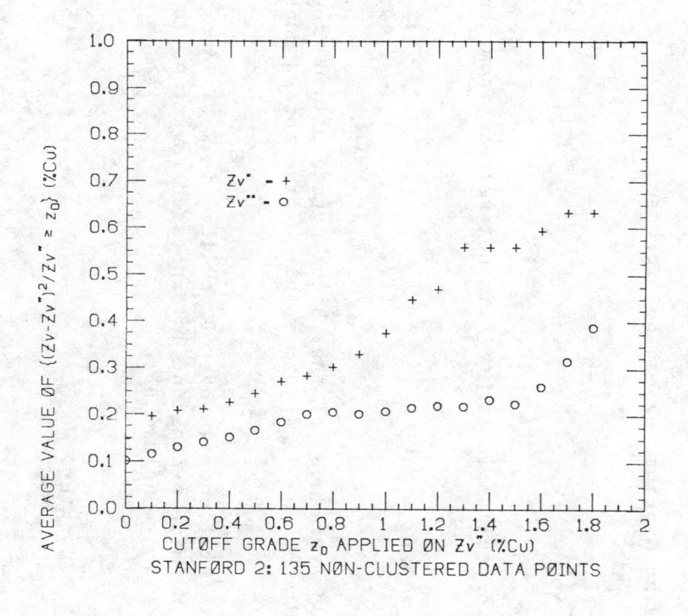

Fig.4 Cumulative conditional squared error given a cutoff z_0; Zv″ holds for the initial estimator Zv* and its corrected estimator Zv**.

STATISTICS	EXHAUSTIVE from 60500 pts	NAIVE from 135 pts	FROM $\gamma_z(h)$ 135 pts	POLYGONAL ESTIMATION		
				Zvm[1]	Zv*	Zv**
Mean Grade	.5816	.6187	–	.5969	.5969	.5955
$D^2(o/D)$	.385	.420	.410	–	–	–
$D^2(v/D)$	.261	–	.286	.286	.396	.239

1: block grade model (cf. relation 3)

Table 1. Non-clustered data set : main statistics of the various grade variables used in the correction model.

STATISTICS	EXHAUSTIVE from 60500 pts	DECLUST from 150 pts	FROM $\gamma_z(h)$ 150 pts	POLYGONAL ESTIMATION		
				Zvm[1]	Zv*	Zv**
Mean Grade	.5816	.6148	–	.5689	.5689	.5677
$D^2(o/D)$	.385	.349	.417	–	–	–
$D^2(v/D)$	.261	–	.274	.274	.324	.237

1: block grade model (cf.relation 3)

Table 2. Clustered data set: main statistics of the various grade variables used in the correction model.

Such a corrected estimator Zv^{**} can be obtained from any initial linear estimator Zv^{*}; the only requisites are the histogram and the variogram of the core data $z(x_{\alpha})$.

IV. EXPERIMENTAL RESULTS

This model of correction is tested using a two-dimensional simulated deposit of copper (quantified in % Cu) with a spatial correlation structure [5]. The exhaustive statistics computed from the 60500 simulated points (table 1, column 1) are used to evaluate the performance of the proposed model.

Two sets of data points are taken from the deposit (with and without clusters). From each data set, 500 mining blocks of equal size v are estimated by the polygonal method: each block receives the grade of the data point closest to its center. Then the two sets of linear block grade estimates are corrected for their conditional bias using the model described above.

Non-clustered data set (table 1)

First, a set of 135 randomly stratified data points is taken. A block grade distribution model (column 4) is built from the naive histogram of these 135 point grades (column 2) expanded into Hermite polynomials (up to order 19). However, the model's mean grade is identified to the overall average grade of the initial poly estimator Zv^{*} which is a better estimate (column 5). The dispersion variance of the model is based on the variography of the 135 data points (column 3) which actually reproduces quite well the true variability of Zv within the deposit (column 1). A corrected estimator Zv^{**} is thereafter derived from this model (cf. relation 8).

The new estimator Zv^{**} thus obtained does well correcting the conditional bias of the initial poly estimator Zv^{*} (figure 3); it also entails a smaller conditional squared error than Zv^{*} (figure 4). Finally, the dispersion variance of this new estimator is considerably reduced relatively to the overestimated one of Zv^{*} (columns 6 and 5).

Clustered data set (table 2)

A set of 150 irregularly spaced data points is taken with preferential sampling mostly from the high grade zones. A block grade distribution model (column 4) is inferred from the declustered histogram of these 150 point grades (column 2) expanded into Hermite polynomials (up to order 19). The mean grade of the model is identified to that of the poly estimator Zv^{*}, while its dispersion variance is deduced from the variography of the data

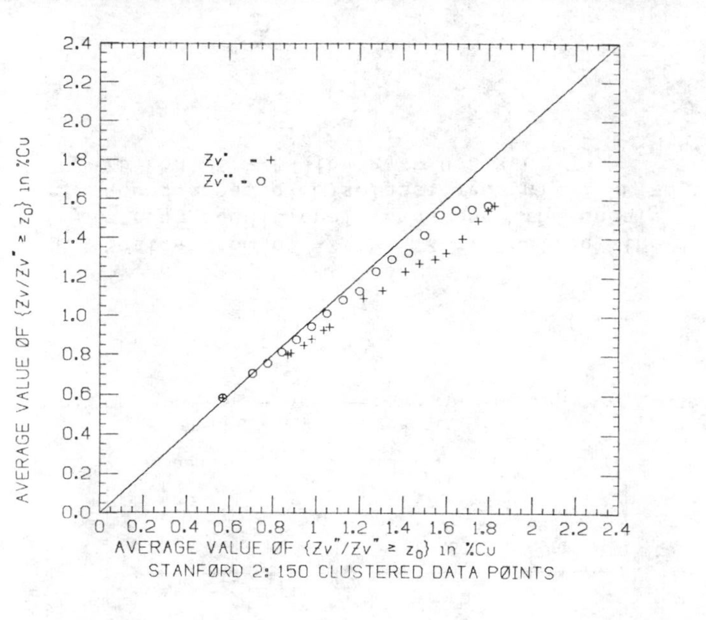

Fig.5 Cumulative conditional bias given a cutoff z_0; Zv″ holds for the initial estimator Zv* and its corrected estimator Zv**. Each symbol '+' or 'o' holds for a cutoff z_0, and $z_0 \in [0.0, 1.6]$, with increment of 0.1 %Cu.

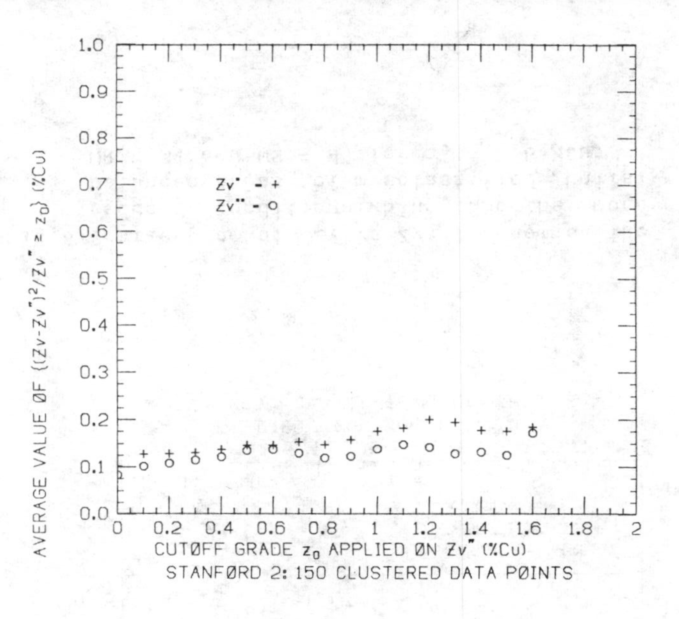

Fig.6 Cumulative conditional squared error given a cutoff z_0; Zv″ holds for the initial estimator Zv* and its corrected estimator Zv**.

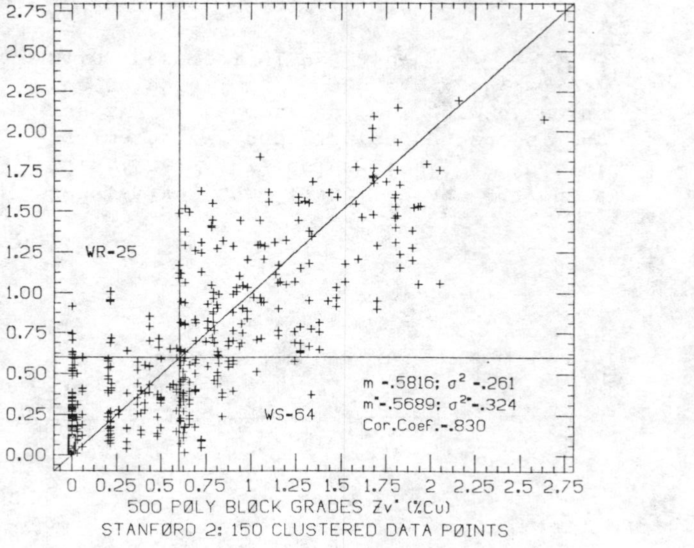

Fig.7 Scattergram of Zv vs Zv*; notice the strong conditional bias and the unbalanced areas of misclassified units: WR = 25 and WS = 64 for z_0 = 0.6 %Cu.

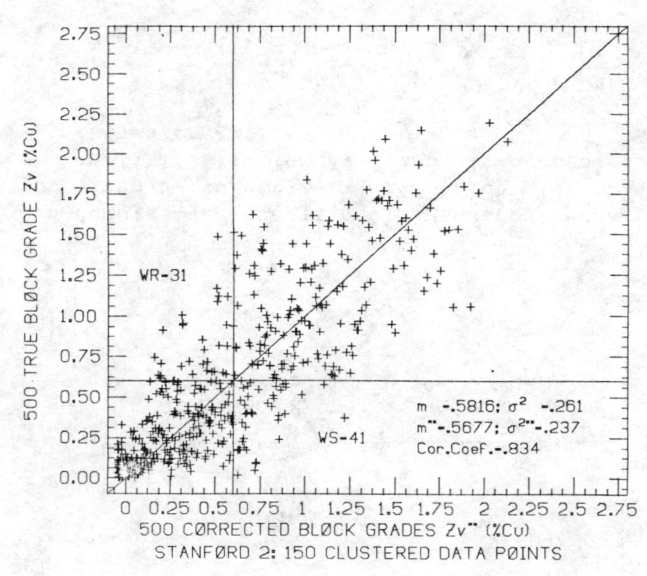

Fig.8 Scattergram of Zv vs Zv**; notice the reduced conditional bias and the more balanced areas of misclassified units: WR = 31 and WS = 41 for z_0 = 0.6 %Cu.

points (column 3); again $\gamma_z(h)$ provides a fair estimate of the variability of Zv (column 1). Then, a corrected estimator Zv** is derived from this model.

The initial Zv* shows an overestimating conditional bias that Zv** eliminates to a considerable extent (figure 5). Zv** also reduces the conditional squared error due to Zv* (figure 6). However, the performance of Zv** decreases as the cutoff becomes high (from 1.4% up) and the number of selected units small (from 36 down).

The overall effect of the correction of Zv* by Zv** is illustrated by the two scattergrams of the true block grade Zv versus Zv* (figure 7) and Zv versus Zv** (figure 8): the last one shows a cloud of points more balanced with regard to the first diagonal, and closer to it. Notice also that the dispersion variance of Zv** is very similar to that of Zv.

These results are satisfactory: in the two cases studied (non-clustered and clustered information), the strong conditional bias of the initial poly estimator Zv* is considerably corrected by Zv** and its conditional squared error is reduced.

Also from these results, the block grade distribution model appears to be robust with regard to the configuration of information as long as a declustered histogram of the data values is used in presence of clusters (cf. table 2, column 2).

On the other hand, it must be stressed that the performance of the correction function is highly dependent on the reliability of the block grade distribution model, mostly through the estimation of its mean grade and dispersion variance (which should be as close as possible to those of the true grade Zv); as an example, a corrected estimator Zv** would carry along a global bias 'built into' the block grade model, reducing the usefulness of conditional unbiasedness.

Añother important aspect in the determination of the correction model consists in an appropriate isofactorial representation of the grade variables that insures a strictly increasing correction function Kv(Zv*).

Finally, although the clustered information in the previously studied case does not invalidate the approximation of a unique correction function based on the overall average covariance of Zv and Zv* (cf.relation 7), it remains that a local correction model would be theoretically required in the presence of such irregulary spaced data points.

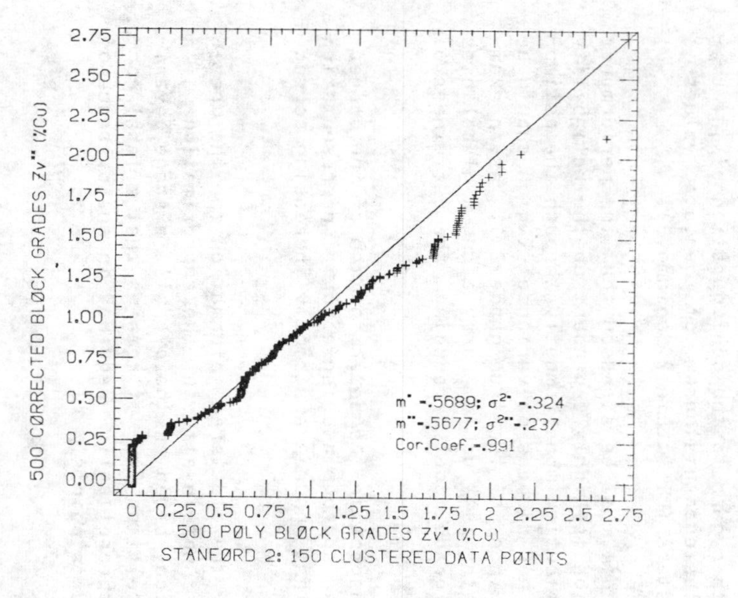

Fig. 9 Correction of Zv* by Zv**. Notice that the unique correction function Zv** = Kv(Zv*) is strictly increasing.

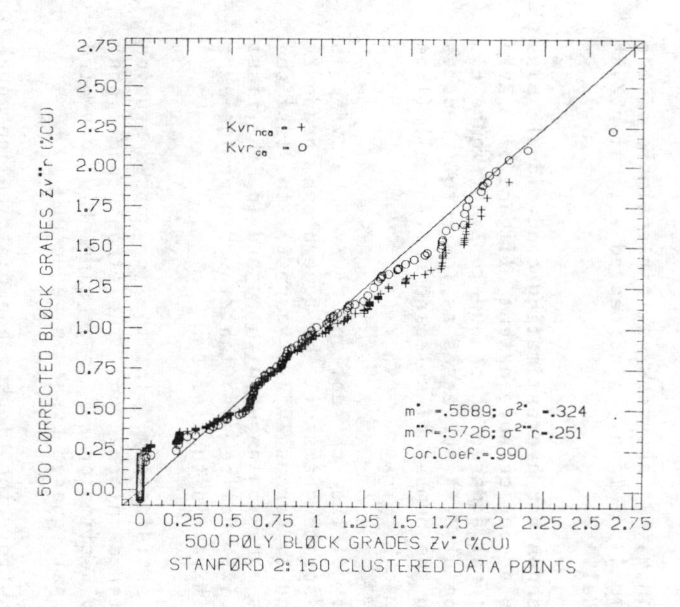

Fig. 10 Correction of Zv* by Zv**r. The '+' holds for the blocks located within the non-clustered areas (nca), the 'o' for those located within the clustered areas (ca). Notice that both regional correction functions Kvr_{nca} and Kvr_{ca} are strictly increasing.

V. LOCAL CORRECTION IN PRESENCE OF CLUSTERS

The model of correction as defined in section III asks for strict stationarity of both random functions Zv(x) and Zv*(x) over the deposit, and thus requires that all the estimated blocks be influenced by an identical configuration of data points. This implies that there should be as many correction functions defined as there are different configurations of data used to estimate the grade of a block. Hence, those blocks influenced by a particular pattern of data points would be corrected in function of their own average covariance $\overline{\mathrm{Cov}}\{Zv,Zv^*\}$.

However, in practice, one might consider only a regional correction corresponding to an averaging of the covariance of (Zv,Zv*) over two sets of blocks: i) the set of blocks influenced by clustered data, and ii) the set of remaining blocks located away from the clusters. Each set would have its own correlation coefficient ρ (cf. relation 7) and therefore its own correction function.

Using again the 150 clustered data points and the initial poly estimator Zv*, such a regional corrected estimator Zv**r is obtained. In terms of conditional bias and squared error, Zv**r is practically equivalent to Zv** (previously defined given a unique overall average covariance, figures 5 and 6); the plots of the global correction function Kv(Zv*) (figure 9) and of the two regional correction functions Kvr(Zv*) (figure 10) are also very much alike.

Such similarities can be explained by the actual small difference in average covariance $\overline{\mathrm{Cov}}\{Zv,Zv^*\}$, going from the well-informed areas (with clusters) to the poorly-informed ones. Moreover, the two regional correction functions being defined over the same range of grades, their combined effect is similar to that of the unique function Kv used in section IV.

Therefore, it appears that a local correction model would be justified only when the configuration of information is very irregular, entailing a significant variation of the covariance of (Zv,Zv*) over the deposit and, consequently, very distinct correction functions from one area to another. In particular, such a model would be required when the areas with different data configurations would have non-overlapping ranges of Zv*-grades; in such a case, the various correction functions defined for specific ranges of Zv* would relay each other, correcting successively all the Zv*-grades.

VI. CONCLUSIONS

The objective of this paper was to provide a correction for conditional unbiasedness that would be theoretically sound and still easy to apply to any linear block grade estimator currently used by mining companies.

The theoretical correction proposed fulfills this objective. Its practical determination is based on an isofactorial representation of the bivariate distribution of Zv and Zv*. Such a model requires only the declustered histogram and the variogram of the core data available.

The first experimental results obtained using such a model are satisfactory. Given information defined on core support and starting from a rather poor block grade linear estimator, the corrected estimator eliminates most of the initial conditional bias and reduces the conditional error, even in the case of clustered information. This last point is important since data available at various stages of development are often irregularly spaced over the deposit; a local adaptation of the correction model can even be considered in presence of very dense clusters.

This model has also been developed in terms of Legendre polynomials; tests made with the same data sets have provided results very similar to those presented in this paper. Finally, a study on a gold deposit is being successfully completed.

REFERENCES

1. Helwick,S.J., 1983, "Geostatistics Versus Conventional Methods: Economical Impact of Misclassification of Ore and Waste", Exxon Minerals Company, presented at 112th AIME Annual Meeting, Atlanta, 1983.

2. Journel,A.G. and Huijbregts,Ch.J., 1978, "Mining Geostatistics", Academic Press, pp.457-459; p.574.

3. Journel,A.G., 1980, "Conditional Bias and its Corrections", Unpublished Research Note, Stanford University.

4. Matheron, G., 1975, "Forecasting Block Grade Distributions: Transfer Functions", Geostat 75, pp.237-251.

5. Verly, G., 1983, "The Multigaussian Approach and its Applications to the Estimation of Local Reserves", Journal of Mathematical Geology, Vol.15, pp.263-290.

mAD AND CONDITIONAL QUANTILE ESTIMATORS

A.G. Journel

Applied Earth Sciences Department
Stanford University, California, USA

Abstract

All krigings are linked to the variance characteristic for spread of errors. The mean absolute deviation (mAD) characteristic leads to median and other quantile-type estimators. The relationships among these criteria and their impact on the corresponding estimators is discussed.

Keywords

Variance - mAD - conditional quantiles - estimation

Introduction

The term "best estimator", or the acronym "BLUE" so often used in the geostatistical literature, appears irritating to some and is certainly misleading for many practitioners. Kriging estimators are only "best" relative to a very particular criterion: the minimization of the estimation variance, or variance of the error.

Statisticians will argue that other characteristics of spread of the error have better resistance properties than the variance, e.g. Tukey (1977). Practitioners may wonder how the unique estimation variance notion can express the variety of their own criteria for goodness. Indeed when they study a map, sometimes they do care about accurate reproduction of local details, but at other times they are prepared to trade local accuracy for global reproduction of the regional trends. How could a map based on a single blind criterion satisfy all these

G. Verly et al. (eds.), Geostatistics for Natural Resources Characterization, Part 1, 261–270.

objectives? Is it thus fair to pretend that any map, kriged or not, is "best", thus implying that all others are poorer?

In the recent past great effort has been given to selling algorithms and software for kriging: whenever a kriged map did not match the expectation of the client, fixes were found or the client was convinced that his expectation was unreasonable and his criteria for goodness inconsistent. When the kriged map showed discontinuities, a whole array of filters were applied or the nugget constant was chopped off the variogram. Conversely, when the kriged map looked overly smooth, additional fake data were introduced or again the variogram was fudged. One may then, and rightfully so, wonder what has happened to the estimation variance after all these manipulations, and whether the final map still deserves the title "BLUE".

Minimization of a criterion is usually obtained at the expense of other criteria. The minimum local estimation variance of the kriged map entails its overall smooth aspect, its exactitude property (honoring the data) can result in nested contour around the data locations, etc. Instead of designing algorithmic patches, it seems to this author more rewarding to study other criteria for goodness of estimation, to evaluate their pros and cons, then try in each particular application to strike a compromise between the adequate criteria: the final map, being thus a compromise, would not carry any such label as "best". Such research should carry great rewards, both theoretical and practical.

In the following, within the class of local accuracy criteria, the mean absolute deviation measure for spread of errors is compared to the variance measure, defining conditional quantile-estimators to be compared to the kriging estimators.

Characteristics of deviation

Without loss of generality, consider a non-negative random variable (R.V.) $Z \geqslant 0$, characterized by its cumulative distribution function (cdf):

$$F(z) = \text{Prob}\{Z \leqslant z\} \qquad (1)$$

The cdf $F(z)$ is such that the moments up to order 2 exists:

$$m = E\{Z\} < \infty, \quad m_2 = E\{Z^2\} < \infty.$$

The usual characteristics of central location and corresponding deviation are:

- the expected (mean) value and variance:

$$m = E\{Z\}, \text{ and: } \sigma^2 = \mathrm{Var}\{Z\} = E\{[Z-m]^2\} \quad (2)$$

- the median and the mean absolute deviation (mAD).

$$M = F^{-1}(.5), \text{ and: } \mathrm{mAD} = E\{|Z-m|\}$$

The expected value m is the value u that minimizes the variance: $E\{|Z-u|^2\}$. The median is the value u that minimized the mAD criterion: $E\{|Z-u|\}$. These results are valid for any distribution and apply in particular to the conditional cdf $F(z/(N))$ of a stationary random function $Z(x)$ conditioned by N surrounding data:

$$F_x(z/(N)) = \mathrm{Prob}\{Z(x) \leq z \,/\, Z(x_\alpha) = z_\alpha \;, \alpha \in (N)\} \quad (3)$$

Thus the best estimator $Z^*(x)$ for the conditional estimation variance criterion, $E\{[Z(x) - Z^*(x)]^2/Z(x_\alpha) = z_\alpha, \alpha \in (N)\}$, is the value that minimizes that variance, that is the mean of the conditional distribution (3) i.e. the conditional expectation:

$$E\{Z(x)/Z(x_\alpha) = z_\alpha, \alpha \in (N)\} = \int_0^\infty z.dF_x(z/(N)).$$

All krigings are estimates of that conditional expectation, and are thus linked to the estimation variance criterion.

Similarly, the best estimate for the mean absolute deviation criterion, $E\{|Z(x) - Z^*(x)|/Z(x_\alpha) = z_\alpha, \alpha \in (N)\}$, is the conditional median: $F_x^{-1}(.5/(N))$.

Skewness and reversal

Conditional expectation and conditional median are identical only if the conditional cdf $F(z/(N))$ is symmetric . Unfortunately, at least in Earth Sciences applications, this is rarely the case. Histograms, $F'(z)$, are usually highly skewed and strictly bounded: $Z \in [0, z_{max}]$. A strictly bounded model for the multivariate cdf of the random function $Z(x)$ leads to asymmetric and bounded conditional cdfs $F(z/(N))$. Depending on the conditioning data values, z_α, $\alpha \in (N)$, skewness of the conditional cdf is either positive or negative. If the data are at the high end of the positively skewed distribution $F(z)$, then the conditional cdf $F(z/(N))$ is negatively skewed entailing a larger risk of over-estimation, with the reverse being true if the N data are at the low end of the initial cdf $F(z)$.

This skewness reversal of conditional distributions has dramatic effects on the prediction intervals attached to estimators, particularly if the initial cdf $F(z)$ is highly skewed. It is not taken into account by estimation techniques that

- either assume a constant shape for conditional distributions, e.g. the lognormal shortcut techniques (David, 1977), or

- only provide an estimation variance (all linear kriging techniques) calling implicitly for a symmetric normal distribution of errors.

Estimation of F(z/(N))

Recent research in Geostatistics, see in particular Sullivan (1983) and Verly (1983) in the present NATO ASI proceedings, has focused on the estimation of conditional distributions of the type $F(z/(N))$ and more precisely on the estimation of spatial distributions which can be seen as spatial composites: $\frac{1}{A} \int_A F_x(z/(N))dx$.

In short, models and techniques are now available to estimate the conditional cdf $F_x(z/(N))$, a function of the argument z. From that function, the conditional mean and median, and also the conditional quantiles, can be deduced, thereby providing the conditional prediction intervals:

$$\text{Prob}^*\{Z(x) \in]z_q, z_{q'}]/Z(x_\alpha)=z_\alpha, \alpha \in (N)\} = q' - q \tag{4}$$

with: $q' > q \in [0,1]$, and z_q being the q-quantile of the estimated conditional cdf $F_x^*(z/(N))$.

If the estimated conditional cdf's $F_x^*(z/(N))$ are asymmetric and account for reversal of skewness, so will the prediction intervals $]z_q, z_{q'}]$.

Expressions for the mAD

Expressions for: $\text{mAD}(u) = E\{|Z - u|\}$ are derived which allow an interesting generalization and the definition of best quantile estimators different from the conditional median. In the course of this analysis, the cdf integral $A(u)$ is defined as the key operator for calculating moments of all orders of the distribution $F(z)$.

Define the truncated mean as the mean of the truncated distribution:

$$m(u) = E\{Z/Z \leqslant u\} = \frac{1}{F(u)} \int_0^u z.dF(z) \tag{5}$$

with :

$$\text{Lim. } m(u) = m = E\{Z\} < \infty \quad , \text{ when } u \to +\infty .$$

The mean absolute deviation of the R.V. Z from any given scalar u is defined as:

$$\text{mAD}(u) = E\{|Z-u|\} = \int_0^\infty |u-z|.dF(z) \tag{6}$$

The previous integral can be developed into

$$\int_0^u (u-z).dF(z) + \int_u^\infty (z-u).dF(z) =$$

$$= u[2F(u) - 1] + m - 2\int_0^u z.dF(z)$$

An integration by parts provides the following:

$$\int_0^u z.dF(z) = zF(z)\Big|_0^u - \int_0^u F(z)dz = uF(u) - \int_0^u F(z)dz$$

The integral $\int_0^u F(z)dz$ has no limit when $u \to +\infty$, so it is better to introduce the following "cdf integral":

$$A(u) = \int_0^u [1-F(z)]dz = u[1-F(u)] + m(u).F(u) \tag{7}$$

Since the moment of order 1 exists, $m = m(\infty) < \infty$ the limit of the cdf integral A(u) is none other than the expected value m:

$$A(\infty) = \int_0^\infty [1-F(z)].dz = m = E\{Z\} \quad . \tag{8}$$

The properties of the cdf integral A(u) are further detailed in the Appendix.

Coming back to the mAD analysis, the previous expression for mAD(u) is rewritten:

$$\text{mAD}(u) = 2[u-m(u)] \cdot F(u) + (m-u) = m + u - 2A(u) \tag{9}$$

Differentiation of the last expression (9) shows that the median is indeed the value u which minimizes mAD(u):

$$\frac{d\text{mAD}(u)}{du} = 1-2A'(u) = 2F(u) - 1 = 0, \text{ For: } u = F^{-1}(.5) = M,$$

the minimum value being:

$$\text{mAD}(M) = m - m(M) = m + M - 2A(M)$$

Generalization

Rewrite the first relation (9) as:

$$mAD(u) \equiv L_{.5}(u) = (m-u) + \frac{F(u)}{.5}\left[u-m(u)\right],$$

then generalize it into the expression:

$$\begin{aligned} L_p(u) &= (m-u) + \frac{F(u)}{p}\left[u-m(u)\right] \\ &= (m-u) + \frac{1}{p}\left[u-A(u)\right], \text{ for all } p \in [0,1] \end{aligned} \quad (10)$$

The minimum of $L_p(u)$ is such that:

$$\frac{\partial L_p(u)}{\partial u} = -1 + \frac{1}{p}\left[1-A'(u)\right] = -1 + \frac{1}{p}F(u) = 0 \quad \Rightarrow \quad u = F^{-1}(p)$$

$$\frac{\partial^2 L_p(u)}{\partial u^2} = \frac{1}{p}F'(u) \geq 0 \quad .$$

Thus the value u which minimizes the criterion $L_p(u)$ is the p-quantile of the distribution F(z). The question is then how that criterion $L_p(u)$ is interpreted.

Asymmetric loss function and quantile-estimators

Consider the estimation of the R.V. Z by the scalar value u, the estimation error being a R.V. (Z-u). Assume then a loss function increasing linearly with the error, but at a different rate depending on whether there is an underestimation, or an overestimation, i.e. Figure 1:

$$L(Z-u) = \begin{cases} \lambda_1(u-z), \text{ For: } z \leq u \text{ (overestimation)} \\ \lambda_2(z-u), \text{ For: } z \geq u \text{ (underestimation)} \end{cases} \quad (11)$$

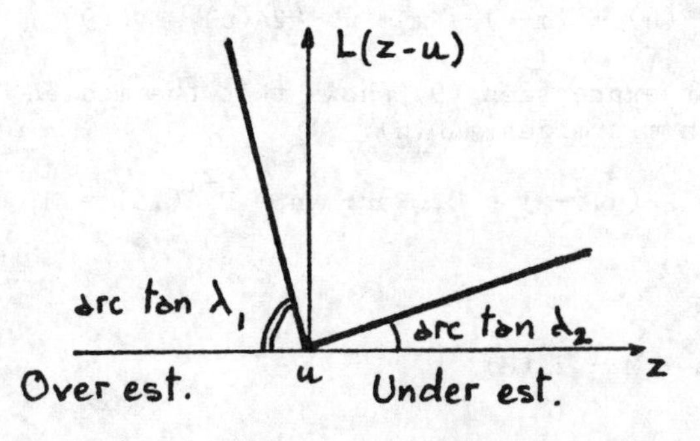

Figure 1. Loss as a function of the error (Z-u)

- The asymmetric linear loss function calls for the p-quantile estimator, with:

$$p = \frac{\lambda_2}{\lambda_1 + \lambda_2} \quad , \quad \lambda_1, \lambda_2 \in [0, +\infty] \; .$$

- The symmetric quadratic loss function calls for the expected value estimator (i.e. all krigings).

The loss function L(Z-u) is a random variable, with expected value:

$$E\{L(Z-u)\} = \lambda_1 \int_o^u (u-z).dF(z) + \lambda_2 \int_u^{\infty} (z-u).dF(z)$$

$$= \lambda_1 uF(u) - \lambda_2 u[1-F(u)] - \lambda_1 \int_o^u z.dF(z) + \lambda_2 \int_u^{\infty} z.dF(z)$$

$$= \lambda_2(m-u) + (\lambda_1 + \lambda_2).F(u).[u-m(u)]$$

Denoting the relative loss rates by:

$$p = \frac{\lambda_2}{\lambda_1 + \lambda_2} \quad , \quad 1-p = \frac{\lambda_1}{\lambda_1 + \lambda_2} \quad , \; p \in [0,1] ,$$

the mean expected loss is:

$$E\{L(Z-u)\} = \lambda_2 \left[(m-u) + \frac{F(u)}{p} [u-m(u)] \right] , \text{ that is:}$$

$$E\{L(Z-u)\} \equiv \lambda_2 \, . \, L_p(u) \quad , \qquad (12)$$

λ_2 being a constant that can be set to one, the estimate u that minimizes the expected loss is the p-quantile $F^{-1}(p)$.

Remarks

- When $\lambda_1 > \lambda_2$, $p < .5$, and the best estimate $F^{-1}(p)$ is less than the median $M = F^{-1}(.5)$. In practical terms, a higher cost of overestimation would call for a more conservative estimate, and the best estimate would be jacked down from the median; this corresponds to the practice of applying a safety factor to an initial estimate.

- Conversely, when $\lambda_2 > \lambda$, $p > .5$, and the best estimate $F^{-1}(p)$ is greater than the median M.

- These results can be applied to the conditional cdf $F_x(z/(N))$ defined in (3):

The best estimate of the unknown outcome of $Z(x)$ given the N surrounding data z_α and a loss function of the type (11) is the p-conditional quantile $F_x^{-1}(p/(N))$.

In mining applications, asymmetric loss functions are encountered in highly selective operations where a wrongly selected unit (overestimation) not only does not pay for its processing costs but also may jeopardize the metal recovery of good units with which it is processed. Therefore the cost of overestimation is proportionally higher than that of underestimation.

Other asymmetric loss functions stem from regulations and penalties triggered by some impurity grade exceeding a given threshold level.

Conclusions

The estimation variance criterion underlying all kriging estimators is but one very particular estimation criterion linked to an error impact proportional to the squared error. The mean absolute deviation criterion corresponds to a more realistic error impact proportional to the absolute error. This error impact can be made asymmetric depending on the sign of the error, thus defining quantile-type estimators. Median and quantile-estimators and their associated, usually non-symmetric, prediction intervals are deduced from estimates of local conditional distributions.

Beyond the class of estimators looking for local accuracy, research is needed to define other classes of estimators aimed towards different criteria corresponding to the diverse uses of a map or a block model. The term "best" should be used with more humility, i.e. be better documented. Also, geostatisticians should accept the fact that, for some specific uses, kriged maps can be inappropriate.

References

David, M. (1977) - "Geostatistical ore reserve estimation", Elsevier Publ., New York, 364 p.

Tukey, J. (1977) - "Exploratory Data Analysis" Addison and Wesley Publ., Reading, Mass., 488 p.

Jones, T.A. (1983) - "Problems in using Geostatistics for Petroleum applications", Proceed. of the 2nd NATO ASI "Geostat Tahoe", Sept. 1983, in this volume, p. 635.

Journel, A.G. (1983) - "The place of non-parametric Geostatistics", ibid.

Sullivan, J. (1983) - "Probability Kriging - Theory and Practice", ibid.

Verly, G. (1983) - "The block distribution given a point multivariate-ϕ-normal distribution", ibid.

Verly, G. (1983) - "The multigaussian approach and its applications to the estimation of local recoveries", in Math. Geology, vol. 15, no. 2, pp. 263-290.

Appendix: The cdf integral A(u)

Recall the definition and relation (7):

$$A(u) = \int_0^u [1-F(z)]dz = u\,[1-F(u)] + m(u).F(u)\text{, with:}$$

$$m(u) = E\{Z/Z \leq u\} = \frac{1}{F(u)} \int_0^u z.dF(z)$$

$$m(\infty) = A(\infty) = m$$

Consider the R.V. $Y \in [0,+\infty]$ with cdf $A(y)/A(\infty)$:

$$\text{Prob}\{Y \leq y\} = A(y)/A(\infty) = \frac{1}{m} \cdot A(y) \tag{13}$$

The following relations link the moments of order (n-1) of Y and the moments of order (n) of the initial R.V. Z:

$$E\{Z^n\} = nm \,.\, E\{Y^{n-1}\}\text{, for all } n \geq 1 \tag{14}$$

$$\sigma^2 = \text{Var}\{Z\} = m.\,[2E\{Y\} - m]$$

$$E\{Z^2\} = 2E\{Z\} \,.\, E\{Y\}$$

The p-quantile of the cdf of Y is linked to the mean of Z through a very simple relation:

$$y_p = A^{-1}(pm) \quad \Leftrightarrow \quad A(y_p) = pm \tag{15}$$

The transform $Z \to Y$ transforms Dirac Z-distributions into uniform Y-distributions. More generally, whatever the skewness sign of the initial Z distribution, the corresponding Y-distribution belongs to a family of positively skewed distributions bounded by a Dirac of parameter $y = 0$ and a uniform distribution within $[0, z_{max}]$.

Generalization

Define the truncated moment of order $n \geq 1$:

$$m_n(u) = E\{Z^n / Z \leq u\} = \frac{1}{F(u)} \int_0^u z^n \, . \, dF(z) < \infty$$

and the order n cdf integral:

$$A_n(u) = \int_0^u z^{n-1} \left[1-F(z)\right] dz, \text{ with: } A_1(u) = A(u) \quad (16)$$

The following relations generalize the properties of the operator A(u):

$$A_n(\infty) = m_n(\infty) = m_n = E\{Z^n\}$$

Y_n being the R.V. with cdf $A_n(y)/A_n(\infty)$, it comes:

$$E\{Z^n\} = n \, m_{n-1} \, . \, E\{Y_{n-1}\} \quad (17)$$

relation to be compared to (14).

$$E\{Z^n\} = (n!) \, m \, . \prod_{j=1}^{n-1} E\{Y_j\} = n \, m \, . \, E\{Y^{n-1}\} \quad (18)$$

The moments of order n of Z appear as the product of the moments of order 1 of the various transforms Y_j. Usage of relation (18) for better inference of the high order moments of Z, starting by the variance and covariance of Z(x), is being investigated.

KRIGING SEISMIC DATA IN PRESENCE OF FAULTS

Alain MARECHAL

ELF-AQUITAINE (PRODUCTION)

ABSTRACT
When processing seismic times, it is of upmost importance to take into consideration the action of faulting, to allow both a correct estimation and display of the data. Kriging with a fault throw as an imposed drift allow to include a very precise model of faults to the standard drift functions. After examining in detail an academic example in 1-D, the routine 2-D practical situation is studied and 2 real examples are shown. KEYWORDS : Seismic data, kriging with faults, contouring.

Introduction

Conventional seismic data appear at a given stage and after convenient processing, as two-ways times to a given horizon, assigned to a series of shot points along a profile. Next step in data processing will consist in plotting the isochrone curves relative to a given area, based on the time values of a certain number of profile (fig.1).

The two successive stages of this operation are :

interpolation, either as an interpolating function or as an interpolating grid.

G. Verly et al. (eds.), Geostatistics for Natural Resources Characterization, Part 1, 271–294.

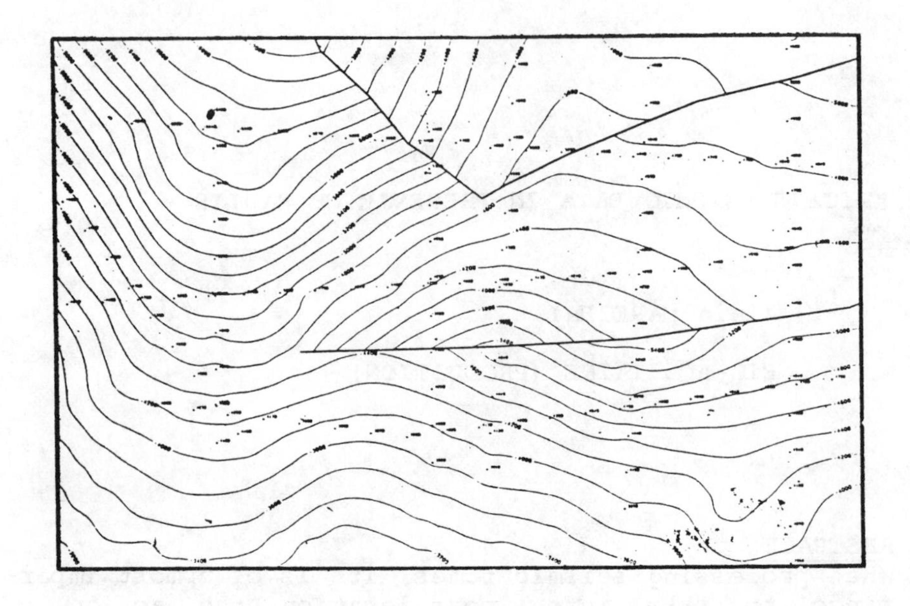

Figure 1. Example of seismic data location (RADIAN data)

- isolines plotting, either directly from the interpolating function, or using the interpolating grid.

Interpolating is actually performed in many cases using Kriging, the result of which is a grid handled by a contouring program : the major problem encountered presently is the convenient treatment of faults.

Existing solutions

Faults as a barrier

Most of the present, commercially available, Kriging programs, KRIGEPACK, BLUEPACK, POLYPAC, are multi-purposed oriented programs, in which faults are considered as limits for the variable under study. This point of view is truly acceptable when the

variable is for instance an ore-grade, the mineralization being often different on both sides of the fault. In the case of seismic reflector depth (or transit time), this point of view is unduly restrictive because the same variable is present on both sides of the fault trace, up to an unknowm fault throw.
Considering faults as barriers for the information on the variable presents usually two important drawbacks :

1. the area under study is splitted in smaller subarea which may contain a too small number of information to allow a good interpolation (see for instance in fig.1 the upper subarea).
2. the contouring programs using a grid as input usually plots the lines within each individual block, the location of the lines being based on the value of the 4 corners of the block. When a block is intersected by the fault trace considered as a barrier (or when the limits of the map do not coincide with the grid), such programs have to extrapolate some values on the grid nodes on the other side of the fault : this extrapolation violates the assumption that the fault is a barrier, and the numerical values obtained are not correct.

Penalty when crossing a fault trace

A first solution to the problem was proposed by J.Pouzet(1980). To overcome the point 1) above, he uses a weighted average interpolator based on all the informations on both sides of the fault trace, penalizing the values on the other side of the fault by considering that the respective samples are located at a longer distance than the true geometrical distance : the method seems to provide reasonable results, but the penalty is totally arbitrary and does not take into account the nature of the regionalization.

Trend surface analysis with a fault indicater

To my opinion a satisfying solution is proposed by Attoh and Whitten (1979) : they use a trend surface interpolator, the trend function incorporating indicator functions of the fault in addition to the usual polynomial functions. The action of the fault is a movement of the unknown surface along a given fault trace, this movement being defined by a 3 components vector : the two first components corresponds to the transcurrent displacement along the discontinuity and the last one to the vertical throw.

From the point of view of the trend model, the method is very satisfactory, although computationaly complex : indeed the amplitude of the horizontal displacement cannot be determined by the usual least square linear system and the authors have to make use of an iterative method to determine all the unknown coefficients.

From a geostatistical point of view however, the method is questionable as is any trend surface analysis method based on a simple least square criterion : the true solution would consist in using a Universal Kriging system based on the same set of trend functions, but using a variogram function (or a generalized covariance) that could be fit by the usual method on sample data. In addition, we know that a method leading to a non-linear system of equations requires a large number of sample values to provide robust results : we shall prefer a method, even less general, which solution will be obtained through the resolution of a linear system.

Splines with discontinuities

An exemple of such a solution is given by P.J. Laurent (1980) in the 1-D case, within the scope of spline interpolation : he shows that provided the classical spline objective is slightly modified, it is possible to incorporate discontinuities within the spline interpolation. P.J. Laurent shows examples of interpolation with discontinuities on the first derivatives in some points : it is not difficult to extend his method to a discontinuity of the function itself, which is exactly what I propose to do within the frame of the Kriging method itself.

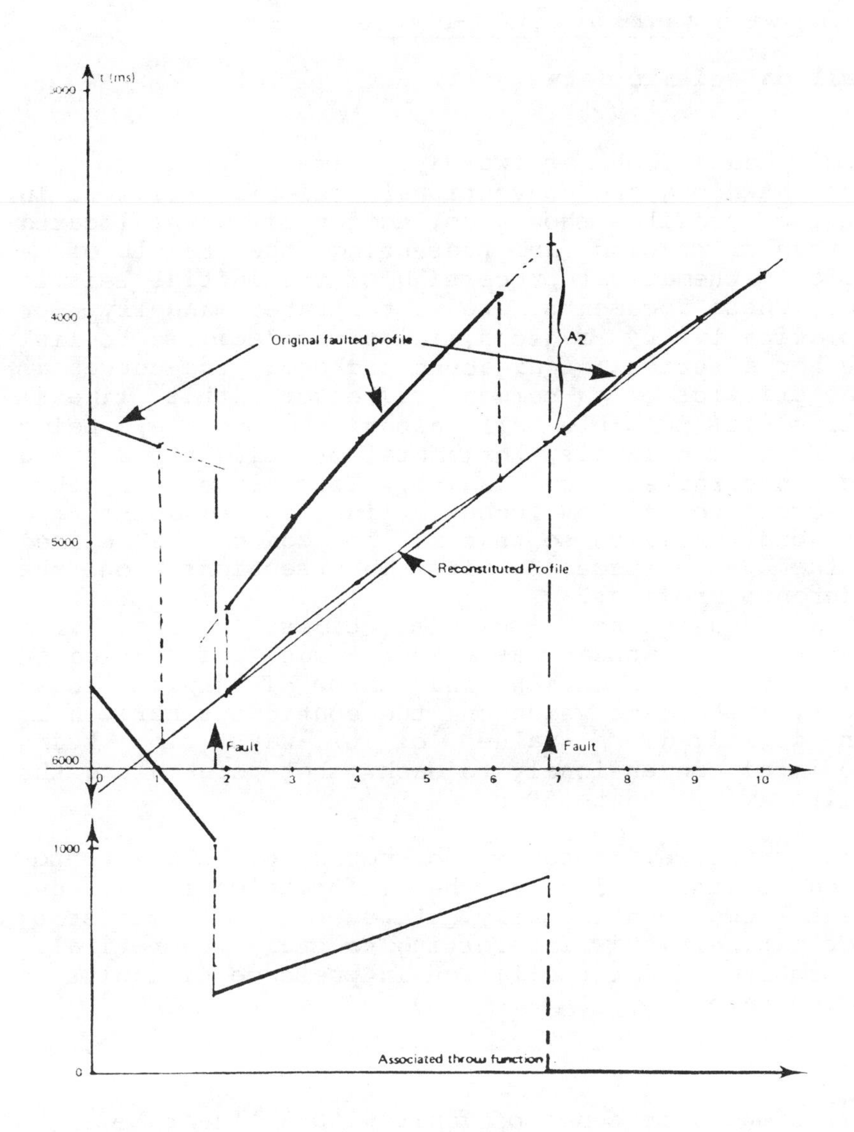

Fig 2 : Seismic Profile N/S n° 6 of Fig 1 .

Kriging with fault in the 1-D case

Recall on seismic data

Seismic data such as two-ways times appear as 1-D functions along the conventional seismic profiles. In fact such profiles show a collection of curves located at regular spacing and presenting the result of a complex mathematical processing of the initial seismic echo. These documents are interpreted manually (or automatically) by a geophysicist who decides to link together a series of adjacent peaks on the curves as caracteristics of a common reflector : this line is followed if possible all along the profile, being usually a continuous, horizontal or slightly dipping line interrupted by faults. This line is then connected to a homologous line in adjacent and orthogonal profiles so that an horizon can be defined in the 2-D space by its intersections on the differents profiles.
As a consequence, seismic times coming from interpretation appear as a continuous information in 1-D, with the location and throw of faults known exactly : the time value on the continuous horizon is then digitized, a value of two-ways time being associated conventionaly to each shot-point along the profile every 25 to 50 m.

So in practice, there is no problem of 1-D interpolation on these data : the information is at one moment known continuously. However, as a matter of illustration, it is interesting to study theoretically the problem of interpollation in presence of faults in the 1-D case.

Estimation in presence of fault with a linear variogram

Figure 2 shows the sequence of values read on the N/S profile n° 6 of figure 1 : according to the complete map, the southern part of the profile corresponds to the undisturbed horizon, the northern part crossing two fault traces. It can be seen easily on this profile that considering only a vertical movement on the faults, the displacement consists in a vertical shift and a change of dip : so in the 1-D case, it seems that a vertical displacement can be modeled by combination of two elementary functions, a step function and a truncated linear function (fig 3).

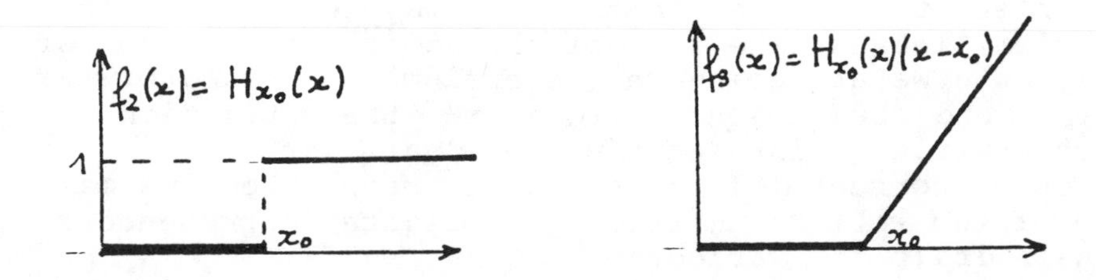

Let us now consider the case of a 1-D variable Z(x) known at a regular spacing ih, i=o, n. We assume the existence of a fault between the two samples positions ph and (p+1)h, at the exact location (p+ε)h. We assume that, after correction of the fault throw, the variable $\tilde{Z}(x)$ is a IRF of order 1 whith a generalized covariance K(h) = - |h|. We look for the kriging estimator of any point x along the line,

$$Z_x^K = \sum_{\alpha=0}^{n} \lambda_x^\alpha z_\alpha$$

It is easy to find the kriging system :

$$\lambda_x^\beta K_{\alpha\beta} + \mu_\ell f_\alpha^\ell = K_{\alpha x} \qquad \alpha = 0, 1, \ldots n$$

$$\lambda_x^\beta f_\beta^\ell = f_x^\ell \qquad \ell = 0, 1, \ldots 3$$

The basic functions f_x^ℓ are, in our case :

$$f_x^0 = 1. \qquad f_x^1 = x$$

$$f_x^2 = H_{x_0}(x) \qquad f_x^3 = H_{x_0}(x)(x - x_0)$$

This system has a unique solution, provided that :

$$\sum_{l=0}^{3} c_l f_\alpha^l = 0 \ , \ \alpha = 0,1,\dots n \implies c_l = 0 \quad l = 0,1,\dots,3$$

In particular, it is necessary that (n+1)> 4.
In order to find an instructive way to resolve this problem, let us make use of the additivity theorem of kriging : we are going to determine the estimators of the trend coefficients, to allow the estimation in each sample point of the residuals (that is the function corrected from the fault throw).Then, we make use of the well known kriging procedure in presence of linear drift and variogram $|h|$.

For the sake of simplicity, let us adopt the point of view of universal kriging. We assume the following decomposition :

$$Z_x = a_0 + a_1 x + a_2 H_{x_0}(x) + a_3 (x - x_0) H_{x_0}(x) + Y_x$$

where Y(x) has locally on the segment (0,nh) a variogram $\gamma(h) = |h|$.

The estimators of the coefficients a_l will be :

$$A_l = \int_L \lambda_l(dx) Z(x)$$

In our case, the measure $\lambda_l(dx)$ will in fact reduce to a linear combination of Dirac measures $\delta_{\alpha h}(dx)$.

The $\lambda_l(dx)$ measures are solution of :

$$\int_L \lambda_l(dx) |x - \beta h| = -\mu_{ls} f^s(\beta h) - \mu_{0l} \qquad \beta = 0,1,\dots n$$

$$\int_L \lambda_l(dx) f^s_x = \delta^l_s(dx) \qquad s = 0,\dots 3$$

The right hand side of the optimality equations of the system is a linear combination of the trend functions so the solution $\lambda_l(dx)$ will be a linear combination of measures solutions of :

$$\int_L \lambda_\ell(dx)|x-y| = f_y^s \qquad s=0,1\,3$$

An easy calculation shows that the following measures correctly combined are solution of the above equation :

$$\nu_0 = \frac{\delta_0+\delta_n}{2}, \qquad \nu_1 = \frac{\delta_n-\delta_0}{2}$$

$$\nu_2 = \frac{\delta_{p+1}-\delta_p}{2} \qquad \nu_3 = \frac{\delta_n-\delta_p}{2}$$

The final results $\lambda_\ell(dx)$ are linear combinations of the above elementary solutions with coefficients determined by the universality equations :

$$\lambda_\ell(dx) = b_{0\ell}\,\nu_0 + b_{1\ell}\,\nu_1 + b_{2\ell}\,\nu_2 + b_{3\ell}\,\nu_3$$

$$\sum_{i=0}^{3} b_{i\ell}\int \nu_i(dx)\, f_x^s = \delta_\ell^s \qquad s = 0,1,..3$$

The estimated coefficients A_ℓ of the trend are :

$$A_0 = z_0 \qquad A_1 = \frac{z_p - z_0}{ph}$$

$$A_2 = z_{p+1} - z_p - \varepsilon\,\frac{z_p - z_0}{ph} - (1-\varepsilon)\,\frac{z_n - z_{p+1}}{n-p-1}$$

$$A_3 = \frac{z_n - z_{p+1}}{n-p-1} - \frac{z_p - z_0}{p}$$

These results are in agreement with the usual results found with a a linear variogram and linear trend :

1. The linear trend is given by the straight line passing through the first and last points of the undisturbed part of the profile.
2. The slope of the fault throw is given by the slope of the line passing through the first and last points of the displaced part of the profile minus the slope of the overall trend.
3. The constant A2 of the throw function is obtained graphically as the vertical distance between the two above mentionned lines at the abcissa of the fault (See a graphical exemple in figure 4) .

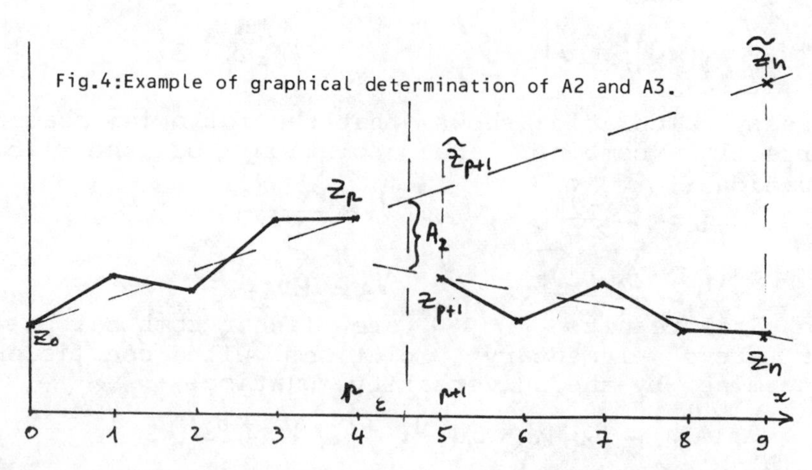

Fig.4:Example of graphical determination of A2 and A3.

4. The two sample values $\hat{z}_{p+1}$ and $\hat{z}_n$ when corrected of the throw coincide with the extrapolation from the last undisturbed point Zp. Indeed, the throw function $D(k)$ for each sample k, k=p+1,n has the following value :

$$\hat{z}_k = z_k - D(k)$$

$$D(k) = z_n - z_p - (k-p)\frac{z_p - z_0}{p} - (n-k)\frac{z_n - z_{p+1}}{n-p-1}$$

$$\hat{z}_n = z_p + (n-p)\frac{z_p - z_0}{p}$$

$$\hat{z}_{p+1} = z_p + \frac{z_p - z_0}{p}$$

Practical case of a succession of faults

The preceding example show how to estimate the parameters of the fault throw in the case of a unique fault, with linear variogram : in practice, it is possible to incorporate to the model many faults points, each fault resulting in two additionnal equations in the kriging system. Indeed, it way be necessary to use an other covariance function especially for seismic data which are very regular : considering that these data are interpreted, it may be justified to use systematically a spline covariance

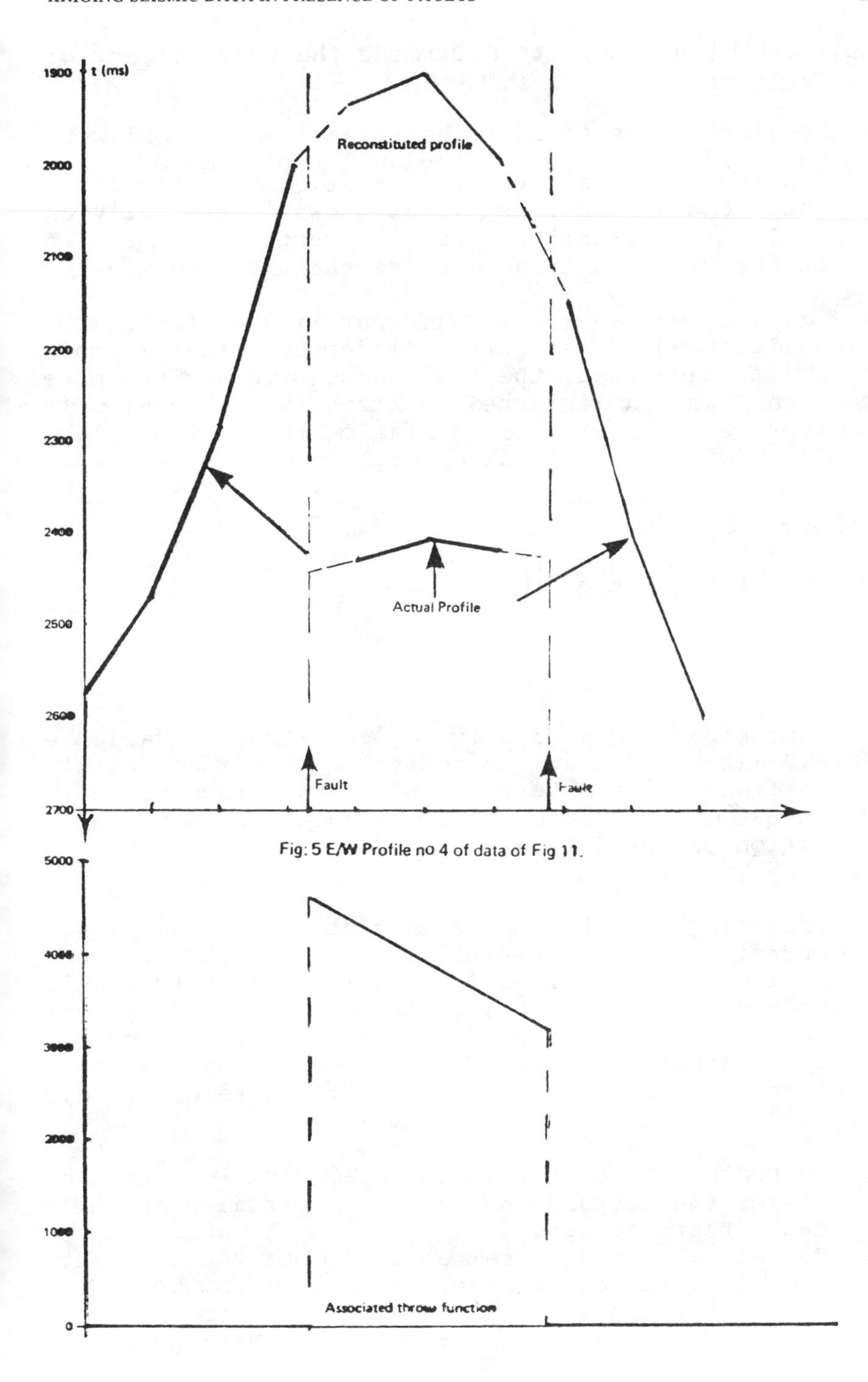

Fig: 5 E/W Profile no 4 of data of Fig 11.

$K(h) = -|h|^3$ at least to determine the coefficients of the fault throw.

In the simple case of $\tilde{Z}(x)$ having a linear trend with linear variogram, the markovian properties of this model allow to make the determination of this fault parameter one by one, as it is shown graphically on figure 2 : the function $\tilde{z}(x)$ corrected of the two faults throws looks in good agreement with the model.

In figure 5, we can see a different kind of tectonics, which deserves a slightly different model : this graben is such that the two successive faults move from the same undisturbed horizon, which obviously requires a model with a quadratic trend. Hence the complete model for this case would be :

$$Z_x = a_0 + a_1 x + a_2 x^2 + a_3 H_{x_1}(x)[1 - H_{x_2}(x)] + a_4 H_{x_1}(x)(1 - H_{x_2}(x))(x - x_1) + Y_x$$

By working graphically on the variable Z(x)-A0-A1x-A2x**2, it is possible to define easily the estimated parameters A3 and A4 and the value of the throw $D(k)$ for each point within the graben.

Conclusion of the 1-D case

The preceding example shows that it is not difficult to determine the caracteristics of the faults movement in the 1-D case, provided it is assumed that the fault points are known, that the fault movement is only vertical, and given an a priori model of covariance for the undisturbed variable. What would be the practical interest of such a calculation in the case of seismic data ?

1. To reconstitute the undisturbed profile, to make easier the comparison with other profiles and help the interpretation.
2. To allow a structural analysis of the complete profile based on a larger number of points : When the faults are many, it is difficult to find long sequences of undisturbed values, which penalizes

the structural analysis. Obviously, the use of estimated throw values instead of real ones will induce a bias (See figure 2 where Z0,Zp and Zn ly on the same straight line), but this bias seems to be less important than for variograms of residuals in universal kriging.

In addition to the above arguments, estimating faults throws in seismic profile would help doing a good contouring in the 2-D space.

Kriging with fault in the 2-D case

Spatial distribution of seismic data

We have recalled previously the caracteristics of the seismic times : very continuous variable, due to a human interpretation, disturbed by a series of faults usually well identified on profiles. At the level of profile interpretation, the variable is known continuously, but later digitized at a regular spacing.

A seismic campaign usually provides two orthogonal sequences of parallele profiles, at regular spacing which varies between 200 m and few kilometres.

From the point of view of estimation and contouring the seismic data present a double difficulty : regular rectangular pattern of sample point with a large ratio of spacing (from 4 to 10), and presence of faults.

Theoretical solution

As in the 1-D situation, we shall assume that a vertical shift of the data provides a satisfactory reconstitution of the undisturbed surface : this approximation is reasonable in practice because most of the faults considered in seismic interpretation are subvertical, so that the horizontal displacement would not be too large. In addition, it has to be considered that the horizon being fairly continuous, the horizontal displacement of the sample points would not change largely the interpolated surface.

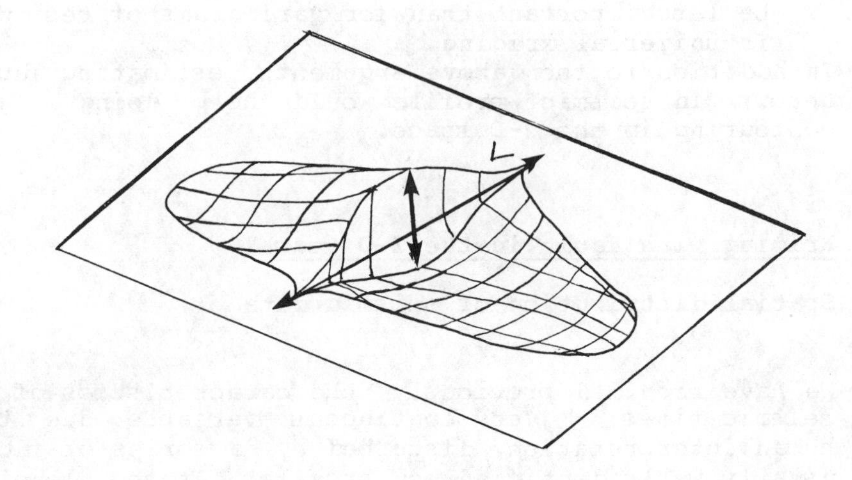

Fig. 6-A : Fault with predefined complex throw

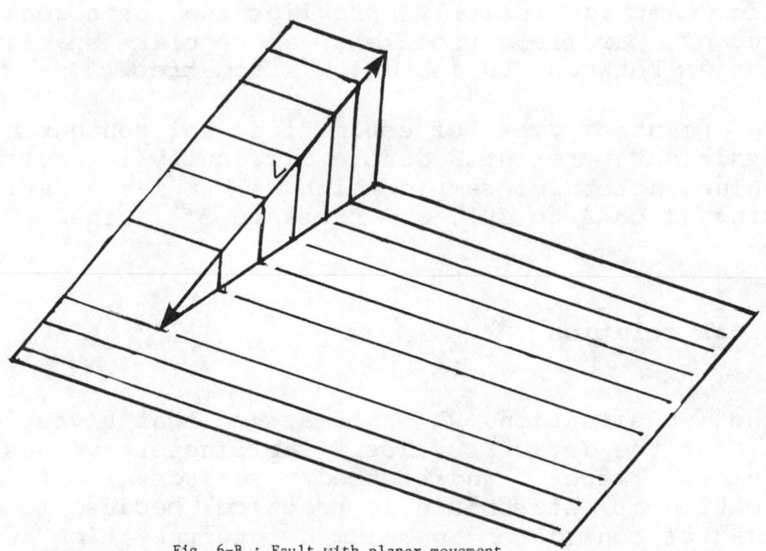

Fig. 6-B : Fault with planar movement

Even when considering only faults with a vertical displacement, there exists a large variety of tectonic accidents :

- .fault trace defined by a unique straight line, or needing a succession of linear segments.
- .fault with a constant throw along the fault trace, or with a variable throw.
- .fault with a planar movement, or with a more complex displacement function (figure 6).

If we want to stay within the scope of linear methods, we need the throw function associated with each accident to be a linear combination of basic trend functions. This can be achieved in two ways :

1. by defining a throw function as a combination of mathematical functions : the example n° 2 of the figure 6 can be generated by a combination of a plane intersected by a step functions $H_L(x,y)$ valid only along the segment L :

$$f(x,y) = a_0 H_L(x,y) + a_1 x H_L(x,y) + a_2 y H_L(x,y)$$

2. by digitizing the shape of a throw function, of unit vertical shift and given lenght of influence along a given fault segment L. The unknown throw will then be considered as proportional to this unit throw function $T_L(x,y)$:

$$f(x,y) = a\, T_L(x,y)$$

If the movement of all the faults in a given area can be represented as a combination of n such basic functions, then it is possible to determine the kriging estimator in this area incorporating these drift functions, under the assumption that the undisturbed variable is an IRF-k function of general covariance K(h) :

$$\lambda^{\beta}_{x} K_{\alpha\beta} + \mu_l f^l_{\alpha} = K_{\alpha x} \qquad \alpha = 1, \ldots N$$

$$\lambda^{\beta}_{x} f^l_{\beta} = f^l_{x} \qquad l = 0, 1, \ldots k, k+1, \ldots k+n$$

Practical solution

As each individual fault in 2-D is described with a much larger number of parameters than in the 1-D case, it is easy to realize that the strict theoretical approach is usually difficult to implement.One can think, however, that such solution could be implemented in a graphical way similar to the one applied in C.A.D. : when defining the fault traces in a given area, the interpreter either would draw a model of the fault movement later digitized by the system, or he could pick up one particular model in an already existing catalogue. At the end of this step, the computer would resolve the above kriging system. In order to test the practicality of the use of a throw function in kriging, the strict theoretical solution given above was abandoned and replaced by an approximation which can be operated on the existing kriging solftwares (KRIGEPACK or BLUEPACK).

The steps are the following :

1. 1-D. Calculation on each seismic profile, for each data point of a throw function as defined in part one of this article.
2. Interpolation on the whole domain of this throw function, in order to get within each fault blocks a smooth throw function computed both on a regular grid covering the domain and on each sample point. It must be emphasized that this throw function will be later included as a trend function in the model, so that it is an element of the model and not an estimated quantity : the interpolation used to define completely this trend function has to be considered as a tool convenient for this purpose, so if the result is not truly satisfactory, there are no inconvenient to modify it arbitrarily (by an additionnal smoothing or by modifying some points of the grid).
3. Standard geostatistical study of the "undisturbed" variables on the sample points using either an automatic structure identification (KRIGEPACK, BLUEPACK) or the computation of a variogram of residuals (if possible). This allows fitting a model of IRF-k for the undisturbed variable
4. Kriging of Z(x) on a regular grid, with a standard IRF-k kriging system : the grid location and spacing will coincide with the grid used to interpolate the throw function of step 2. It is possible to check graphically the undisturbed

function by contouring and see whether it is now a regular and smooth function or if some faults would have been omitted.

5. Reconstitution of the final function Z(x) by adding $\widetilde{Z}(x)$ on the grid nodes to D(x).

$$Z_i^* = \widetilde{Z}_i^* + D_i$$

The whole method would work well in a interactive graphic system : display of each profile to check the fault location and throw, reunion of the fault points along a given fault traces, calculation, display and possibly correction of the 2-D throw function, interpolation and display of the undisturbed surface, display of the final result. The use of such a system would allow an easy modification of the assumptions concerning the faults, with the possibility of a quick visual check.

Example of use of the method

The two following examples were processed with the standard software in use at Elf Aquitaine (Production) :

1. KRIGEPACK, in use since 1974 for automatic structure identification, kriging and generation of the results on a grid.
2. MIXGRID, for operations on grid values
3. An experimental contouring program, part of the GEOL package lend to us for tests by J.L.MALLET

For reasons of confidentiality and to allow an easy comparison with other methods, the examples shown come from data published in publicity brochures : example 1 (figures 1, 2, 7, 8) is taken from a brochure of Radian Company for the software CPS-1, Example 2 (Figures 5, 8, 9) is taken from a brochure of Sattleger Ing. for their sýstem SISPP.

In both examples, the above mentioned method was applied using the existing software KRIGEPACK, MIXGRID without any change. A clear improvement in the final result would be obtained by smoothing the trend function obtained by KRIGEPACK using a cubic covariance function. It can be seen, however, that the reconstituted horizon is very likely, justifying the method : it was obtained by kriging with the parameters defined by the automatic structure indentification.

Application of the method

There are two applications to surface interpolation :

1. gridding of the variable to allow some mathematical applications such as volumetrics, boundary conditions in reservoir models,ray-trace calculation, migration of seismic times.
2. display of the variable through contouring or block diagram.

Our method for processing faults provides a clear improvement in the two above-mentioned applications :

A major difficulty for mathematical calculation in volumetrics for instance is occasioned by absent values in the grid. Indeed kriging programs requires the presence of a minimum number of sample values in the vecinity of each point to interpolate, so that the output grid may contain a fairly large number of voids whenever the fault blocks contain too few samples (see figure 10). This is no longer the case when kriging the undisturbed surface with all the samples considered together : for instance in a unique neighbourhood it is possible to interpolate the result on the complete grid of values.
In addition, if kriging is done incorporating fault functions as described in a paragraph before the kriging system will provide a significant kriging variance, assuming the hypothesis of continuity of the surface up to a fault shift. In particular this allows the practice of conditional simulation based on fixed fault traces.

The method of throw function determination allows an improvement in contouring near the fault trace : any contouring method have to "invent" an extrapolation of the gridded values through the fault trace. Such extrapolation can be done more exactly using the grid of undisturbed values and an extrapolation of the throw function accross the fault trace (figure 15) :

The domain to contour is divided in individual fault blocks limited either by fault trace or by a dummy limit : Within each such block the throw function is gridded and slighly extrapolated over the fault trace, so that by addition to the undisturbed surface, a "natural" extrapolation of the surface can be obtained and contoured up to the fault traces limit : the advantage appears clearly on figure 14, compared with a similar figure (figure 13) obtained by a standard contouring method.

Conclusion

The solution for dealing with faults proposed in this paper does not require any new developpement of the kriging theory : It is an illustration of the concept of "external drift" proposed long time ago in geostatistics as a way to include in the model some additional information other than sample values . Such concept has not been much exploited until now, the geostatistician prefering models with undiferenciated trend functions, such as polynomials.
It is my opinion that the "Man of the Art" in different disciplines (a geologist, a geophysician, etc...) should be asked to provide a more precise information about the trend of the phenomenon under study than the usual list of polynomial functions : Such a practice is obviously more risky than the practice of automatic determination of standard basic trend functions, but allows the incorporation in the numerical interpolation procedure of some amount of qualitative knowledge. Existing examples are :

- inclusion of interpolated seismic data as a trend function for the estimation of depth based on few wells only (DELHOMME et al,1981).

- definition of a circular symetry for some geological phenomenon such as volcanic pipe or a porphiry-type deposit.

- definition of the direction of trend of a phenomenon by eliminating in the model some trend functions (Example of a phenomenon with a vertical trend only)

As a conclusion, much work was done to increase the number of admissible covariance functions and develop the use of some of them (such as the spline covariance) : it is time now to investigate the use of newer trend functions allowing a better representation

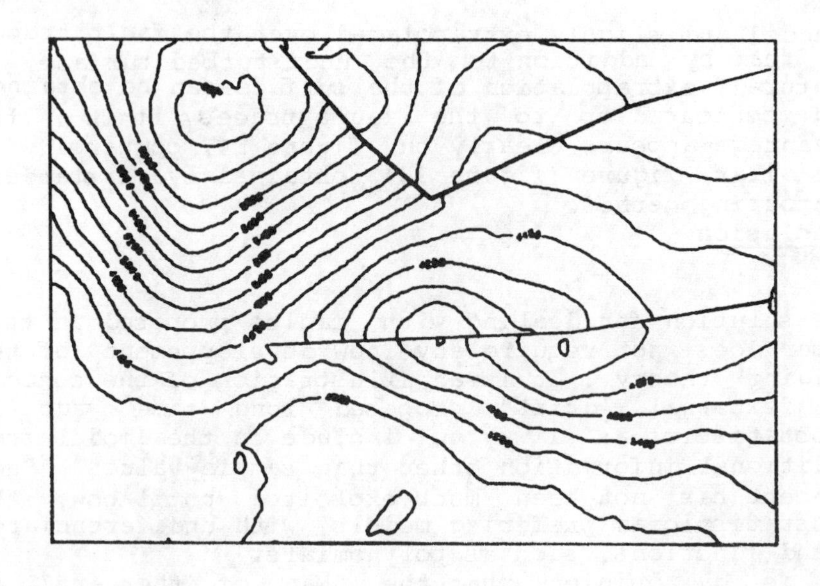

Fig. 7 : Radian Data : Standard krigepack processing considering faults as barriers

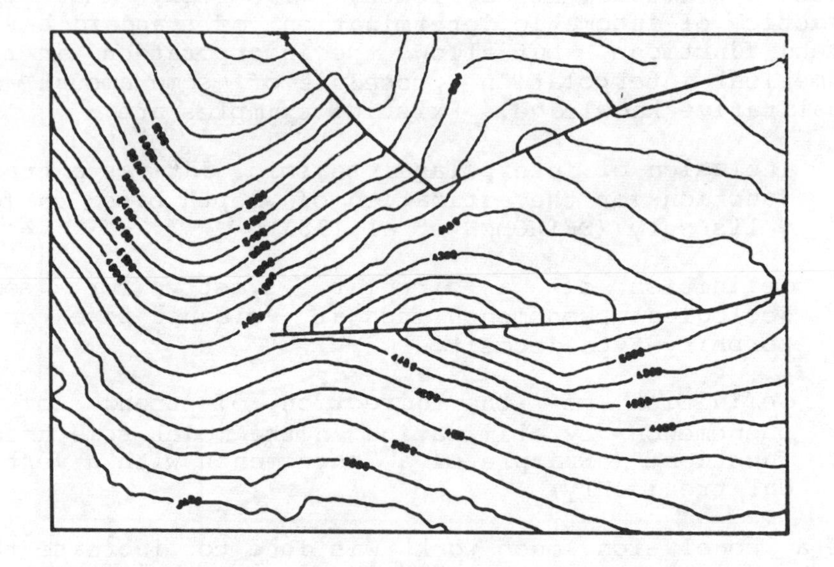

Fig. 8 : Radian Data : kriging using a throw function

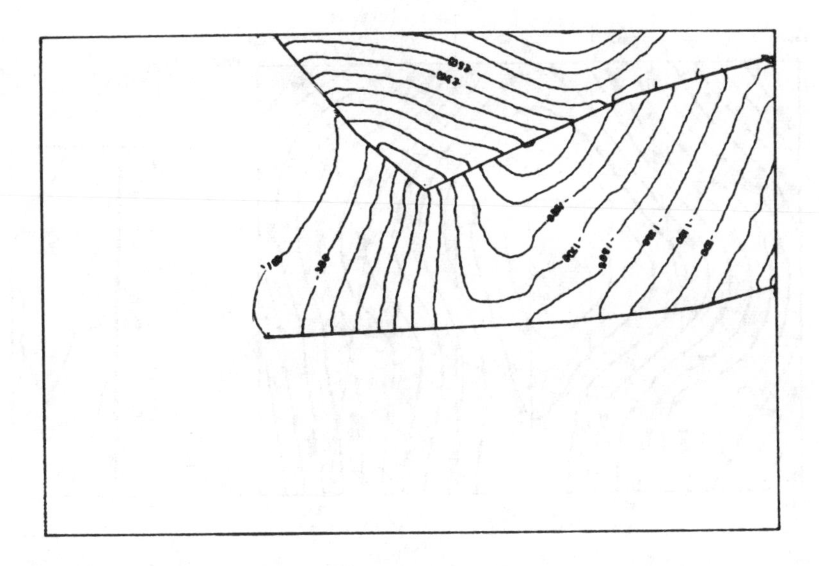

Fig. 9 : Radian Data : Throw function generated by krigepack from throw sample values.

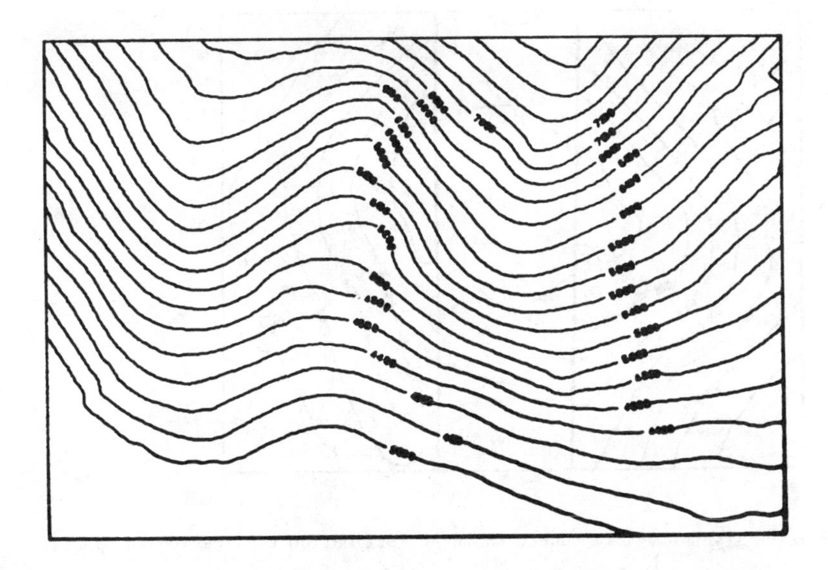

Fig. 10 : Radian Data : Cartography of reconstituted surface by krigėpack, with automatic structure identification.

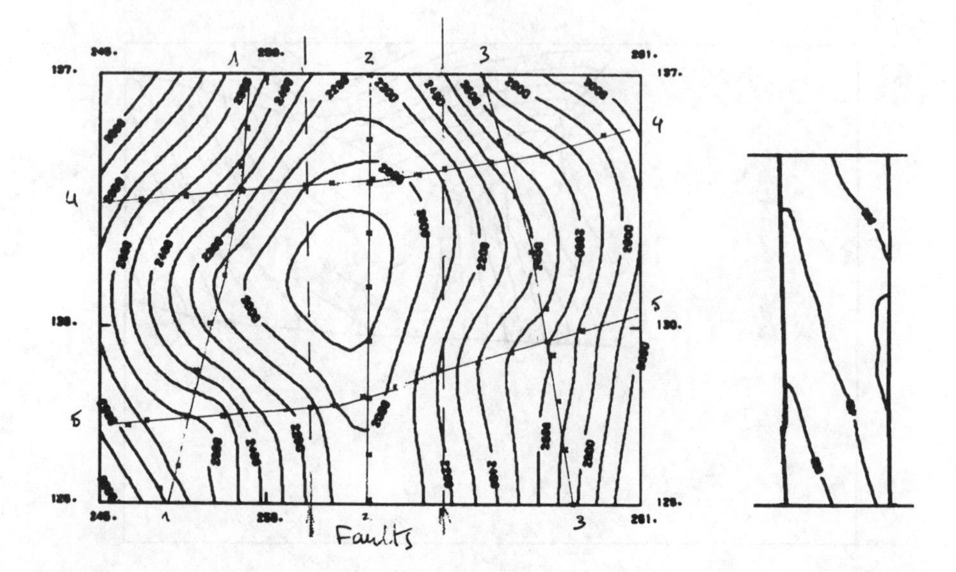

Fig. 11:Sattlegger data:Display of sample location and cartography of the reconstituted surface.

Fault throw function

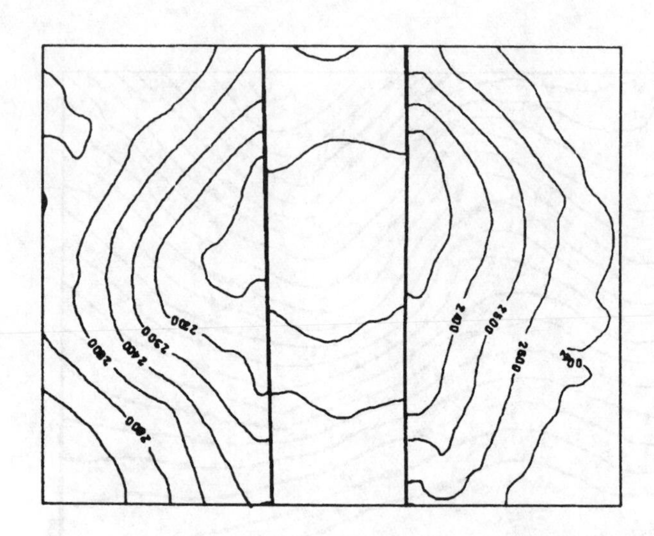

Fig.12:Sattlegger data.Standard KRIGEPACK estimation considering the faults

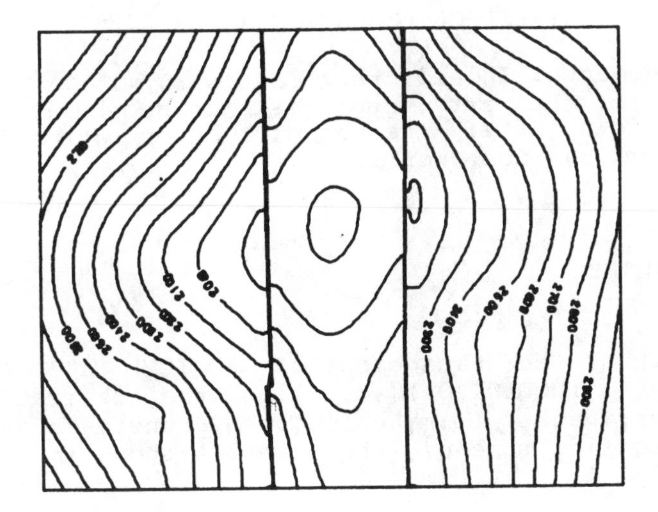

Fig. 13 : Sattlegger data : Final grid with fault throw function contoured directly by a contouring with fault program.

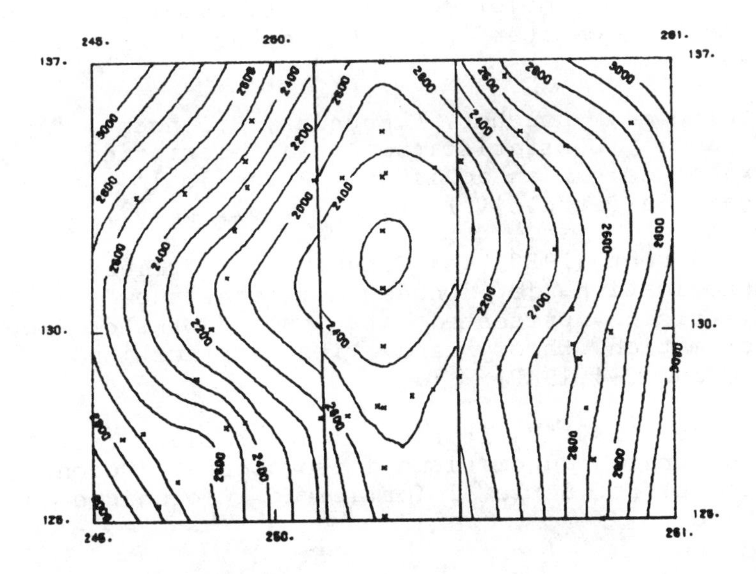

Fig.14:Sattleger data.Final grid contoured by fault block to ensure a good continuity to the fault trace.

of some phenomenon. Indeed, the last word will remain to the "Man of the art", geologist or geophysician, the only who can justify the use of such or such model for describing the real world.

ACKNOWLEDGEMENTS

I wish to thank the Management of SOCIETE NATIONALE ELF AQUITAINE (PRODUCTION) ,to allow the publication of this paper.Thanks to my collegue B.Brunet for his help in processing some of the examples shown in this paper.

BIBLIOGRAPHY

K. Attoh, E.H.T. Whitten (1979) "Computer program for regression model for discontinous structural surfaces" (Computer and Geosciences, Vol. 5 pp47-71)

J.P.Delhomme,M.Boucher, G.Meunier, F.Jensen : "Apport de la Geostatistique à la description des stockages de gaz en aquifère" (Revue de l'institut Français du Petrole,Vol 36, n°3 1981, pp309,327)

P.J. Laurent (1980)"Inf-Convolution Spline pour l'approximation de données discontinues" (May 1981.Article presented at the "Workshop on approximation theory and its application" Haïfa,June 3-4,1980)

J. Pouzet (1980)"Estimation d'une surface faillée pour un tracé automatique d'isovaleurs" (Sciences de la terre. Série Informatique Géologique n° 14.,Nancy)

CO-KRIGING - NEW DEVELOPMENTS

Donald E. Myers
Department of Mathematics
University of Arizona
Tucson, Arizona 85721

Abstract

Co-Kriging or Joint Estimation utilizes data from correlated variables to improve the estimation of all variables or to compensate for missing data on some variables. The general formulation of Co-Kriging in matrix form was given by the author. The matrix form emphasizes the analogy with Kriging of one variable utilizing only spatial dependence. General conditions are obtained for covariance matrix functions and variogram matrix functions. The extension to block co-kriging is delineated including the Co-Kriging variance. A simple algorithm is given for obtaining the "under-sampled" case from the general matrix formulation. Finally a method for reducing the size of the system of equations is given and a iterative method provided which allows solution of even singular systems in which entries are matrices.

INTRODUCTION

In (2), (3) the author has given in matrix form the general formulation of co-kriging and shown that it is a direct extension of kriging in that form. In this paper we extend those results in several ways. Section one is concerned with the general problem of the variance of vector linear combinations and conditions necessary for a matrix function to be a covariance or a variogram.

G. Verly et al. (eds.), Geostatistics for Natural Resources Characterization, Part 1, 295–305.

Although the extension to block estimation does not present any significant difficulties, it was not included in (2), (3), and for the sake of completeness is covered in Section two.

The "under sampled" case is the one that has received the most attention in mining applications but only for a few variables and a few sample locations. Section three describes a simple algorithm for obtaining the system of equations in matrix form for any "under-sampled" problem from the general form of Co-Kriging.

Finally Section four introduces methods for solving the large systems of equations that are generated in the use of Co-Kriging.

NOTATION

The notation used in (3) will be followed here. Recall that Tr denotes the trace, i.e. the sum of the diagonal entries and A^T denotes the transpose of A .

VARIANCE OF VECTOR LINEAR COMBINATIONS

One of the singular characteristics of Ordinary Kriging, i.e., the use of variograms and IRF-0's is that certain linear combinations can have finite variance even though the random function does not have finite variance. One of the properties of a variogram is the following

$$-\sum\sum \gamma(x_i - x_j)\lambda_i\lambda_j \geq 0 \qquad (1)$$

for all weights $\lambda_1, \ldots, \lambda_n$ such that $\lambda_1 + \ldots + \lambda_n = 0$. That is, $-\gamma$ is conditionally positive definite. In the case of a second order stationary random function the covariance must be positive definite. For Co-Kriging it is then appropriate to ask what kind of matrix functions can be covariance matrices or variogram matrices.

Let K(h) be an m x m matrix with entries $k_{ij}(h)$. In the case of second order stationary random functions K(h) should satisfy

$$\mathrm{Tr}\ E[\sum_{i=1}^{n}\sum_{j=1}^{n} \Gamma_i^T \bar{Z}(x_i)^T \bar{Z}(x_j)\,\Gamma_j]$$

$$= \mathrm{Tr}[\sum_{i=1}^{n}\sum_{j=1}^{n} \Gamma_i^T K(x_i - x_j)\,\Gamma_j] \geq 0 \qquad (2)$$

where we assume without loss of generality that

$$E[Z(x)] = [0, \ldots, 0]$$

If the Z_i's are IRF-0's then $K(h)$ should satisfy

$$\mathrm{TrE}\,[\sum_{j=1}^{n}\sum_{i=1}^{n} \Gamma_i{}^T\bar{Z}(x_i)^T\bar{Z}(x_j)\Gamma_j]$$

$$= - \mathrm{Tr}[\sum_{i=1}^{n}\sum_{j=1}^{n} \Gamma_i{}^T K(x_i - x_j)\Gamma_j] \geqslant 0 \qquad (3)$$

for all $\Gamma_1, \ldots, \Gamma_n$ with

$$\sum_{j=1}^{n} \Gamma_j = \begin{bmatrix} 0 \ldots 0 \\ \vdots \quad \vdots \\ 0 \ldots 0 \end{bmatrix} \qquad (4)$$

Both cases may be considered at the same time. Let Γ_{st}^j be an m x m matrix whose only non-zero entry is λ_{st}^j. Then

$$\Gamma_j = \sum_{j=1}^{n}\sum_{t=1}^{n} \Gamma_{st}^j \qquad (5)$$

Substituting in (2) or (3) and recalling that (2) must be satisfied for all Γ's or (3) must be satisfied for all Γ_j's satisfying (4), it is necessary and sufficient that

$$\sum_{i=1}^{n}\sum_{j=1}^{n} \lambda_{ss}^i\, k_{ss}(x_i - x_j)\, \lambda_{ss}^j +$$

$$\sum_{i=1}^{n}\sum_{j=1}^{n} \lambda_{ss}^i k_{st}(x_i - x_j)\, \lambda_{tt}^j +$$

$$\sum\sum \lambda_{tt}^i\, k_{tt}(x_i - x_j)\, \lambda_{tt}^j$$

$$\geqslant 0 \qquad (6)$$

$$\leqslant 0 \qquad (7)$$

for all s, t . (6) corresponds to (2) and (7) to (3). This can be described somewhat simpler.

$$\lambda_{ss}^1, \lambda_{ss}^2, \ldots, \lambda_{ss}^n$$

are diagonal entries in $\Gamma_1, \ldots, \Gamma_n$ respectively. $\lambda_{tt}^1, \ldots, \lambda_{tt}^n$ are also diagonal entries. Let

$$A_s = [\lambda^1_{ss}, \ldots, \lambda^n_{ss}] \tag{8}$$

$$B_t = [\lambda^1_{tt}, \ldots, \lambda^n_{tt}] \tag{9}$$

$$K_{uv} = \begin{bmatrix} k_{uv}(x_1-x_1) \ldots k_{uv}(x_1-x_n) \\ \vdots \qquad\qquad \vdots \\ k_{uv}(x_1-x_n) \ldots k_{uv}(x_n-x_n) \end{bmatrix} \tag{10}$$

then (16) can be written

$$A_s K_{ss} A_s^T + A_s K_{st} B_t^T + B_t K_{tt} B_t T \geqslant 0 \tag{11}$$

and (17) as

$$A_s K_{ss} A_s^T + A_s K_{st} B_t^T + B_t K_{tt} B_t^T \leqslant 0 \tag{12}$$

$$\sum_{i=1}^{n} \lambda^i_{ss} = 0 \ , \ \sum_{i=1}^{n} \lambda^i_{tt} = 0 \tag{13}$$

It is clearly necessary that

$$A_s K_{ss} A_s^T \geqslant 0$$

$$B_t K_{tt} B_t^T \geqslant 0 \tag{14}$$

which are consequences of (21). (24) is the usual positive definite condition for covariance functions.

BLOCK CO-KRIGING

In (2), (3), (4) the author indicated that the formulation of Co-Kriging given would extend to estimation of block values, the results were not given. For the sake of completeness they are icluded herein. Let V be a volume, area or length and let

$$\bar{Z}_V = [\frac{1}{V} \int_V Z_1(x)dx, \ \ldots, \ \frac{1}{V} \int_V Z_m(x)dx] \tag{15}$$

If $\bar{Z}^*_V = \sum_{k=1}^{n} Z(x_k) \Gamma_k$, the problem as before is to determine the Γ_k's so that Z^*_V is unbiased and the estimation error has minimum variance.

Stationary/Covariance case

If $Z_1, Z_2, \ldots, Z_m$ are second order stationary and

$$\bar{C}(h) = \begin{bmatrix} C_{11}(h) \ldots C_{1m}(h) \\ \cdot \quad\quad \cdot \\ \cdot \quad\quad \cdot \\ \cdot \quad\quad \cdot \\ C_{m1}(h) \ldots C_{mm}(h) \end{bmatrix}$$

then

$$E[\bar{Z}(x_i)]^T[\bar{Z}_V] = = \frac{1}{V}\int_V \bar{C}(x_i - x)dx = \bar{C}(x_i, V) \qquad (16)$$

and

$$E[\bar{Z}_V]^T(\bar{Z}(x_j)] = \frac{1}{V}\int_v \bar{C}(x - x_j)dx = \bar{C}(v, x_j) \qquad (17)$$

$$E[\bar{Z}_V)^T[\bar{Z}_V] = \frac{1}{V^2}\int_V\int_V \bar{C}(x-y)dxdy = \bar{C}(V,V) \qquad (18)$$

The Kriging equations become

$$\sum_{j=1}^{n} \bar{C}(x_i x_j)\,\Gamma_j + \bar{\mu} = \bar{C}(x_i, V), \quad \sum_{j=1}^{n} \Gamma_j = I \qquad (19)$$

with Kriging variance

$$\sigma^2_{CK} = \text{Tr}\,\bar{C}(V,V) - \text{Tr}\sum \bar{C}(x_j, V)\,\Gamma_j - \text{Tr}\bar{\mu} \qquad (20)$$

and as with punctual Co-Kriging the component corresponding to each Z_i may be selected out.

The Intrinsic Case

If the Z_i's are IRF-0's then we may write

$$\bar{Z}_V - Z^*_V = \sum_{k=1}^{n} \frac{1}{V}\int_V [\bar{Z}^* x(- \bar{Z}(x_k)]dx \qquad (21)$$

and $E\,[\bar{Z}_V - \bar{Z}^*_V]^T[\bar{Z}_V - \bar{Z}^*_V]$ becomes

$$= \sum\sum \frac{\Gamma_i^T}{V^2} \int\int E[\bar{Z}(x) - \bar{Z}[x_i)]^T[\bar{Z}(y) - \bar{Z}(x_j)]dxdy\,\Gamma_j$$

$$= \sum\sum \Gamma_i^T \frac{1}{V^2}\int_V\int_V [\bar{\gamma}(x-x_j) + \bar{\gamma}(x_i - y) - \bar{\gamma}(x-y)$$

$$- \bar{\gamma}(x_i x_j)]dxdy\ \Gamma_j \tag{22}$$

From (22) it is easily seen that the Kriging equations are

$$\sum_{j=1}^{m} \bar{\gamma}(x_i - x_j)\ \Gamma_j + \bar{\mu} = \bar{\gamma}(x_i, V) \quad , \quad \sum \Gamma_j = I \tag{23}$$

$$\text{where} \quad \bar{\gamma}(x_i, V) = \frac{1}{V} \int_V \bar{\gamma}(x - x_i)dx \tag{24}$$

If the samples have nonpunctual support then $\bar{\gamma}(x_i, x_j)$ is replaced by

$$\frac{1}{V_i V_j} \int_{V_i} \int_{V_j} \bar{\gamma}(x-y)dxdy \tag{25}$$

V_i, V_j being the supports at x_i, x_j.

The Co-Kriging Variance is

$$\sigma^2_{CK} = \text{Tr} \sum_1^m \bar{\gamma}(x_i, V)\ \Gamma_i - \text{Tr}\ \bar{\mu} - \text{Tr}\ \frac{1}{V^2} \int_V \int_V \bar{\gamma}(x-y)dxdy \tag{26}$$

Since the off-diagonal entries in

$$\frac{1}{V^2} \int_V \int_V \bar{\gamma}(x-y)dxdy \tag{27}$$

are not required otherwise the last term in σ^2_{CK} is simply

$$\sum_{i=1}^{n} \frac{1}{V^2} \int_V \int_V \gamma_{ii}(x-y)dxdy \tag{28}$$

THE "UNDER-SAMPLED" CASE

The following example is found in (1) Journel.

1	A	$S_0^{\bullet}$	B	Data $Z_1(0)$, $Z_2(0)$, $Z_2(A)$, $Z_2(B)$, $Z_2(C)$
	C		D	$Z_2(D)$

V is the square with vertices A,B,C,D. It is desired to estimate Z_V where

$$Z_V = \frac{1}{V} \int_V Z_1(x)dx$$

using all the data. In (1), the problem was simplified by using the symmetry i.e.
$Z_2(S_R) = \frac{1}{4}[Z_2(A) + Z_2(B) + Z_2(C) + Z_2(D)]$. The system of equations given in (1) is a sub-system of the following

$$\begin{bmatrix} \bar{\gamma}(S_0,S_0) & \bar{\gamma}^*(s_0,S_R) & I \\ \bar{\gamma}_*(S_0,S_R & \bar{\gamma}_*^*(S_R,S_R) & I_* \\ I & I^* & 0 \end{bmatrix} \begin{bmatrix} \Gamma_1 \\ \Gamma_2 \\ \bar{\mu} \end{bmatrix} = \begin{bmatrix} \bar{\gamma}(S_0,V) \\ \gamma_*(\bar{S}_R V) \\ I \end{bmatrix} \quad (29)$$

where

$$\bar{\gamma}^*(S_0,S_R) = \begin{bmatrix} 0 & \gamma_{12}(S_0,S_R) \\ 0 & \gamma_{22}(S_0,S_R) \end{bmatrix} \quad (30)$$

$$\bar{\gamma}_*^*(S_R,S_R) = \begin{bmatrix} 0 & 0 \\ 0 & \gamma_{22}(S_R,S_R) \end{bmatrix} \quad (31)$$

$$\bar{\gamma}_*(S_0,S_R) = \begin{bmatrix} 0 & 0 \\ 0 & \gamma_{22}(S_R,S_R) \end{bmatrix} \quad (32)$$

$$I_* = \begin{bmatrix} 0 & 0 \\ 0 & 1 \end{bmatrix}, \quad I^* = \begin{bmatrix} 0 & 0 \\ 0 & 1 \end{bmatrix} \quad (33)$$

This system also provides for the estimation of Z_{2V} along with Z_{1V} . Moreover λ^2_{11} , λ^2_{12} will be arbitrary and hence may be taken to be zero, that is $Z_1(S_R)$ is not used in the estimation of Z_{1V} or Z_{2V} . In a geometric/graphical way then it is easy to see how the general Co-Kriging system is changed to the under-sampled version.

$$\begin{bmatrix} \bar{\gamma}(x_1-x_1) \cdot\cdot\cdot \bar{\gamma}(x_1-x_n) \\ \vdots \\ \bar{\gamma}(x_1-x_n) \cdot\cdot\cdot \gamma(x_n-x_n) \\ I \cdot\cdot\cdot\cdot\cdot I \end{bmatrix} \begin{bmatrix} I & \Gamma_1 \\ \vdots & \vdots \\ I & \Gamma_n \\ 0 & \bar{\mu} \end{bmatrix} = \begin{bmatrix} \bar{\gamma}(V,x_1) \\ \vdots \\ \bar{\gamma}(V,x_n) \\ I \end{bmatrix} \quad (34)$$

Each column/row in (35) corresponds to a sample location. Each column/row within a $\bar{\gamma}$ corresponds to a variable. For each variable that is not sampled at all locations, locate the column/row corresponding to the location, then in each $\bar{\gamma}$ (and I) change all of the column/row entries corresponding to the variable. In the case of the rows the deletion is applied on the right side of (35) as well. To see that constraining the variogram matrices is equivalent to constraining certain weights to be zero is easily seen.

For simplicity suppose Z_{i_0} is not sampled at x_{j_0}. Then Γ_{j_0} has additional constraints imposed namely

$\lambda_{i_0 1}^{j_0}, \ldots, \lambda_{i_0 m}^{j_0}$ are all zeros.

Let $\hat{\Gamma}_{j0}$ be the modified Γ_{j0}. Similarly let $\bar{\gamma}_*(x_{j0}-x_p)$ be the same as $\bar{\gamma}(x_{j0}-x_p)$ except that in the i_0 row all entries are zeros and $\bar{\gamma}_*(x_p-x_{j0})$ is the same as $\bar{\gamma}(x_p-x_{j0})$ except that all entries in the i_0 column are all zeros, that is, $\bar{\gamma}*(x_p-x_{j0}) = \bar{\gamma}_*(x_{j0}-x_p)^T$. Let I_* be an identity matrix except that the i_0 row is all zeros and $(I*)^T = I_*$.

If $\bar{Z}$ is a vector IRF-0 then the estimation variance in the under-sampled case is

$$
\begin{aligned}
&\mathrm{Tr}[\sum_{j\neq j0} \Gamma_j^T \bar{\gamma}(x_j-x) + \hat{\Gamma}_{j0}\,\bar{\gamma}(x_{j0}-x)] \\
&+ \mathrm{Tr}[\sum_{j\neq j0} \bar{\gamma}(x-x_j)\Gamma_j + \bar{\gamma}(x-x_{j0})\hat{\Gamma}_{j0}] \\
&+ \mathrm{Tr}[\sum_{i\neq j0}\sum_{j\neq j0} \Gamma_i^T \bar{\gamma}(x_n-x_j)\,\Gamma_j \\
&+ \sum_{i\neq j0} \Gamma_i^T \bar{\gamma}(x_i-x_{j0})\,\hat{\Gamma}_{j0} \\
&+ \sum_{j\neq j0} \hat{\Gamma}_{j0}^T \bar{\gamma}(x_{j0}-x_i)\,\Gamma_j \\
&+ \hat{\Gamma}_{j0}^T \bar{\gamma}(x_{j0}-x_{j0})\,\Gamma_{j0}]
\end{aligned}
\qquad (35)
$$

The universality conditions can be written

$$\sum_{j \neq j0} I\,\Gamma_j + I\,\hat{\Gamma}_{j0} = I \tag{36}$$

Consider the following identities

$$\text{(i)} \qquad I\,\hat{\Gamma}_{j0} = I_* \Gamma_{j0} \tag{37}$$

$$\text{(ii)} \qquad \hat{\Gamma}_{j0}^{T-} \gamma(x_{j0}-x) = \Gamma_{j0}^{T-} \gamma_*(x_{j0}-x)$$

$$\text{(iii)} \qquad \bar{\gamma}(x-x_{j0})\,\hat{\Gamma}_{j0} = \bar{\gamma}*(x-x_{j0})\,\Gamma_{j0}$$

$$\text{(iv)} \qquad \hat{\Gamma}_{j0}^{T-} \gamma(x_{j0}-x_j)\,\hat{\Gamma}_j = \Gamma_{j0}^{T-} \gamma_*(x_{j0}-x_j)\,\Gamma_j$$

$$\text{(v)} \qquad \hat{\Gamma}_{j0}^{T-} \gamma(x_{j0}-x_{j0})\,\hat{\Gamma}_{j0} = \Gamma_{j0} = \Gamma_{j0}^{T-} \gamma_*^*(x_{j0}-x_{j0})\,\Gamma_{j0}$$

$$\text{(vi)} \qquad \Gamma_j^{T-} \bar{\gamma}(x_j-x_{j0})\,\hat{\Gamma}_{j0} = \Gamma_j^{T-} \bar{\gamma}*(x_j-x_{j0})\,\Gamma_{j0}$$

Then the estimation variance can be written in terms of the modified $\bar{\gamma}$'s instead of in terms of the modified Γ_j's. All arbitrary entries in the Γ_j's will be set equal to zero.

NUMERICAL METHODS

Because of the size, of the system of equations to be solved in Co-Kriging, makes it formidable for more than a few variables and a few sample locations. We describe first a method for reducing the system and then an iterative method that requires less core memory to solve the system.

Utilizing the matrix form given by (35) let

$$K = \begin{bmatrix} \bar{\gamma}(x_1-x_1) & \cdots & \bar{\gamma}(x_1-x_n) \\ \vdots & & \vdots \\ \bar{\gamma}(x_n-x_1) & \cdots & \bar{\gamma}(x_n-x_n) \end{bmatrix} \tag{38}$$

$$E = [I \cdots I]^T \tag{39}$$

$$\bar{\Gamma} = [\Gamma_1 \cdots \Gamma_n]^T \tag{40}$$

$$K_0 = [\bar{\gamma}(V,x_1) \cdots \bar{\gamma}(V,x_n)]^T \tag{41}$$

Then (35) may be written as

$$\begin{bmatrix} K & E \\ E^T & 0 \end{bmatrix} \begin{bmatrix} \bar{\Gamma} \\ \bar{\mu} \end{bmatrix} = \begin{bmatrix} K_0 \\ I \end{bmatrix} \tag{42}$$

$$\text{i.e.} \qquad K\,\bar{\Gamma} + E\,\bar{\mu} = K_0 \quad E^T\,\bar{\Gamma} = I \tag{43}$$

If $KU = K_0$, $KV = E$ then (44) becomes

$$K(\bar{\Gamma} + V\bar{\mu}) = KU \quad \text{or} \quad \bar{\Gamma} = U - V\bar{\mu} \tag{44}$$

If moreover

$$(E^T V)W = EU - I \quad \text{then} \quad \bar{\Gamma} = U - VW \tag{45}$$

This not only reduces the size of the system but also avoids the possibility that (43) is ill-conditioned.

Kacmancz and Tanabe (6) have given an iterative method for solving systems of equations even when the system is singular: the author (5) has extended this to systems such as (35) or (43). For any two n x 1 matrices X, Y whose entries are m x m matrices let

$$\langle X,Y\rangle = Y^T X \quad \text{and} \quad (X,Y) = \operatorname{Tr} \langle X,Y\rangle \tag{46}$$

(X,Y) is an inner-product and $\| X \| = \sqrt{(X,X)}$ is a norm. For a system

$$AX = B \tag{47}$$

A a p x n matrix, X n x 1, B p x 1 whose entries are m x m matrices let $\bar{A}_i$ be the ith column in A^T. Assume $\| \bar{A}_i \| > 0$ for all i. Let f_i be defined as follows

$$f_i(X) = X - \frac{1}{\alpha_i} [\bar{A}_i < X, \bar{A}_i > - A_i B_i] \tag{48}$$

and $F(X) = f_1 \circ f_2 \circ \dots \circ f_p(X)$. If X_0 is any initial element and

$$X_{i+1} = F(X_i) \tag{49}$$

then $X_0, X_1, \dots, X_q, \dots$ is a sequence converging to the solution of $AX = B$.

By writing

$$B = [B_1 \cdots B_p]$$

the original system $AX = B$ can be written

$$\langle X,\bar{A}_i\rangle = B_i \quad ; i=1, \ldots, p \tag{50}$$

If X is the solution of the system then $f_i(X)$ is the projection of X onto the hyperplane given by (51) even if A is invertible this method is useful since only one row of A need be in core at one time. Gaussian reduction would require all of A in core at one time. This projection method can be used for the under sampled case and also in conjunction with the reduction technique given earlier in this section.

A computer program has been written at the University of Arizona by J. Carr for Co-Kriging and which utilizes the projection algorithm for solving the system.

References

(1) Journel, A. G. 1977 "Geostatistique Minier" Centre de Geostatistique Fontainebleau, 737 p.

(2) Myers, D. E. 1981 "Joint Estimation of random functions: the matrix form" Research Report, Department of Mathematics, University of Arizona 32 p.

(3) Myers, D. E. 1982 "Matrix Formulation of Co-Kriging" Mathematical Geology, Vol. 14, No. 3, 249-257.

(4) Myers, D. E. 1983 "Estimation of Linear Combinations and Co-Kriging" to appear in Mathematical Geology Vol. 15, No. 5

(5) Myers, D. E. 1983 "On solving large scale linear systems", submitted for publication.

(6) Tanabe, K. 1971 "Projection method for solving a singular system of linear equations and its application", Numerische Math. 17, 203-214

THE PLACE OF NON-PARAMETRIC GEOSTATISTICS

A.G. Journel

Applied Earth Sciences Department
Stanford University, California, USA

Abstract

By varying the search space of service variables data onto which the unknown is projected, a large category of krigings including disjunctive and indicator krigings are put in frame. The non-parametric technique of indicator and probability krigings are then developed towards estimation of spatial distributions.

Keywords

Non-parametric Geostatistics - Indicator Kriging - Probability Kriging -Disjunctive Kriging - Conditional Distribution

Introduction

Estimation of spatial distributions, and its application to estimation of recovery functions and other excursions of random functions over a threshold level, represents possibly the greatest challenge for geostatisticians.

The very recent introduction of the Indicator approach has given rise to an extremely stimulating debate. The newly developed non-parametric approaches are set in frame with the distribution-dependent techniques, allowing a better appreciation of their pros and cons.

It is shown that most krigings, including disjunctive kriging and the various forms of indicator krigings, fall into

G. Verly et al. (eds.), Geostatistics for Natural Resources Characterization, Part 1, 307–335.

the category of service-variable approaches and differ by the level of modelling required. As for the support effect correction, the block distribution model can be introduced either before or after the kriging process.

The enhanced indicator approach, called probability kriging (PK), is developed extensively. It makes more complete use of the experimental structural information, and does in practice yield better results.

I. A Continuous Line of Thought

Since the inception of their discipline in the early 1950's under a wide variety of words in a large spectrum of application fields, geostatisticians have been dealing with essentially one basic problem with one set of tools. That of estimating a given function of an unknown realization $z(x)$ at location x, from a set of N spatially correlated data $\{z(x_\alpha), \alpha \in (N)\}$. The set of tools belongs to the category of regression analysis, and received the generic name of "kriging".

Whether the location x is replaced by a time t, the realization $z(x)$ is replaced by a spatial average $z_v(x)$ over an area or block of size v, whether $z(x)$ represents a core grade, a water table elevation or a crop yield, the basic problem remains the same: a risk-qualified estimation of some given function h of the unknown $z(x)$.

The crucial problem of estimating recoverable reserves within a panel V is no exception. It consists of estimating two particular known functions h_1 and h_2 of the selective mining unit (smu) grades:

$$h_1(z_v(x)) = j_v(x; z) = \begin{cases} 1, & \text{if } z_v(x) > z \\ 0, & \text{if not} \end{cases} \tag{1}$$

$$h_2(z_v(x)) = j_v(x;z) \cdot z_v(x;z)$$

with:

$z_v(x) = \frac{1}{v} \int_{v(x)} z(u)\, du$, being the mean grade of the smu $v(x)$,

z, being the cut-off grade. These functions, once averaged over the K smu's constituting the panel V, provide the so-called "recovery functions" over panel V:

- tonnage recovery factor: $t_{v/V}(z) = \frac{1}{K} \sum_{k=1}^{K} j_v(x_k;z)$

- quantity of metal recovery factor $q_{v/V}(z) = \frac{1}{K} \sum_{k=1}^{K} j_v(x_k;z) \cdot z_v(x_k)$

- ore grade: $m_{v/V}(z) = q_{v/V}(z)/t_{v/V}(z)$ (2)

In pollution control, a similar problem is that of monitoring the number of days (support v) within a year (V), days during which acidic deposition has exceeded a threshold level z, and estimating the total acidity then deposited.

More generally, this is the classical problem of characterizing the "excursions" of the process $Z_v(t)$ over a certain level z, during a particular period of time V informed by N data $\{z(x_\alpha), \alpha \in (N)\}$.

The kriging estimation technique consists of performing a regression of the unknown, $h(z(x))$ or $h(z_v(x))$, on either the data themselves $\{z(x_\alpha), \alpha \in (N)\}$, or on some transform of them $\{\phi(z(x_\alpha)), \alpha \in (N)\}$. The bewildering array of specific krigings (UK, OK, SK, IK, PK, DK, SLnK, OLnK, MG. . .) should not mask a common and very simple principle, that of regression, i.e. minimization of a distance defined as:

- either the non-conditional variance of errors:

$$E\{[h(Z_v(x)) - h^*(Z_v(x))]^2\} \quad (3)$$

- or the conditional variance of errors:

$$E\{[h(Z_v(x)) - h^*_v(Z_v(x))]^2 / Z(x_\alpha) = z_\alpha, \alpha \in (N)\} \quad . \quad (4)$$

One should be aware that the estimation variance, a moment of order 2, might not be the most appropriate criterion to be minimized, see Journel (1984) in this volume. Other criteria include:

- the mean absolute error:

$E\{|h(z_v(x)) - h^*(Z_v(x))|\}$, calling for median regression instead of mean regression. Active research is being conducted in this avenue.

- truncated moments, whether of order 1 or 2:
$E\{[h(Z_v(x)) - h^*(Z_v(x))]^2 / Z_v(x) > z\}$, for the important units are those that will be recovered, i.e. those around and above the cut-off.

- more generally the minimization of the expected value of a loss function L characterizing the impact of an error value on the project at hand: $E\{L(h(Z_v(x)) - h^*(Z_v(x)))\}$.

II. The conditional distribution formulation

Consider then the basic problem of estimating any measurable function h of an unknown argument $z_V(x)$ from a set of neighboring data $\{z(x_\alpha), \alpha \in (N)\}$

The optimization criterion considered hereon is the classical estimation variance, previously defined in (3).

Classically, the regionalization $\{z(x), x \in D\}$ within the area D covered by the N data is interpreted as a particular spatial outcome of a random function (in short R.F.) Z(x). In addition, the R.F. model Z(x) is assumed to be stationary.

Under the estimation variance criterion (3), the best estimate of $h(Z_V(x))$ is the conditional expectation:

$$[h(Z_V(x))]^*_{CE} = E\{h(Z_V(x))/Z(x_\alpha) = z_\alpha, \alpha \in (N)\} \tag{5}$$

classically defined by a Stieltjes integral of the conditional cumulative distribution function (in short cdf) of $Z_V(x)$:

$$[h(Z_V(x))]^*_{CE} = \int_0^\infty h(u) \,.\, dF_V(u/(N)), \text{ with:}$$

$$F_V(u/(N)) = \text{Prob } \{Z_V(x) \leq u/ Z(x_\alpha) = z_\alpha, \alpha \in (N)\}, \tag{6}$$

the range of outcomes of $Z_V(x)$ being $[0, +\infty]$.

Knowledge of the conditional distribution $F_V(z/(N))$ not only provides the conditional expectation (5), but also the corresponding conditional estimation variance (4) which is none other than the variance attached to the cdf. $F_V(z/(N))$.

Also, since the function h is known, the conditional distribution of the random variable $h(Z_V(x))$ can be retrieved from that of $Z_V(x)$, i.e. from $F_V(z/(N))$:

$$H_V(z/(N)) = \text{Prob } \{h(Z_V(x)) \leq z/ Z(x_\alpha) = z_\alpha, \alpha \in (N)\} \tag{7}$$

For example if the function $h(z_V)$ is strictly increasing and inversible, it comes:

$$H_V(z/(N)) = \text{Prob } \{Z_V(x) \leq h^{-1}(z)/ Z(x_\alpha) = z_\alpha, \alpha \in (N)\} = F_V(h^{-1}(z)/(N))$$

The quantiles z_p, z_{1-p} of the conditional cdf. $H_V(z/(N))$ provide estimates of the confidence intervals for $h(Z_V(x))$:

$$\text{Prob}^*\{h(Z_v(x)) \in]z_p, z_{1-p}] / Z(x_\alpha) = z_\alpha, \alpha \in (N)\} \tag{8}$$
$$= H_v(z_{1-p}/(N)) - H_v(z_p/(N)) = (1-p)-p = 1-2p$$

Of course these confidence intervals do not account for the error of estimation of the conditional cdf $H_v(z/(N))$.

Hence, it appears that knowledge of the conditional cdf. $F_v(z/(N))$ would fully solve the problem of risk-qualified estimation of the unknown $h(Z_v(x))$.

Before proceeding on the estimation of $F_v(z/(N))$, it may be worthwhile going through an example of development of formulae (5) to (7).

An example: Consider the problem of estimation of recoverable reserves and the two corresponding functions h_1 and h_2 defined in relations (1).

- The conditional distribution of the R.V. $h_1(Z_v(x)) = J_v(x; z)$ is Bernouilli distributed and fully characterized by its mean:

$$[J_v(x;z)]^*_{CE} = E\{J_v(x;z)/Z(x_\alpha) = z_\alpha, \alpha \in (N)\} = 1-F_v(z/(N))$$

- The best estimate of the unknown $h_2(z_v(x) = j_v(x;z).z_v(x)$ is the conditional expectation:

$$[h_2(z_v(x))]^*_{CE} = E\{J_v(x;z) . Z_v(x)/Z(x_\alpha) = z_\alpha, \alpha \in (N)\} =$$
$$\int_z^\infty u \cdot dF_v(u/(N)).$$

This Stieltjes integral is approximated by the discrete sum:

$$\int_z^\infty u\, dF_v(u/(N)) \simeq \sum_{k=1}^{K} u'_k \cdot [F_v(u_{k+1}/(N)) - F_v(u_k/(N))],$$

the u'_k are central characteristics of the K intervals $[u_k, u_{k+1}]$ discretizing the interval $[z, +\infty]$. For further discussion on the critical choice of the central values u'_k, and in particular of the last one u'_K, refer to Sullivan (1984 and thesis work).

The conditional distribution of $h_2(Z_v(x)) = J_v(x;z) . Z_v(x)$ is the truncated distribution defined as:

$$H_v(u/(N)) = \text{Prob}\{J_v(x;z) . Z_v(x) \le u/ Z(x_\alpha) = z_\alpha, \alpha \in (N)\}$$

$$= \begin{cases} F_v(z/(N)), \text{ for all } u \in [0,z] \\ F_v(u/(N)), \text{ for all } u \ge z \end{cases}$$

III. Estimating the conditional distribution

It has been shown that the key to the problem of risk-qualified estimation of any function $h(z_V(x))$ of the unknown value $z_V(x)$ is the estimation of the conditional cdf of $Z_V(x)$. This conditional cdf, defined in relation (6), can be written as the conditional expectation of an indicator R.F.

Let:

$$I_V(x;z) = 1 - J_V(x;z) = \begin{cases} 1, & \text{if } Z_V(x) \le z \\ 0, & \text{if not} \end{cases} \text{, then:} \quad (9)$$

$$E\{I_V(x;z)/Z(x_\alpha) = z_\alpha, \alpha \in (N)\} = F_V(z/(N)) \quad . \quad (10)$$

Thus, estimation of the cdf $F_V(z/(N))$ amounts to estimation of the conditional expectation of the indicator R.F.'s $I_V(x;z)$. Note that there are as many R.F.'s $I_V(x;z_k)$ as there are bounds z_k needed to discretize the interval of variability $[0, +\infty]$ or in practice $[0, z_{max}]$.

By definition of a conditional expectation, cf. for example Rao (1965), the value $F_V(z/(N))$ appears as the projection of the corresponding R.V. $I_V(x;z)$ onto the vectorial space H_N of all possible measurable functions of the N data $Z(x_\alpha)$, see Figure 1.

Estimation of the distribution $F_V(z/(N))$ can also be obtained by projection of $I_V(x;z)$ onto any subspace $L_N \subset H_N$, thus defining an "L-kriging" of both $I_V(x;z)$ and $F_V(z/(N))$. Indeed by virtue of the theorem of the 3 perpendiculars, the projection of $I_V(x;z)$ onto L_N coincides with the projection of $F_V(z/(N))$ onto L_N, see Figure 1; this last projection being the best estimator of $F_V(z/(N))$ that can be found in the space L_N.

Requisites for the projection

Consider the vectorial subspace $L_N \subset H_N$ constituted by a certain family of functions ℓ of the N data:

$$L_N: \quad \{\ell(Z(x_\alpha), \alpha \in (N))\}$$

If the functions ℓ are allowed to be any measurable function, then $L_N = H_N$. If the functions ℓ are restricted to linear combinations of the N data plus the constant 1:

$$L_N = \zeta_N + 1: \quad \left\{ \sum_{\alpha=1}^{N} \lambda_\alpha Z(x_\alpha) + \lambda_o \cdot 1 \right\} \subset H_N$$

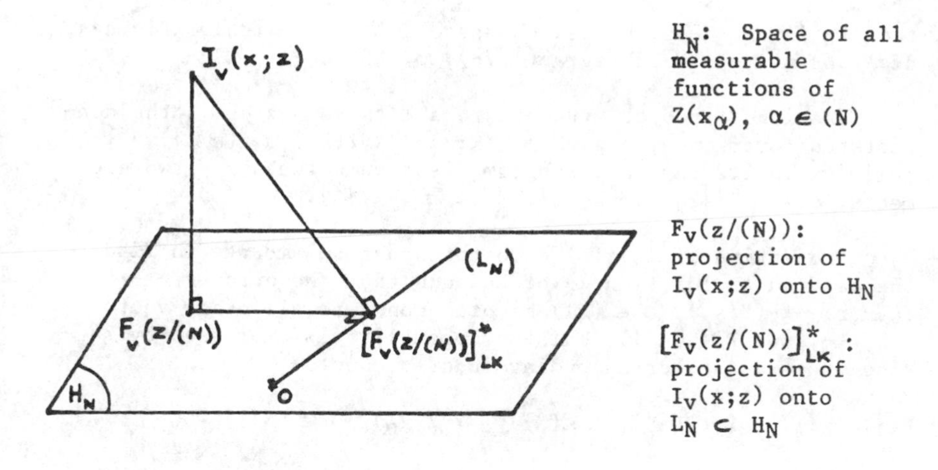

Figure 1. Conditional expectation and L-Kriging estimator

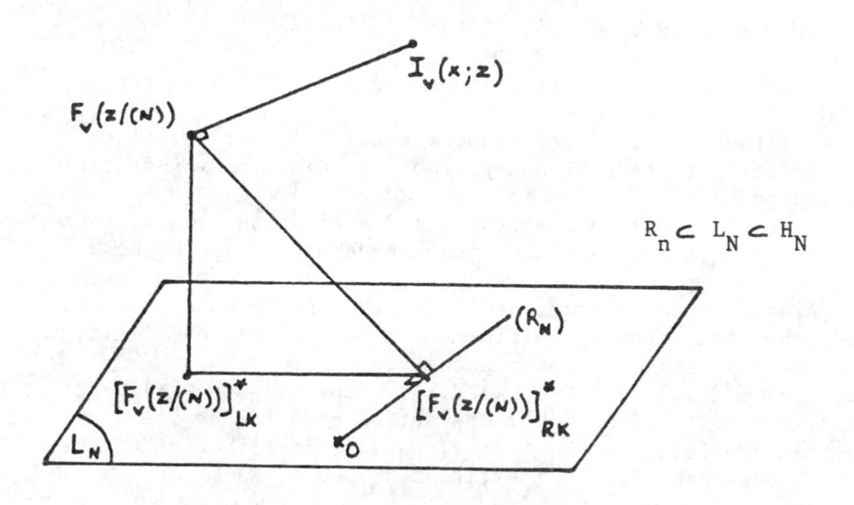

Figure 2. Search space and hierarchy of krigings.
(The larger the search space, the closer the projection).

then L_N is equal to the search space $\zeta_N + 1$ of linear krigings, cf. Journel and Huijbregts (1978), p. 559.

The inner product attached to all these spaces is the non-centered covariance, e.g. $E\{I_v(x;z) \,.\, \ell(Z(x_\alpha),\alpha \in (N))\}$, defining in its turn a variogram distance between any two elements, e.g., $E\{[I_v(x;z) - \ell(Z(x_\alpha),\alpha \in (N))]^2\}$.

Projection of $I_v(x;z)$ onto the space L_N amounts to find the particular function ℓ_K of L_N such that the error vector $I_v(x;z) - \ell_K(Z(x_\alpha), \alpha \in (N))$ be orthogonal to all other vectors ℓ of L_N, see Figure 1. This orthogonality is expressed by setting to zero the corresponding inner products:

$$E\{[I_v(x;z) - \ell_K(Z(x_\alpha), \alpha \in (N))] \,.\, \ell(Z(x_\alpha), \alpha \in (N))\} = 0,$$

for all $\ell \in L_N$.

It thus appears that the determination of the projection:

$$\ell_K(Z(x_\alpha), \alpha \in (N)) = [I_v(x;z)]^*_{LK} \equiv [F_v(z/(N))]^*_{LK} \text{ , of } I_v(x;z)$$

onto L_N requires knowledge of the following two sets of covariances:

$$E\{I_v(x;z) \,.\, \ell(Z(x_\alpha),\alpha \in (N))\}, \text{ and}$$

$$E\{\ell_i(Z(x_\alpha),\alpha \in (N)) \,.\, \ell_j(Z(x_\alpha), \alpha \in (N))\} \tag{11}$$

for all functions ℓ, ℓ_i, $\ell_j \in L_N$

These sets of covariances need to be

- either inferred from experimental information, thus defining a class of non-parametric, distribution-free approaches
- or deduced from multivariate distribution models, thus defining a class of distribution-dependent approaches.

Also it appears that the larger the search space $L_N \subset H_N$, the richer the structural information needed. At one limit, if $L_N = \zeta_N + 1$, the only covariances needed are: $E\{I_v(x;z) \,.\, Z(x_\alpha)\}$ and $E\{Z(x_\alpha).Z(x_\beta)\}$, for all $\alpha, \beta \in (N)$. At the other limit, if $L_N = H_N$, the set of covariances needed requires knowledge of the $(N + 1)$-variate distribution of the $(N + 1)$ dependent R.V.'s $I_v(x;z)$, $Z(x_\alpha)$, $\alpha \in (N)$.

Thus and clearly, distribution-free approaches will be limited to small search spaces such as $\zeta_N + 1$ (Linear Krigings), whereas model-dependent approaches can afford considering larger search spaces, up to H_N for the multi-variate Gaussian (MG) approach, see Verly (1983).

All krigings are projections of an unknown R.V. onto a search sub-space $L_N \subset H_N$. The larger the subspace, the more structural information or modelling is needed, and the closer is the kriging estimator to the "optimum optimorum" that is the projection onto the large space H_N. Thus the inclusions $R_N \subset L_N \subset H_N$ of Figure 2 implies the following order relation between the estimation variances, defining a hierarchy of krigings:

$$E\{[I_v(x;z) - [F_v(z/(N))]^*_{RK}]^2\} \geq E\{[I_v(x;z) - [F_v(z/(N))]^*_{LK}]^2\} \geq$$

$$\geq E\{[I_v(x;z) - F_v(z/(N))]^2\} \quad .$$

Beware though that an inaccurate structural analysis or an inadequate modelling in the larger, but "riskier", space H_N, may reverse the practical hierarchy. In other terms, the MG estimate corresponding to a projection onto H_N is not necessarily the best estimate in practice.

IV. The service variable approach

Coming back to the estimation of the conditional distribution $F_v(z/(N))$ by projection of the R.V. $I_v(x;z)$ onto H_N or any subspace $L_N \subset H_N$, the two set of required covariances are, cf. (11):

$E\{I_v(x;z).\ell(Z(x_\alpha), \alpha \in (N))\}$ for all functions $\ell \in L_N$

$E\{\ell_i(Z(x_\alpha), \alpha \in (N)).\ell_j(Z(x_\alpha), \alpha \in (N))\}$,

for all functions ℓ_i, $\ell_j \in L_N$.

The problem of estimating the first set of covariances will be dealt with in the next section "The support effect".

If ℓ_i and ℓ_j are functions of all N data at a time then there is only available one realization, $\ell_i(z_\alpha, \alpha \in (N))$ and $\ell_j(z_\alpha, \alpha \in (N))$, for each of the R.V.'s $\ell_i(Z(x_\alpha), \alpha \in (N))$ and $\ell_j(Z(x_\alpha), \alpha \in (N))$. Therefore experimental inference of the corresponding covariance is not possible.

One way around this problem is to consider a smaller space $D_H \subset L_N$, for example the space of all measurable functions

$\{\ell(Z(x_\alpha)), \alpha \in (N)\}$ of one datum at a time. Then, and because of stationarity, for each particular function ℓ there are available N realizations: $\ell(z_\alpha)$, $\alpha \in (N)$, of the R.V. $\ell(Z(x))$, thus allowing experimental inference of the covariance function $C_{\ell_i \ell_j}(h) = E\{\ell_i(Z(x+h)) \cdot \ell_j(Z(x))\}$.

An even smaller space is defined by restricting the function ℓ to a single type ℓ_k:

$$S_N\{\ell_k\}: \quad \{\sum_{\alpha=1}^{N} \lambda_\alpha \cdot \ell_k(Z(x_\alpha))\} \subset D_N \subset L_N \subset H_N.$$

The R.V.'s $\ell_k(Z(x))$ are called "service-variables", the search space $S_N\{\ell_k\}$ is the space of all linear combinations of the N service-variable data. Again the covariance function $C_{\ell_k}(h) = E\{\ell_k(Z(x+h)).\ell_k(Z(x))\}$ can be inferred directly from the N spatially correlated data: $\ell_k(z_\alpha), \alpha \in (N)$.

A slightly larger search space can be defined by allowing the function ℓ of a single datum to take two different expressions ℓ_k and $\ell_{k'}$:

$$S_N\{\ell_k, \ell_{k'}\} = \{\sum_{\alpha=1}^{N} \lambda_\alpha \, \ell_k(Z(x_\alpha)) + \sum_{\alpha=1}^{N} \nu_\alpha \, \ell_{k'}(Z(x_\alpha))\} \subset D_N$$

Again, the covariance functions needed:

$$C_{\ell_k}(h) = E\{\ell_k(Z(x+h)) \cdot \ell_k(Z(x))\},$$

$$C_{\ell_{k'}}(h) = E\{\ell_{k'}(Z(x+h)).\ell_{k'}(Z(x))\} \text{ , and :}$$

$$C_{\ell_k \ell_{k'}}(h) = E\{\ell_k(Z(x+h).\ell_{k'}(Z(x))\},$$

can be inferred directly from the two sets of service variable data: $\{\ell_k(z_\alpha), \alpha \in (N)\}$ and $\{\ell_{k'}(z_\alpha), \alpha \in (N)\}$.

Some examples

- The function ℓ_k is the identity function: $\ell_k(Z(x)) = Z(x)$, and the search space is:

$$\zeta_N + 1 = S_N(Z) + 1: \quad \{ \sum_{\alpha=1}^{N} \lambda_\alpha \, Z(x_\alpha) + \lambda_o \cdot 1\} \; .$$

All linear krigings are defined within $\zeta_N + 1$, or linear manifolds of it if conditions on the weights $\lambda_\alpha, \lambda_o$ are introduced, cf. Universal Kriging (UK) in Journel and Huijbregts (1978), p. 560.

- The function ℓ_k is the indicator function:

$$\ell_k(Z(x)) = I(x;z) = \begin{cases} 1, & \text{if } Z(x) \leq z \\ 0, & \text{if otherwise} \end{cases}$$

and the search space is:

$$S_N(I) + 1: \quad \left\{ \sum_{\alpha=1}^{N} \lambda_\alpha(z) \cdot I(x_\alpha;z) + \lambda_o(z).1 \right\} \quad .$$

This search space is that of the "Indicator Kriging" (in short IK) approach, see hereafter.

- The function ℓ can take the two previous forms: $\ell_k(Z(x)) = Z(x)$, and: $\ell_{k'}(Z(x)) = I(x;z)$, or better

$$\ell_k(Z(x)) = F(Z(x)) = Un(x), \text{ with: } F(z) = \text{Prob}\{Z(x) \leq z\}$$

$$\ell_{k'}(Z(x)) = I(x;z) \; , \tag{12}$$

defining the search space:

$$S(Un,\, I) + 1: \left\{ \sum_{\alpha=1}^{N} \nu_\alpha(z)\, F(Z(x_\alpha)) + \sum_{\alpha=1}^{N} \lambda_\alpha(z).I(x_\alpha;z) + \lambda_o(z) \right\} \tag{13}$$

This search space is that of the "Probability Kriging" (in short PK) approach, see Sullivan (1984) and hereafter Section VI.

Note that the transform $un(x_\alpha) = F^*(z(x_\alpha))$ is essentially a rank order of the datum $z(x_\alpha)$. Thus PK amounts to a cokriging using the spatial variability of the data rank order.

- The function ℓ_k takes the form of an Hermitian polynomial of order n of the gaussian transform Y(x) of the initial R.F. Z(x):

$$\ell_k(Z(x)) = h_n(Y(x)), \text{ for any n integer} \geq 0 \tag{14}$$

with: Y(x), being the normal score transform of Z(x), defined by: $Y(x) = G^{-1}(F(Z(x))$, $G(y) = \text{Prob}\{Y(x) \leq y\}$ being the standard normal cdf.

The corresponding search space is:

$$S_N(h_n): \quad \left\{ \sum_{\alpha=1}^{N} \lambda_\alpha(n).h_n(Y(x_\alpha)) \right\} \tag{15}$$

and is the search space considered for linear kriging of the unknown $h_n(Y(x))$.

These linear estimators $[h_n(Y(x))]^*_{SK}$ for n = 0 to n_o, are then linearly recombined to provide the "Disjunctive kriging" (in short DK) estimator, of order n_o, of the R.V. $I_V(x;z)$, cf. Matheron (1976). More precisely, the inverse normal score transform is developed into a Hermitian expansion:

$$Z(x) = \phi(Y(x)) = \sum_{n=0}^{\infty} \phi_n.h_n(Y(x)),$$

the ϕ_n being known coefficients. The estimators $[h_n(Y(x))]^*_{SK}$ are recombined into the DK estimator of Z(x):

$$[Z(x)]^*_{DK} = \sum_{n=0}^{\infty} \phi_n.[h_n(Y(x))]^*_{SK}$$

IK, PK, or DK?

It thus appears that all three, IK, PK, DK, are service-variable approaches, considering different transforms $\ell(z_\alpha)$ of the initial N data. The question is then to determine which transform is the most adequate for the particular goal at hand, which is the estimation of the conditional cdf $F_V(z/(N)$ through that of the indicator R.V. $I_V(x;z)$, cf. relations (9) and (10) and Figure 1.

Since the final goal is to estimate a cdf value $F_V(z/(N))$, it makes good practical sense to use available cdf-type data, such as the indicator data $I(x_\alpha; z)$, for all z, and the uniform transform data $Un(x_\alpha) = F(Z(x_\alpha))$; that is to consider the previously defined PK search space, cf. relations (12) and (13).

Note that the normal R.F. Y(x) considered by DK is none other but an inverse-gaussian transform of the previous uniform transform: $Y(x) = G^{-1}(F(Z(x))) = G^{-1}(Un(x))$, cf. the definition (14). From a strict non-parametric approach, this additional transform (G^{-1}) seems useless, and even possibly harmful for it transforms well behaved bounded R.F.'s., $Z(x) \in [0, z_{max}]$, or $U_n(x) \in [0,1]$, into an unbounded R.F. $Y(x) \in [-\infty, +\infty]$, see Beckman (1973).

But the true difference between PK and DK lies elsewhere. The structural information, i.e. the set of covariances, required by IK or PK is inferred directly from the corresponding service-variables data. Whereas the structural information required by DK is deduced from a distribution model; more precisely the R.F. Z(x) is supposed to be bi-ϕ^{-1}-normal distributed, that is its univariate gaussian transform $Y(x) = \phi^{-1}(Z(x))$ is supposed to be also bivariate normal distributed. Hence, it can be said that:

- IK and PK are distribution model-free, service variable approaches to the cdf, $F_V(z/(N))$

- DK is a distribution-dependent, service variable approach to the same cdf, $F_V(z/(N))$.

Experimental inference of the covariances: $E\{h_n(Y(x+h)) \cdot h_n(Y(x))\}$, for all n = 1 to n_o, required for the DK approach would call for inference of moments of order 2n of the R.F. Y(x). For evident reasons of robustness, such an inference has never been attempted in practice even for n as low as 2. In the DK approach, all these covariances are deduced from the binormal model.

On the other hand, the IK and PK approaches do not require moments of higher order than 2.

It might be argued that DK can be performed using an orthogonal basis different from the Hermitian, e.g. the Legendre basis corresponding to the uniform transform Un(x) defined in (12). The corresponding DK would still be bivariate distribution dependent, e.g. bi-F^{-1}-uniform dependent.

V. The support effect

There remains to investigate the inference or modeling of the first set of covariances (11), that is of:

$$E\{I_V(x;z) \cdot \ell\,(Z(x_\alpha),\alpha \in (N))\} \qquad (16)$$

for all functions ℓ belonging to the search space L_N.

Unless the support v of the R.V. $I_V(x;z)$ and the conditional cdf $F_V(z/(N)$ defined in relations (9) and (10) is the same as the support of the N data $\{Z(x_\alpha), \alpha \in (N)\}$, experimental inference of the previous covariances is not possible. For there is usually no information available on block v-support. This is the old devil of Mining Geostatistics appearing every time statistics on non-data support are needed.

From theory, it is clear that determination of the block v-support distribution function $F_v(z) = \text{Prob}\{Z_v(x) \leq z\}$ requires knowledge of the multivariate distribution of the point-support R.F. $Z(x)$, whose spatial average is

$$Z_v(x) = \frac{1}{v} \int_{v(x)} Z(u)du.$$

Similarly, this multivariate distribution is required to determine covariances of the type (16) and the conditional cdf $F_v(z/(N))$ defined in (6). Just as clearly, such multivariate distribution is not accessible from a finite number N of quasi point-support data: $Z(x_\alpha)$, $\alpha \in (N)$.

Thus it appears that data-support Geostatistics is the absolute frontier of a distribution-free approach. Any step beyond this frontier must draw not from the data but from a model that, by definition, cannot be refuted until block support data become available.

The question is then to investigate block distribution models and if needed models for the covariances of the type (16).

Conditions on the block distribution

The fact that a block distribution model cannot be refuted short of block-support data does not mean that it can be anything. It must verify at least the following conditions, under the hypothesis of stationarity for the point support R.F. $Z(x)$.

(i) All v-support distributions have same mean:
$$E\{Z_v(x)\} = E\{Z(x)\} = m \text{ , for all size } v \tag{17}$$

(ii) The v-support variance is given from the Z-variogram by the classical dispersion variance relation:

$$\text{Var}\ \{Z_v(x)\} = \sigma_v^2 = \text{Var}\ \{Z(x)\} - \bar{\gamma}(v,v) = \sigma^2 - \bar{\gamma}(v,v),$$

with: $\gamma(h) = \frac{1}{2} E\{[Z(x+h) - Z(x)]^2\}$,

and: $\bar{\gamma}(v,v) = \frac{1}{v^2} \int_v du \int_v \gamma(u-u')du'$

(iii) The block distribution bounds are within those of the point distribution:

If: $Z(x) \in [0, z_{max}]$, then: $Z_v(x) \in [z_{v,min}, z_{v,max}] \subset [0, z_{max}]$

(iv) If $F(z)$ and $F_V(z)$ are the cdf's of the point-support and block v-support R.F.'s $Z(x)$ and $Z_V(x)$, then these cdf's must verify the following inequality:

$E\{\ell(Z_v(x))\} \leq E\{\ell(Z(x))\}$, for all convex and measurable functions ℓ,

i.e. $\int_0^{\infty} \ell(z) \, . \, dF_v(z) \leq \int_0^{\infty} \ell(z) dF(z)$.

A convex function ℓ is a function such that:

$\ell(\lambda z + (1-\lambda)z') \leq \lambda\ell(z) + (1-\lambda).\ell(z')$, for all $\lambda \in [0,1]$ and all z,z', implying that:

$\ell(E\{Z\}) \leq E\{\ell(Z)\}$, for all R.V.'s Z.

The preceding condition (17-iv) ensures that the selectivity of an operation decreases as the block size (support of the selection) increases; indeed the spatial averaging within the block size does reduce selectivity.

All classically used distribution models do verify all conditions (17), from the simple and widely used affine correction of variance to the more sophisticated discrete Gaussian model introduced by Matheron (1976).

Post-kriging correction for support effect

It has been shown that, provided the search space is restricted to spaces such that of PK defined in (13), a distribution-free, service-variable type kriging can provide an estimate of the point-support conditional distribution
$F(z/(N)) = \text{Prob}\ \{Z(x) \leq z/\ Z(x_\alpha) = z_\alpha,\ \alpha \in (N)\}$.
From that result is it possible to deduce the required block-support conditional distribution:

$F_v(z/(N)) = \text{Prob}\ \{Z_v(x) \leq z/\ Z(x_\alpha) = z_\alpha,\ \alpha \in (N)\}$?

The idea is to apply to the point-support cdf $F(z/(N))$ a correction for support effect. All such corrections require knowledge of the conditional variance attached to the cdf. $F_V(z/(N))$. This conditional variance is written, cf.(17)(ii):

$$\text{Var}\{Z_V(x)/\ Z(x_\alpha) = z_\alpha,\ \alpha \in (N)\} =$$

$$\text{Var}\{Z(x)/\ Z(x_\alpha) = z_\alpha,\ \alpha \in (N)\} - \bar{\gamma}(v,v/(N)).$$

An estimate $\text{Var}^*\{Z(x)/\ Z(x_\alpha) = z_\alpha, \alpha \in (N)\}$ of the point conditional variance is obtained straightforwardly from the estimate of the point conditional cdf $F(z/(N))$. There remains the term $\bar{\gamma}(v,v/(N))$ calling for the non-stationary conditional semivariogram.

$$\gamma_x(h/(N)) = \frac{1}{2} E\{[Z(x+h) - Z(x)]^2 / Z(x_\alpha) = z_\alpha,\ \alpha \in (N)\},$$

which unfortunately is inaccessible experimentally.

Two possible approximations for the conditional variance follow:

(i) if the block size v is small with regard to the range of the Z-variogram, that is if the absolute variance correction term $\bar{\gamma}(v,v)$ is small, approximate $\bar{\gamma}(v,v/(N))$ by the known non-conditional value $\bar{\gamma}(v,v)$. The consequent approximation for the block conditional variance is then:

$$\text{Var}\ \{Z_v(x)/Z(x_\alpha) = z_\alpha,\ \alpha \in (N)\} \simeq \tag{18}$$

$$\text{Var}^*\ \{Z(x)/Z(x_\alpha) = z_\alpha,\ \alpha \in (N)\} - \bar{\gamma}(v,v) \geq 0$$

This approximation amounts to ignore the conditioning effect on the absolute correction of variance, i.e. on the short scale variability of point grades within the block size v. Note that $\bar{\gamma}(v,v)$ must be small enough to ensure positiveness of the approximation (18).

(ii) identify the ratio of conditional variances to the known ratio of non-conditional variances, yielding the following approximation for the block conditional variance:

$$\text{Var}\ \{Z_v(x)/Z(x_\alpha) = z_\alpha,\ \alpha \in (N)\} \simeq$$

$$\text{Var}^*\{Z(x)/Z(x_\alpha) = z_\alpha,\ \alpha \in (N)\}\ .\ \left[1 - \frac{\bar{\gamma}(v,v)}{\sigma^2}\right] \quad . \tag{19}$$

This approximation amounts to ignore the conditioning effect on the relative correction of variances.

Once the block conditional variance is approximated by either relation (18) or (19), an affine correction or any other correction for support effect is proceeded upon to get an estimate of the block conditional cdf. $F_v(z/(N))$.

Remarks

There is little place here to discuss the pros and cons of the numerous techniques available for support effect correction. Short of some information about the block grade distribution, all of them carry some degree of arbitrariness and none of them can be considered as appropriate in all cases.

For highly selective operations with variable small size smu's, preferentially located to separate blobs of high grade material, the affine correction of variance performs well, for it does not symmetrize the initial data distribution and does preserve the main modes.

Conversely, for less selective operations within a mineralization where high and low grades are more intermingled, a strong symmetrization and smoothing of the initial data distribution is required. The discrete Gaussian model, or for that purpose a straightforward and much simpler normal or Student t model for the block model may be used.

The correction is post-kriging in the sense that the kriging step is not affected by the block distribution model. Sensitivity analysis to various block models can be easily performed. Conversely, the disjunctive kriging process of $F_V(z/(N))$ depends directly on the block distribution model used to determine the covariances of type (16) required by the DK system.

VI. The Indicator Approach (IK)

It has been shown that data support Geostatistics is the absolute frontier of distribution-free approaches, but that a post-kriging correction for support effect may provide estimates of the required block-support conditional cdf. $F_V(z/(N))$.

Therefore consider the problem of distribution-free estimation of the data-support conditional cdf.

$$F(z/(N)) = \text{Prob}\ \{Z(x) \leq z/\ Z(x_\alpha) = z_\alpha,\ \alpha \in (N)\}.$$

This section is an extension of a previously published work on IK, cf. Journel (1983), and its reading should be backed by the more practical presentations made during the present NATO ASI "Geostat Tahoe", see in particular the paper from J. Sullivan (1984).

The conditional cdf. $F(z/(N))$ appears as the projection of the R.V. $I(x;u)$ onto the PK search space defined in (13). The estimator is written:

$$[I(x;z)]^*_{PK} = \sum_{\alpha=1}^{N} \lambda_\alpha(z).I(x_\alpha;z) + \sum_{\alpha=1}^{N} \nu_\alpha(z).F(Z(x_\alpha)) + \lambda_o(z) \tag{20}$$

That is the PK estimator is a linear combination of the two sets of service-variables $\{Un(x_\alpha) = F(Z(x_\alpha)), \alpha \in (N)\}$ and $\{I(x_\alpha;z), \alpha \in (N)\}$. The first set corresponds to the data rank orders, the second to indicator data for the particular cut-off z considered:

$$un_\alpha = F^*(z_\alpha) = \text{Prob}^*\{Z(x) \leq z_\alpha\} \tag{21}$$

$$i(x_\alpha;z) = \begin{cases} 1, & \text{if } z_\alpha \leq z, \\ 0, & \text{otherwise} \end{cases}, \quad \alpha \in (N).$$

Both sets of data appear as cdf-type data used for kriging a cdf value $F(z/(N)$, hence the name "probability kriging" (PK).

The projection of $I(x;z)$ onto the search space $S_N(Un, I) + 1$ amounts to a very simple and classical cokriging of the R.V. $I(x;z)$ from the two interdependent and spatially correlated sets of data (21).

Unbiasedness conditions

Classically, unbiasedness conditions are investigated first:

$$E\{[I(x;z)]^*_{PK}\} = \sum_{\alpha=1}^{N} \lambda_\alpha(z).E\{I(x_\alpha;z)\} + \sum_{\alpha=1}^{N} \nu_\alpha(z).E\{Un(x_\alpha)\} + \lambda_o(z)$$

The uniform transform R.F. $Un(x)$ is stationary and uniformly distributed within $[0,1]$, hence: $E\{Un(x)\} = .5$, for all x.

The indicator R.F. $I(x;z)$ is stationary and Bernoulli distributed with mean:

$$E\{I(x;z)\} = \text{Prob}\{Z(x) \leq z\} = F(z), \text{ for all } x.$$

It comes:

$$E\{[I(x;z)]^*_{PK}\} = F(z). \sum_{\alpha=1}^{N} \lambda_\alpha(z) + \tfrac{1}{2} \sum_{\alpha=1}^{N} \nu_\alpha(z) + \lambda_o(z)$$

which should be set equal to: $E\{I(x;z)\} = F(z)$. Thus various alternative sets of unbiasedness conditions can be considered:

$$\lambda_o(z) = \left[1- \sum_{\alpha=1}^{N} \lambda_\alpha(z)\right].F(z) - \tfrac{1}{2} \sum_{\alpha=1}^{N} \nu_\alpha(z) \tag{22}$$

$$\begin{cases} \sum_{\alpha=1}^{N} \nu_\alpha(z) = 0 \\ \lambda_o(z) = \left[1- \sum_{\alpha=1}^{N} \lambda_\alpha(z)\right].F(z) \end{cases} \tag{23}$$

$$\begin{cases} \sum_{\alpha=1}^{N} \nu_\alpha(z) = 0 \\ \sum_{\alpha=1}^{N} \lambda_\alpha(z) = 1 \quad , \quad \lambda_o(z) = 0 \end{cases} \tag{24}$$

$$\begin{cases} \nu_\alpha(z) = 0, \text{ for all } \alpha \in (N), \text{ and all } z \\ \lambda_o(z) = \left[1- \sum_{\alpha=1}^{N} \lambda_\alpha(z)\right].F(z) \end{cases} \tag{25}$$

Going from (22) to (25) the unbiasedness conditions becomes more and more stringent:

- The last set (25) amounts to ignore the uniform data altogether and corresponds to the straightforward IK approach described in Journel (1983), and used by B. Davis (1984) and C. Lemmer (1984).

- The penultimate set (24) corresponds to the PK version adopted by J. Sullivan (1984). It has the advantage not to call for the datum $F(z)$, with its related stationarity problems, in the expression of the estimator (20); hence reducing the kriging smoothing effect. Of course, the cdf $F(z)$, in practice its estimate, is still needed for definition of the uniform transform data: $un_\alpha = F^*(z_\alpha)$.

- The second set corresponds to a simple cokriging system (SCK) using explicitly the mean value $E\{I(x;z)\} = F(z)$ in the expression of the estimator.

- The unique unbiasedness condition (22), the less stringent of all, is deemed to be "lax" for it allows pooling the two sets of service-variable data without a specific condition on each set of weights. More practice is needed though to check out that guess.

The PK system

Depending on the set of unbiasedness conditions retained, the PK system would change slightly. The system developed here corresponds to the classical simple cokriging and the set (23) of unbiasedness conditions.

The second condition (23) can be built-in in the expression of the PK estimator:

$$[I(x;z)-F(z)]^*_{PK} = \sum_{\alpha=1}^{N} \lambda_\alpha(z)[I(x;z)-F(z)] + \sum_{\alpha=1}^{N} \nu_\alpha(z).Un(x_\alpha) \tag{26}$$

The only unbiasedness condition left is: $\sum_{\alpha=1}^{N} \nu_\alpha(z) = 0$

The co-kriging system is then written:

$$\sum_{\beta=1}^{N} \lambda_\beta(z)\, C_I(x_\alpha-x_\beta;z) + \sum_{\beta=1}^{N} \nu_\beta(z)\, C_{UI}(x_\alpha-x_\beta;z) = C_I(x-x_\alpha;z) \quad \text{for all } \alpha \in (N) \tag{27}$$

$$\sum_{\beta=1}^{N} \lambda_\beta(z)\, C_{UI}(x_\alpha-x_\beta;z) + \sum_{\beta=1}^{N} \nu_\beta(z)\, C_U(x_\alpha-x_\beta) + \mu = C_{UI}(x-x_\alpha;z) \quad \text{for all } \alpha \in (N)$$

$$\sum_{\beta=1}^{N} \nu_\beta(z) = 0$$

that is, in matrix form: $[PK] \cdot [X] = [B]$, with:

$$[PK] = \begin{bmatrix} [C_I(x_\alpha-x_\beta;\ z)] & [C_{UI}(x_\alpha-x_\beta;\ z)] & [0] \\ [C_{UI}(x_\alpha-x_\beta;\ z)] & [C_U(x_\alpha-x_\beta)] & [1] \\ [0]^T & [1]^T & 0 \end{bmatrix}$$

being the left-hand side matrix (2N + 1, 2N + 1),

$$[X]^T = [\lambda_1(z) \ldots \lambda_N(z)\ \nu_1(z)\ \nu_2(z) \ldots \nu_N(z)\ \ \mu]$$

$$[B]^T = \left[\ [C_I(x-x_\alpha;z)]\ [C_{UI}(x-x_\alpha;\ z)]\ 0\right]$$

$[X]$ is the column matrix of (2N + 1) unknowns including the Lagrange parameter μ ,

$[B]$ is the right-hand side column matrix (2N + 1).

Required structural information

The matrix covariances required for this cokriging consists of 3 covariances:

- Indicator covariance:

$$C_I(h;z) = E\{I(x+h;z)\ I(x;z)\} - [F(z)]^2 = S^2(z) - \gamma_I(h;z) \tag{28}$$

with: $S^2(z) = \text{Var}\{I(x;z)\} = F(z)[1-F(z)] = C_I(o;z)$

and: $\gamma_I(h;z) = \frac{1}{2} E\{[I(x+h;z) - I(x;z)]^2\}$

- Uniform transform covariance:

$$C_U(h) = E\{Un(x+h)\ .\ Un(x)\} - \frac{1}{4} = \frac{1}{12} - \gamma_U(h) \tag{29}$$

with: $\text{Var}\{Un(x)\} = \frac{1}{12} = C_U(o)$

and: $\gamma_U(h) = \frac{1}{2} E\{[Un(x+h) - Un(x)]^2\}$

- Cross-covariance, after symmetrization in (h, -h):

$$C_{UI}(h;z) = E\{Un(x+h).I(x;z)\} - \tfrac{1}{2}F(z) = C_{UI}(0;z)-\gamma_{UI}(h;z) \leq 0 \tag{30}$$

with:

$$C_{UI}(o;z) = E\{Un(x).I(x;z)\} - \tfrac{1}{2}F(z) = -\tfrac{1}{2}F(z).[1-F(z)] = -\tfrac{1}{2}\, S^2(z)$$

since: $E\{Un(x).I(x;z)\} = \int_o^z F(u).dF(u) = \frac{F^2(z)}{2}$

and: $\gamma_{UI}(h;z) = \frac{1}{2} E\{[Un(x+h) - Un(x)]\ .\ [I(x+h;z) - I(x;z)]\}$.

Consider the h-scattergram of Figure 3, representing the bivariate distribution of Z(x+h), Z(x). The difference $[I(x+h;z) - I(x;z)]$ is different from zero only in the two areas denoted A and B, where one variable is greater than z and the other is less than z.

- within A: $[I(x+h;z) - I(x;z)] = -1$, and $Un(x+h) \geq Un(x)$
- within B: $[I(x+h;z) - I(x;z)] = +1$, and $Un(x+h) \leq Un(x)$

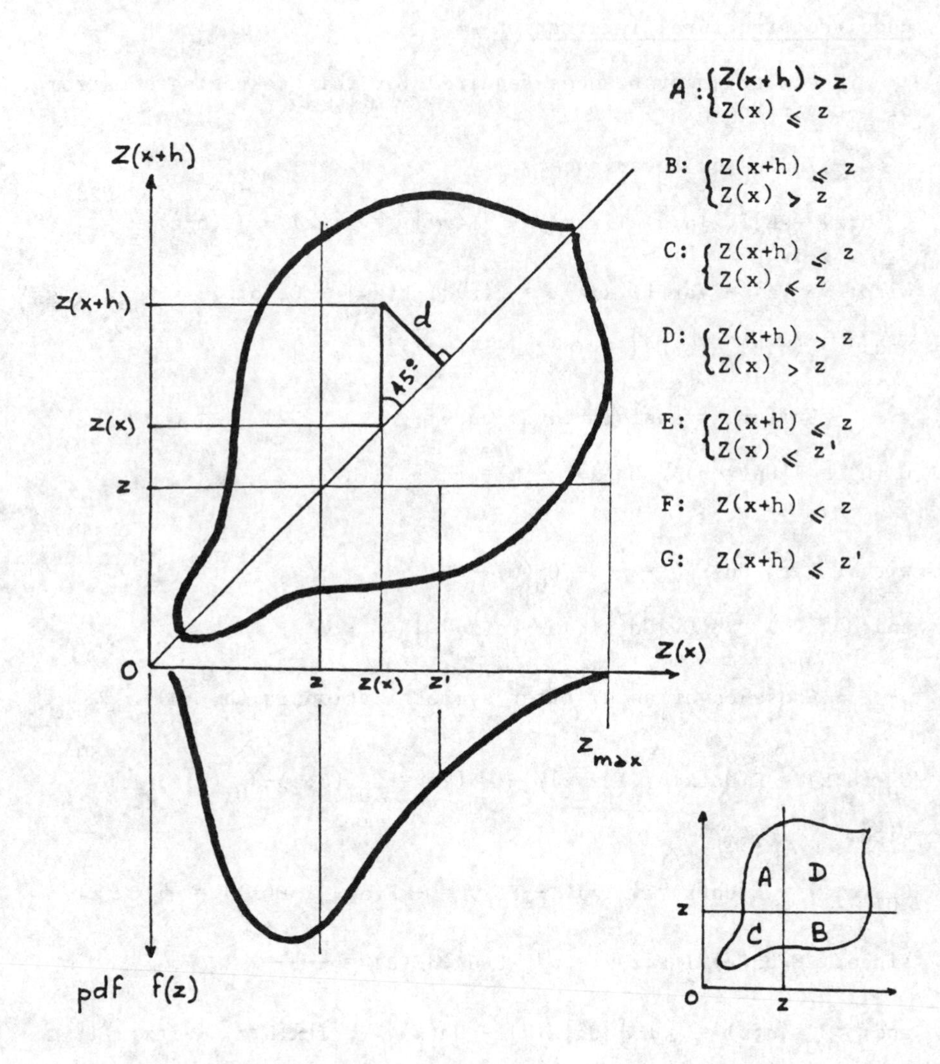

Figure 3. Bivariate probability distribution function (h-scattergram) of Z(x), Z(x+h).

That is, within either A or B:

$$[Un(x+h) - Un(x)] \cdot [I(x+h;z) - I(x;z)] = - | Un(x+h) - Un(x)| \leq 0$$

Moreover since the h-scattergram is made symmetric around the first diagonal, the cross-variogram is written:

$$\gamma_{UI}(h;z) = -E\{|Un(x+h) - Un(x)| \cdot [1-I(x+h;z)] \cdot I(x;z)\} \leq 0 \qquad (31)$$

Remarks

- The cross semi-variogram (31) appears linked to the conditional moment of order 1 of the transform R.F. Un(x). More precisely:

$$\gamma_{UI}(h;z) = -E\{|Un(x+h) - Un(x)|/Z(x+h) > z, Z(x) \leq z\}$$

$$\cdot \text{Prob } \{Z(x+h) > z, Z(x) \leq z\} =$$

$$= -E\{|Un(x+h) - Un(x)|/ Z(x+h) > z , Z(x) \leq z\} \cdot \gamma_I(h;z) .$$

This linkage bears very promixing seeds for further research on the usage of the structural moment of order 1 as criterion of optimization.

The corresponding correlogram is written:

$$\rho_{UI}(h;z) = \frac{2\sqrt{3}}{S(u)} \cdot C_{UI}(h;z) \leq 0, \text{ yielding:}$$

$$\rho_{UI}(0;z) = - \sqrt{3} \cdot S(z) \in \left[- \frac{\sqrt{3}}{2}, 0\right] .$$

The highest negative value for correlation between U(x) and I(x;z) is obtained for the median cut-off:

$z = M = F^{-1}(.5)$ entails: $S^2(M) = \text{Max } (S^2(z)) = ¼$, and:

$$\text{Max } |\rho_{UI}(0;z)| = |\rho_{UI}(0;M)| = + \frac{\sqrt{3}}{2} = + .866 \qquad (32)$$

Since the geometric configuration of data is identical for all R.F.'s Un(x), I(x;z) for all z, a unique run of a variogram subroutine can provide all required variograms for all cut-offs z. Similarly, implementation of a PK software is immediate from an Ordinary Kriging (OK) software.

The PK structural analysis is likely to be robust with regard to outlier data values z_α, for $\gamma_I(h;z)$ and $\gamma_U(h)$ are variograms of rank order type data, and $\gamma_{UI}(h;z)$ is a moment of order 1.

Order relations problems

The PK estimates (20) of the conditional cdf. $F(z/(N))$ may not verify the order relations of a cdf that is:

$$\begin{cases} F(z/(N)) \text{ is non a decreasing function of } z \\ F(x/(N)) \text{ is valued in } [0,1]. \end{cases} \quad (33)$$

This problem is shared by all projection-estimators in true sub-spaces of H_N. The DK estimator is no exception: $[F(y/(N))]^*_{DK}$ is not a strict cdf, even if the corresponding Hermitian developments were extended to $n_o = +\infty$. Only the projection of $I(x;z)$ onto H_N, i.e. the actual conditional cdf $F(z/(N))$ does verify these conditions, see Figure 1.

These order relations problems are linked to the non-convexity of the kriging process: the kriged estimator is not necessarily within the interval defined by the highest and lowest data of the same type used. In ordinary kriging, such non convex estimates are linked to negative weights.

An engineering solution around this problem is provided by quadratic programming, cf. Sullivan (1984). More precisely, given a series of initial estimates $[F(z_k/(N))]^*$, k = 1 to K, which might not verify the order relations (33), a series of corrected estimates $[F(z_k/(N))]^{**}$, k = 1 to K, is obtained such that the amount of correction:

$$\sum_{k=1}^{K} w_k \left[[F(z_k|(N))]^* - [F(z_k)/(N))]^{**} \right]^2 \text{ is minimum.}$$

The weights w_k can be chosen so as to reduce the amount of correction done in any particular critical interval of the cut-off z_k. Of course:

$$\sum_{k=1}^{K} w_k = 1, \text{ and: } w_k \geq 0.$$

Practice has shown that, provided the variograms used in the PK or IK systems do not present hole-effect, i.e. values greater than the sill value, the amount of order relations problems observed on the corresponding estimates is minimal.

The h-scattergram interpretation

The h-scattergram, i.e. the representation of the bivariate probability distribution function (in short pdf) of $Z(x + h)$, $Z(x)$, given in Figure 3 provides a useful insight to the structural relations between IK, PK and DK.

Consider any point within this scattergram of coordinates $[z(x+h), z(x)]$. The distance d of that point to the first diagonal is: $d = |z(x+h) - z(x)| \cdot \cos 45^o$, hence: $d^2 = \frac{1}{2} [z(x+h) - z(x)]^2$, and the semi-variogram: $\gamma_Z(h) = \frac{1}{2} E\{[Z(x+h) - Z(x)]^2\}$ appears as the inertia moment of the h-scattergram with respect to the first diagonal. Although an important characteristic of the bivariate pdf, the semi-variogram value $\gamma_Z(h)$ does not usually characterize this bivariate pdf. In other words, two bivariate pdfs. with the same variogram function $\gamma_Z(h)$ may be widely different.

The bivariate cdf of $Z(x+h)$, $Z(x)$ is written: $F_h(z, z') = \text{Prob}\{Z(x+h) \leq z, Z(x) \leq z'\} = E\{I(x+h; z) . I(x;z')\}$ and appears as the non centered cross-covariance between the two indicator R.F.'s $I(x;z)$ and $I(x;z')$. Hence knowledge of the indicator variogram matrix $[\gamma_I(h;z,z'), \text{ for all } z,z']$ is dual to that of the bivariate cdf. $F_h(z,z')$. Recall that:

$$\gamma_I(h;z,z') = \tfrac{1}{2}E\{[I(x+h;z) - I(x;z)].[I(x+h; z') - I(x;z')]\} .$$

This indicator variogram matrix thus carries a much richer structural information than the mere Z-semivariogram $\gamma_Z(h)$. Thus a kriging technique drawing from this variogram matrix must lead to more powerful estimates than a γ_Z-based linear kriging. This was precisely the starting idea of both DK and IK.

DK amounts to perform an indicator cokriging making full use of the previous indicator variogram matrix, i.e. of the bivariate cdf. $F_h(z,z')$. This would be a fantastic achievement were this variogram matrix experimentally inferred. Unfortunately, this essential structural information is mostly drawn from a bivariate distribution model, namely a bi-ϕ^{-1}-normal model for $F_h(z,z')$, cf. a previous remark made at the end of section IV. The only bivariate feature of that model which is experimentally inferred is a joint moment of order 2, more precisely the Y-semivariogram $\gamma_Y(h)$ of the normal transform $Y(x) = \phi^{-1}(Z(x))$. In other terms, if the two normal R.V.'s $Y(x+h)$ and $Y(x)$ are not also binormal distributed, the DK model is inappropriate.

On the other hand, the various forms of IK attempt to infer experimentally parts of the previous indicator variogram matrix $[\gamma_I(h;z,z')]$:

(i) In the "multiple" IK approach, so called because it considers K cut-offs z_k, only the K diagonal terms of that matrix are inferred:

$$\gamma_I(h;z_k) = E\{[I(x+h;z_k) - I(x;z_k)]^2\} = \text{Prob }\{A\} = \text{Prob }\{B\},$$

cf. areas A and B on Figure 3. This amounts to estimate the bivariate cdf. $F_h(z,z')$ by a series of K values $F_h(z_k, z_k)$, k = 1 to K.

(ii) The PK approach, in addition to the previous K diagonal terms $[\gamma_I(h;z_k)$, k = 1 to K$]$, retains:

- a series of K conditional moments of order 1: $[\gamma_{UI}(h;z_k)$, k = 1 to K$]$, cf. definition (31)

- and an overall joint moment of order 2:

  $\gamma_U(h) = \frac{1}{2}E\{[Un(x+h) - Un(x)]^2\}$, this last structural function being in all points similar to the gaussian transform semi-variogram $\gamma_Y(h)$ inferred for the DK process.

From a strict distribution-free approach, PK is superior to DK in the sense that it draws more from the data: not only a joint moment of order 2 but also the direct indicator variograms and joint conditional moments of order 1. But if the gaussian transform $Y(x) = \phi^{-1}(Z(x))$ is proven to be indeed bi-normal-distributed, then DK has a definite superiority. Consequently, in DK applications, some emphasis should be given to checking for bi-ϕ^{-1}-normality of the data $z(x_\alpha)$.

The median IK approximation

In the earlier applications of OK, cf. Journel (1983), a direct attempt to ensure the order relations (33) consisted in assuming an intrinsic coregionalization model for the indicator variogram matrix $[\gamma_I(h;z,z')]$. More precisely, the K direct variograms $[\gamma_I(h;z_k)$, k = 1 to K$]$ required by the IK approach were assumed to be proportional one to another:

$$\gamma_I(h,z_k) = \frac{S^2(z_k)}{S^2(z_{k'})} \cdot \gamma_I(h;z_{k'}), \text{ for all } k, k'$$

with the variance factors $S^2(z) = \text{Var }\{I(x;z)\} = F(z)[1-F(z)]$ being known from the univariate cdf $F(z)$.

Consequently, a single direct indicator variogram needed to be inferred: a good minimax-type choice being the median indicator variogram corresponding to the cut-off

$z_k = M = F^{-1}(.5)$, providing an equal split of the corresponding indicator data $i(x_\alpha;M)$ in values (0,1), the maximum value for the sill: $\text{Max}\,\{S^2(z)\} = S^2(M) = .25$, and the maximum correlation (32).

The distinct advantage of this median IK approximation is that it reduces the number of indicator kriging systems of the type (27) from K to 1, thus reducing considerably the order relations problems. The excellent solution brought to order relations problems by quadratic programming has since removed much of the focus on median IK.

Matheron (1982) has shown that the intrinsic coregionalization model underlying the median IK approximation is characteristic of the following bivariate cdf:

$$F_h(z,z') = \rho_Z(h) \cdot F(\min(z,z')) + [1-\rho_Z(h)] \cdot F(z) \cdot F(z'), \tag{34}$$

with:

$$\rho_Z(h) = 1 - \frac{\gamma_Z(h)}{\text{Var}\,\{Z(x)\}}$$ being the correlogram of $Z(x)$.

Interestingly, this model was introduced independently at Stanford in an effort to approximate the cross values $F_h(z,z')$ from the values $F_h(z,z)$ estimated from the direct indicator variograms $\gamma_I(h;z)$. Indeed the affine correction type relation:

$$\frac{F_h(z,z')}{F_h(z,z)} = \frac{\text{Prob } E}{\text{Prob } C} \simeq \frac{F(z')}{F(z)} = \frac{\text{Prob } G}{\text{Prob } F}$$, cf. Figure 3

leads to the model (34).

The bivariate cdf (34) corresponds to a probability $\rho_Z(h)$ for the two R.V.'s $Z(x+h)$ and $Z(x)$ to be identical, and a probability $[1-\rho_Z(h)]$ for them to be independent. Under this model, it can be shown that the correlogram of any measurable function $h(Z(x))$ is identical to $\rho_Z(h)$. In particular all indicator correlograms are identical, entailing that all indicator variograms are proportional one to another.

Thus it appears that the median IK approximation amounts to adopt the bivariate model (34) characterized by a correlogram $\rho_Z(h) = \rho_I(h;M)$ and a univariate cdf. $F(z)$, both inferred experimentally; a modelling level equivalent to that of the DK model. It may be argued though that a bi-ϕ^{-1}-normal is more "classic and realistic" than the model (34).

Conclusions

The large array of kriging species, sometimes obscured by cumbersome notations and mathematics, may mask their very simple common line: they are all regression techniques and differ only by the particular functions of the data being recombined into an estimator.

When these functions are function of one datum at a time, they are called "service-variables". Most kriging falls into that service-variable category, including linear krigings, disjunctive kriging and indicator krigings. The variograms characterizing the spatial variability of these service-variables are either inferred experimentally (OK, IK, PK) or deduced from a bivariate distribution model (DK, median IK approximation).

A kriging process applied to the indicator R.F. I(x;z) allows estimation of the conditional cdf.

$$F(z/(N)) = E\{I(x;z)/ Z(x_\alpha) = z_\alpha,\ \alpha \in (N)\} .$$

A block-support correction which, in absence of block data is necessarily distribution-dependent, then provides an estimate of the block-support conditional cdf.

$$F_V(z/(N)) = \text{Prob}\ \{Z_V(x) \leq z/\ Z(x_\alpha) = z_\alpha,\ \alpha \in (N)\} .$$

This last estimate is then used to provide an estimate of any known function h of an unknown block value $z_V(x)$:

$$[h(Z_V(x))]^* = E^*\{h(Z_V(x))/Z(x_\alpha) = z_\alpha,\ \alpha \in (N)\} =$$

$$= \int_0^\infty h(z).dF_V(z/(N)) \quad .$$

The critical problem of estimating local recovery functions, or more generally spatial distributions of block values $Z_V(x)$ within an area $\{x \in V\}$, is solved by considering for particular functions h the ones defined in relations (1).

Also an estimate of the conditional cdf. of the transform $h(Z_V(x))$ can be deduced from that of $F_V(z/(N))$. The corresponding conditional quantiles provide estimates of the prediction intervals for $[h(Z_V(x)]^*$.

The h-scattergram visualization of the bivariate structural information available shows that there is much more to it than the mere Z-variogram, hence allowing for further developments in non-parametric Geostatistics.

References

Beckmann, P., 1973, "Orthogonal polynomials for engineers and physicists", Golem Press, Boulder, 280 p.

Davis, B., 1984, "Indicator Kriging as applied to an alluvial gold deposit", in this volume, p. 337.

Journel, A. and Huijbregts, Ch., 1978, "Mining Geostatistics", ed. Academic Press, 600 p.

Journel, A., 1983, "Non-parametric estimation of spatial distributions" in Math. Geology, Vol. 15, no. 3, pp. 445-468.

Journel, A., 1984, "mAD and Conditional quantile estimators", in this volume , p. 261

Lemmer, C., 1984, "Estimating local recoverable reserves via IK", in this volume.

Matheron, G., 1976, "A simple substitute for conditional expectation: disjunctive kriging" in Proceedings of the 1st Geostat. NATO ASI, Rome (1975), D. Reidel, p. 221-236.

Matheron, G., 1982, "La destructuration des hautes teneurs et le krigeage des indicatrices" Report N-761, CGMM, Fontainebleau.

Rao, C.R. 1965, "Linear statistical inference and its applications" ed. Wiley and Sons, 625 p.

Sullivan, J. (1984), "Probability Kriging - Theory and Practice", in this volume, p. 365.

Verly, G., 1983, "The multigaussian approach and its applications to the estimation of local recoveries", in Math. Geology, Vol. 15, no. 263-290.

Verly, G., 1984, "The block distribution given a point multivariate ϕ-normal distribution", in this volume, p. 495.

INDICATOR KRIGING AS APPLIED TO AN ALLUVIAL GOLD DEPOSIT

Bruce M. Davis

St. Joe American Corporation

A case study of indicator kriging (I.K.) used to estimate in-situ grade in an alluvial gold deposit is presented.

The deposit and sampling present problems which may cause difficulties in other linear and nonlinear estimation procedures. These problems are either mitigated or avoided by using I.K.

INTRODUCTION

The deposit under study is a gold placer of the gold bearing Tertiary gravel deposits of California. The auriferous gravels are the result of deposition by the Tertiary Yuba River system. Bedrock is composed of metasedimentary and metavolcanic rocks. Three distinct gravel units exist at the property; however, this study is restricted to the lower, coarser, compositionally less mature unit. The lower gravel contains the most significant amounts of gold, and it is most crucial for economic reasons to estimate the quantity of gold contained in this unit.

The area being considered for possible mining operations is intersected by 53 vertical drill holes. These 53 holes are made up of 36, 6 inch diameter churn drill holes, drilled between 1917 and 1939 and 17, 36 inch diameter "body" drill holes drilled between 1981 and the present.

The drilling presents two problems. The first is that the total number of holes available to make an evaluation is small. The second related problem is that the difference in volumes sampled by the two different types of drills raises the possibility

G. Verly et.al. (eds.), Geostatistics for Natural Resources Characterization, Part 1, 337–348.

of important differences in support. These differences could further reduce the amount of data available for estimation.

In addition to the support problem, the deposit presents other potential geostatistical problems. A few very large grade values occur, but there is no reason to believe that the values are in error and should be discarded. The frequency distribution of the sample grades is well represented by a lognormal distribution, but the panels which will eventually be considered are many times larger than the samples. Therefore, an assumption of lognormality for panel grades is questionable given the result of the m-dependent central limit theorem.

These problems put an unique burden on any method used to estimate grade. The method used must mitigate any possible support effect differences. It must not be greatly effected by the departure from normality or lognormality, and it must allow for the presence of anomalous grade values without requiring they be trimmed or removed.

The only presently available method which adequately handles each of the above problems is the technique of indicator kriging (5). This case study describes how indicator kriging is used to estimate in-situ grade when faced by problems that would cause serious difficulties in other geostatistical methods.

SUPPORT DIFFERENCES

Because of the paucity of data available to estimate reserves, it would be very helpful to use both pre-and post-1980 drilling together. The relationship between volume sampled and grade was, therefore, examined.

To begin, the volume of material from the lower gravel at each drill hole was available from the drill logs. The amount of gold recovered from washing that volume was also given. The linear correlation coefficient of grade (amount of gold recovered divided by the volume of material washed) and volume was calculated as -0.156. This means approximately 2.5 percent of the variation in the grade-volume data could be explained by a linear relationship.

Figure 1 shows a scatter plot of grade versus volume. The distinction between the pre-and post-1980 drilling is readily apparent from the two clusters of values along the volume axis. However, the scatter of grades for each group of volumes is very similar except for the one extreme grade from a churn drill hole. The scatter of grades versus volume on this plot gives no compelling evidence that a nonlinear relationship exists between grade and volume.

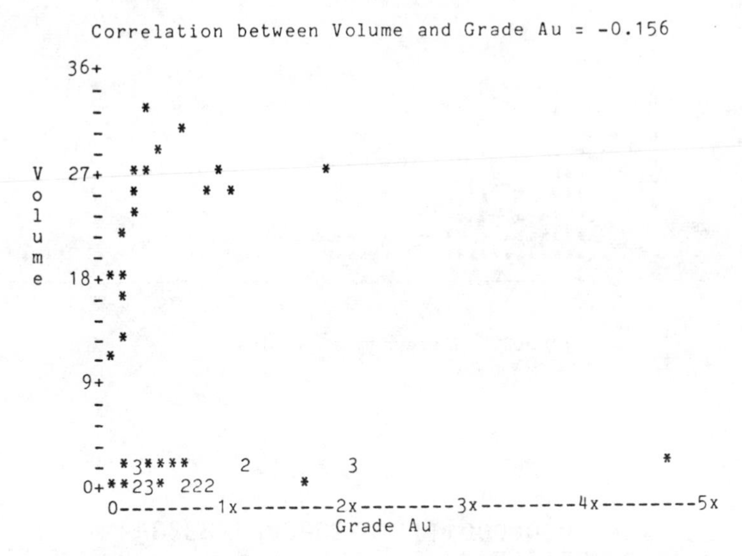

Figure 1 - Scatterplot of Sample Volume Versus Grade

The nature of indicators also mitigates the effect of change in support on the grade estimates. This is the result of the indicator being associated with a range of values rather than a single value. Indicators are, therefore, less sensitive to possible inaccuracies in the specification of the grade because of inadequate sample volume. While examination of the data provides no evidence of a relationship between grade and volume, this property of indicators further insures the integrity of the estimates to be produced for this property.

PROBLEMS OF OUTLIERS AND DISTRIBUTION

Figure 2 shows the histogram of grade. There are several values which are anomalous. One value is extremely far removed from the remaining values while five values are clearly anomalous but less removed from the main body of data. The stem and leaf display (7) (Figure 2.a) of the grade data gives a similar picture of the distribution of anomalous values.

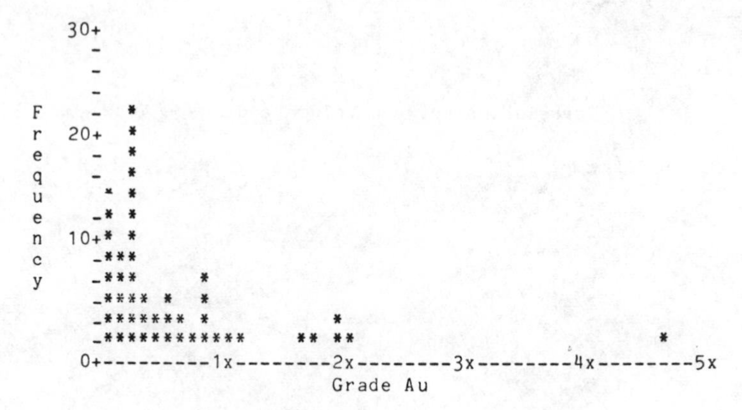

Figure 2 - Histogram of Au Grade

```
+0* 00000001111222222222233344
+0. 55566788889
 1* 01
 1. 6799
 2* 0
 2.
 3*
 3.
 4*
 4. 6
```

Figure 2.a - Stem and Leaf Display of Grade

It should be noted that large values considered here are designated as anomalous as opposed to outliers (in spite of the heading). The term outlier connotes error in the data. Careful examination of the data used here gives no indication that any of the values reported is in error. However, the sampling method (churn drilling) is fraught with potential for generating errors in the grade of the sample. It is necessary, therefore, that the estimation process use the large values while protecting against the possibility of error.

Figure 3 shows the normal score plot (6) of the natural logarithm of grade. Also shown is the correlation between the normal scores and the log-grade values which is a measure of the straightness of the line. The high correlation and visual examination of the plot suggest that the data are "lognormally distributed".

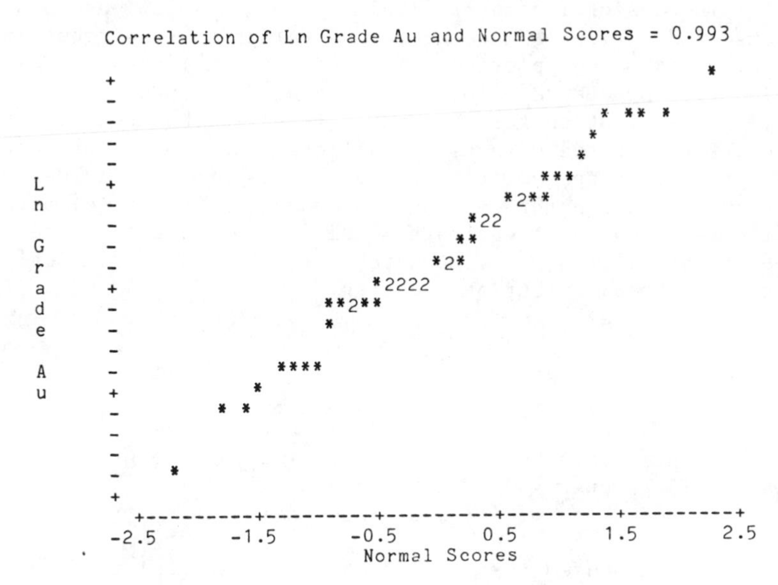

Figure 3 - Normal Frequency Plot of the Natural Logarithm of Grade

One method proposed to the author for dealing with the anomalous values was rejection of the data above an arbitrary cutoff. Another method involved the adjustment of all values above an arbitrary cutoff (e.g., 2.5 standard deviations above the mean) to the mean value of the histogram class containing the grades greater than the cutoff value. After either rejection or trimming it was suggested that a natural log transformation be made to deal with the skewed distribution of the data.

These proposals were rejected. The reduction in the amount of data required by the first proposal was deemed unacceptable, though Hampel (2) showed such a rejection rule performs well in resisting the effects of outliers.

Both proposals presented a problem that is often overlooked. While the procedures are robust, robustness and nonparametric methods are not the same (4). Like most robust procedures, these are parametric in nature. Other procedures such as "robust kriging" (3), are also very distribution dependent and may require several passes through the data. In this case where the point support data are well represented by a lognormal model and the proposed mining panel size is large in relation to the point data, making an assumption of lognormality for the frequency of grade in the panels seems highly inappropriate.

The use of indicator kriging provides solutions to the problems of distribution and anomalous values that no robust parametric method can provide. The indicator kriging is a distribution free procedure eliminating the need to specify an underlying distribution for the data. It, also, has the property of resistance to the effects of outliers/anomalies, a property not inherent to nonparametric methods. This property is derived from the rank order nature of the procedure. As Huber (4) puts it, "The performance of estimates derived from rank tests tends to be robust...., this a fortunate accident, not intrinsically connected with distribution-freeness."

GRADE ESTIMATION

Because of the problems, indicator kriging is the geostatistical procedure best suited to estimating the grade at this particular property. Lack of data still limits the estimation to the production of in-situ grade estimates.

The data used for estimation purposes consist of the grade for the entire thickness of the lower gravel and the thickness of the lower gravel.

Selection of Indicator Cutoffs

From the histogram of grade (Figure 2) and the scattergrams of grade versus gravel thickness (Figure 4), five cutoff values of grade were selected. These cutoffs were also selected to have a value near the median of the grade data, one or two cutoff values below the median and two or three cutoffs above the median.

The indicator for each cutoff is defined as

$$i_c(\underline{x}) = \begin{cases} 0, & \text{if } z(\underline{x}) > z_c \\ 1, & \text{if } z(\underline{x}) < z_c \end{cases} \tag{1}$$

where $z(\underline{x})$ is the grade at location $\underline{x}$ and z_c is the cutoff value. This definition is recommended by Journel (5) to allow interpretation of indicator estimates as steps in the cumulative frequency histogram of grade over the area estimated.

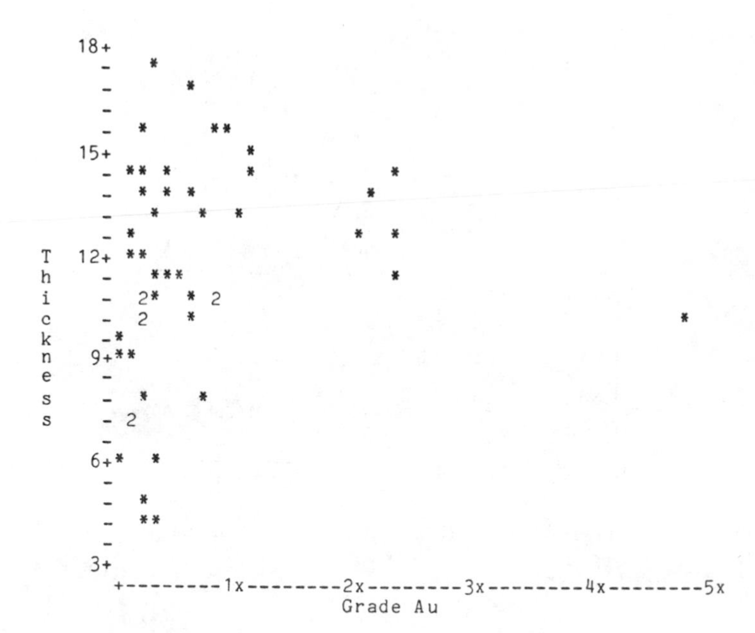

Figure 4 - Scatterplot of Gravel Thickness Versus Grade

Indicator Variograms

For each cutoff, variograms of the indicators were calculated in two directions. One along fences of drill holes, and one across the fences of drill holes. Sparse data prevented meaningful sample variograms to be calculated in any other directions. From the set of five variograms, three were chosen for purposes of estimation. The variograms for the rejected cutoffs were not appreciably different from one of the selected variograms. Therefore, use of these cutoffs in the estimation would not materially alter the estimate produced from only three cutoffs.

The pair of indicator variograms for the greatest cutoff is shown in Figures 5.a-5.b. The small number of holes accounts for the small number of pairs at most lags and the "noisy" sample variogram values. The sills of the models are fitted based on the sample variogram values and the value of the sample variance of the data used to calculate a variogram.

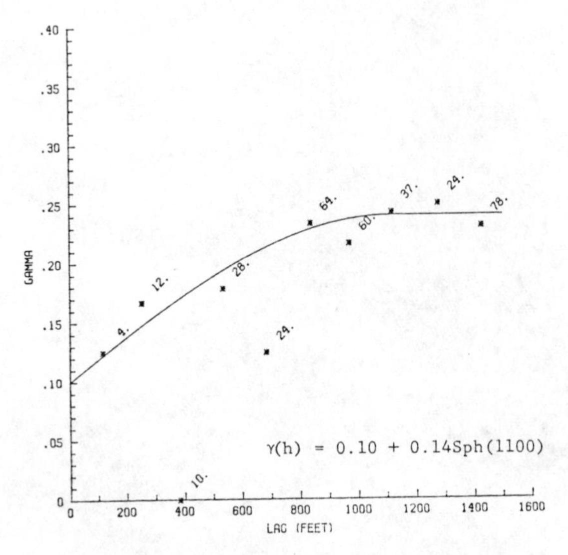

Figure 5.a - Cutoff 3 Along Channel Indicator Variogram

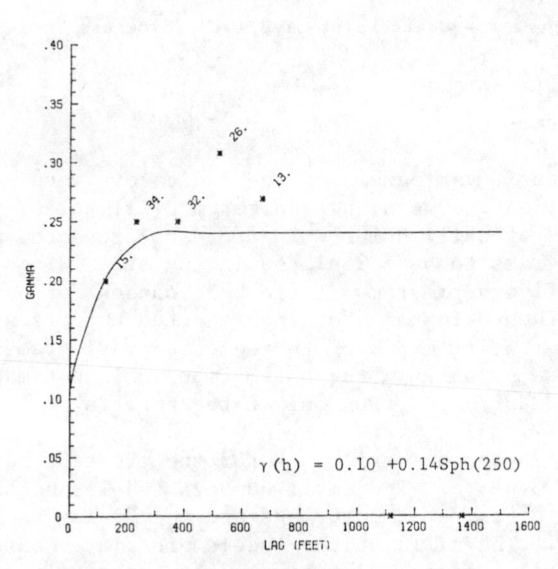

Figure 5.b - Cutoff 3 Across Channel Indicator Variogram

It should be stated that the first cutoff value corresponds to the 0.23 quantile of the grade data; the second cutoff value corresponds to the 0.62 quantile. These quantile values are consistent with the sill values of the fitted models for each of the three indicator variograms.

The variogram models are also consistent with the geology. The along fence variograms are variograms taken across the channel. These have the same or shorter ranges than the along channel or across fence variograms. Since it is believed the amount of gold tends to be more continuous along the channel than across it, this set of indicator variograms agrees with the geologic interpretation. The nugget effect of the third indicator variogram is larger than for the variograms of the lower cutoffs. If higher grade is related to greater gold nugget size or more intense local concentration, the nugget effect for this variogram should be larger.

Estimation of the In-Pit Grade Histogram

Grade in a pit designed on the basis of a polygonal reserve estimate was estimated by indicator kriging. Point estimates were made on a regular grid covering the pit. The point estimator was the ordinary kriging estimator (using the notation of Davis and Grivet (1)

$$\phi^*(\underline{x}_o;z_c) = \underline{\lambda}'\underline{i} + (1 - \underline{\lambda}'\underline{e})F^*(z_c) \qquad (2)$$

where $\underline{x}_o = (x_o, y_o)$ is the location of the grid node, $\underline{\lambda}$ is the vector of simple (unconstrained) kriging weights, $F^*(z_c)$ is an estimate of the global value of the indicator, $\underline{e}$ is a vector of ones, and $\underline{i}$ is a vector of the indicator values at the drill hole locations. Point estimates, 2, falling inside the pit limits were then averaged to produce an estimated in-pit histogram of grade of the lower gravel. That histogram is given in figure 6.

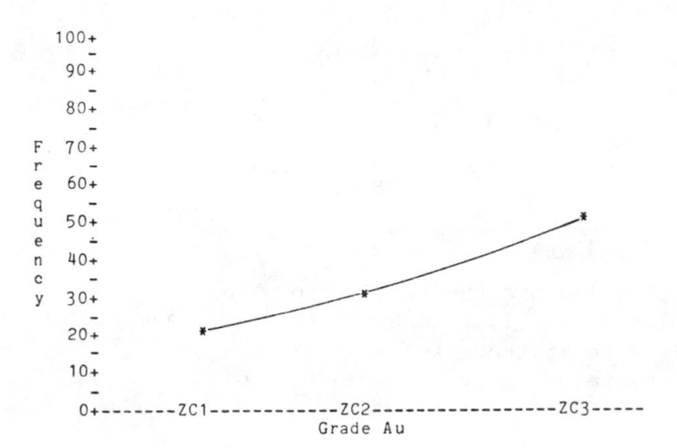

Figure 6 - Estimated Cumulative Frequency Histogram of Grade Inside the Pit

Written in the form, 2 above several aspects of the kriging estimator may be observed. In areas of sparse data, weights in the vector will be small so the estimate will be dominated by the global estimate of the indicator. Similarly when the point to be estimated is near to drill hole data, the estimate will be less effected by the global value and more the result of the local values. Also, 2, shows that because kriging does not preclude the generation of negative weights, it is possible to generate negative values of an indicator or otherwise misrepresent the order relationship of the estimated histogram.

The estimation in this study produced 26% of the point estimates of the indicator of cutoff 1 greater than for cutoff 2. Similarly, 29% of the point estimates of cutoff 2 were greater than those for cutoff 3. While less than 1% of the estimates for cutoff 1 were greater than those for cutoff 3.

These order relationship problems were corrected using the method suggested by (5). The corrections were made sequentially according to

$$\begin{aligned} \phi^{**}(\underline{x}_o;z_{c1}) &= \mathrm{Max}\{\phi^{*}(\underline{x}_o;z_{c1}),0\} \\ \phi^{**}(\underline{x}_o;z_{c2}) &= \mathrm{Max}\{\phi^{**}(\underline{x}_o;z_{c1}),\phi^{*}(\underline{x}_o;z_{c2})\} \\ \phi^{**}(\underline{x}_o;z_{c3}) &= \mathrm{Max}\{\phi^{**}(\underline{x}_o;z_{c2}),\phi^{*}(\underline{x}_o;z_{c3})\} \\ \phi^{***}(\underline{x}_o;z_{c3}) &= \mathrm{Min}\{\phi^{**}(\underline{x}_o;z_{c3}),1\} \end{aligned} \tag{3}$$

The histogram given in figure 6 is the result of the corrected estimates.

The estimated in-situ mean grade in the pit was calculated from the formula

$$\begin{aligned} m^{*}(p) = {} & z_{c1}/2\{\phi^{**}(p;z_{c1})\} + (z_{c2} + z_{c1})/2 \times \\ & \{\phi^{**}(p;z_{c2}) - \phi^{**}(p;z_{c1})\} + (z_{c3} + z_{c2})/2 \\ & \times \{\phi^{***}(p;z_{c3}) - \phi^{**}(p;z_{c2})\} \\ & + z_m\{1 - \phi^{***}(p;z_{c3})\} \end{aligned} \tag{4}$$

where z_{ci}, i = 1-3 are the three cutoff values and z_m is the average value of the last class of the raw grade histogram. The value z_m is used instead of the usual $(z_{max} + z_{c3})/2$ to resist the anomalous value z_{max}.

CROSS-VALIDATION

Cross-validation of the estimate is hampered by the lack of production data. All comparisons must involve the drilling data only.

The cross-validation was performed by calculating the estimated histogram over windows containing at least seven drill holes and then comparing the estimated proportion with the proportion of drill hole grades below each cutoff. It was realized that because of spatial correlation the true proportion might deviate from that calculated from the drill hole grades.

Figure 7 gives the comparison of the true with the estimated proportions. For the three cutoffs, the estimates are greater than the true values by from 3 to 5 percent. This means the estimated grade may be slightly conservative; however, it is impossible to discriminate between the values given the large variability in the data. A discrepancy of 5 percent is very reasonable and lends support to the validity of the method. Of course, the check does not strictly validate the method because drill hole data rather than actual production data was used.

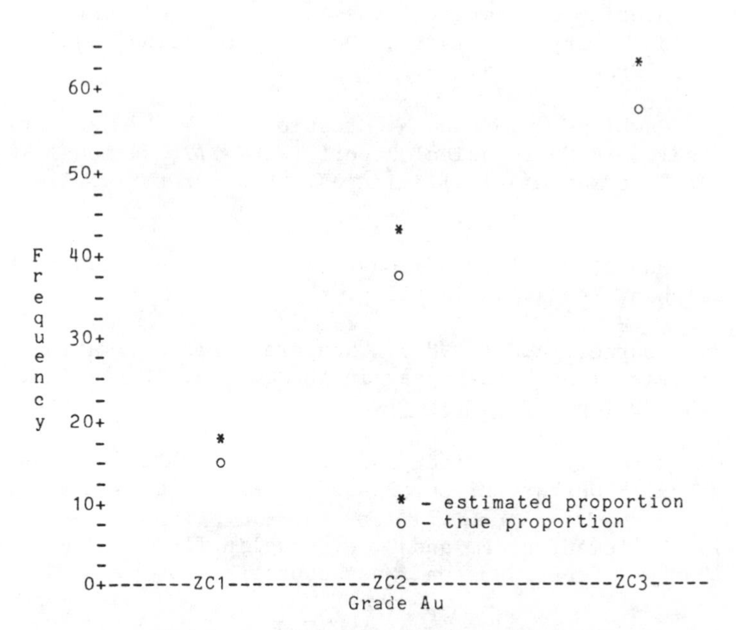

Figure 7 - Comparison of True and Estimated Proportions

CONCLUSIONS

Problems with the support of the samples, anomalous grades, and distribution of the grade within a panel present a unique aspect to estimating grade in this deposit. The potential for intractability exists in parametric geostatistical methods under these conditions despite their robustness. Distribution free geostatistical methods such as ordinary kriging are not resistant to the presence of anomalous grades and in the presence of sparse data may over value the deposit.

Only the rank order procedure of indicator kriging is rich enough to correctly handle each of the problems presented by the deposit.

REFERENCES

1. Davis, M.W. and C. Grivet (1983), "Kriging in a global neighborhood," to appear in the Journal of the Int. Assoc. for Math. Geology.

2. Hampel, F. (1974), "Rejection rules and robust estimates of location: An analysis of some Monte Carlo results," Proc. European Meeting of Statisticians and 7th Prague Conference on Information Theory, Statistical Decision Functions and Random Processes, Prague, 1974.

3. Hawkings, D.M. and N. Cressie (1982), "Robust kriging - a proposal," NRIMS Technical Report TWISK 217, National Research Institute for Mathematical Sciences, CSIR, Pretoria, South Africa, 28 p.

4. Huber, P. (1977) "Robust statistical procedures," SIAM, Philadelphia, Pennsylvania, 55 p.

5. Journel, A.G. (1983), "Nonparametric estimation of spatial distributions," to appear in the Journal of the Int. Assoc. for Math. Geology.

6. Ryan, T., B. Joiner and B. Ryan (1981), "MINITAB reference manual," Duxbury Press, Boston, Massachusetts, 154 p.

7. Velleman, P. F. and D. C. Hoaglin (1981), "ABC's of EDA," Duxbury Press, Boston, Massachusetts, 354 p.

ESTIMATING LOCAL RECOVERABLE RESERVES VIA INDICATOR KRIGING

I.C. Lemmer

Gold Fields of South Africa, Johannesburg

ABSTRACT

The planning of selective mining operations requires knowledge of local recovery functions, i.e. how recovered tonnage and ore grade vary with cutoff grade for each mining panel under consideration. Theoretically this requires estimating local grade distributions within such panels.

Indicator kriging suggests a linear way of estimating a local distribution directly. The relationship between grade and indicator correlograms in general is derived, and its implications examined. A generalized indicator function is proposed and the corresponding estimator of local distributions formulated.

The results generated for a computer-simulated orebody are very encouraging and compare well with non-linear methods used on the same orebody.

1. PRELIMINARIES

In practice the problem of recoverable reserves is usually formulated as follows: given a large mining block or 'panel', area A, with N drillhole samples having grades z_α, α = 1 to N, in and in the immediate vicinity of A, what is the proportion of selective mining units within A with average grades higher than a cutoff grade z and what will be the corresponding average grade of all the units so selected in A. The drilling density at this stage is usually of the order of one hole per

G. Verly et al. (eds.), Geostatistics for Natural Resources Characterization, Part 1, 349–364.

panel. For the purposes of this study, it will be assumed that selective mining units have core-size volumes like the data and selection will thus be performed on their local grade distribution in A. Changing the volume or support of units constitutes an additional problem not considered here.

An indicator kriging (IK) procedure has recently been proposed by Journel (1) for estimating such local panel distributions from the available sample data. This approach introduces indicator step functions in the grade $z(\underline{x})$ at $\underline{x}$,

$$i(z(\underline{x});z) = i(\underline{x};z) = \begin{cases} 1, & \text{if } z(\underline{x}) \leqq z \\ 0, & \text{if } z(\underline{x}) > z \end{cases} \quad (1.1)$$

and their mean value over panel area A (within deposit D),

$$\phi(A;z) = 1/A \int_A i(\underline{x};z)\, d\underline{x} \quad (1.2)$$

as functions of the cutoff grade z. Since $\phi(A;z)$ amounts to the grade cumulative histogram, knowledge of it allows a direct calculation of the core recovery functions within A.

The proportion of tonnage recovered and the corresponding recovered quantity of metal after applying cutoff z are given by:

$$\begin{aligned} t(A;z) &= 1 - \phi(A;z) \\ q(A;z) &= \int_z^{\infty} u \frac{\partial \phi(A;u)}{\partial u}\, du \end{aligned} \quad (1.3)$$

with relevant average ore grade $m(A;z) = q(A;z)/t(A;z)$.

The IK approach consists of constructing linear estimators of $\phi(A;z)$ via some kriging procedure in which the estimate $\phi^*(A;z)$ is a linear function of the $i(\underline{x}_\alpha;z)$ defined for the N closest data values, $\alpha = 1, \ldots, N$ in and around area A. To implement such a linear kriging procedure one elevates the i's to random functions $I(Z(\underline{x});z)$ of the random grade functions $Z(\underline{x})$. Then $I(Z(\underline{x});z) = I(\underline{x};z)$ can be assigned an expected value

$$E\{I(\underline{x};z)\} = F(z) = \text{Prob}\,\{Z(\underline{x}) \leqq z\} \quad (1.4)$$

for all $\underline{x}$ ε stationary deposit D, a centered covariance

$$\sigma_I(h;z) = E\{I(\underline{x}_\alpha;z).I(\underline{x}_\beta;z)\} - F^2(z) \quad (1.5)$$

between the random variables at $\underline{x}_\beta$ and $\underline{x}_\beta + h$, and a variance

$$\text{Var}\{I(\underline{x};z)\} = \sigma_I^{\,2}(z) = F(z)\left[1 - F(z)\right]$$

The notation emphasizes that all these functions are parametrized by the cutoff grade z. Unbiased linear estimators $\phi^*(A;z)$ of $\phi(A;z)$ are provided by the standard expressions

$$\phi^*(A;z) = \begin{cases} \sum_\alpha \nu_\alpha(z).i(\underline{x}_\alpha;z) + [1 - \sum_\alpha \nu_\alpha(z)].F(z), & \text{simple kriging (SK)} \quad (1.7a) \\ \sum_\alpha \lambda_\alpha(z).i(\underline{x}_\alpha;z), & \text{ordinary kriging (OK)} \quad (1.7b) \end{cases}$$

with associated kriging systems

$$\left.\begin{array}{ll} \text{SK:} & \sum_{\beta=1}^{N} \nu_\beta(z).\rho_I(h_{\alpha\beta};z) \\ \text{OK:} & \begin{cases} \sum_{\beta=1}^{N} \lambda_\beta(z).\rho_I(h_{\alpha\beta};z) - \mu \\ \sum_{\beta=1}^{N} \lambda_\beta(z) = 1 \end{cases} \end{array}\right\} = 1/A \int_A \rho_I(h_{\alpha o};z)dx_o = \overline{\rho}_I(\alpha;z) \qquad \alpha = 1, \ldots, N \qquad (1.8)$$

where we have substituted the indicator correlogram value ρ_I for the indicator covariance value via

$$\sigma_I(h;z) = \sigma_I^2(z).\rho_I(h;z) \qquad (1.9)$$

Implementation of either of the above estimates for $\phi(A;z)$ entails a knowledge of the indicator correlogram of the deposit <u>for each cutoff value z.</u>

2. INDICATOR COVARIANCE

Let us now investigate how the indicator covariance depends on the cutoff grade by assigning an isofactorial bivariate density (2)

$$f_{\alpha\beta}(z_\alpha,z_\beta) = f(z_\alpha).f(z_\beta) \sum_{n=o}^{\infty} T_n(\alpha,\beta)\ \chi_n(z_\alpha)\ \chi_n(z_\beta) \qquad (2.1)$$

to the grades. We recall here that the orthogonal polynomials χ_n, having the marginal density f as weight function over their orthogonality interval, are uniquely determined by this weight function, provided that certain rather general integrability conditions are met (3). From this fact it follows that the first two polynomials are

$$\chi_0(z) = 1, \qquad \text{and} \qquad \chi_1(z) = (z - m)/\sigma \qquad (2.2)$$

irrespective of the actual form of f(z), where m and σ^2 are the mean and variance of the grade. Construction of higher order polynomials naturally depend on knowing the higher moments of f(z) too.

We employ the expansion (2.1) to obtain the following expression for the indicator covariance,

$$\sigma_I(h;z) = \sum_{n\geq 1}^{\infty} T_n(h).F_n^{\,2}(z) \tag{2.3}$$

with

$$F_n(z) = \int_{-\infty}^{z} f(z').\chi_n(z')\,dz' \tag{2.4}$$

In particular, $F_0(z)$ is identical with the grade distribution value at z:

$$F_0(z) = \int_{-\infty}^{z} f(z').\chi_0(z')\,dz' = F(z) \tag{2.5}$$

Note that, with the exception of F_0, all the F_n's vanish for large z due to the orthogonality of χ_n with $\chi_0 = 1$. Consequently σ_I becomes small at high cutoffs (the destructurization effect of high grades (4)). One can also invoke the assumed completeness of the χ_n's to evaluate (2.3) for h = 0 and retrieve the previous expression (1.6) for the indicator variance,

$$\sigma_I^{\,2}(z) = \sum_{n\geq 1}^{\infty} F_n^{\,2}(z) = F_0(z)\left[1 - F_0(z)\right] \tag{2.6}$$

The result (2.3) establishes the connection between $\sigma_I(h;z)$ and the covariances $T_n(h_{\alpha\beta}) = E\{\chi_n(Z_\alpha).\chi_n(Z_\beta)\}$ of polynomial functions of the grade. In particular, we see in view of (2.2) that

$$T_0(h) = 1, \quad \text{and} \quad T_1(h) = E\{(Z_\alpha - m)(Z_\beta - m)\}/\sigma^2 = \rho(h) \tag{2.7}$$

where $\rho(h)$ is the grade correlogram of the deposit. It is known that the particular case of a Gaussian bivariate distribution for $f_{\alpha\beta}$ in (2.1) gives $T_n(\alpha,\beta) = \rho^n(h_{\alpha\beta})$ in (2.3). Thus the term $T_1 = \rho(h)$ proportional to the grade covariance is seen to dominate the T_n's in this case for small ρ. If we assume this to be a common feature of any reasonable distribution $f_{\alpha\beta}$, then, isolating the coefficients a_n $(n \geq 2)$ of terms proportional to ρ in $T_n(\alpha,\beta) = E\{\chi_n(Z_\alpha).\chi_n(Z_\beta)\}$ by multiplying out the two polynomials, one can rearrange (2.3) to read

$$\sigma_I(h;z) = \rho(h).\{F_1^{\,2}(z) + a_2F_2^{\,2}(z) + \ldots\} + \text{more complicated grade correlation terms} \tag{2.8}$$

The actual values of the a_n's (some or all of which may be zero as in the Gaussian case) are unimportant for our further discussion. The important point is that the indicator covariance becomes proportional to the grade covariance $\sigma(h) = \sigma^2\rho(h)$ in the approximation shown, while the cutoff dependence is now completely carried by the factor $F_1{}^2 + a_2F_2{}^2 + \ldots$ in curly brackets.

One can hope that breaking off σ_I at the first term will be good when $\rho(h)$ is small. Anticipating such an approximation comes from practical experience. One knows empirically that, due to the nugget effect, $\rho(h)$ exhibits a very sharp drop from unity as h increases. Thus, excepting $h = 0$, this empirical circumstance suggests the linear form (2.8) as a realistic expression for $\sigma_I(h;z)$. As with the full series, this approximation still contains the destructurization effect in the vanishing of the cofactor of $\rho(h)$ at large z.

We have tested the proportionality between $\rho_I(h;z)$ and $\rho(h)$ empirically on a simulated tabular deposit of nickel laterite origin, called Stanford II. It is 'perfectly known' through 60500 'core' values covering a square metre each. It was thus possible to construct the true 'local' grade and indicator variograms. The experimental omnidirectional relative variograms for grade and indicators corresponding to three cutoffs are shown in Figure 1a. Next, the indicator correlogram values $\rho_I(h;z)$ for various cutoffs were plotted versus the grade correlogram $\rho(h)$ as shown in Figure 1b.

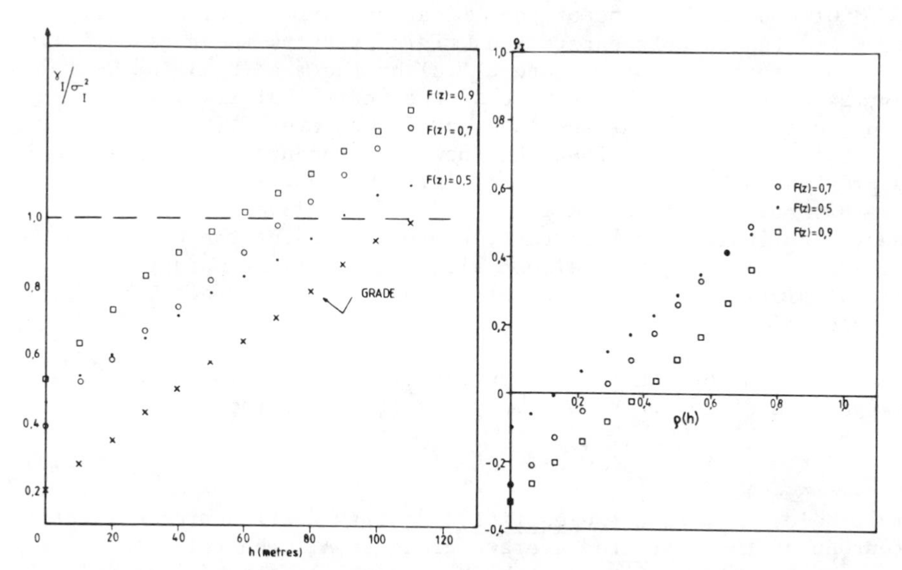

Figure 1a. Based on the 60500 grade values of Stanford II: relative variograms for grade and indicators at 3 cutoffs.

Figure 1b. Indicator correlogram values at the 3 cutoffs plotted versus grade correlogram values to illustrate proportionality.

Observe that this latter plot relates two empirical quantities. One sees that the expected linear relation between ρ_I and ρ is excellently obeyed, except for $\rho(h)$ near one, where the entire approximation breaks down. The occurrence of negative values for $\rho_I(h;z)$ in this case arises from the fact that, at the range, the experimental indicator variogram overshoots the theoretical sill, see Figure 1a.

A linear relation between indicator and grade correlograms has an interesting consequence for kriging indicator weights. Inserting $\rho_I(h;z) = b(z).\rho(h)$ (or, for that matter, $\rho_I(h;z) = a(z) + b(z).\rho(h)$) into either of the kriging systems (1.8) simply replaces ρ_I by ρ everywhere. Thus the indicator weights λ_α become independent of cutoff and identical with the kriging weights that estimate the average grade in A. This in turn means that, as long as ρ_I depends linearly on ρ, the destructurization of the indicator correlogram at high cutoff grades has no effect on the estimates of the indicators themselves. In practice this will hold for a deposit where the nugget effect of the grade variogram constitutes roughly one third or more of the sill value. The mere fact that selective mining necessitates estimating local densities, signifies a highly variable mineralization, which in turn signifies a substantial nugget effect. This causes one to suspect that the linear relationship between ρ_I and ρ will hold for most relevant practical cases.

Of course, outside of the linear approximation, destructurization at high cutoffs enters the kriging systems in an essential way and causes SK to estimate $\phi(A;z)$ by the mean $F(z)$ and OK to estimate $\phi(A;z)$ by the average of the individual $i(\underline{x}_\alpha;z)$. One can see these two limiting cases arise whenever $\rho_I(h;z) << 1$ for $h \neq 0$ in eqs. (1.8). Then the (now z-dependent) weights are approximated by $\lambda_\alpha \approx \rho_I(\underline{x}_\alpha;z) << 1$ in SK and $1/N$ in OK. This feature suggests that OK is to be preferred over SK when destructurization is important, thereby avoiding the undesirable feature of SK of giving almost all of the kriging weight to the global distribution $F(z)$ in estimating the high-cutoff behaviour of each $\phi(A;z)$.

We also note that the identity of grade and indicator kriging weights in either OK or SK imply that the relation

$$Z_A^* = \int_0^\infty z^* \frac{\partial \phi^*(A;z)}{\partial z^*} dz^* \tag{2.9}$$

must hold, i.e. that the estimated density distribution correctly reproduces the estimated average grade in A. The relation (2.9) has already been pointed out by Matheron (4) for the 'mosaic' model for which a linear relation between ρ_I and ρ holds by construction.

3. INDICATOR KRIGING

In this section we compare the performance of simple and ordinary indicator kriging with reference to the true recovery functions for Stanford II. We also compare the performance of another procedure that introduces the idea of a generalized indicator function.

3.1 Simple and Ordinary Kriging

Stanford II can be divided into 500 11 × 11 metre panels, each with a central 'drillhole' value. The 9 closest holes were used in estimating each of 384 panels, excluding border ones. Of the 500 drillhole data, one third has zero values and the rest have a positively skewed distribution. Grade and indicator omni-directional variograms for 18 cutoffs were calculated using the 500 data (as opposed to the 'true' variograms of Figure 1a) and the mean F(z) was taken to be their experimental distribution $\hat{F}(z)$. Since the data is so regular, the kriging weights per panel remain the same. Examples of weights corresponding to the same discretization point for grades and indicators at a few cutoffs, using SK or OK, are the following (expressed as percentages rounded to one decimal).

Cutoff z where	9 SK weights									Weight of $\hat{F}$
F(z)=0.3	7.3	8.1	7.3	4.8	13.7	13.8	14.1	13.7	8.1	9.2
F(z)=0.5	7.1	8.0	7.1	4.3	14.5	14.6	15.0	14.5	8.0	6.9
F(z)=0.7	7.2	8.1	7.2	4.5	14.3	14.3	14.7	14.3	8.1	7.5
F(z)=0.9	5.6	6.0	5.6	1.7	16.9	15.7	19.3	16.9	6.0	6.3
F(z)=0.98	-0.1	-0.0	-0.1	0.0	3.5	2.1	5.8	3.5	-0.0	85.2
	9 OK weights									
F(z)=0.3	8.2	9.3	8.2	6.0	14.6	15.0	14.7	14.6	9.3	
F(z)=0.5	7.8	9.0	7.8	5.3	15.2	15.5	15.4	15.2	9.0	
F(z)=0.7	7.9	9.1	7.9	5.5	15.0	15.4	15.2	15.0	9.1	
F(z)=0.9	6.2	7.0	6.2	2.7	17.5	16.7	19.5	17.5	7.0	
F(z)=0.98	9.4	9.5	9.4	9.6	13.0	11.6	15.1	13.0	9.5	
Grade(OK)	4.9	5.5	4.9	1.0	19.0	17.4	22.7	19.0	5.5	

Note the constancy of the weights, and, since they are a hundred times smaller as fractions and then multiplied by values of 0 or 1, using only the median cutoff variogram for all cutoffs will make very little difference to the result, except at extreme cutoffs, using SK, where $\hat{F}$ gets a weight of 85% in every panel.

The estimated local densities tend to be 'spiky' (see Figures 4a and 4b). The estimated global distributions and recovery functions are practically identical for SK and OK, as can be expected. However, the recovery functions tend to have an oscillating behaviour (see Figure 5).

3.2 Generalized Indicator Kriging (GIK)

The oscillations in the estimated global recovery functions are perhaps not too surprising, given the discontinuous nature of the individual indicators making up the estimator $\phi^*(A;z)$. One smoothing procedure to counteract this effect is suggested by the following heuristic argument. Suppose we discretize the area A into a number of points $\underline{x}$ and use IK to estimate the point probability distributions $i^*(\underline{x};z)$ as a preliminary to constructing $\phi^*(A;z)$. If the grades z_α in the vicinity of A happen to be equal, then all the $i^*(\underline{x};z)$ will be boxlike as functions of the cutoff z. So $\phi^*(A;z)$ will be boxlike too. We also know that the kriged estimates $Z^*(\underline{x})$ of the actual grades at the points $\underline{x}$ in A will have their average Z_A^* correctly reproduced by $\phi^*(A;z)$ according to eq. (2.9) whenever the indicator kriging system is cutoff independent. However, the associated grade kriging variances σ_K^2 of the estimates $Z^*(\underline{x})$ are not incorporated into the estimators $i^*(\underline{x};z)$ as our hypothetical example shows. This circumstance suggests that we introduce a revised estimator

$$i_G^*(\underline{x};z) = \sum_\alpha \lambda_\alpha i_G(\underline{x}_\alpha;z) \tag{3.1}$$

where i_G is a generalized indicator function, that has the shape of a distorted step function at $z_\alpha = z$. If the extent of distortion is controlled by the kriging variance σ_K^2 of the grade estimate at $\underline{x}$, then i_G must depend on σ_K in such a way that

$$\lim_{\sigma_K \to 0} i_G(\underline{x};z) = i(\underline{x};z) \tag{3.2}$$

or, equivalently, that

$$\lim_{\sigma_K \to 0} \frac{\partial i_G}{\partial z} = \lim_{\sigma_K \to 0} f_G[z(\underline{x}) - z] = \delta[z(\underline{x}) - z] \tag{3.3}$$

in addition to $i_G \to 0$ and 1 for $z \to \pm\infty$.

There are many functions available in the mathematical literature that satisfy these conditions. A particularly convenient one for our purposes is provided by the Fermi-Dirac distribution (FD) (5)

$$F_D(Z;z) = \tfrac{1}{2} - \tfrac{1}{2}\tanh[\pi/(2\sqrt{3})(Z - z)/\sigma] \tag{3.4}$$

with the associated density function

$$f_D(Z - z) = \frac{\partial F_D}{\partial z} = \pi/(4\sqrt{3}\sigma).\mathrm{sech}^2[\pi/(2\sqrt{3})(Z - z)/\sigma] \tag{3.5}$$

Some other properties of the FD distribution that are particularly appropriate and useful for geostatistics are noted in the Appendix. Eq. (3.4) provides one possible parametrization of i_G if we take $\sigma = \sigma_K$, i.e. we 'smear out' the indicator function at each data point first in the shape of an FD and by an amount dictated by the relevant grade kriging variance. Then the result of using eq. (3.1) with equal grades at $\underline{x}_\alpha$ would be an indicator at $\underline{x}$ distorted by exactly the kriging variance of the estimated grade at $\underline{x}$.

Let us suppose next that the λ's in eq. (3.1) have been determined by OK. The next question is: with respect to estimating what quantity is the estimator $i_G^*(\underline{x};z)$ unbiased? To answer this question, consider i_G^* and i_G as particular realizations of random functions I_G^* and I_G in the usual way. Then if F(z) is the distribution of actual grades,

$$E\{I_G(\underline{x};z)\} = \int_{-\infty}^{+\infty} i_G(\underline{x};z)\ dF(z(\underline{x}))$$

$$= \int_{-\infty}^{+\infty} F(z(\underline{x})).f_G(z(\underline{x}) - z)\ dz(\underline{x})$$

or

$$E\{I_G(\underline{x};z)\} = \int_{-\infty}^{+\infty} E\{I(\underline{x};z')\}.f_G(z' - z)\ dz' \tag{3.6}$$

Hence $i_G^*(\underline{x};z)$ is an unbiased estimator of the folded distribution function

$$F_G(z) = \int_{-\infty}^{+\infty} F(z').f_G(z' - z)\ dz' \tag{3.7}$$

since from (3.1) and (3.6)

$$E\{I_G^*(\underline{x};z)\} = \sum \lambda_\alpha E\{I_G(\underline{x};z)\} = F_G(z) \tag{3.8}$$

assuming a stationary deposit.

Note that $F_G(z)$ can be constructed from an indicator kriging run over the deposit after deciding on a parametrization for f_G, calculating $i_G^*(\underline{x};z)$ and $\phi_G^*(A;z)$ for discretization points and panels in the process. Subsequently one moves to an estimator (call it i* again) that is unbiased with respect to the global distribution function F(z) by simply replacing each i_G in eq. (3.1) by

$$i_G'(\underline{x};z) = i_G(\underline{x};z).[F(z)/F_G(z)] \quad (3.9)$$

Then

$$i^*(\underline{x};z) = \sum \lambda_\alpha i_G'(\underline{x}_\alpha;z) = i_G^*(\underline{x};z).[F(z)/F_G(z)] \quad (3.10)$$

is an unbiased estimator of F(z). As a result,

$$\phi^*(A;z) = 1/A \int_A i^*(\underline{x};z)\, d\underline{x} = \phi_G^*(A;z).[F(z)/F_G(z)] \quad (3.11)$$

The step subsequent to the indicator kriging run therefore simply amounts to multiplication by a set of constant factors.

Thus far the considerations are free of the actual parametri-

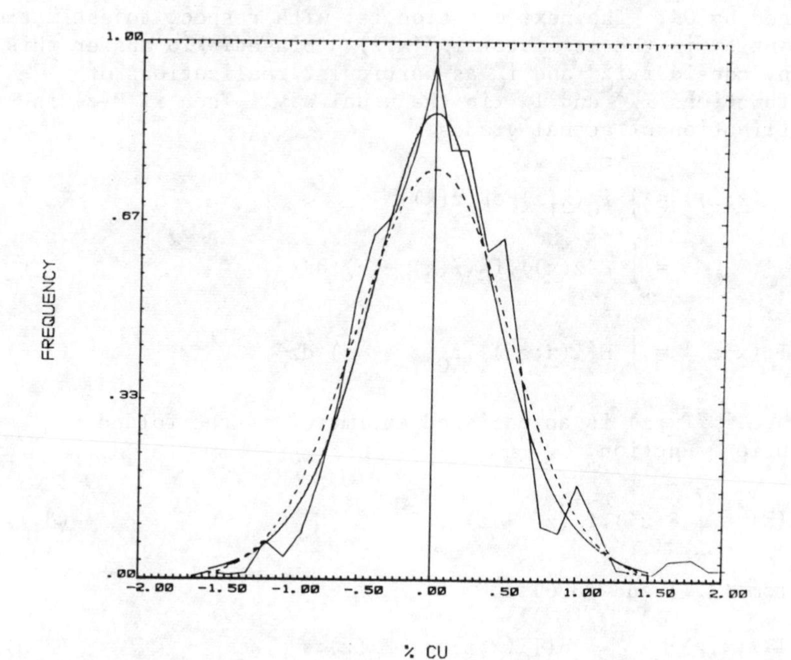

Figure 2. Fermi-Dirac density function (solid line), m = 0, σ^2 = 0.273, superimposed on an illustration of the histogram of kriging errors of the Chuquicamata deposit, m = 0.01, σ_E^2 = 0.273, and normal density function (dotted line), m = 0, σ^2 = 0.273, taken from Journel and Huijbreghts, 'Mining Geostatistics', 1978, p. 58.

zation employed for f_G. In actual computations, we have used F_D and f_D from eqs. (3.4) and (3.5) for i_G and f_G with $\sigma^2 = \sigma_K{}^2$, the corresponding discretization point kriging variance. The philosophy is thus that an ordinary indicator value at a data point $\underline{x}_\alpha$ is applicable at discretization point $\underline{x}$, but with a certain amount of uncertainty. This amount of uncertainty is taken to be given by the grade kriging variance that results when estimating a grade value at $\underline{x}$ using the specific data configuration. The <u>same</u> amount of uncertainty is applied at all data points within the configuration. The choice of form for the shape of the distorted indicator function is ad hoc. However, it is of interest to note that f_D of eq. (3.5) with $\sigma^2 = \sigma_E{}^2$, the experimental error variance over a deposit, gives an excellent description of the distribution of errors $[Z^*(\underline{x}) - Z(\underline{x})]$, see Figure 2. Since the generalized indicator estimator $i_G{}^*(\underline{x};z)$ also has to reflect in some way the uncertainty $(Z(\underline{x}) - z)$ relative to z with which the grade $Z(\underline{x})$ at $\underline{x}$ is recorded/not recorded, it is perhaps a very reasonable model to use for i_G.

The indicator covariance calculations for $i_G{}'$ present no problem, and will in fact factor whenever the ordinary indicator covariance does. The very fact that the new indicator data are 'smeared out' and that their values do not jump abruptly from one to zero, of course means that there is far more continuity, even

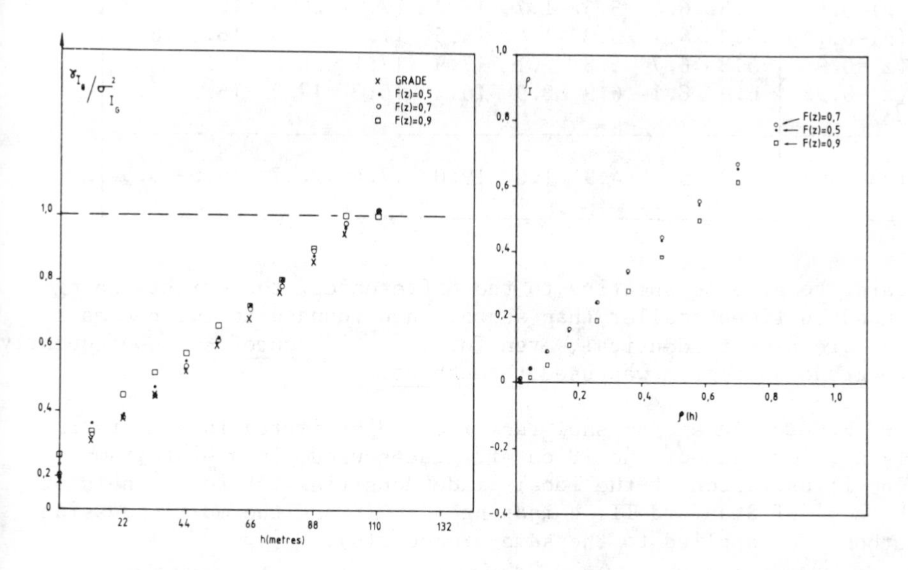

Figure 3a. Based on the 500 grade values of Stanford II: relative variograms for grade and generalized indicators.

Figure 3b. Generalized indicator correlogram values at the 3 cutoffs plotted versus grade correlogram values, based on fitted models.

for the highest cutoff grades, and the corresponding variograms demonstrate that there is practically no destructurization present.

To illustrate, the average grade kriging standard deviation over the deposit was inserted into eq. (3.4) to build generalized indicators at the previous 500 data points for variogram calculations. The experimental omnidirectional relative variograms for grade and generalized indicators corresponding to three cutoffs are shown in Figure 3a. The new indicator relative variograms, although based on only 500 data values as opposed to 60500 used for Figure 1a, are stable and very similar to the grade relative variogram. Figure 3b shows the linear relationship between the grade and generalized indicator correlograms, this time based on the spherical models fitted to the relative variograms in Figure 3a. For comparison purposes, the following are some OK weights based on these models, given again as percentages rounded to one decimal.

Cutoff z where	9 OK weights								
F(z)=0.3	5.9	6.9	5.9	2.3	17.7	17.1	19.8	17.7	6.9
F(z)=0.5	5.5	6.4	5.5	1.7	18.2	17.2	20.9	18.2	6.4
F(z)=0.7	5.3	6.1	5.3	1.4	18.5	17.3	21.5	18.5	6.1
F(z)=0.9	5.8	6.7	5.8	2.1	17.9	17.1	20.2	17.9	6.7
F(z)=0.98	6.9	8.1	6.9	3.9	16.3	16.3	17.2	16.3	8.1
Grade(OK)	4.9	5.5	4.9	1.0	19.0	17.4	22.7	19.0	5.5

Again, to give perspective to the differences, the weights being a hundred times smaller than shown, when rounded to two decimals they are almost identical, even for <u>very high cutoffs</u>. Consequently the <u>grade</u> variogram was used <u>throughout</u>.

Figures 4a and 4b show the true and estimated increments in $\phi(A;z)$, for each of the 18 cutoff grades used, in a histogram-type illustration of the local grade densities for four panels or 'blocks' of Stanford II, comparing IK, GIK and the multi-gaussian method (MG) applied to the same orebody (6).

The estimated GIK local densities are remarkably close to the true ones and do practically as well as the MG estimates. At the same time, only linear kriging is involved, it is applicable to any distribution, and there is no need to transform the data values to a standard distribution.

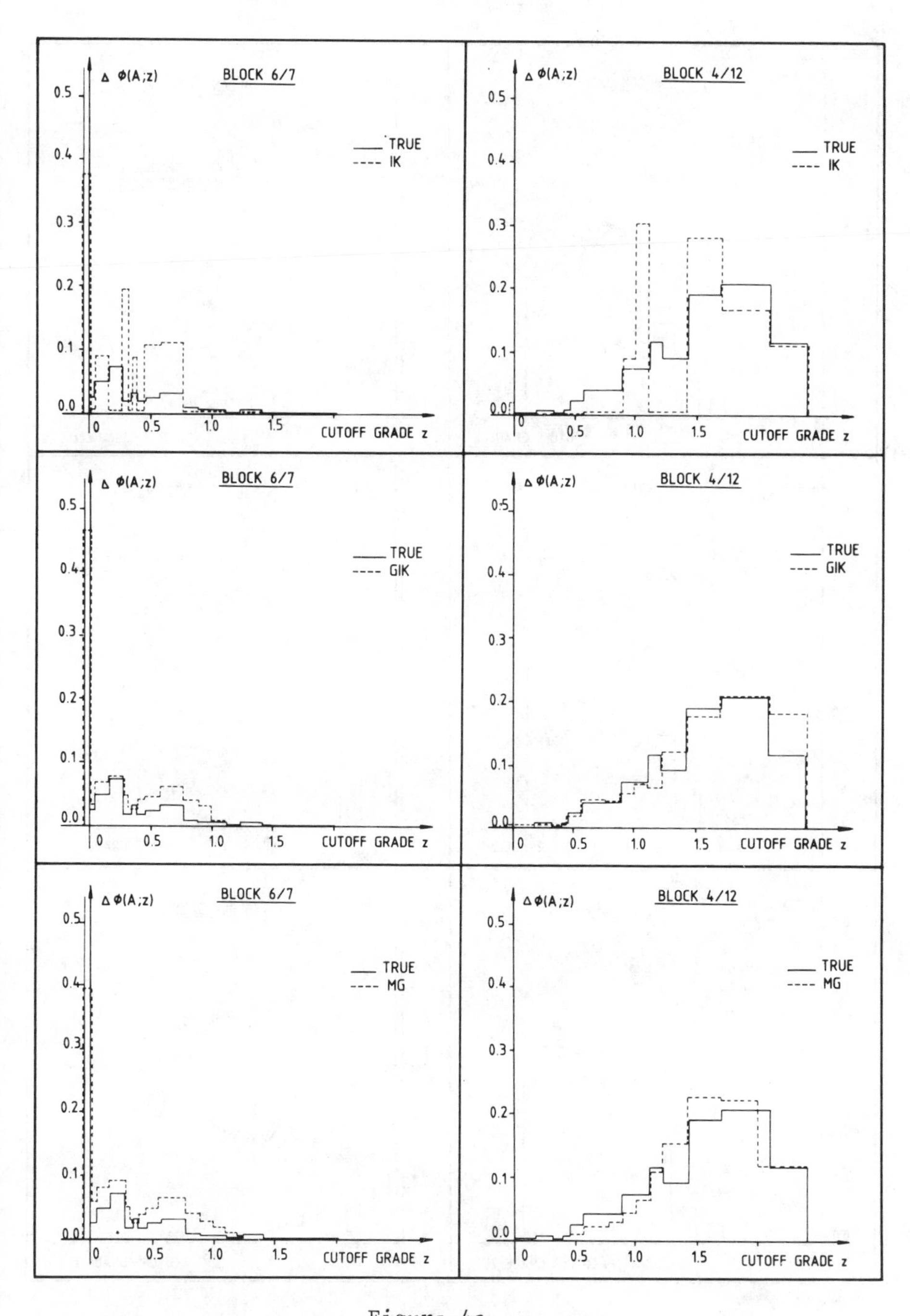

Figure 4a

Figures 4a and 4b (next page) show the true and estimated increments in $\phi(A;z)$ corresponding to 18 chosen cutoff grades, comparing IK, GIK and MG for four panels or 'blocks' of Stanford II.

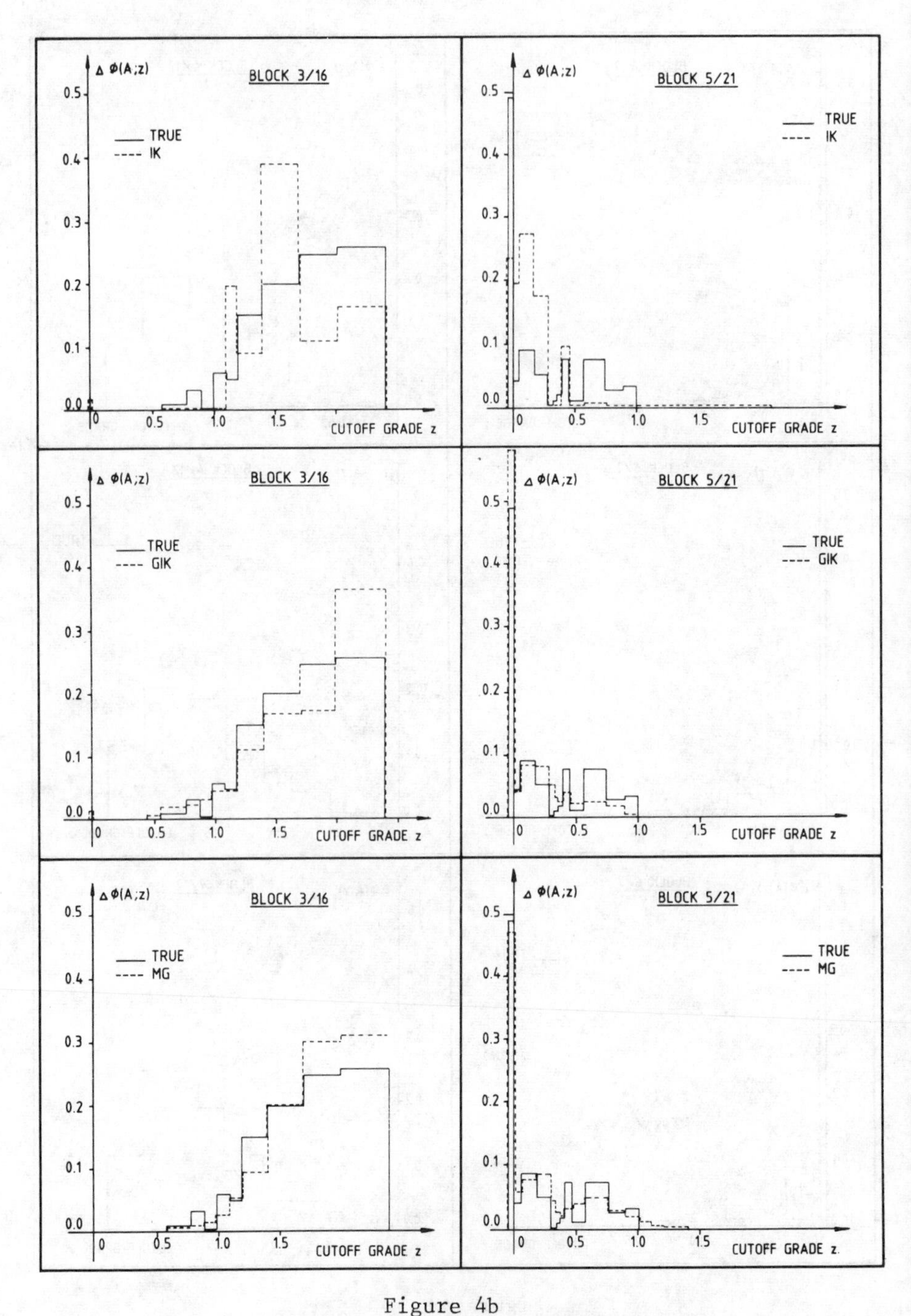

Figure 4b
In each estimation only the 9 closest drillholes were used. A 'spike' for zero cutoff grade (representing waste in the panel) is shown just to the left of zero cutoff grade.

Figure 5 shows the global recovery functions for Stanford II for IK and GIK, based on the estimation of 384 panels, excluding the border panels. Where the correction factor in (3.9) causes i* to be slightly larger than 1, it is set to 1, which in turn causes $\hat{F}(z)$ to be slightly underestimated globally for high cutoffs. The cutoff for which q(A;z) in (1.3) is taken as zero was calculated such that the estimated global distribution reproduces the mean grade. The ore grade recovery function is estimated well (see Figure 5) and compares favourably with those produced by non-linear methods for the same orebody (6).

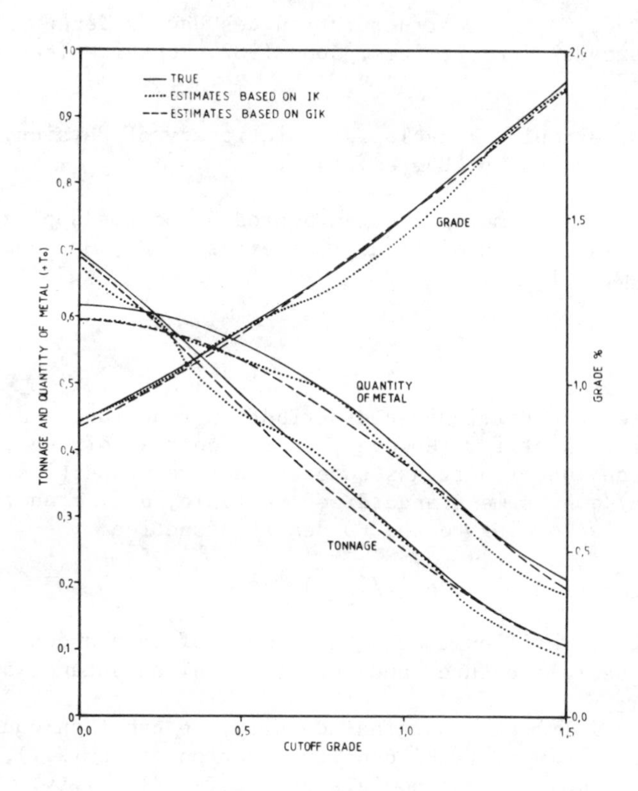

Figure 5. Global recovery functions obtained by ordinary kriging of generalized indicator functions.

ACKNOWLEDGMENTS

I wish to express my gratitude to Prof. A.G. Journel for suggesting this problem and for his guidance in many phases of the work. I would also like to gratefully acknowledge the study opportunity afforded by Gold Fields of South Africa which allowed the research reported here to be carried out.

REFERENCES

1. Journel, A.G., Non-parametric Estimation of Spatial Distributions, 1982 Apcom Proceedings.

2. Beckmann, P., Orthogonal Polynomials for Engineers and Physicists, 1973, The Golem Press, Boulder, Colorado.

3. Abramovitz, M. and Stegun, I., Handbook of Mathematical Functions, 1972, Dover Publications, New York, p. 771.

4. Matheron, G., La Destructuration des Hautes Teneurs et le Krigeage des Indicatrices, June 1982, Fontainebleau Series No. N-761.

5. See for example, Encyclopedic Dictionary of Physics, Pergamon Press, p. 85, eq. (4).

6. Verly, G., The multigaussian approach and its applications to the estimation of local recoveries, 1983, Mathematical Geology, Vol. 15.

APPENDIX

We used the Fermi-Dirac distribution function as a possible parametrization of i_G. However, in the context of, say, a grade distribution function it has some attractive properties. Let $y = (Z - z)/\sigma$ be some standardized variable, e.g. grade, distributed according to the FD density function

$$f_D(y) = \pi/(4\sqrt{3})\ \mathrm{sech}^2[\pi/(2\sqrt{3})y] \tag{A.1}$$

It is readily verified that y has a mean of zero and a variance of one. Some useful features and computational advantages of $f_D(y)$ are:

(i) Using $f_D(y)$ as a marginal density, we can then construct an isofactorial bivariate FD density function as in (2.1), where the first few normalized polynomials are $\chi_0(y) = 1$, $\chi_1(y) = y$, $\chi_2(y) = \sqrt{5/16}\,(y^2 - 1)$, $\chi_3(y) = 1/36\ \sqrt{7/3}\,(5y^3 - 21y)$, etc.

(ii) By construction the distribution expression is simple,

$$F(y) = \int_{-\infty}^{y} f_D(y')\,dy' = \tfrac{1}{2} + \tfrac{1}{2}\tanh[\pi y/(2\sqrt{3})] \tag{A.2}$$

(iii) The quantity of metal recovered at cutoff y_c would be

$$Q(y_c) = \int_{y_c}^{+\infty} y' f_D(y')\,dy' = y_c[1 - F(y_c)] - \sqrt{3}/\pi\ \ln\,[F(y_c)]$$

(iv) It can be used to express confidence limits for OK estimates.

CONDITIONAL RECOVERY ESTIMATION THROUGH PROBABILITY KRIGING - THEORY AND PRACTICE

JEFF SULLIVAN

Department of Applied Earth Sciences
Stanford University
School of Earth Sciences
Stanford, California

ABSTRACT

The probability kriging technique is an improvement on the distribution free indicator kriging technique for obtaining conditional recoverable reserves. Probability kriging is similar to indicator kriging in that both techniques utilize indicator data and no assumption concerning the shape of the conditional distribution is made. Indicator kriging however does not utilize some easily obtainable information which causes, in certain cases, the indicator kriging estimator to be smoothed, conditionally biased, and in general a poor local estimator. The cases where indicator kriging performs poorly will be identified and it will be shown that by including additional information, through the probability kriging estimator, that the quality of the estimator will be improved. The probability kriging technique is then tested on a gold deposit and the results are presented.

ACKNOWLEDGEMENTS

The author would like to thank the Freeport Gold Company for graciously supplying data from their Jerritt Canyon deposit.

G. Verly et al. (eds.), Geostatistics for Natural Resources Characterization, Part 1, 365–384.

INTRODUCTION

The field of conditional recovery estimation has recently been expanded due to the introduction of the non-parametric and distribution free indicator kriging technique. This technique has expanded the field since it represents the first technique which does not make some type of bivariate or multivariate distribution hypothesis. Since indicator kriging does not make any distribution hypothesis, it must rely exclusively on the information provided by the data. Because this technique relies so strongly on the data, it is imperative that as much information as possible be extracted from the data and included in the estimation procedure. As will be shown, the indicator kriging technique does not include some easily obtainable information concerning the data. This omission can cause the indicator kriging estimate to be smoothed and strongly conditionally biased. The probability kriging technique, which will be introduced, utilizes more of the available information and hence represents an improvement over the indicator kriging estimator. Probability kriging incorporates the favorable points of indicator kriging: it is distribution free, nonparametric, unbiased, and easy to apply. Probability kriging, however, represents an improvement over indicator kriging since it has smaller estimation variance and less smoothing. To understand how probability kriging can improve on indicator kriging it is necessary to examine the basis of indicator kriging.

WHAT IS INDICATOR KRIGING?

Indicator kriging (IK) provides an estimate of the spatial distribution of a given support (v), taken to be point support in this case, within a panel V. The indicator data, used in this estimation, are simple transforms of the grade Z(x) which are defined as

$$i(x;z) = \begin{cases} 1 \text{ if } Z(x) \leq z \\ 0 \text{ if } Z(x) > z \end{cases}$$

where z is any cutoff of interest.

An important and useful property of the indicator data is that when the cutoff, z, is varied i(x;z) can be seen as as a crude distribution function.

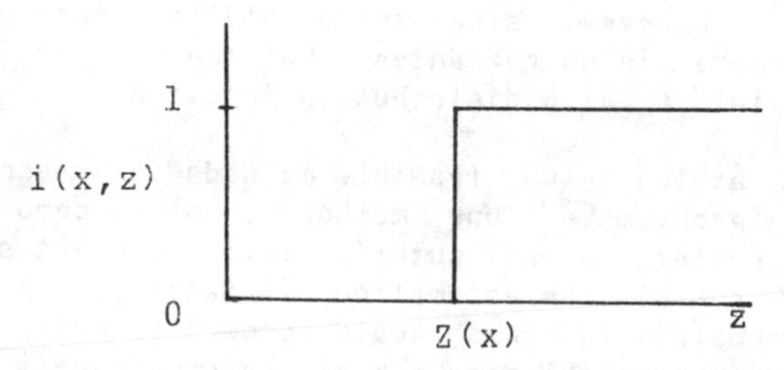

Given this interpretation of the data as distribution functions it is natural that these data be used to estimate the distribution of points within a panel V. This estimation is performed through simple kriging at a sufficient number of cutoffs. The structural function used in this kriging is the indicator variogram defined as

$$\gamma_i(h,z) = .5\ E[(I(x+h,z)-I(x,z))^2].$$

The simple kriging system is expressed as:

$$\sum_1^n \lambda_\alpha \gamma_i(x_\alpha - x_\beta, z) = \overline{\gamma}_i(x_\beta, V, z) \qquad \beta = 1 \text{ to } n$$

$$\sigma^2_{ik} = \sum_1^n \lambda_\alpha \overline{\gamma}_i(x_\alpha, V, z) - \overline{\gamma}_i(V,V,z)$$

The estimated distribution function for a given z is then

$$\phi^*(V,z) = \sum_1^n \lambda_\alpha \gamma_i\ i(x_\alpha,z) + (1-\sum_1^n \lambda_\alpha)\hat{F}(z)$$

where $\hat{F}(z)$ is the mean of all indicator data for cutoff z. The recovered tonnage at cutoff z is simply:

$$(1-\phi^*(V,z))(\text{tonnage factor}).$$

The IK estimator provides an unbiased estimate of the recovered tonnage at any cutoff of interest. One difficulty of this method is the possibility of order relation problems. Order relation problems occur when the distribution function estimated by IK; is decreasing, has negative values, or has values greater than 1. In short, an estimated distribution has order relation problems if it is not a valid distribution function. These order relation problems can occur because the IK simple kriging system at each cutoff is solved independently from the IK systems at all other cutoffs. That is, the IK system provides an optimal solution, in the sense that it minimizes the estimation variance, at each cutoff; however, since these solutions are arrived at independently there is no guarantee that these solutions taken together will yield a valid distribution function.

There are at least two feasible methods for resolving the order relation problems. One method involves combining the simple kriging systems for all cutoffs into one giant system and minimizing the sum of the estimation variances. This system would contain constraints which would force the order relations to hold. The fundamental drawback of this method is that the system of equations would be very large, hence the solution would be expensive.

The second method, which closely approximates the results of method 1, takes the results given by the IK algorithm and fits a distribution function which minimizes the weighted sum of squared deviations from the optimal IK solution. For instance suppose the following distribution function was estimated by IK (order relation problems of this magnitude are extremely rare). The distribution function which minimizes the squared deviation from the IK solution is shown as a dashed line.

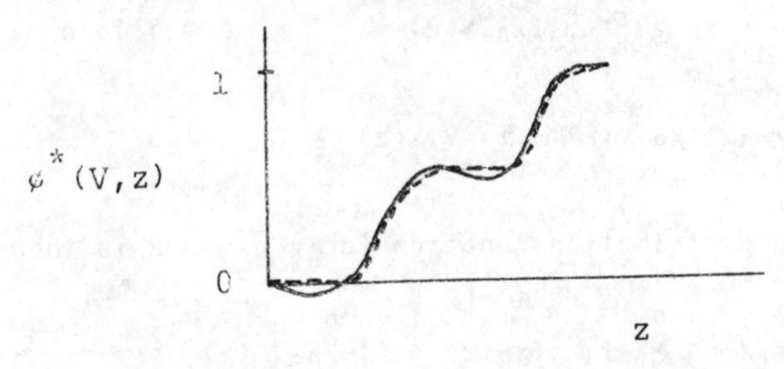

The algorithm used in obtaining the dashed line can be expressed as follows

$$\text{Min} \quad W_i \sum_{1}^{n} (F_i - I_i)^2$$

subject to

$$F_{i+1} > F_i \qquad \text{for} \quad i = 1, n-1$$
$$F_1 > 0$$
$$F_n < 1.$$

where F_i is the final solution (dashed line)
I_i is the initial solution given by IK ($\phi^*(V,z)$)
and W_i is a weight given to each cutoff, if $W_i = 1/n$ for all i the the objective amounts to minimizing the average squared deviation.

Notice that this system of equations involves minimizing the sum of quadratic terms subject to linear constraints. The solution of such a system can be obtained through quadratic programming (Gill,et al,1982). A very nice property of this quadratic programming solution is that the optimal, IK solution is unchanged in regions where the estimated distribution function is nondecreasing; that is, the quadratic programming algorithm only takes effect when the order relations are not verified.

DEVELOPMENT OF PROBABILITY KRIGING

PK represents an improvement over IK since it is able to utilizes more of the information present in the bivariate distribution of Z(x) and Z(x+h). To see how PK is able to utilize more of this information it is first necessary to see what information is used by IK.

The indicator kriging estimate is based on the indicator variogram which can easily be interpreted as the $Pr(Z(x) \leq z$ and $Z(x+h) > z)$, hence it is directly related to the bivariate distribution of Z(x) and Z(x+h). Since the IK estimate is directly related to the variogram of I(x;z), if more information concerning the bivariate distribution of Z could be extracted, the quality of the estimator would increase. The question is then: how can additional information concerning the bivariate distribution be obtained? The most obvious way of obtaining this information is through co-indicator kriging (CIK). CIK utilizes not only the indicator variograms but also the co-indicator variograms: $Pr(Z(x) \leq z1$ and $Z(x+h) > z2;\ z1 \neq z2, \forall z1, z2)$. Thus CIK, through the indicator cross variograms, provides complete knowledge of the bivariate distribution. This knowledge, although obtainable, has a very high price since a very large number of cross indicator variograms must be modeled. Additionally the co-kriging system which must be solved is enormous. Hence CIK will not be used, in spite of its excellent properties, because of the extremely large computational costs.

Since we cannot, practically, include all of the information concerning the bivariate distribution within the IK estimate through CIK (without resorting to some bivariate distribution model), a simpler method of obtaining more information must be found. Examining a representation of the bivariate distribution of Z(x) and Z(x+h) gives a clue for the manner in which this information can be obtained. Consider a possible bivariate distribution:

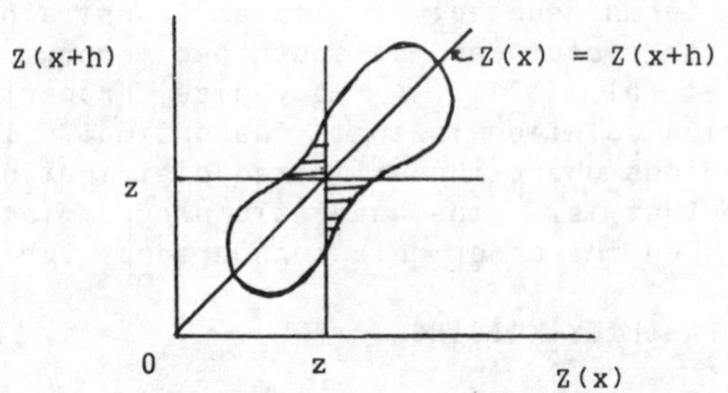

The shaded area is the value of the indicator variogram for cutoff z This value, although very useful does not give any indication of how the mass is distributed within the shaded area. For instance, a bivariate distribution which is closely packed about the line Z(x)= Z(x+h) and a distribution which is loosely packed could easily have identical indicator variogram values. Compare distributions A and B below.

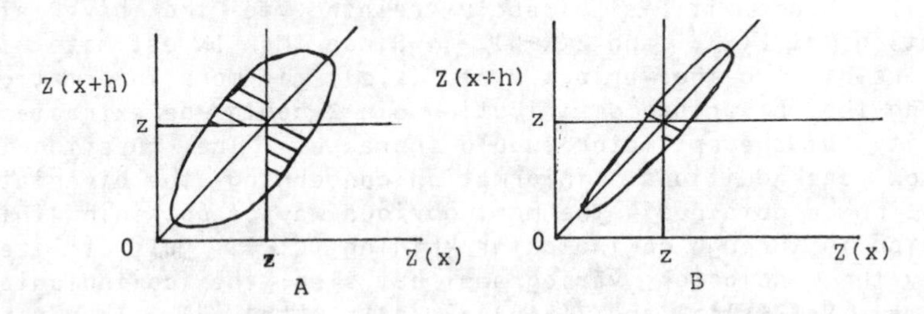

Given the indicator variogram information these distributions could appear identical when in reality they are very different. Hence a piece of information which distinguishes these distributions would be very useful and ultimately would improve the IK estimate.

It can be shown that the semivariogram is the expected distance of all pairs {Z(x),Z(x+h)} from the line Z(x)=Z(x+h) (Journel, 1984). Thus if the variogram is calculated only over the region of the bivariate distribution for which the indicator variogram is non-zero, it would easily show the difference between case A and case B shown above. An estimator which incorporated such information would outperform IK. A method of incorporating exactly this information has not been developed, however similar information can be included utilizing the conditional variogram of order 1: $E(|Z(x+h)- Z(x)|/Z(x)>z$ and $Z(z+h)\leqslant z)$.

The conditional variogram of order 1 arises in the following manner. Suppose that both the indicator data and the grade Z(x) are used to provide an estimate of $\phi(V,z)$. This would then be a type of cokriging. The cross structure γ_{zi} is defined as

$E((I(x+h,z)-I(x))(Z(x+h)-Z(x)))$. Consider the possible values of this quantity:

If $Z(x+h)>z$ and $Z(x)>z$ or $Z(x+h)\leq z$ and $Z(x)\leq z$ then $\gamma_{zi}=0$

If $Z(x+h)>z$ and $Z(x)\leq z$ then $\gamma_{zi}=-(Z(x+h)-Z(x))$

If $Z(x+h)\leq z$ and $Z(x)>z$ then $\gamma_{zi}=(Z(x+h)-Z(x))$

$\gamma_{zi}=-|Z(x+h)-Z(x)|$

So $\gamma_{zi}=.5\ E(-|Z(x+h)-Z(x)|/Z(x+h)>z$ and $Z(x)\leq z$ or $Z(x+h)\leq z$ and $Z(x)>z)$ and 0 when $Z(x)$ and $Z(x+h) > z$ or $Z(x)$ and $Z(x+h) < z$. This is simply the previously mentioned conditional variogram of order 1 and gives the expected semi-distance between all pairs, $\{Z(x),Z(x+h)\}$, for which $Z(x)>z$ and $Z(x+h)<z$ and the line $Z(x)=Z(x+h)$. Given the information provided by this cross structure the cokriging estimator

$$\phi^*(V,z) = \sum_1^n \lambda_\alpha (x_\alpha\ z) + \sum_1^n \nu_\alpha U(x_\alpha)$$

will improve on the IK estimate.

One difficulty with this estimator is the potential great difference in the units of I and Z. The indicator data are either 0 or 1 while the Z values can, in theory, take on values between 0 and 100%. The difference in units could cause the estimate to be unstable. So it is natural to transform the Z data so that they are comparable in magnitude to the indicator data. Such a transformation is readily available; simply replace $Z(x)$ by its experimental cumulative distribution function $F(Z(x))$. That is, a variable $U(x)$ will be considered where

$$U(x_\alpha)=F(Z(x_\alpha)).$$

An estimator which utilizes both the $U(x)$ and $I(x;z)$ data can be obtained from a standard co-kriging system

$$\sum_1^n \lambda_\alpha \rho_i(x_\alpha - x_\beta,z) + \sum_1^n \nu_\alpha \rho_{ui}(x_\alpha - x_\beta,z) + \mu_1 = \overline{\rho}_i(x_\beta,V,z)$$

$$\sum_1^n \lambda_\alpha \rho_{ui}(x_\alpha - x_\beta,z) + \sum_1^n \nu_\alpha \rho_u(x_\alpha - x_\beta,z) + \mu_2 = \overline{\rho}_{ui}(x_\beta,V,z)$$

$$\sum_1^n \lambda_\alpha = 1$$

$$\sum_1^n \nu_\alpha = 0$$

$$\sigma^2_{pk} = \overline{\rho}_i(V,V,z) + \mu_1 - \sum_1^n \lambda_\alpha \overline{\rho}_{ui}(V,x_\alpha,z)$$

Where ρ_i, ρ_{ui}, and ρ_u are the correlograms associated with γ_i, γ_{ui}, and γ_u(Journel,1984). The corresponding estimator of is:;;

$$\phi^*(V,z) = \sum_1^n \lambda_\alpha i(x\ z) + \sum_1^n \nu_\alpha U(x_\alpha)$$

This is termed the probability kriging estimate because of the probabilistic interpretations of the i(x;z) and U(x) data.

The estimation variance of the PK estimator is certainly less than the estimation variance of the IK estimator. This can be easily shown through projection theory (Journel,1984).

COMPARISON OF IK AND PK

The PK estimator is a better estimator than IK under the estimation variance criterion. However there are other factors which must be considered when comparing two estimators. Two important factors which will be examined are smoothing and conditional bias.

The smoothing of the PK or IK estimator can be expressed as

$$\mathrm{var}(\phi(V,z)) - \mathrm{var}(\phi^*(V,z)).$$

In words, smoothing is the difference between the variance of the true recovered tonnage and the variance of the estimated recovered tonnage.

The smoothing of the IK estimator can, in certain cases, be large. The effect of high smoothing on the estimator is that the estimated tonnage recovered is underestimated in high grade panels and overestimated in low grade panels. The smoothing of the IK estimate at a given cutoff is approximately

σ^2_{ik} (appendix).

Thus, as continuity, as measured by the indicator variogram, decreases, smoothing increases. The smoothing of PK, on the other hand, is approximately

σ^2_{pk} (appendix).

Since the estimation variance associated with the PK estimator is less than the estimation variance associated with the IK estimator, the smoothing of PK will be less than the smoothing of IK.

An example for which the smoothing of IK is large is provided by a simulated deposit entitled Stanford 2B. The indicator variograms for this simulation have fairly high nugget effects (fig 1) which as expected lead to high smoothing of the IK estimate (fig 2). When the PK estimator is applied to this simulation the smoothing is greatly reduced (fig 2).

The conditional bias of an estimator, defined as $E[(\phi^*(V,z) - \phi(Vz)) | \phi^*(V,z)=p]$, can cause very large over or under estimation at a specific proportion p. If for instance at a give cutoff, the estimated tonnage is always greater than the true tonnage when the estimated tonnage is (.9)(tonnage factor). There is a conditional bias at this estimated tonnage. Such a conditional bias will cause an overestimation of the tonnage recovered from high grade blocks.

An example of the extreme conditional bias which can occur with the IK estimator is given by a second simulated deposit entitled Stanford 2C. This simulation is extremely continuous and this fact is reflected in the indicator variograms which show low nugget effects (fig 1). The kriging plan for the IK estimation of this simulation was such that a datum was located at the center of the panel to be estimated while the remainder of the data were located outside of the panel. Due to the low nugget effect, the datum located at the center of the panel received a high kriging weight. Because of the high weight given to this single datum, the value of the estimator is highly influenced by the value of this datum. As this datum is an indicator datum, its only possible realizations are 0 and 1. Hence the estimator will tend to be approximately 0 or 1 while the true recovered tonnage is continuous between 0 and 1. This will result in high conditional bias for very high or very low tonnage estimates (fig 3).

The PK estimator, because of the inclusion of the U(x) data, can take on any value between 0 and 1. In addition the inclusion of the cross variogram between U(x) and I(x;z) causes the weight given to the central datum to be decreased. Hence the PK estimate for the Stanford 2C deposit shows much less conditional bias than the IK estimate (fig 3).

Another problem which the IK estimator can encounter is also related to the binary nature of the indicator data. Consider, for example, a possible kriging plan for which only four data surround a panel. Assume that these data are located at the corners of the panel and that the indicator variogram model is isotropic. Given this configuration each datum will receive the

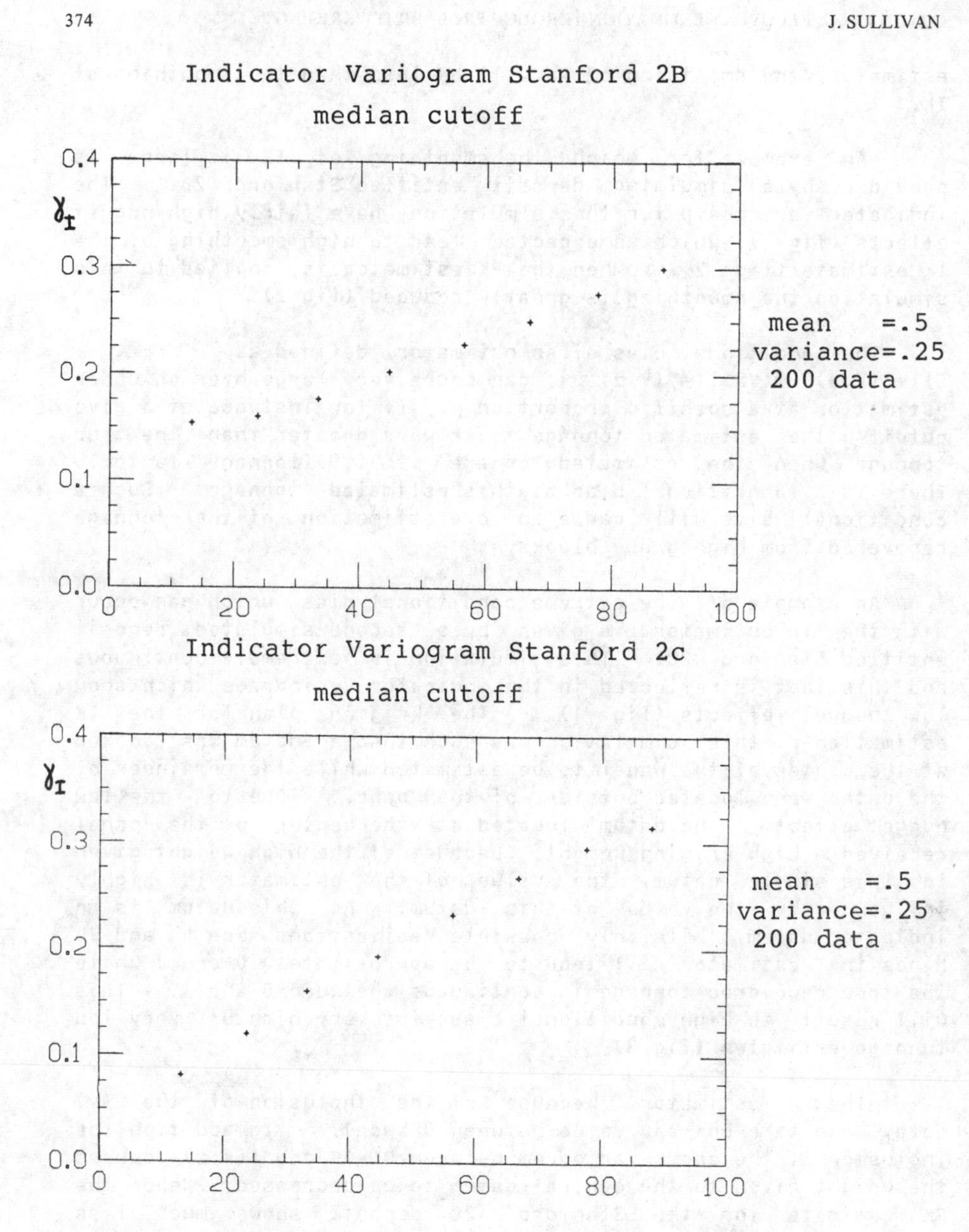

Figure 1. Behavior of indicator variograms near the origin.

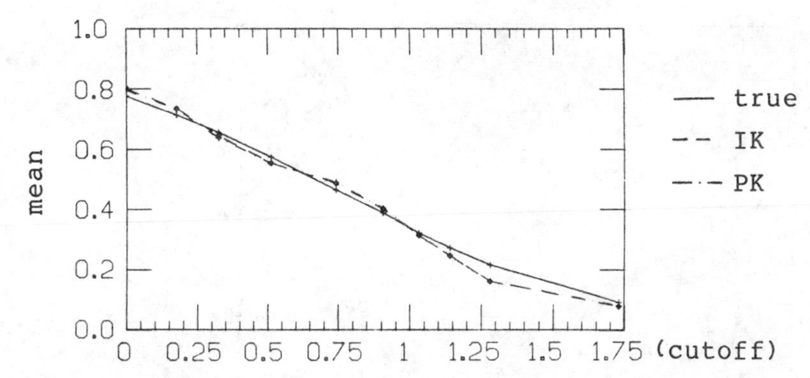

Mean recovered tonnage as a function of cutoff.

This plot shows that both IK and PK are nearly unbiased over the entire range of cutoffs.

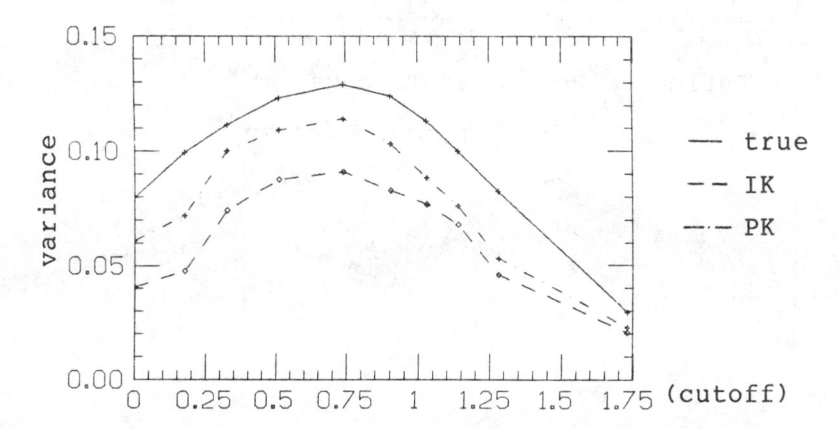

Variance of recovered tonnage as a function of cutoff.

The difference between the variance of true recovered tonnage and the variance of estimated recovered tonnage is the smoothing of the estimator. Notice that at all cutoffs the smoothing of PK is less than the smoothing of IK.

Figure 2. Mean and variance of IK and PK estimators.

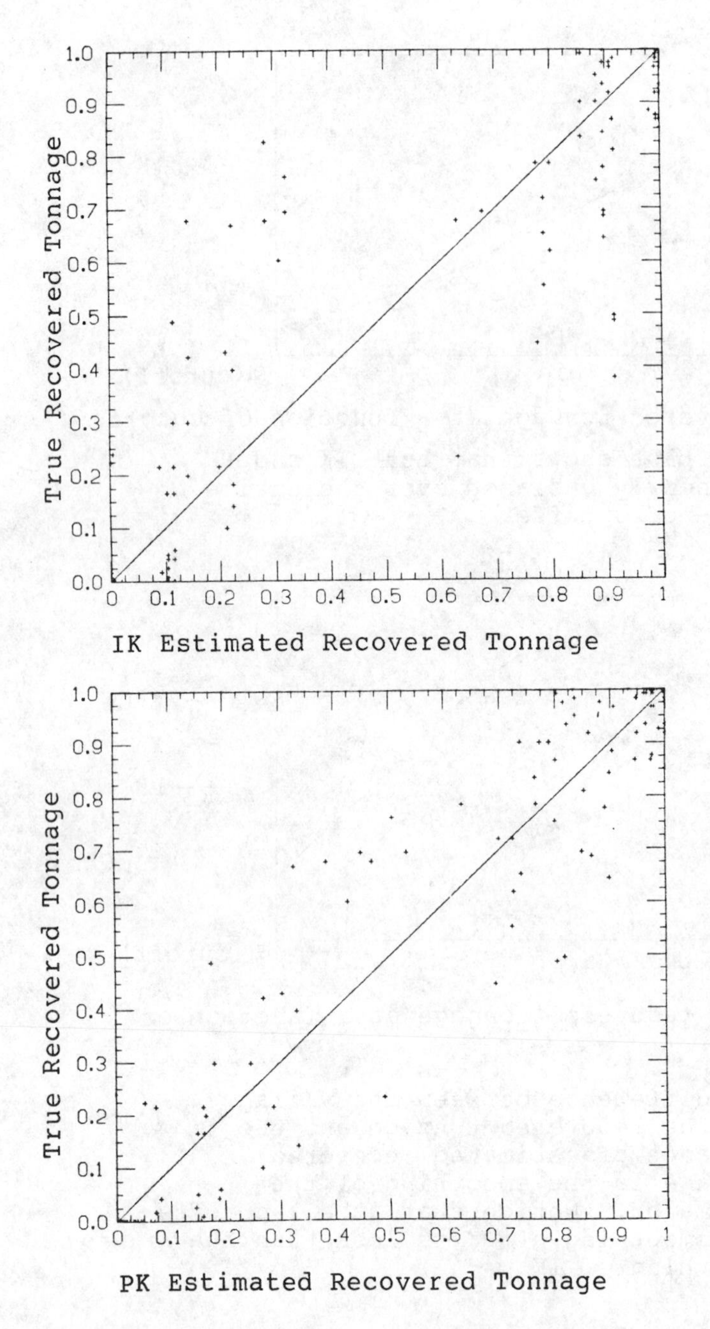

Figure 3. Scattergrams of estimated vs. true recovered tonnage.

same weight. Because the indicator data are either 0 or 1, there are only five possible estimates of the percentage of the total block recovered: 0,.25,.5,.75,1.0. Thus, in cases where there are only a few data surrounding a panel which receive significant weight, the IK estimator has a severe resolution problem. The PK estimator does not share this problem, because the U(x) data allow the PK estimate to take on any value between 0 and 1.

APPLICATION OF PK TO THE JERRIT CANYON DEPOSIT

The Jerrit Canyon deposit is a Carlin type gold deposit located in north central Nevada near the Oregon border. It is currently the one of the largest gold producers in the United States. The mine is owned and operated by the Freeport Gold Company.

The data base provided by Freeport consists of approximately 8000 blasthole fire gold assays located on approximately 12 foot centers. From blast hole location maps 119 well informed 100ft by 100ft. panels were defined. Each of these panels contains 60 to 70 blast hole assays. This information provides the true distribution of points within a panel against which the PK results will be compared.

To simulate an exploration drilling program, data were chosen at the corners of the panels on a regular 100ft grid to provide an exploration data base of 193 composites. In order to obtain an idea of the short scale variability of the indicator variograms 20 of the blast holes, located randomly throughout the deposit, were twinned. The mean of the 193 exploration data is .105 oz/ton and the coefficient of variation is 2.0. This compares well with the mean and coefficient of variation of the 8000 data: .105 and 1.86.

After examining the data base and checking that there were no obvious data entry errors, variography was performed. The variogram of grade (fig 4) shows no structure beyond 100ft., however, the twin hole information indicates a fairly low nugget effect. Given this variogram model; the spacing of the data; and the fact that the smu's are much smaller than the size of the panel, it would be impossible to accurately krige the grade of the selective mining units (smu's) due to the smoothing of ordinary kriging. Hence, the necessity of a conditional distribution approach such as PK which will ultimately determine the distribution of smu's within each panel. At this stage, however, only point recovery functions will be estimated since the true smu recovery functions are not available for comparison.

To perform PK both indicator variograms and conditional variograms of order 1 must be determined. These variograms were

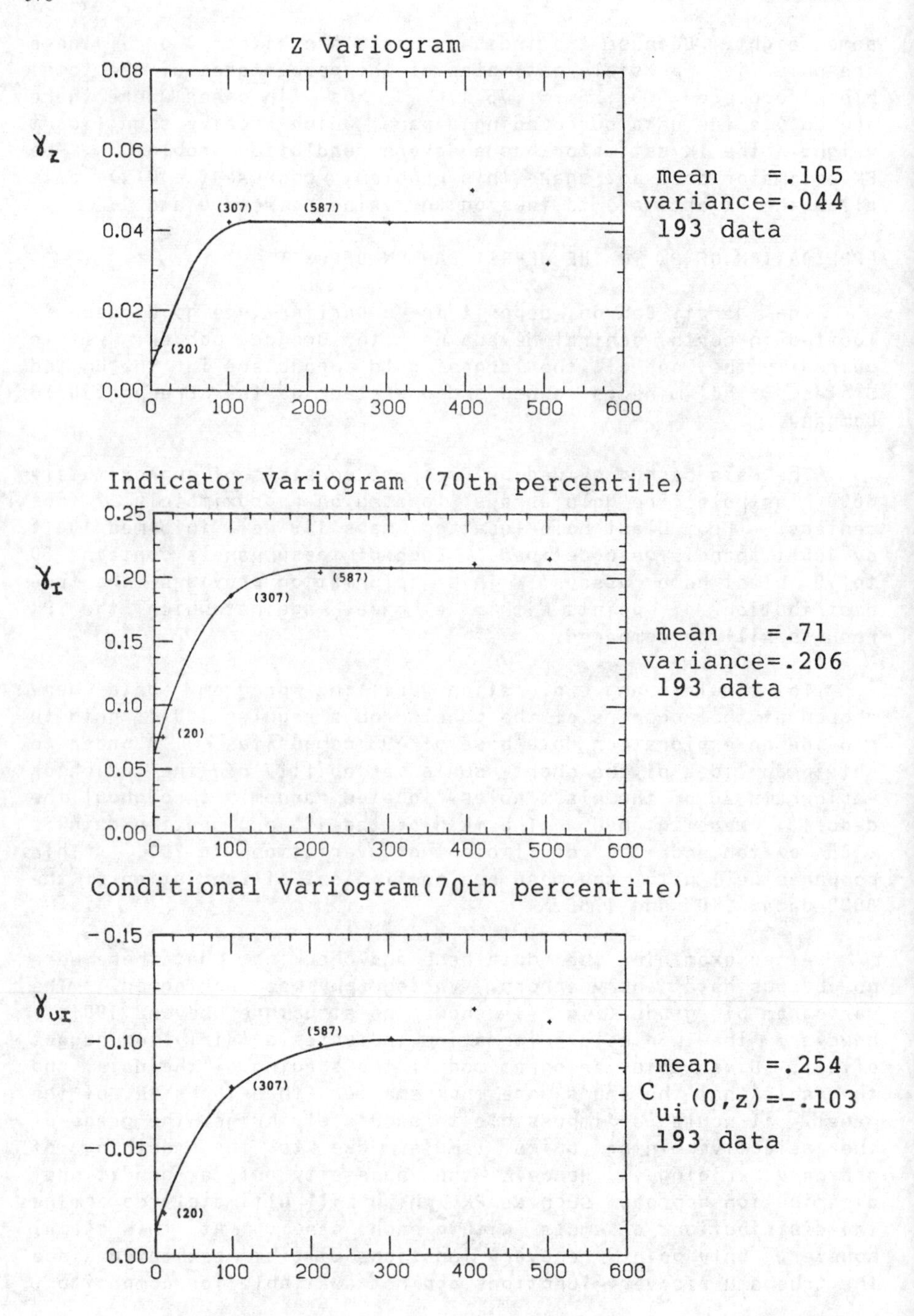

Figure 4. Variograms from the Jerrit Canyon deposit.

calculated for 10 cutoffs. The first cutoff corresponds to the detection limit of the fire assay. The remaining 9 cutoffs correspond to deciles of the initial distribution of the 193 data. Both types of variograms (fig 4) show low nugget effects and ranges which decrease systematically as cutoff increases. At the lowest cutoff the range is 400ft while at the highest cutoff the range is 130ft. This decease in range with increase in cutoff indicates that the high grade material tends to occur in pods while the low grade material is scattered throughout the deposit. Significant anisotropies were not observed in either the indicator variograms or the conditional variograms of order 1 at any cutoff.

In kriging a panel, the data used in the estimate were the four data at the corners of the panel plus the next eight closest data for a total of 12 data. Due to the range of the variogram only the 4 closest data receive significant weight, hence the remaining 8 data have little effect of the estimate. Given the fact that only the closest 4 data receive significant weight, the problem at hand is to estimate a conditional distribution using only 4 data without resorting to any distribution hypothesis. This appears to be quite a difficult task, however the results are quite good.

PK was performed at the 10 previously mentioned cutoffs, however, due to lack of space, the results from only 3 cutoffs, corresponding to the 20th, 50th, and 80th percentiles of the initial distribution, will be presented. The results at these 3 cutoffs are summarized in the scattergrams of true vs. estimated recovery and their associated statistics (figs 5,6,7). The results show that PK is an unbiased estimator of tonnage over the entire range of cutoffs; in addition, the smoothing of the estimator is small. Furthermore the estimates are very nearly conditionally unbiased, so there is no significant overestimation of the tonnage recovered from low grade panels or underestimation of tonnage recovered from high grade panels. The correlation between the estimated and true recoveries as measured by the correlation coefficient, ranges from .79 to .86 indicating that PK is a good local estimator of tonnage. Notice that the mean squared error is largest at cutoff #2, the median cutoff. This is consistent with the fact that the estimation variance is largest at the median cutoff (Journel, 1982).

These results indicate that PK is a reliable estimator throughout the range of the distribution both globally and locally.

CONCLUSION

Probability kriging represents a further development in the

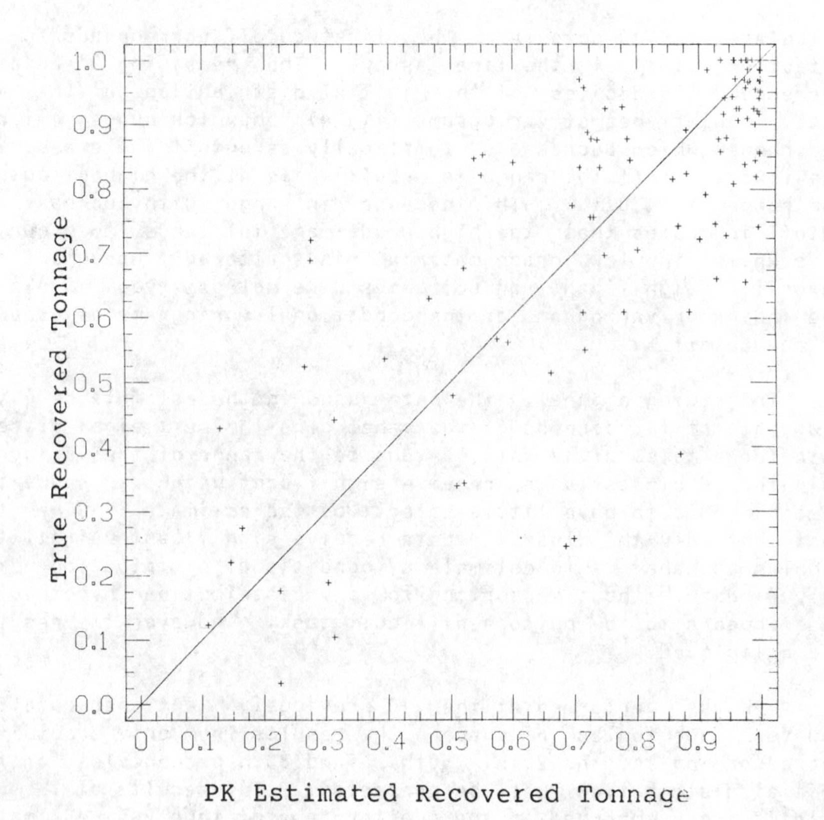

Summary Statistics

	mean	variance
true	.791	.0573
estimated	.811	.0604

mean error = .0196

mean squared error = .0249

correlation coefficient = .792

Figure 5. PK results 20th percentile.

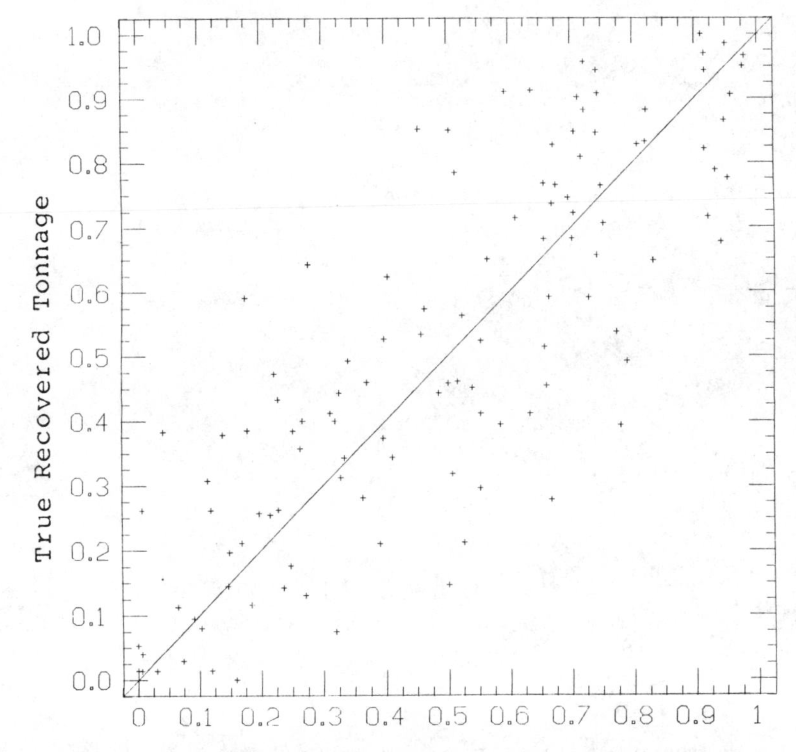

Summary Statistics

	mean	variance
true	.499	.0890
estimated	.512	.0778

mean error = .0134
mean squared error = .0276
correlation coefficient = .836

Figure 6. PK results 50th percentile.

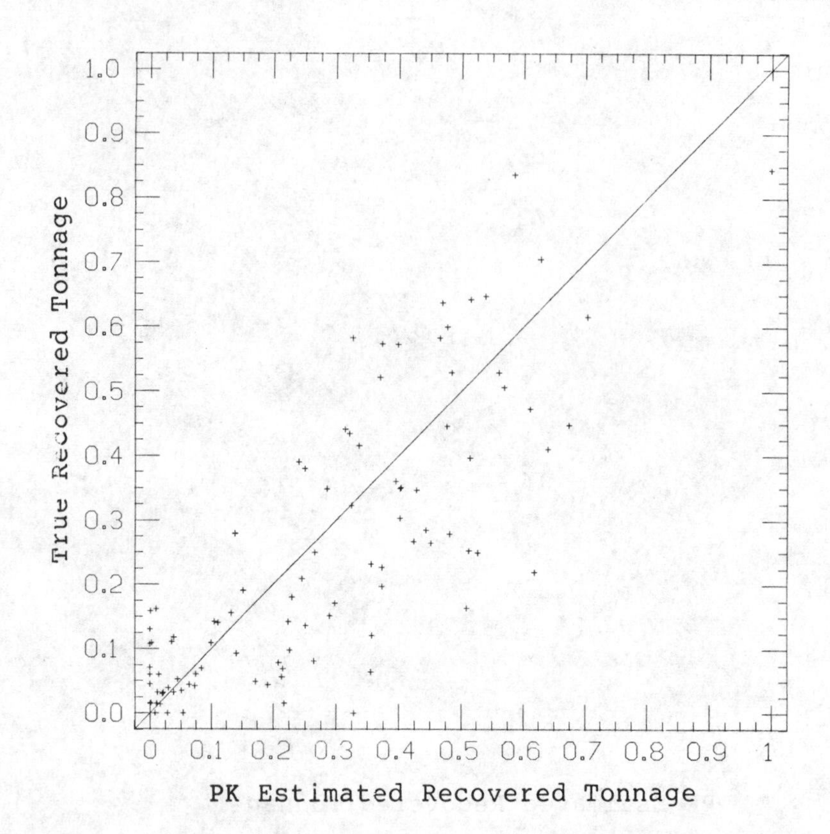

Summary Statistics

	mean	variance
true	.201	.0454
estimated	.219	.0513

mean error = .0177

mean squared error = .0137

correlation coefficient = .863

Figure 7. PK results 80th percentile.

field of non-parametric distribution free conditional distribution estimation. It is an improvement over the theoretically sound indicator kriging technique since it provides a lower estimation variance and eliminates some of the smoothing and conditional bias problems that can plague IK in certain situations. Furthermore the Probability Kriging technique has proven to be an accurate and efficient estimator on an actual deposit with a relatively high coefficient of variation.

APPENDIX

Smoothing of IK and PK

True variance of $\phi(V,z)$ within the deposit D

$$S^2(z) = 1/k \sum_1^n (\phi(V_k,z) - \overline{\phi}(V,z))^2$$

$$\overline{\phi}(V,z) = 1/k \sum_1^n \phi(V,z) = \phi(D,z)$$

$$S^2(z) = 1/k \sum_1^n \phi^2(V_k,z) - \phi^2(D,z)$$

$$E(S^2(z)) = 1/k \sum_1^n E(\phi^2(V_k,z) - E(\phi^2(D,z))$$

$$E(\phi^2(V,z)) = 1/V^2 \iint E(I(x,z)I(x',z))\, dx\, dx'$$

$$= \overline{C}(V,V,z) + F^2(z)$$

$$E(\phi^2(D,z)) = \overline{C}_i(D,D,z) + F^2(z)$$

$$E(S\hat{k}(z)) = \overline{C}_i(V,V,z) - \overline{C}_i(D,D,z)$$

The spatial variance of the IK estimator $\phi^*(V,z)$ within D is

$$\phi^*(V,z) = \sum_1^n \lambda_\alpha(z)[i(x_\alpha,z) - \hat{F}(z)] + \hat{F}(z)$$

the sk system is:

$$\sum_1^n \lambda_\alpha C_i(x_\alpha - x_\beta,z) = \overline{C}_i(A,x_\beta,z) \qquad \beta = 1 \text{ to } n$$

$$\sigma^2{}_{ik} = \overline{C}_i(V,V,z) - \sum_1^n \lambda_\alpha(z)\overline{C}_i(V,x_\alpha,z)$$

$$S^{*2}(z) = 1/k \sum_1^n (\phi^*(V,z) - \phi^*(D,z))^2$$

$$= 1/k \sum_{1}^{n}(\phi^*(V,z) - \hat{F}(z))^2$$

$$E(S^{*2}(z)) = 1/k \sum\sum\lambda_\alpha(z)\lambda_\beta(z)\ C_i(x_\alpha - x_\beta,z)$$
$$= 1/k \sum\sum\lambda_\alpha(z)\overline{C}_i(V,x_\alpha,z)$$
$$= \underline{1/k} \sum(\overline{C}_i(V,\underline{V},z) - \sigma^2{}_{ik}(V,z))$$
$$= \overline{C}(V,V,z) - \overline{\sigma^2}{}_{ik}(V,z)$$

as $D \rightarrow \infty$ $\quad E(S^2(z)) = \overline{C}(V,V,z)$

Hence the smoothing of the IK estimate is simply:

$$E(S^2(z) - S^{*2}(z)) = \overline{\sigma^2}{}_{ik}(V,z)$$

By similar logic the smoothing of the PK estimator is:

$$E(S^2(z) - S^{*2}(z)) = \overline{\sigma^2}{}_{pk}(V,z) + 2\overline{\mu}_1$$

since $\overline{\mu_1}$ << than $\overline{\sigma^2}{}_{pk}(V,z)$

the smoothing is $\sigma^2{}_{pk}(V,z)$.

REFERENCES

Gill, P.; Murray, W; Saunders, M; and Wright, M, (1982). Users Guide for SOL/QPSOL: A FORTRAN Package for Quadratic Programming, Technical Report SOL 82-7. Systems Optimization Laboratory, Stanford University, Stanford, California.

Journel, A; (1984), The Place of Non-parametric Geostatistics, proceedings of 2nd NATO ASI, Sept 1983.

RECOVERY ESTIMATION : A REVIEW OF MODELS AND METHODS

Alain MARECHAL

ELF-AQUITAINE (PRODUCTION)

ABSTRACT

This paper is a review of the "State of the Art" in recovery estimation, from the theoretical and practical point of view. After a recall on early methods, some of them being still usefull, the main methods currently employed are compared in the case of point recovery : namely indicator kriging, disjunctive kriging, conditional distribution. The pratice of each method is reviewed then in the important case of block recovery, including a comparison of the change-of-support methods involved.

Keywords : recovery estimation, indicator, kriging, disjunctive kriging, conditional distribution, change of support.

1. INTRODUCTION

1.1. Recall on definitions

In the natural resources industries, there is a clear distinction between resources and recoverable reserves. The resources present in a given orebody cannot be extracted completely for two kinds of reasons :

. Technical reasons : equipment size, strength of materials, organizational problems.

. Economical reasons : the cost of extraction is uneconomical.

The term "recovery estimation" adopted in geostatistics refers to a very particular situation :

- . the technical environment of the exploitation can be summerized in the definition of a minimum selection volume,
- . the economical environment can be summerized in the definition of a selection cut-off applied to the selection volume average grade.

G. Verly et al. (eds.), Geostatistics for Natural Resources Characterization, Part 1, 385–420.

Recovery estimation, as termed just above, happens to apply nearly exactly to the real conditions of mining big open-pit orebodies : the methods reviewed here have been applied to metal mines such as copper, iron, uranium, molybdenum, gold, silver etc. The recovery estimation methods were developped during the 70's at the cost of much research work in the geostatistician community : indeed, identifying the problem of recovery estimation and providing a workable solution has been a major break-through in the field of ore-reserve evaluation.

This paper is a theoretical review of methods, hence no particular case study will be detailed.

1.2. Recall on terminology

The purpose of recovery estimation is the evaluation of the tonnage and mean grade recovered by selection at a given cut-off in a given orebody.

. The basic selection step applies to a unit selection volume v, called selection block.

. The selection criterion is a cut-off zc applied to an estimated mean grade Zv* of the selection block v. In this case, the selection is called "indirect". If it happens that the true block grade Zv is known, or if it is considered to be estimated nearly exactly, the selection is called "direct".

. If a block is selected without any geometrical constraints of access, the selection is called "free".

In resource evaluation, the orebody is divided into regular panels V, the size of which corresponds usually to the sampling interval.

In recovery estimation, the orebody will be too divided into regular panels V, and we assume the ore to be selected :

. on unit blocks of size v,

. without geometrical constraints (free selection),

. by applying a cut-off criterion on a block grade estimate Z*v,

Consequently only part of each panel V will be selected as ore, and so only a fraction of tonnage and metal in place will be recovered. For any panel V, we define the two recovery function as :

. T (zc) : estimated proportion in the panel V of ore tonnage recovered at cut-off value zc,

. Q (zc) : estimated quantity of metal recovered per unit tonnage in place in V at the cut-off value zc.

The recovery function will be termed "direct" or "indirect". In practice, the result will be presented as follow, for a panel V of tonnage TV in place :

. estimated recovered tonnage in V : T = TV T(zc)

. estimated quantities of metal recovered in V : Q = TV Q(zc)

. estimated grade of the ore recovered in V : $M = Q(zc)/T(zc)$

Notice that the two functions $T(zc)$, $Q(zc)$ for a given panel, depend on :

. The size of the unit selection block v : if v reduces at a point, $T(zc)$ and $Q(zc)$ will be called "point recovery functions",
. The amount of information available when <u>performing the estimation</u> and incorporated in the estimators T and Q,
. The amount of information <u>available when mining</u> an incorporated in each block estimator Zv^*.

These elements are known as the volume effect and the information effect : for a good presentation, see JOURNEL, HUIJBREGTS (1978).

1.3. Recall on the volume effect and information effect

Let us consider all the blocks v_i to be selected in a given orebody, Zv_i their grade, Z^*vi their estimated grade from any set of information, $F(z)$ and $F^*(z)$ their respective cumulative histogramm.

a. If the selection were performed on the true grade Zv_i (perfect information) the recovered tonnage, quantity of metal and benefit above a cut-off zc would be respectivly :

$$T(zc) = T_M\,(1 - F(zc)) \qquad T_M : \text{tonnage of the orebody}$$

$$Q(zc) = T_M \int_{zc} z\, F(dz) \tag{1}$$

$$B(zc) = Q(zc) - zc\, T(zc)$$

These three functions depends on the distribution $F(z)$ of the block grade : it is well known to any miner (and geostatistician) that the bigger the block v_i the smoother their grade distribution. Consequently, for high (resp. low) cut-off, the bigger the selection blocks, the lower (resp. higher) the recovered tonnage. The influence of the size of the selection blocks is called the volume effect. G. MATHERON (1982) showed mathematically that for any grade distribution, the bigger the selection support, the smaller the operating benefit $B(zc)$.

b. In the case of indirect selection on block v_i, the recovered tonnage, metal and operating benefit will be :

$$T(zc) = T_M \;\; 1 - F^*(zc)$$

$$Q(zc) = T_M \int_{zc} z(z^*)\, F^*(dz^*) \tag{2}$$

The recovered tonnage will depend directly on the histogramm of the estimated block Z^*v : whatever the mathematical estimator involved in Z^*v, less sample information will result in a smoother distribution for the estimated grades Z^*_V.

In the case of the estimator being kriging, its dispersion variance is always smaller than the variance of the true grade (smoothing effect fromula).
The influence of information on the estimator is called the "information effect". MATHERON (1982) showed that the smaller the information, the smaller the operating benefit.
In conclusion, to be realistics, any recovery estimator will have to take into account both the volume and information effect.

2. EARLY SOLUTIONS

2.1. Recovery estimation in the early 70's

As soon as the practice of 3-D block kriging was made possible on computer around 1970, a block by block estimation of big open-pit mines was performed : selecting the blocks which kriged grade was above the cut-off was supposed to provide an estimator of the recovered tonnage in the open-pit. MARECHAL (1972), DAVID (1972) showed by comparing these results with the actual recovery that the method was biased. The smoothing effect of kriging (Information effect) reduced the variability of estimated grade, hence changing the proportion of blocks above the cut-off grade.

Estimating the grade of each small block was unrealistics and moreover without real importance. Which block was to be selected as ore was a matter of mining practice, when the real estimation problem was : which proportion of all the blocks to be mined in the pit will be selected as ore ? This is equivallent to estimate, locally or globally the histogramm of blocks grades.

Only large panels beeing correctly estimated, is it possible to derive from the data an estimator of the distribution of block grades within such panel ? Two kinds of solutions were proposed, according to whether actual recovery figures were available or not.

2.2. Experimental regression on real or estimated grades

MARECHAL (1972), WILLIAMSON and MUELLER (1976) had access to the actual results of recovery in two open-pit mines : Algorrobo (iron ore, Chile) and Cyprus Pima Mine (porphyry-copper, Arizona). It was possible to get by regression the average proportion P(zv) of ore recovered at the operating cut-off as a function of the real panel's grade zv (computed from the blast holes). Actually, two variants of the method were used :

. WILLIAMSON and MUELLER worked on a regression of the recovery as a function of the estimated panel grade : they superimposed for each bench the ore inventory based on DDH samples and the final ore-blasts outline to compute statistically the average proportion of ore (> 0.4 %), low grade ore (0.2 - 0.4 %) and waste for a 100 x 100' panel for various classes of inventory ore grade.
. MARECHAL and FLORES computed the same kind of experimental

curves by quadratic regression for ore, stockpile material and waste, giving the proportion and grade recovered in a panel as a function of the panel grade (computed from the blast holes),

$$P_o(zv) = a_o^o + a_o^1 \, zv + a_o^2 \, z^2v$$

For new panels, zv is not known, but can be estimated by kriging z^k. An unbiased estimator Po* of Po (zv) is then :

$$P_o^* = a_o^o + a_o^1 zv_k + a_o^2 (zv_k)^2 + a_o^2 \sigma^2_{kv}$$

In practice, it was found necessary to define as a conditional estimation variance, that is to correct the standard kriging variance by a proportional effect term.

Advantages of both methods are :

. The results are fully representative of the past mining practice : if nothing changes (cut-off, bench height, mineralization) the forecast based on these production figures will be very accurate.

. Once the regression curves are established, the recovery estimation is obtained directly from ordinary kriging.

Disadvantages of both methods are :

. They are attached to a given mine practice.

. The already mined area must be representative of the whole orebody.

. The estimated recovery for a given panel relies only on the estimated panel's grade (and possibly the corresponding estimation variance) : so the individual sample surrounding information is not completely used.

. The method is applicable only in already mined orebodies. In the case of unmined orebodies, a distribution model for the blocks grade in the panels has to be assumed to allow a prior calculation of the curves P(zv), Q(zv) : this is the principle of the alternative method proposed later by MARECHAL (1974) and DAVID (1977).

2.3. Normal and lognormal local distribution

It has been shown above that the recovery in a given panel V depends only on the local histogramm in V of the future estimators z*vi of each block v_i to be selected : this is exactly true for the tonnage and approximately true for the metal, provided that the estimator is approximately conditionaly unbiased. MARECHAL (1974) studied the case of a (multivariate) gaussian distribution of grade, showing that the estimator of each block grade was a gaussian variable with a variance function of the location of the block in the panel : then the distribution of blocks estimators z*vi in a panel V is a mixture of gaussian distribution of different means and variances.

As an approximation, it is possible to replace this mixed gaussian distribution by an unique gaussian distribution of same mean and variance :

$$G(m, \sigma^2, dz)$$ with :

$$m = \lambda_v^a Z_a$$ (kriging estimator of the panel V)

$$\sigma^2 = \sigma^2(v^*/V) + \sigma^2_{KV}$$

derived from the dispersion variance of the future block estimators v* in V and the kriging variance of V: σ^2_{KV}

A similar approximation can be done in the lognormal case : the distribution of the future block estimates will be considered as lognormal, with an arithmetic mean equal to the kriged grade Z*v of the whole panel and a logaritmic variance given by :

$$\sigma^2 = \mathrm{Log}\left(\frac{M^2+\Sigma^2_{v^*}}{M^2+\Sigma^2_{v^*}+\Sigma^2_{KV}}\right)$$

M. DAVID (1977) proposes an approximation of the above lognormal distribution by an normal distribution.

2.4. Conclusion on early methods

- They are based on standard geostatistical methods : linear kriging, dispersion variance calculation, kriging variance.
- These methods, though imperfect, provided a definite improvement in recovery estimation, compared to the application of a cut-off selection on the blocks estimators. They can be considered as still attractive whenever the conditions of their use are met : access to a large amount of mine operation results, or reasonable validity of the normal/lognormal assumption.

3. EXISTING METHODS FOR POINT-RECOVERY ESTIMATION

3.1. Interest of point-recovery estimation

Considering first the estimation methods defined for point-recovery allows separating two important problems :

- the problem of local histogram estimation,
- the change-of-support problem.

The first problem has various exact mathematical solutions, while the second has generally only approximate solutions.

From the practical point of view, point-recovery estimation is important in a study of an already operating open-pit mine : it is possible to compare an estimated local distribution whith the blast-holes histogram for each panel.

3.2. Introduction to indicator cokriging and disjunctive kriging

3.2.1. Purpose of disjunctive kriging

As recalled previously, estimating the point recovery in a given panel V is equivallent to estimating the local histogram of point grades Z(x) in V, which in turn is equivallent to estimating the indicator function of point-grades for all the points of panel V.

Let Z(x) be a stationary R.F. with p.d.f F(dz), bivariate p.d.f between two points x, x + h, F_h (dz, dz) its convariance function. Z(x) is known on N sample points. The indicator function for cut-off grade z associated to the numerical fonction z(x) is defined by :

$$Iz(x) = \begin{cases} 1 \text{ if } z(x) > z \\ 0 \text{ if } z(x) < z \end{cases} \tag{3}$$

Considering now Z(x), a stationary R.F., Iz (x) defines a new stationary R.F. with the following characteristics :

$$E \quad Iz(x) = 1 - F(z) \tag{4}$$

$$VAR \quad Iz(x) = F(z)\ (1 - F(z) \tag{5}$$

Iz(x) has a stationary convariance function

$$\sigma_z(h) = \int_z^{+\infty}\int_z^{+\infty} F_h(du, du') - [1 - F(z)]^2$$

Then, the indicator function appears as a particular case of service-variable, for long time in use in geostatistics (see a good presentation of indicator-function in BOUCHIND'HOMME, 1980). In particular, the idea of using indicator functions is the basis of MATHERON'disjunctive kriging (1972). It has been again proposed by A. JOURNEL(1982) as a practical method for point-recovery estimation.
The need for a new estimator appears clearly when dealing with highly skewed distributions : kriging will give the same weight to a given sample point, the observed grade being 1 g/T or 1000 g/T for instance. So instead of using an estimator like :

$$Z_x^* = \lambda_\alpha^* Z_\alpha$$

it would be nice in some cases to use an estimator like :

$$Z_x^* = \lambda_x^*(Z_\alpha) Z_\alpha \qquad \text{equivallent to } \sum_\alpha f_\alpha(Z_\alpha) \tag{7}$$

As any continuous function can be represented by a combination of indicator functions, the new estimator proposed by MATHERON can be approximated by a linear combination of indicator functions of each sample.

$$Z_x^* = \sum_\alpha \sum_i \mu_{i\alpha}^* I_{z_i}(x_\alpha) \tag{8}$$

zi, i = 1, k is a sequence of cut-off values allowing a good representation of the variable Z.

Then, it is easily seen that the m.q optimum estimator of the above form, for the unknown value Z_{xo} is the complete cokriging of Z_{xo} using the indicator values I_{zi} of each sample points : because Zx is represented in disjoint classes through the use of indicator functions, this estimator was named disjunctive kriging (D.K.).

3.2.2. Discrete and continuous disjunctive kriging

a. The variable Z(x) can be approximated in a discrete way in a sequence of classes limited by the values z_{i+1}, z_i : for each class i, Z is represented by a value $\bar{z}_i$ (centre of the class or sample mean in the class) and the R.F Z(x) can be written :

$$Z(x) = \sum_{i=1}^{n} \bar{Z}_i [I_{z_i}(x) - I_{z_{i+1}}(x)]$$

with the additional conventions : $z_o = o$, $Z_{n+1} = 0$
we have

$$Z(x) = \sum_{i=1}^{n+1} (\bar{Z}_i - \bar{Z}_{i-1}) I_{z_i}(x) \tag{9}$$

So, in each point x, Z(x) appears as a linear combination of the indicator functions $I_{zi}(x)$, and the complete cokriging of Z(xo) using the indicator values will appear as a linear combination of indicator cokriging estimators :

$$Z^*_x = \sum (Z_i - Z_{i-1}) I^*_{z_i}(x)$$
$$I^*_{z_i}(x) = \sum_\alpha \sum_j \lambda^\alpha_x (z_i, z_j) I_{z_j}(x_\alpha) \tag{10}$$

$\lambda^\alpha_x(z_i, z_\gamma)$ cokriging coefficient of $I_{z_\gamma}(x_\alpha)$ in the estimation of $I_{zi}(x)$.

b. The variable Z(x) is represented in a continuous way, for instance as an integral of its indicator function. In the case of a positive variable Z(x), we have :

$$Z(x) = \int_0^{+\infty} I_z(x) dz \tag{11}$$

Let $I^*_z(x)$, $\forall z$, be the complete cokriging of the function $I_z(x)$; by linearity Z*(x) will be :

$$Z^*(x) = \int_0^{+\infty} I^*_z(x) dz \tag{12}$$

$$I^*_z(x) = \sum_\alpha \int \lambda^\alpha_x (z, z') I_{z'}(x_\alpha) dz' \tag{13}$$

Writing again (12) : $Z^*(x) = \sum_\alpha \int \lambda^\alpha_x (z, z') I_{z'}(x_\alpha) dz dz'$

defining $\mu^x_\alpha(z') = \int \lambda^x_\alpha (z, z') dz$ it comes :

$$Z^*(x) = \sum_\alpha \int \mu^x_\alpha (z') I_{z'}(x_\alpha) dz'$$

Each term $\int \mu^x_\alpha (z') I_{z'}(x_\alpha) dz$ is the definition of a function $h_\alpha(z_\alpha)$ through the indicator functions. We are back to the usual presentation (7) of the D.K estimator.

Presenting the D.K. estimator as an indicator cokriging allows an easy understanding of the model specifications necessary to determine a D.K estimator in a discrete form (10) or continuous form (13).
By definition of cokriging, the cokriging coefficients $\lambda_x^a(z, z')$ of formula (13), $\lambda_x^a(z_i, z_j)$ of relation (10) are solutions of systems like (ordinary cokriging) :

$$\lambda_x^\beta(z_i, z_j)\sigma_{z_j z_k}(x_a - x_\beta) = \sigma_{z_k z_i}(x_a - x) + \mu$$

$$\Sigma \lambda_x^\beta(z_i, z_j) = 1 \qquad \Sigma \lambda_x^a(z_i, z_j) = 0 \qquad (14)$$

Then both the complete cokriging and disjunctive kriging require the whole set of cross-covariance functions :

$$\sigma_{zz'}(h) = E[I_z(x+h) I_{z'}(x)] - (1-F(z))(1-F(z'))$$

between the indicators I_z and I_z', for all z, z' (in the continuous case) or for all z_i, (discrete case).
Modeling $\sigma_{zz'}(h)$ is equivallent to defining the point probability distribution of Z(x+h). Z(x)

$$\sigma_{zz'}(h) = P(Z(x+h) \geqslant z, Z(x) \geqslant z') - [P(Z(x+h) \geqslant z][P(Z(x) \geqslant z' \qquad (15)$$

Consequently, the complete cokriging of indicator functions, called in the continuous case disjunctive kriging, supposes the definition of the bivariate probability distribution of the R.F Z(x) for any pair of points (x + h, x) : in particular fitting a certain number of covariances $\sigma_{zz'}(h)$ to experimental indicator variograms is equivallent to adopting an underlying model of bivariate distribution

3.2.3. Characteristics of the disjunctive kriging estimator
The theoretical properties of the D.K. estimator are those of any cokriging estimation, and depend on the variant of kriging used.

a. $I_{z_i}^*(x) = 1 - F(z) + \sum_a \sum_j \lambda_x^a(z_i, z_j)[I_{z_j}(x_a) - 1 + F(z_j)]$ (16)

This is the simple kriging estimator (S.K.) based on an explicit model of p.d.f. for Z(x) : F(z). As any simple kriging tied to a known mean 1 - F(z), the validity of this estimation is based on a strict stationarity over the sampled field : in particular, whenever the information is sparse, the estimation will be equal to the mean 1 - F(z).

b. $I_{z_i}^*(x) = \sum_a \sum_j \lambda_x^a(z_i, z_j)[I_{z_j}(x_a)$ (13)

This is the ordinary kriging (O.K.), mainly used in a local neighbourhood : it is known that such a kriging requires only a local stationarity of the covariances $\sigma_{zz'}(h)$

c. $I^*_z(x)$ could be defined by an universal kriging system, based on a set of spatial trend functions for $I_z(x)$ (MATHERON 1977). In fact, increasing the number of universality conditions increases the number of negative kriging weights, a major nuisance for indicator cokriging.
Indeed, negative values weighting 0 or 1 values (see formula 13) may provide a negative estimate for Iz(x), or allows $I^*_{z_{i+1}} > I^*_{z_i}$ wich $z_{i+1} > z_i$.
Stated simply, the cokriging estimator of the p.d.f of Z(x) on an unsampled point x is not a p.d.f, and the number of abnormal values increases when the number of universality conditions increases : in practice simple cokriging is recommended.

4. INDICATOR ESTIMATION BY CONDITIONAL DISTRIBUTION

4.1. Definitions

Consider N sample points $x_a, a = 1 . N$ and a target point x. We define the (N + 1) vectorial R.V $(Z_x, Z_a, a = 1 . .$ by its (N + 1) p.d.f $F(z, z_i \ldots z_n) = P[Z_x < z, Z_1 < z_1 \ldots]$

We can derive from this vectorial p.d.f :

. $fx(z), fa(z), \ldots$: marginal densities of Z_x, Z_a
. $fx(z|z_1, \ldots z_a, \ldots z_N)$: conditional density of Zx when each Z(x) is fixed to the value Z_α

To estimate φ(Zx), φ being any measurable function, the estimator must be any function h $(Z_1 \ldots Z_N)$ of the sample value. h $(Z_1 \ldots Z_N)$ minimizing the variance of error is the conditional expectation of Z given $Z_1 \ldots Z_N$.

$$\phi^*(Zx) = h(z_1 \ldots z_a \ldots \quad \text{such that} \quad E(\phi^* - \phi)^2 \quad \text{minimum}$$

$$\phi^*(Zx) = \int \phi(z) fx(z|z_1 \ldots z_N) dz \tag{17}$$

In particular if the function Φ(Zx) to estimate is the indicator function $I_z(x)$, it appears that the optimum estimator is the conditional p.d.f. itself :

$$I^*_z(x) = \int_z^{+\infty} fx(z|z_1 \ldots z_N) dz = 1 - Fx(z|z_1 \ldots z_N) \tag{18}$$

Notice that :

. by construction, this estimator $I_z{}^*(x) = h_x(Z_1 \ldots Z_N)$ has the good properties, 1 - I*z being a p.d.f.
. there is no need for Z(x) to be stationary : whenever one can specify a (N+1) - p.d.f for $(Z_x, Z_1 \ldots Z_N)$ the conditional expectation is implicitly defined.

In mining applications, it is frequent to use above 10 sample values in one estimator : the corresponding model of p.d.f.

$F(z, z_1 \ldots z_n)$ would be very complex in the general case. In practice, the only multivariate distribution used is the multivariate gaussian. For any continuous R.V. there exist a transformation to a gaussian R.V. : Applying such transformation to $Z_{x1}, Z_1 \ldots Z_N$, each tranformed variable Y_α can be considered individually as a gaussian standard N(0,1). The decisive additional hypotheses consists in assuming that the point (N + 1) distribution of $(Y_x, Y_1 \ldots Y_N)$ is gaussian standard multivariate of mean o, variance 1 and covariance matrix $\rho_{\alpha\beta}$

4.2. Multivariate gaussian model

The major advantage of the gaussian multivariate distribution lies in the great simplicity of the conditional distribution : It is still a gaussian distribution. Indeed, the conditional variable Y_x $Y_1 \ldots Y_N$ has the following law :

$$Y_x \mid Y_1 \ldots Y_N \doteq m_x + \sigma_x \upsilon \qquad : N(O, 1)$$

$$m_x = \lambda_x^a y_a ; \qquad \lambda_x^a \text{ solution of: } \lambda_x^\beta \rho_{\alpha\beta} = \rho_{\alpha x}$$

$$\sigma_x^a = 1 - \lambda_x^a \rho_{\alpha x}$$

. The above system coincides with the simple kriging system of Yx by the Y . Then :

m_x is the kriging of Yx

σ_x^a is the variance of kriging of Y_x

It is very important to realize the pratical consequences of a multigaussian model on the properties of the estimator $I^*_y(x)$:

. The whole conditioning information $(Y_1 \ldots Y_n)$ is synthesized within one figure $\lambda_x^a y_a$, the only to appear in the conditional distribution.

. The dispersion of the conditional distribution is independant of the sample values Y_x, so that Y_x $Y_1 \ldots Y_n$ shows a perfect absence of "proportional effect".

Consequently, in such a model, the indicator function estimator as the following definition :

$$I_y^*(x) = 1 - G\left(\frac{y - \lambda_x^a y_a}{\sigma_x}\right) \tag{19}$$

Adopting a complete multivariate model fixes strongly the mathematical form of the estimator, the experimental information appearing as a simple parameter in a function.

5. INDICATOR KRIGING

5.1. Simplifying the complete indicator cokriging

JOURNEL (1982) proposes to simplify the complete indicator cokriging shown in (10), replacing it by ordinary kriging with indicator values at the same cut-off z. In other term, he considers the R.F. $I_z(x)$ on different cut-off z as different service-variables estimated independantly by the sample service-variable $I_z(x_\alpha)$, using their corresponding variogram JOURNEL emphasizes some of the difficulities of this approach :

. necessity to compute and fit a model for the indicator variogram at each cut-off.

. No garantee that the estimators I*z(x) for various cut-off will show the expected order relations.

JOURNEL (1982) proposes to overcome these difficulties by adopting a unique variogram model for the $I_z(x)$, in practice the model fitted to the median indicator variogram. His arguments are :

. for most cut-off values, the relative variograms of indicators are experimentally very close to the median variogram, which is generally well defined experimentally.

. keeping a unique variogram model for the different z leads to a constant set of kriging weights λ^a , hence unsuring the order relation for I*z(x), as long as the number of negative weights λ^a is small (which is the case with the usual variogram models).

5.2. Advantages and disadvantages

What are the advantages of indicator kriging ?

a. simplicity of the method, totally based on the practice of kriging,

b. the validity of ordinary kriging in the case of local stationarity only, makes the method applicable even when the grade function is not strictly stationary,

c. when O.K. is used, no prior model of p.d.f. F(z) is needed.

d. processing of grade distribution with a spike (e.g. with a large proportion of zero values) is possible, either by considering the zeros as a separate population or by including them in the lower class of cut-off.

The disadvantages are of two kinds :

a. numerical problems when the caracteristics of the estimator $I^*_z(x)$ are not satisfacying :

. negative results or violated order relations,

. discontinuous aspect of the local distribution estimator : the density distribution associated to $I^*_z(x)$ considered as a p.d.f. is a weighted average of Dirac measures assigned to the sample points. So, when they are few samples in the vicinity of x, the estimator of the local distribution corresponds in fact to a discrete distribution with few possible values (the sample values observed in the vicinity of x). A classical solution in statistics to smooth such histogram is to replace the Dirac measures by some smoother Kernel (see the use of Histospline in MARECHAL 1976).

b. Necessity to compute and fit a variogram for each cut-off : for some cut-off, it will be difficult to find a satisfying model. BOUCHIND'HOME (1980) uses as model at various cut-off the functions derived theoretically by the bigaussian relations.

c. The major drawback of indicator kriging is related to the abandon of cokriging : because of the strong correlation between indicator functions of various cut-off, a large amount of information is lost when disregarding these other indicator functions. Indeed, only in exceptional situations will the indicator kriging be equivallent to the complete cokriging.

5.3. Theoretical validity of indicator kriging

Replacing indicator cokriging by indicator kriging is theoretically justified only in two situations :

a. Indicator functions for various cut-off $Iz(x)$, $Iz'(x')$ are uncorellated : it can be seen on (15) that the covariance function $\sigma_{nm}(h)$ is null only if $Z(x + h)$ and $Z(x)$ are independant whatever h. Apart from this case (of little practical interest) there are always strong correllations between the indicator functions at various cut-off, especially for the indicator functions at the same location x.

b. Cokriging reduces theoretically to kriging because of intrinsec correlations between the $Iz(x)$, $Iz'(x')$. MATHERON (1982) has shown that such a situation occurs only for a very special model of bivariate distribution, called by him "Mosaïc model", where the direct and cross-correlogram of indicator function I_z are all identical : in particular, high grade indicator function have the same variogram as low grade indicator functions. (20)

$$F_\rho(dz, dz') = \rho(h)\delta_z(dz')F(dz) + (1 - \rho(h))F(dz)F(dz')$$

In this model, $Z(x + h)$ and $Z(x)$ are equal with the probability $\rho(h)$, or independant with same p.d.f $F(z)$ with the probability $1-\rho(h)$.

This is the only model for which the indicator functions are a set of "autokrigeable" functions, that is functions for which cokriging reduces theoreticaly to kriging. It can be seen that such a model is in contradiction with most of the experimental observations, whereas the "destructurization of high grade" typical of a bigaussian distribution seems to correspond much better to the reality. Then, to quote MATHERON (1982), "in the bigaussian case, substituting indicator kriging to dis- junctive kriging must result in a very substantial loss of information".

Using indicator functions as service-variable is not recommended because they do not constitute in general an "autokrigeable" family. Conversely for some bivariate models, there exists an autokrigeable family made of orthogonal polynomials : it is then appropriate to use these polynomials as service-variables in place of the indicator functions.

6. DISJUNCTIVE KRIGING OF ISOFACTORIAL MODELS

6.1. General case

Let us see how the complete cokriging of indicator reduces to a series of kriging of service variables in the case of isofactorial models. In the general continuous presentation of D.K., we look for an estimator of R(Zx), R being any continuous function, the estimator being defined by :

$$R^{DK} = \sum_{\beta} f_{\beta}(Z_{\beta})$$

In practice, it will be often convenient to make a one-to-one tranform of Zx through an anamorphosis function ϕ, such that :

$$Z(x) = \phi(Y_x) \quad ; \quad Y_x = \phi^{-1}(Z_x)$$

The problem is then stated in equivallent terms :

Estimate $T(Y_x) = R[\ (Y_x)]$ by $T^{DK} = \sum_{\beta} f'_{\beta}(Y_{\beta})$

Provided the p.d.f. G(y) associated with Yx is enough regular to have a whole set of moments $E(Y^h x)$, $\forall$ n, then there exist a family of orthogonal polynomials Xn(y) such that :

$$\int X_n(y) X_m(y) g(y) dy = \delta_{n,m} \quad \forall n, m \tag{21}$$

Consequently any function T(Yx) or $f_{\beta}(Y_{\beta})$ can be expressed as an expansion of such polynomials :

$$T(Y_x) = \sum_{n=0}^{\infty} \tau_n X_n(Y_x)$$

$$f_{\beta}(Y_{\beta}) = \sum_{n=0}^{a} f_n^{\beta} X_n(Y_{\beta})$$

By linearity of the D.K. estimator of T (Yx), this one can be expressed as :

$$T^{DK} = \sum_{n=0}^{\infty} \tau_n X_n^{DK}$$

X_n^{DK} being the D.K. estimator of the function Xn(Yx)

$$X_n^{DK} = \sum_{p} \sum_{\beta} \lambda_{p,n}^{\beta} X_p(Y_{\beta}) \tag{22}$$

Then D.K. appears as a cokriging of each $X_n(Y_x)$ function by all the service-variables $X_p(Y_{\beta})$. This cokriging will reduce exactly to a sequence of kriging system when the bivariate distribution $F_{\alpha\beta}(dY_{\alpha}, dY_{\alpha})$ of two sample values located in x_{α} and

X_β is assumed to be isofactorial (MATHERON, 1972) :

$$f_{\alpha\beta}(Y_\alpha, Y_\beta) = \sum_{n=0}^{\infty} \rho_n(x_\alpha - x_\beta) X_n(X_\alpha) X_n(Y_\beta) g(Y_\alpha) g(Y_\beta) \quad (23)$$

Using the Correspondance Analysis terminology, the X_n are the factors associated with the bivariate law : writing the bivariate density as in (23) shows that the family of factors are the same with regard to Y_α and Y_β , hence the term "isofactorial" (P. BECKMAN 1973). From (23), we derive :

$$E[X_n(Y_\alpha) \mid Y_\beta = y] = \rho_n(x_\alpha - x_\beta) X_n(y) \quad (24)$$

Hence :

$$E[X_n(Y_\alpha) X_m(Y_\beta)] = \delta_{n,m} \quad (25)$$

So the service-variables $X_n(Y_\alpha)$, $X_m(Y_\beta)$ are uncorrellated as soon as n ≠ m, and the function $\rho_n(h)$ is the correllogram function of the service variable $X_n(Y_x)$. Then, the cokriging estimator (22) reduces to a simple kriging estimator using the service-variable $X_n(Y_x)$, of covariance function $\rho_n(h)$.

$$X_n^{OK} = \sum_\beta \lambda_n^\beta X_n(Y_\beta) \quad (26)$$

$$\sigma_{Kn}^2 = 1 - \lambda^a \rho_n(a\, x)$$

$$\lambda_n^\beta \rho_n(x_\alpha - x_\beta) = \rho_n(x_\alpha - x) \quad \forall a = 1, n \quad (27)$$

Then, whatever be the function T(Yx) known by an expansion given above, we have :

$$T^{OK} = \sum_{n=0}^{\infty} \tau_n \sum_\beta \lambda_n^\beta X_n(Y_\beta) \quad (28)$$

$$\sigma_{OK}^2(T) = \sum_{n=1}^{\infty} \tau_n^2 \sigma_{Kn}^2 \quad (29)$$

6.2. Bigaussian particular case

A very popular model for bivariate distribution is the bigaussian :

$$g_\rho(Y_\alpha, Y_\beta) = \sum_{n=0}^{\infty} \frac{\rho^n(h)}{n!} H_n(Y_\alpha) H_n(Y_\beta) g(Y_\alpha) g(Y_\beta)$$

Hn(y) are the non-normalized Hermite polynomials and the correllogram of Hn(y) is simply $\rho^n(h)$, $\rho(h)$ being the covariance function of the gaussian equivallent Yx.

As an example, the indicator function in x above the cut-off $z=\phi(y)$ will be obtained through the Hermite expansion of the function Iz(x) :

$$I_z^{OK}(x) = 1 - G(y) - \sum_{n=1}^{\infty} \frac{g(y) H_{n-1}(y)}{n!} \left(\sum \lambda_n^\beta H_n(Y_\beta)\right) \quad (30)$$

written slightly differently, this formula reads :

$$I_z^{DK}(x) = 1 - F(z) + \sum_{n=1}^{\infty} \sum_{\beta} \mu_n^{\beta} H_n [\phi^{-1}(Z_\beta)]$$

The estimator looks very similar to an indicator kriging with known mean 1 - F(z) for which the indicator functions of each Z_β have been replaced by the service-variable $H_n[\phi^{-1}(Z_\beta)]$ Formula (30) can be written again as :

$$I_z^{DK}(x) = \int_z^{+\infty} f_x^{DK}(y)\,dy$$

$$f_x^{DK}(y) = \sum_{n=0}^{\infty} \frac{H_n^{DK}(x) H_n(y)}{n!} g(y) \qquad (31)$$

$I_z^{DK}(x)$ is obtained by integration from a density function as is the D.K. estimator of any function :

$$\phi_x^{DK}(Y) = \int \phi(y) f_x^{DK}(y)\,dy$$

Unfortunately, f_x^{DK} resulting from a complete cokriging, is not a true density function and may take negative values for some values of y.

In practice, the steps in a D.K. study will be :

. Transforming of the sample values $Z_\alpha \to Y_\alpha$ using either a mathematical transformation (e.g. logarithmic) or the "graphical transformation" attached to the ordered sequence of Zx.

. Variogram calculation and fit of γ, hence fit of $\rho(h)$

. Resolution of n kriging systems with covariance ρ^n .

These systems are usually about 10, the high order ones being easily resolved because of their dominant diagonal. In fact, these systems refer directly to the mean indicator function $\overline{I_z(V)}$ in a complete panel V, so that one can estimate directly the point grade distribution in the panel V, through the estimated mean density f_v^{DK} . As recalled before, the point-recovered tonnage and point-recovered metal are linked to this distribution.

$$T_v^{DK}(z) = \int_y^{+\infty} f_v^{DK}(u)\,du \qquad (y = \phi^{-1}(z))$$

$$Q_v^{DK}(z) = \int_y^{+\infty} \phi(u) f_v^{DK}(u)\,du$$

It is possible, to compute the grade of co-products recovered when the main grade is selected at a cut-off z. (MARECHAL, 1982) : the grade of the recovered co-product in the panel is derived from the estimated co-product grade of the panel, modified by an expansion of terms including the H_n^{DK} of the panel and the correlation coefficient between the transformed main and coproduct grade.

6.3. Other isofactorial models

Transforming the grade Zx to a gaussian distribution is very frequent for the following reasons :

. many grade distributions are experimentaly close to lognormal type distibutions. Then, it seems natural to model the grade as a transformed gaussian variable.

. it is frequent to undertake successively on a given orebody a recovery estimation study and a conditional simulation study : as it is well known, all the R.F. simulation methods currently in use in stationary geostatistics generate gaussian variables. Generally, it will be found convenient to adopt a unique gaussian tranformed model for both the recovery estimation study and the simulation study, to find consistent recovery results in both studies. See the example shown on (fig. 6 and 8, MARBEAU, MARECHAL 1980).

So, in many examples of continuous distribution, the use of bigaussian transformed model was found satisfying. Examples of checks of the basic property of the model, i.e.

$$E[H_n(Y_{x+h}) \mid Y_x = y] = \rho^n(h) H_n(y)$$

have been published GUIBAL, REMACRE 1983, BOUCHIND'HOMME 1980) variograms of the Hn (Yx) have been checked experimentally (BOUCHIND'HOMME, 1980).
In the case of the above checks not be found corrects, there remain the possibility of adopting a more general bivariate model based on gaussian marginal distributions, the Hermitian model :

$$f_n(Y_{x+h}, Y_x) = \sum_{n=0}^{\infty} \frac{T_n(h)}{n!} H_n(y_{x+h}) H_n(Y_x) g(Y_{x+h}) g(Y_x) \quad (32)$$

The sequence of functions Tn(h), correllogram of the service-variables Hn(Yx) is no longer of the type $\rho^n(h)$, but is necessarily of the type :

$$T_n(h) = \int \rho^n(h) F(d\rho) \quad (33)$$

In other term, these models are obtained as a mixture of bigaussian model of different covariance functions. Such models can be obtained by simulation through the combination of various complete gaussian simulations : MATHERON (1982) proposes an interesting example of such simulation :

$$Z(x) = Z_1(x) \cos X(x) + Z_2(x) \sin X(x) \quad (34)$$

Z_1, Z_2 are two independant gaussian standards R.F. of same covariance $\rho(h)$, Xx an independant stationary R.F.

A much more important (in practice) class of isofactorial model is related with discrete distributions, allowing a discrete representation of the R.F. Zx similar to the one used in indicator kriging : such a representation is very suitable to a variable having an important proportion of zero values. To remain close to the gaussian case, the continuous variable Zx is replaced by a discrete binomial Nx with (N+1) values, from o to N and parameter p. (MATHERON, 1980).

Consider a serie of N + 1 classes of values :

$$Ci = (z_i < Zx < z_{i+1}) \quad i = 0,1 \dots N$$

with possibly the first class Co containing the zero values of Zx. The limits of class (z_i, z_{i+1}) are chosen so that the frequency of Z_α observed in each class Ci is equal to the theorical probability Wi for a binomial variable Nx to be equal to i :

$$Wi = P(Nx = i) = C^i_n \, p^i \, (1p)^{N-i} \qquad (35)$$

the parameter p is fixed in the model. If Zx has zero values in proportion po, p will be related to po by :

$$p_o = (1-p)^N \quad ; \quad p = 1 - p_o^{1/N} \qquad (36)$$

If there are no zero value, p will be fixed arbitrarily so that the classes Ci represent correctly the histogram of Zx. We replace Zx by the discrete variable $\tilde{Z}_x$ according to :

$$\tilde{Z}_x = \bar{Z}_i \quad \text{if} \quad z_i \leq Z_x < z_{i+1}$$

Then, we have the following binomial anamorphosis :

$$\tilde{Z}_x = \phi(N_x) \quad , \quad N_x = \phi^{-1}(\tilde{Z}_x)$$

$$z_i \leq z_\alpha < z_{i+1} \Longrightarrow N_\alpha = \phi^{-1}(\tilde{Z}_\alpha) = i$$

The isofactorial model associated with the binomial distribution can be found in MATHERON (1980). With the notations :

$$Wi = C^i_N \, p^i (1-p)^{N-i}$$

$$wi = C^i_N \, p^i \, ^i_N \, (1-p)^i$$

$$\rho_{\alpha\beta}(i, j) = P(N_\alpha = i, N_\beta = j) = \sum w_n \, \rho^n_{\alpha\beta} \, H_n(i) H_n(j) W_i \, W_j \qquad (37)$$

The orthogonal integer polynomials (unnormed) Hn(i) follow a recurrence relation : (38)

$$(N-n)p(1-p)H_{n+1}(i) + [i - Np + n(2p-1)] H_n(i) + nH_{n-1}(i) = 0$$

$$Ho(i) = 1 \qquad H_1(i) = (Np - 2p + 1 - i)/w_1$$

As in the bigaussian model, the correllogram of the Hn(Nx) is of the type : $\rho^n(h)$ where $\rho(h)$ is the correllogram of the associated binomial function Nx, which can be fit experimentally.

Following the standard steps in D.K estimation, a D.K estimation of the probability distribution of Nx in an unsampled point x will be given by :

$$P^{DK}(N_x = j) = W_j \sum_{n=0}^{N} w_n H_n^{DK}(N_x) H_n(j) \tag{39}$$

$$H_n^{DK}(N_x) = \sum_{\beta} \lambda_n^{\beta} H_n(N_\beta)$$
$$\lambda_n^{\beta} \rho_{\alpha\beta}^{n} = \rho_{\alpha}^{n} x \qquad \forall \alpha = 1, \nu \tag{40}$$

The binomial anamorphosis of Zx is the vector $\phi(i) = \{\bar{Z}_i\}$ This function will be expressed as a Hn expansion of the type :

$$\phi(i) = \sum_{n=0}^{N} w_n \psi_n H(i) \tag{41}$$

with $\psi_n = \sum_{i=0}^{N} W_i \, \phi(i) H_n(i)$ $\qquad \phi(i) = \{\bar{Z}_i\} \implies \psi_n = \sum_{i=0}^{N} W_i \, \bar{Z}_i \, H_n(i)$

So, the proportion of ore and the metal recovered above the cut-off z_i will be :

$$T_{z_i}(x) = \sum_{j=1}^{N} P_j^{DK}$$

$$Q_{z_i}(x) = \sum_{j=1}^{N} \bar{Z}_j \, P_j^{DK} \tag{42}$$

In practice, discrete D.K. implies about the same kind of calculation as continuous D.K. : the main advantage of the method is the possibility to handle without problems a grade distribution with a spike (in particular the bi-binomial model for point and block grades provides a change of support formula for the proportion of zero grades in blocks). MATHERON has studied ultimately (1983) an isofactorial model with nega- tive binomial distribution which provides a more flexible change of support formula and seems well appropriate to such discrete distribution as the number of stones/m^3 in a diamond orebody.

6.4. Multivariate gaussian conditional distribution

The use of the conditional distribution to estimate the point recovery was suggested early (MARECHAL 1974, MATHERON 1974, PARKER, SWITZER 1976) and actually put in practice in the lognormal case by PARKER (1975) and then by PARKER, JOURNEL, DIXON (1979), JOURNEL (1980). The theoretical formula for the general case can be found in MARECHAL (1982), VERLY (1982).

After a suitable transformation $Y_x = \phi^{-1}(Z_x)$ the point grade gaussian equivallent R.F Yx is assumed to be gaussian multivariate. Then, in a given point x, the recovered tonnage and metal are :

$$T_z(x) = 1 - G\left(\frac{y - \lambda^\beta_x y_\beta}{\sigma^x_\kappa}\right) \quad ; \quad y = \phi^{-1}(z) \quad u = \frac{y - \lambda^\beta_x y_\beta}{\sigma^x_\kappa}$$

$$Q_z(x) = \int_u^{+\infty} \phi(\lambda^\beta_x y_\beta + \sigma^x_\kappa v)\, g(v)\, dv \qquad (43)$$

At first glance, a major disadvantage of this solution is that Tz(x) and Qz(x) do not depend linearly of x, as whith Indicator kriging or D.K, so that the recovery in a panel should be evaluated point by point : it was shown (JOURNEL 1980, VERLY 1982) that the integral of the functions of formula (43) within a panel V can be approximated in a very satisfying way by the arithmetic mean of such functions computed on a small number of sample points. If a set of 4 x 4 points in the panel V is used, resolving a unique kriging system provide directly the quantities $\lambda^\beta x$ and σ^x_κ for all these points ; Tz(x) and Qz(x) are then computed for each of the 16 points and averaged to provide the recovery estimation of the panel. A way to avoid the computation of Tz(x) and Qz(x) on these 16 points was proposed by MATHERON (1974) and tested by GUIBAL and REMACRE (1983) under the name of uniform conditioning : it consists in using the conditional distribution of a point x at random in V, conditioned by the kriging estimator of the mean of the gaussian equivallent in the panel V. This local distribution is gaussian, so that the method appears as a generalization of the lognormal local distribution approach, whith the same pros and cons.

The use of any gaussian model is based on the definition of the anamorphosis function Zx = ϕ(Yx). A standard practice to transform Z_a in Y_a is to use the rank r_a of each Z_a (with possibly a special treatment for Max Z and Min Z).

$$Y_a = G^{-1}\left(\frac{r_a}{N}\right) \qquad (44)$$

We see that the tonnage recovered Tz(x) can be computed directly from the Y_a, y_c and the ordered sequence of Z_a. Let ri be the rank of data Z_i, such that its transformed value Yi is :

$$Y_i = \lambda^\beta_x y_\beta + y\,\sigma^x_\kappa$$

Then $\quad T^*_z = \frac{r_i}{N}$

In the same way, the mean recovered grade Q_z / T_z will be estimated as the mean of all the samples Z_a such that Z_a > Zi

An alternative solution consists in fitting an Hermite expansion to the point anamorphosis function Φ and then make use of this representation to compute exactly Q_Z as given in (13). The exact formula can be found in (MARECHAL 1982). Such a practice is apparently less convenient than the above one : however it is necessary in the case of estimation of block recovery using a change-of-support formula acting on the anamorphosis expansion.

As a conclusion, the property of the multigaussian conditional distribution to be a shifted gaussian distribution allows a very simple determination of the recovery in each point which then can be easily averaged over the panels V.

6.5. The problem of non-stationarity

All the estimators reviewed above are based on the knowledge of the p.d.f. in any two points : such model can be non-stationary, which means that the p.d.f. will be a function of the location points x, x + h.

Such models are not used in practice because of the impossibility to infer their caracteristics. However, there are practical situations when it is not possible to assume the stationarity. Two such situations remain practically in the scope of the methods seen above :

$Z(x) = m(x) + Z_1(x)$ the trend m(x) has a clear physical meaning and its mathematical form $a_l f^l x$ can be precisely estimated by universal kriging. Then the trend is included in the model and the recovery estimation is performed on the variable $Z_1(x)$

$$I_z(Z_x) = I_{z-m(x)}(Z_1(x))$$

. Z(x) is locally stationary, within a convenient vicinity of each point. This definition applies not only to the covariance function, but also to the marginal distribution $F_x(dz)$. In such a case, MATHERON (1978) has shown that the D.K. recovery estimator can be made unbiased whatever the unknown distribution F(dz) or anamorphosis Φ by adjoining universality conditions to the D.K kriging system. Unfortunately, it was found (MARECHAL, TOUFFAIT 1980) that such additional conditions increased the abnormal aspect of the D.K density, making it unpracticable for estimating a density : it can be usefull as a non-linear estimator of the grade itself.

In the frame of the multigaussian model, it is possible to replace the S.K estimator $Y^*_{SK} = \lambda^\alpha y_\alpha$ of the mean of the conditonal distribution by an O.K estimator Y^*_{OK} defined in the immediate vicinity of x. Unfortunately the relation * is no longer valid, the variable $Y_{OK} + \sigma_{OK} U$ when deconditioned having a variance larger than 1 : the method can be used however

$* \; Y|_{Y_1 \ldots Y_N} \overset{L}{\sim} Y^*_K + \sigma_K U$

if the μ_K parameter of the O.K system is small (the information is close to the panel). The various recovery estimators listed above have good numerical properties only when they are close to a true conditional distribution. Then, the solution for recovery estimation in the non-stationary case must be searched in fitting local stationary models and not in adding universality conditions to the kriging systems.

An interesting example of such practice is proposed by PARKER, JOURNEL, DIXON (1979) : they assume that the grade is locally lognormal, within each subarea of the orebody.

$$Z(x) = \exp(m_A + \sigma_A Yx)$$

The caracteristics m_A, σ_A of the model for each subarea are fitted using cross-validation on the local data.

The method can be generalized following a two-steps transform :
. a first transform $\phi^{-1}(Z_x)$ makes the Y globally gaussian : ϕ^{-1} may be for instance simply log Z_α.
. within each subarea of the orebody, Y_x is assumed to be a non-standard normal R.F such that : $Y_x = m_A + \sigma_A Y'_x$

The local model could be easily fitted using indicator kriging as an estimator of the histogram within the subarea or directly as an estimator of m_A and σ_A. The Y'_α are then supposed to be multivariate gaussian or bivariate according to whether a conditional distribution or a D.K. estimator is used.

The interest of the method is that both the conditional distribution parameters $\lambda^\alpha y_\alpha$, σ_K and the Hermite expansion of ϕ behave well with respect to a linear shift of the argument. For instance, if we use O.K. for determining the conditional distribution, then the recovery functions given in formula (43) remained unchanged with this model, $\lambda^\beta_x y_\beta$ being the O.K estimator based on the y_β and $\sigma_K = \sigma_A \rho_K$ where ρ^2_K is the kriging variance derived from the use of a correllogram function

As a consequence, the use of local parameters m_A and σ_A will only imply a "proportional effect" change in the final formula, without implying any modification of the kriged estimate Y^*_x. In the case of D.K estimation, the sample gaussian equivallent will have to be modified localy before computing the H_n^{DK}

$$Y'_\alpha = \frac{Y_\alpha - m_A}{\sigma_A} \Longrightarrow H_n^{DK} = \lambda^\beta_n H_n(Y'_\beta)$$

The expansion coefficients of the local anamorphosis $\phi'(Y') = \phi(m_A + \sigma_A y')$ are easy to compute (MARECHAL 1982, formula 3).

7. BLOCK RECOVERY ESTIMATION AND THE CHANGE-OF-SUPPORT PROBLEM

7.1. Recalls on definition

The ore/waste selection is done when mining by applying the cut-off criterion to an estimate Z*v of each unit selection block. It has been shown (MATHERON 1976) that provided this estimator is (approximatively) conditionally unbiased, the recovery functions Tz and Qz (termed "indirect" in this case) can be derived from the conditional distribution (or an estimate of) of the estimator Z_v^*: $F_v^* = P(Z_v^* > z | z_1 \ldots)$. For instance, the indicator function of Z_v^* will be estimated as :

$$I_z^*(v) = \int_z^{+\infty} f(z_v^* \; z \ldots z) \, dz_v^* \tag{44}$$

The exact determination of f $(z^*v \; z_1 \ldots z_n)$ is impossible in practice, for the way the estimator Z_v^* will be computed at the mining stage is not known at the exploration stage : an approximation have to be done, considering z_v^* as a kriging estimator of given estimation variance. (In other term, one must anticipate the amount of information available at he mining stage).

For the brevity of this review, the case of direct recovery will be mainly examined.

7.2. Model for computing a block conditional distribution

For determining the conditonal distribution of the block grade Zv, we may consider the formula seen above in section 3.3 : A (N + 1) p.d.f is given for the R.V. $(Zv, Z_1 \ldots Z_N)$ which includes implicitly the conditional distribution $F_v^N(dz/z_1 \ldots z_N)$. Notice that the determination of the conditional distribution calls for not only for a block grade model Fv(dz), but also for the joint p.d.f. between Zv and the information $Z_1 \ldots Z_N$.

Deriving a model Fv(dz) for block grade from the sample distribution of point grade is resolving the change-of-support problem. The resulting distribution Fv(dz) can be used for defining the global recovery functions of the whole orebody whenever the information is evenly covering the orebody. However, for local recovery estimation, it is necessary to define more elements of the model than only Fv(dz).

7.3. Model needed for block indicator cokriging

Following the same arguments as in section 3.2., block recovery estimation amounts to estimate the indicator function :

$$\begin{aligned} I_z(v) &= 1 \qquad Z_v \geq z \\ I_z(v) &= 0 \qquad z > Z_v \end{aligned} \tag{45}$$

This function can be estimated by a complete cokriging, using the indicator function of the sample grades : a minimum prerequisite is the expectation of the target R.V, 1 - $F_v(z)$:

$$I_z^*(v) = 1 - F_v(z) + \sum_i \sum_\beta \lambda_v^\beta(z_i) [I_{z_i}(x_\beta) - (1 - F(z_i))] \tag{46}$$

As usual, the cokriging system is based , on the cross-covariances $\sigma_{z_i z_j}(h)$ of point-indicator and on the cross-covariances $\sigma_{z_i z}(v, x_a)$

$$\sigma_{z_i z}(v, x_a) = P(Z_v \geq z, Z_a \geq z_i) - [P(Z_v \geq z)]\,[P(Z_a \geq z_i)]$$

For local block recovery estimation, a change-of-support model must be found together with a joint probability model for the block grade Zv and each sample grade Z_α.

7.4. The change-of-support problem

In theory, $F_v(z)$ can be derived from the spatial law of Z(x) for all the points within the block v. Hence in general, $F_v(z)$ cannot be linked to the point p.d.f F(z) only, and deriving Fv(z) requires a complete probability model. Defining such models allows exceptionally a mathematical determination of Fv(z), for mere reasons of calculus : MATHERON(1981) has studied some particular distributions and made numerical comparisons between the approximate formula in use in geostatistics and the real theoritical result.

A complete gaussian tranformed model allows the determination of Fv (dz) at least after discretization of v in N blocks. This allows to check, for instance, that the block grade in a point lognormal distribution in no longer strictly lognormal. (Although experimentally the lognormal permanence is remarkably checked in many cases).

In practice, the change-of-support operation will consist in defining a relation between the point grade p.d.f. $F_x(dz)$ and the block grade p.d.f Fv(dz). This operation must satisfy a certain number of constraints derived from the stationary hypothesis and the Cartier therorem for p.d.f (MATHERON 1981b).

a. Fv(dz) and $F_x(dz)$ have the same mean, m.
b. The variance σ_v^2 associated with Fv(dz) must be equal to the value given by integration of the variogram of Zx :

$$\sigma_v^2 = \sigma^2 - \gamma(v, v) \tag{46}$$

c. The interval of definition of Zv must be identical to that of Zx.
d. Defining the operating benefit yielded by a grade distribution F(dz) selected at the cut-off level zc as :

$$B_{z_x}(z_c) = \int_{z_c}^{+\infty} (z - z_c)\,F(dz)$$

The benefit yielded by a selection on block must be lesser or equal to the benefit yielded by a selection on point grades, for all zc (support effect).

$$B_{Z_v}(z_c) \leq B_{Z_x}(z_c) \quad \forall z_c \tag{47}$$

7.5. General method for the change-of-support problem

The basic method for finding change-of-support relations consists in starting from the following condition (MATHERON 1976). Let us consider a block v and any point x inside (they could be also smaller block v' covering v). Conditionally to Zv having a given value z, the mean of all point grades of v must be z.

$$\frac{1}{V}\int_v E(Z_x \mid Z_v = z) = z \tag{48}$$

If we randomize uniformly the location x in V and define Zx as the R.V Zx when x is at random, (48) reads :

$$E(Z_{\bar{x}} \mid Z_v) = z_v \tag{48 bis}$$

Different change-of-support relations can be derived by the following method by considering different isofactorial models : For any continuous p.d.f G(dy) of density g(y), both Zx and Zv can be transformed to a R.V of p.d.f G(dy) using two different anamorphosis functions :

$$Z_x = \phi(Y_x) \qquad\qquad Z_v = \phi_v(X_v) \tag{49}$$

These two functions can be expanded with regard to the orthogonal family $\chi_n(y)$ related to g(y).

$$\begin{aligned} \phi(y) &= \sum_0^\infty \phi_n \chi(y) \\ \phi_v(y) &= \sum_0^\infty \phi_n d_n(v) \chi_n(y) \end{aligned} \tag{50}$$

We assume the existence of two isofactorial models attached to g(y). One links the R.V observed in any pair of points x_1, x_2 :

$$f(Y_{x_1}, Y_{x_2}) = \sum_{n=0}^\infty T_n(x_1, x_2)\chi_n(Y_{x_1})\chi_n(Y_{x_2})g(Y_{x_1})g(Y_{x_2}) \tag{51}$$

The second one links the R.V observed (after tranformation) in any point x and block V :

$$f(Y_x, X_v) = \sum_{n=0}^\infty T_n(x, v)\chi_n(Y_x)\chi_n(X_v)g(Y_x)g(X_v) \tag{52}$$

If x is considered at random in V, then (52) reads :

$$f(Y_{\bar{x}}, X_v) = \sum_{n=0}^\infty \overline{T_n(v, v)}\chi_n(Y_{\bar{x}})\chi_n(X_v)g(Y_{\bar{x}})g(X_v)$$

So, the conditional distribution of $Y_{\bar{x}} \mid X_v$ comes :

$$f(Y_{\bar{x}} \mid X_v) = \Sigma \overline{T_n(v, v)}\, \chi_n(X_v)\, \chi_n(Y_{\bar{x}})\, g(Y_{\bar{x}}) \qquad (53)$$

Applying the relation (48) and equating identical terms in the expansions leads to the following result :

$$d_n(v) = \overline{T_n(v, v)}$$

By expressing directly $Z_v = \phi_v(X_v)$ by integration in V, we find a second relation :

$$d_n(v)\, T_n(v, x) = \frac{1}{v} \int_v T_n(x, y)\, dy$$

Hence finaly two results of importance :

$$d_n(v) = \sqrt{\overline{T_n(v, v)}} \qquad (54)$$

$$T_n(x, v) = \overline{T_n(v, v)} / d_n(v) \qquad (55)$$

(54) provides a change-of-support formula, while (55) links any bivariate distribution (Zx, Zv) to the initial joint bivariate distributions (Zx, Zy) : by construction of these two relations, the change of support conditions seen above are satisfied :

$$E(Z_v) = E(Z_x)$$

$$Var(Z_v) = \Sigma(0) - \overline{\Sigma(v, v)} \qquad \Sigma(h) = Cov(Z_{x+h}, Z_x)$$

$$\Sigma(h) = \sum_1^{\infty} \phi_n^2 T_n(h) \quad T_n(h) = Cov(Y_{x+h}, Y_x) \qquad (56)$$

. Unfortunately, it cannot be proved that $\phi_n d_n(V)$ will -necessarily be an anamorphosis function, nor Tn (x, V) corresponds always to a true p.d.f.

However, two popular change-of-support formula are derived from this formulation, namely the "affine correction" and the Hermite change-of-support.
Indeed, if we assume that Zx is totally gaussian, we have then :

$$Z = m + \sigma Y \qquad (57)$$

$$z = m + \sigma_v X$$

The change-of-support formula reduces to do = 0, $d_1 = \sigma_v/\sigma$, dn = 0 n > 1. The distribution (Zx, Zv) is bigaussian, with covariances parameters given by the standard geostatistical formula. From (57), it comes :

$$\frac{Z_v - m}{\sigma_v} \approx \frac{Z - m}{\sigma} \qquad (58)$$

The two normalized variables Zv and Zx have the same p.d.f, indeed a standard gaussian p.d.f.
Now, we assume that Zx an Zv are transformed to gaussian :

$$Z_x = \phi(Y_x) \qquad Z_v = \phi_v(X_v)$$

Then, the above formula (52), (53) define Hermitian bivariate distributions and (54) define the Hermitian change-of-support. Some unsatisfying aspects of the Hermtian model noticed above prompted MATHERON (1976, 1981, 1983) to imagine a slightly different method : it consists in providing from the start both types of joint p.d.f, i.e the point-point distribution and the point-block distribution. A famous example of such practice is found in the Discrete Gaussian model (1976).(in the sequel D.G. model).
The basic R.V considered here are block grades Z_{vi} of all the selection blocks of the orebody, and the sample grades Zx. We assume that both types of variable are related to gaussian transformed equivallent variables :

$$Z_x = \phi(Y_x) = \sum_{n=0}^{\infty} \frac{\psi_n}{n!} H_n(Y_x)$$

$$Z_v = \phi_v(X_w) = \sum_{n=0}^{\infty} \frac{\psi_n \, d_n(v)}{n!} H_n(X_v)$$

The basic hypothese now is that, any pair Y_α, Y_β are bigaussian and also any pair Y_α, X_{v_i}, with correlation coefficients $\rho_{\alpha\beta}$ and $r_{\alpha i}$ respectively. For such an assumption to be valid, one has to consider each point sample value Z_α as located at random within a small block v_i, so that the dependance between any point transformed grade Y_α and its "companion block" transformed grade X_{v_i} will be fixed by a unique parameter r = cov (Yx, Xv).
Applying condition (48) on the above model leads directly to the D.G. change-of-support formula :

$$d_n(v) = r^n$$

and the parameter r can then be fixed by the condition of variance of Zv :

$$\mathrm{Var}[\phi_v(X_v)] = \overline{\Sigma(v,v)}$$

$$\sum_{n=1}^{\infty} \frac{(\psi_n r^n)^2}{n!} = \overline{\Sigma(v,v)} \tag{60}$$

The relation (60) easy to resolve numerically, defines a unique value $o < r < 1$.
As recalled above, a by-product of the basic hypothesis is to define completely the whole set of bivariate laws existing between the R.V Y_α, Y_β , Y_α, X_{v_i} , and even X_{v_i}, X_{v_j} .

It can be shown that the whole model is derived from the following quantities :

. the anamorphosis function ϕ .

. the change-of-support parameter r.

. the covariance function R(h) between the block transformed grade (Xv, Xv+h).

The D.G. model can be considered too as a complete multigaussian model between all the Y_α, X_{vi} allowing the computation of block-recovery by conditional distribution. Two other examples of this mecanism of block model determination can be found in MATHERON (1981) and (1983) for discrete distributions, allowing to handle a change-of-support for distributions with a large proportion of zero values.

7.6. Comparison between the practice of different change-of-support methods

The two major change-of-support formula currently in practice are the Affine Correction and the D.G. (otherwise termed generalized lognormal permanence). Various checks have been done with production data (MARECHAL 1975, GUIBAL, REMACRE 1983, DAVID, 1977) but few direct comparisons exist between both methods (MUGE 1982), and only one exists comparing theoreticaly exact results and change-of-support methods (MATHERON 1981). In general, for global recovery estimation purposes, for small change-of-support and not heavy-tailed distributions, the results of recovered tonnage obtained by the two methods do not differ considerably, but difference may appear in recovered metal.

. The major advantage of Affine Correction is simplicity : the method can be used without any mathematical modeling of the point grade histogram. Global recovery estimation can be obtained directly form the ordered sequence of sample values.

As $\frac{Z_v - m}{\sigma_v} \simeq \frac{Z_x - m}{\sigma}$

Hence

$$P(Z_v > z) = P(Z_x > \frac{\sigma}{\sigma_v}(z-m) + m) \qquad (61)$$

So $I_z^*(v) = \frac{r_\kappa}{N}$, r_κ rank of Z_κ such that

$$Z_a = \frac{\sigma}{\sigma_v}(z-m) + m$$

. The Affine Correction honors the condition of decreasing benefits for selection on large blocks.

The Affine Correction is exact for gaussian distribution and good for quasi-gaussian distributions. It preserves lognormality, but in a rather strange way. Let Zx be any 3 parameters lognormal.

$$Z_x = z_M + \epsilon \exp(m + \sigma Y_x)$$

$$Z_v = M + (z_M - M)\Sigma_v/\Sigma \cdot (\Sigma_v/\Sigma)\exp(m + \mathrm{Log}\,\Sigma_v/\Sigma + \sigma Y_x) \qquad (62)$$

This example emphasizes one of the drawbacks of the method, which is to alter the domain of definition of Zv. In particular, it gives a probability zero for block grades to be zero even when point grades have a spike for zero. Last, the affine correction should be avoided for large variance reduction, the distribution not tending to a gaussian limit distribution.

The D.G. permanence requires an Hermite expansion of the anamorphosis function (except in the normal/lognormal case) : this expansion is currently computed on the ordered sequence of data, which allows a good restitution of the histogram and of the experimental variance. The determination of the parameter r is very easy too, so that the whole calculation can be done with a table calculator. The transformed distribution fulfills all the above mentioned conditions : in addition, it preserves lognormality without altering the limits of definition. At the end, we may include in the D.G. permanence practice the non-continuous version presented above, based on binomial or negative binomial model : this allows processing any type of distribution continuous with spike or discrete.

Comparing the D.G. approximation with the real results for some special cases, MATHERON (1981) shows :

- The D.G. approximation is good as soon as the point grades within the block are in good correlation (a large nugget-effect would damage the quality of the fit).
- A distribution with a spike for zero can be correctly processed provided the frequency attached to zero values is small (say 10 %).

As a conclusion, the D.G. formula (or the discontinuous equivallent) is a convenient and precise approximation for the global recovery estimation. In many situations, the Affine Correction performs well too, although presenting some theorretical inconvenients : then the latter is a good solution for a quick global calculation. However, in practice, recovery estimation will be performed locally and for this problem the affine correction does not provide a satisfactory solution.

7.7. Change-of-support for local recovery estimation

We recalled previously the two possible approaches for estimating the local block recovery : either determining the conditional block grade distribution through a complete probabilistic model or estimating such distribution by a complete cokriging.

Determining the conditional block grade distribution presumes that a change-of-support formula was used to define the block-grade model and to define the point probabilities ($Zv, Z_1 \ldots Z_n$). The final result would be for a given block, a conditional p.d.f :

$$F_v(z|z_1 \ldots z_N) = P(Z_v < z | Z_1 = z_1 \ldots Z_N = z_N)$$

What are the caracteristics of this p.d.f ? (63)

. The expectation is $E(Z_v | z_1 \dots z_N) = \frac{1}{V} \int_V E(Z_x | z_1 \dots z_N)\, dx$
The mean of this distribution is the average in the block of the point condititional expectation.

. The variance is the conditional "estimation variance" of Z_v by the sample values $z_1 \dots z_N$: $\sigma^2(Z_v | z_1 \dots z_N)$

It is appealling to the mind to replace, as an approximation, the unknown p.d.f. $Fv(dz\ z_1 \dots z_N)$ by $F_x(dz\ z_i \dots z_N)$ conveniently transformed to get the right variance, that is using a "local" affine correction : this method is proposed by JOURNEL (1983). Indeed, $Fx\ (dz/\ z_i \dots z_N)$ average point conditioned distribution for x varying in v has the correct expectation : there remains only to approximate the variance term

. It is easy in the gaussian case (only case where the affine correction is exact) to determine the value of $\sigma^2(Z_v | z_1 \dots$

$$\sigma^2(Z_v | z_1 \dots z_N) = \sigma_v^2 - \lambda_v^a \sigma_{av} \text{ close to:} \tag{64}$$
$$\sigma^2(Z_{\bar{x}} | z_1 \dots z_N) = \sigma^2 - \overline{\lambda_x^a \sigma_{ax}}$$

JOURNEL proposes to derive $\sigma^2(z_v | z_1 \dots z_N)$ from the equivallent term computed in point recovery $\sigma^2(z_{\bar{x}} | z_1 \dots z_N)$ which in the gaussian case amounts to approximate $\lambda_v^a \sigma_{av}$ by $\overline{\lambda_x^a \sigma_{ax}}$:

$$\sigma^2(Z_v | z_1 \dots z_N) \# \sigma_v^2 - \sigma^2 + \sigma^2(Z_{\bar{x}} | z_1 \dots z_N)$$

with : $(\overline{\lambda_x^a})(\overline{\sigma_{ax}}) \# \overline{\lambda_x^a \sigma_{ax}}$ (65)

In the gaussian case, the quantity $\sigma^2(Z_{\bar{x}} | z_1 .$ is independant of the $(z_1 \dots z_n)$ and is easily computed from the variogram : in the general case, $\sigma^2(Z_{\bar{x}} | z_1 .$ must be computed from the estimated point recovery function and will show large numerical variations so that it is not sure that the variance computed in (65) will always be positive.

Recalling that $\sigma^2(Z_v | z_1 .$ can be considered as the conditional estimation variance of Zv, this quantity can be approximatly derived from the non-conditional variance through a "proportioned effect" relation : this is basically the way a conditional estimation variance was computed in the early methods reviewed above (MARECHAL 1972 a, DAVID 1977). In a different way, JOURNEL (1983) proposes to modify the point conditional variance $\sigma^2(Z_{\bar{x}} | z_1 .)$ by σ_v^2/σ^2 the ratio of the a priori variances (this at least ensures that the result is positive). Then the conditional block distribution is given by :

$$Z_v^* | z_1 \dots z_n \simeq Z_v^* + \sigma_{\kappa x} \frac{\sigma_v}{\sigma}(Z_{\bar{x}}^* - Z_v^*) \tag{66}$$

Formula (66) shows that the substitute $Z^*_v | (z_1 \dots z_n)$ for $Z_v | (z_1 \dots z_n)$ when deconditioned has a correct mean and variance but the deconditioned distribution cannot be equivallent to the global affine correction : $Z_v \simeq m + \frac{\sigma_v}{\sigma}(Z_x - m)$

Then the affine correction is inconsistant when applyed at the global level or at the local level : the recovered tonnage and metal estimated in both ways may not be equal.

On the other side, the D.G. model provides a consistent model to derive, either by D.K. or conditional distribution, an estimation of the local block grade histogram. In addition, the nature of the D.G. model allows to derive the histogram of various block sizes from a unique point histogram : indeed, the two basic parameters of the model, namely r and R(h) are tied to the size of v, but in practice R(h) is nearly invariant within a given class of sizes for v, especially in presence of a nugget effect for $\gamma(h)$.

Let us detail the example of the conditional distribution. In terms of transformed variable, we look for the distribution :

$$X_v | y_1 \dots y_N \doteq \lambda_v^\beta y_\beta + \rho_{Kv} U$$

with

$$\lambda_v^\beta \rho_{\alpha\beta} = r R_{\alpha i}$$

$$\rho_{Kv}^2 = 1 - r \lambda_v^\beta R_{\alpha i} \quad (67)$$

We know that $\rho_{\alpha\beta} = \delta_{\alpha\beta} + (1 - \delta_{\alpha\beta}) r^2 R_{\alpha\beta}$ let $S^{\alpha\beta}$ be the inverse matrix of $\rho_{\alpha\beta}$. For the reasons quoted above and because $\rho_{\alpha\beta}$ is diagonal dominant the vector $S^{\alpha\beta} R_{\alpha\beta}$ is nearly invariant to changes of r. Then, defining $\lambda_i^\beta = S^{\alpha\beta} R_{\alpha i}$ it comes :

$$X_v | y_1 \dots y_N \doteq r(\lambda_i^\beta y_\beta) + (1 - r^2 + r^2 \rho_{Ki})^{1/2} U \quad (68)$$

The recovery functions for various block sizes will be obtained from the same basic kriging results $\lambda_i^\beta y_\beta$, ρ_{Ki}^2, changing only the value of r : a similar simplification exists in the case of D.K. estimation.

There exists an other method to relate the block recovery to the point recovery estimation, different of the affine correction : it consists in drawing from the D.G.. model a regression formula between the block grade and the point grade (BOUCHIND'HOMME 1980) By definition of the block-point grade and point grade in the D.G. model, we have :

$$Z_v | Z_x = y \doteq \phi_v(ry + \sqrt{1-r^2}\,U) \quad (69)$$

So we can attach to each sample grade z_α a new service-variable which will be $U_\alpha = E(Z_v | Z_\alpha) = E[\phi_v(rY_\alpha + \sqrt{1-r^2}\,U)$ (70)

The function defined in (70) can be computed for each sample point, so that all the block recovery problem can be stated in term of indicator estimation of the new service variable U_α : indeed when production data are available, it is possible to fit experimentally the regression curve defined in (70), avoiding the use of any model.

7.8. Indirect block recovery estimation

All the methods used to compute indirect block recovery are based on the following two assumptions :

- the final block estimator Z^*_v is assumed to be conditionally unbiased.
- it is possible at the exploration stage to assign an estimation variance $\sigma^2(Z^*_v)$ to this estimator.

These hypothesis are sufficient when using an affine correction, but have to be completed in the case of the D.G. model whith an hypothesis about the correlation between Z^*_v and the present information. However in practice, nothing is changed to the estimation methods seen above, a part from changing the fit of such parameters as r* in the D.G. model.

It must be emphasized that considering an indirect recovery is very important in practice. Because the information drawn from blast holes is often not precise, it is impossible to consider that the selection is done on real grades. Whenever operational results are known, it will be important to calibrate the parameter linked to Z^*_v using the observed recovery : in a study of a massive orebody already partialled mined, JACKSON and MARECHAL (1979) had to make vary the parameter r of the D.G. model to find a value giving a good match between estimated and actually recovered tonnages : the value r thus fitted corresponds to volumes larger than the real selection unit.

7.9. Estimation variance of recovery estimators

The situation is very different for global and local recovery estimation.

The tonnage and quantity of metal in point recovery can be considered as the average, for the whole ore-body, of the service variables Iz(x) and $Z_x I_z(x)$: after fitting the experimental variograms of these two variables, the estimation variances can be computed without difficulties using the standard approximation for global estimation variances. In the case of block recovery, we are faced with the need for adopting a change of support formula, the precision of which cannot be judged theoritically. With an affine correction, Iz(v) and ZvIz(v) can be linearly related to the point variables, so that the estimation variances will be derived from the variances computed in the point recovery case. With a D.G. model, $U_\alpha = E(Z_v | Z_\alpha)$ can be computed for each sample point, so that the service-variable giving directly the tonnage and metal recovered on block can be studied, their variogram fitted and estimation variances computed.

- Experimental checks have shown that usually a local recovery estimation is fairly imprecise at the level of each panel, especially for the quantity of metal at high cut-off values. The precision is better expressed in term of conditional variance : this quantity is theoritically accessible only when using the conditional distribution approach in the D.G. model. Other estimators (D.K., I.K., etc.) provide only as unconditional variance, which would have to be modified by an "ad-hoc" method such as proportional effect, etc.

8. CONCLUSION

Let us recall two major observations :

. Estimating recoverable reserves is of uppermost economical importance, because it is the only way to account for the two major factors influencing the selectivity : the support effect and the information effect. It is only by acknowledging these effects that geostatistics was able to resolve correctly the "problem of vanishing tons" (to quote M. DAVID).

. In many orebodies, the average grade estimation of each individual panel is not very precise : what about estimating the local point grade histogram !

It results from the two above points that performing an effective recovery estimation deserves a special effort :

. necessity to use somewhat sophisticated methods based on well defined probability models. Such methods exist and have been reviewed in this paper : they are not difficult to understand if correctly taught to the right people, and the softwares necessary for applying them exist or can be easily developped.

. necessity to devote more human time and thinking to fit and check the models. It has been noticed that, the more complete the model, the more satisfying the estimator : but it is known too, that the more precise the model, the more likely the model is wrong !

Consider the example of the Petroleum Industry : for such important problems as seismic data processing or reservoir production simulations, physically complex models are used, involving high level mathematics and resulting in large softwares of high C.P.U. consumption. There are no reasons why one of the major challenge of the modern Mining Industry, namely estimating, efficiently planning and operating big selective mine, would not be considered with the same attention as oil men consider the acquisition and processing of exploration and production data. Consequently, the future for recovery estimation methods lies in developping more or less complete probability models adapted to various particular cases, together with developping the methods for fitting and validating these models. Example of such cases are the negative binomial model adapted to diamonds estimation, the 3 parameters lognormal adapted to the south-african gold orebodies, or the gaussian transformed model largely used for estimating disseminated massive orebodies. The estimation methods corresponding to theses models exist : they are the conditional distribution and disjunctive kriging. It would be very important, in addition, to develop the simulation methods corresponding to theses new non-gaussian models.

There is a need, at the exploration stage, to appreciate globaly for large units the result of selective mining at different fixed cut-off grades. Indicator kriging based on the panel's grade kriging weights will provide a quick, unexpensive initial estimation of the recovered reserves : the method is already routinely used at the global level for declusterized histogram determination and it could be systematically associated with any panel's kriging program. An important by-product of this practice would be to provide information for fitting local models in case of nonstationarity. At this first level of global estimation, when simple unsphosisticated methods are required, the change of support can be equivallently performed using an affine correction or a discrete gaussian method.

Finally, one must not forget that some selection methods are fairly complex and cannot be idealized as a free direct selection : there are no alternative than, simulating the mining method on a simulation of the grade distribution. As emphasized above, there is a close link between the models to be used in recovery estimation and in conditional simulation.

ACKNOWLEDGMENTS

A grateful acknowledgment is made to the management of the Societe Nationale ELF-AQUITAINE (Production) for allowing time to the author for writing this paper, the matter of which was mainly gathered during the last ten years when the author participated to the develop; ment of recovery estimation together with Professor Georges MATHERON at the Centre de Geostatistique de FONTAINEBLEAU.

REFERENCES

NB : CGMM refers to Centre de Geostatistique et Morphologie Mathématiques, 35 rue Saint Honoré - 77300 FONTAINEBLEAU
APCOM refers to the various conferences named : "Application of Computer methods to the Mining Industry".

1. BECKMANN P. (1973) : "Orthogonal polynomials for engineers and physicist" Cohen Press, Boulder, Co.
2. BOUCHIND'HOMME JF (1980) : "Estimation de l'Uranium récupérable sur les gisements sédimentaires stratiformes exploitables à ciel ouvert" Doctor Ing. Thesis University of Nancy, INPL.
3. DAVID M. (1972) : "Grade tonnage curve, use and misuse in ore reserve estimation. Trans. Inst. Min. Metall. p : 129, 132.
4. DAVID M.(1973), DAGBERT M. and BESLILES JM. (1977) : "The practice of porphyry copper deposit estimation for grade and ore-waste tonnage demonstrated by several case studies". Pro. 15th APCOM, AUSTRALIAN I.M.M. Parkville.

5. DAVID M. (1977) : "Geostatistical ore reserve estimation" Elsevier 1977. p : 313, 320.
6. GUIBAL D., REMACRE A. (1983) : "Local estimation of recoverable reserves : comparing various methods witch the reality on a porphyry copper deposit" to be published on the proceeding, NATO-ASI TAHOE-83.
7. JACKSON M., MARECHAL A. (1979) : "Recoverable reserves estimated by disjunctive kriging : a case study" Proceeding, 16th APCOM, S.M.E-AIME, New-York.
8. JOURNEL A. (1973) : "Le formalisme des relations ressources-reserves, Simulations de gisements miniers", Revue de l'Industrie Minérale, section Mines 4. p : 214, 226.
9. JOURNEL A. (1980) : "The lognormal approach to predicting local distributions of selective mining unit grades" Mathematical Geology, Vol 13.
10. JOURNEL A. (1982) : "Indicator approach to spatial distributions" Proc. of 17th APCOM, Denver, published by AIME
11. JOURNEL A. (1983) : "Nonparametric estimation of spatial distributions" Math. Geology, Vol 15 n° 3.
12. JOURNEL A, HUIJBREGTS Ch. (1978) : "Mining geostatistics" Academic Press, London.
13. MARBEAU JP., MARECHAL A. (1980) : "Geostatistical estimation of uranium ore reserves". Invited review paper in "Uranium evaluation and mining techniques" IAEA Vienna 1980
14. MARECHAL A. (1972a) : "El problema de la curva tonelage-ley de corte y su estimation" Boletin de geostatistica, Vol 1 (May 1972). Universidad de Chile, Satiago.
15. MARECHAL A. (1972b) : "El problema de la estimation de la distribution local de leyes, Bol Geostatistica Vol 4, n° 72. p : 55, 69.
16. MARECHAL A. (1974) : "Généralités sur les fonctions de transfert" Internal Report M-384 CGMM.
17. MARECHAL A. (1975) : "Analyse numérique des anamorphosees gaussiennes. Internal Report CGMM 1975.
18. MARECHAL A. (1975): "Forecasting a grade-tonnage distribution for various panel sizes" Proceedings, 13th APCOM Clausthal, R.F.A.
19. MARECHAL A., TOUFFAIT Y. (1980) : "Recovery estimation of non-stationary orebody using disjunctive kriging". Proceedings, Computer methods for the mining industry Meeting Moscou 1980.
20. MARECHAL A. (1982) : "Local recovery estimation for co-products by disjunctive kriging". Proceeding of 17th APCOM, Denver, AIME.
21. MATHERON G. (1972) : "Le krigeage disjonctif". Internal report CGMM.
22. MATHERON G. (1974) : "Les fonctions de tranfert des petits panneaux". Internal report N-395. CGMM.
23. MATHERON G. (1976) : "Forecasting block grade distributions : the tranfer functions". Proceedings 15th NATO-ASI "Advanced geostatistics in the mining industry".

24. MATHERON G. (1978) : "Peut-on imposer des conditions d'universalité au krigeage ?". Internal report n° CGMM.
25. MATHERON G. (1980) : "Modeles isofactoriels pour l'effet zéro". CGMM Internal Report N-659 June 1980.
26. MATHERON G. (1981a) : "Remarques sur le changement de support". Internal Report n° N-690 Feb. 1981. CGMM.
27. MATHERON G (1981b) : "La selectivité des distributions". Internal Report n° N-686 Feb. 1981. CGMM.
28. MATHERON G. (1982) : "La destructuration des hautes teneurs et le krigeage des indicatrices". Internal Report N-761. CGMM.
29. MATHERON G. (1983) : "Isofactorial models and change of support" to be published in Proceedings, NATO-ASI. TAHOE-83
30. VERLY G. (1983) : "The multigaussian approach and its applications to the estimation of local recoveries". Mathematical Geology, Vol. 15. n° 2.
31. MUGE F. (1982) : "The role of recuperation functions in the early stage of the mine planning of an iron orebody". Proceedings of 17th APCOM, AIME Publications.
32. PARKER H. (1975) : "The geostatistical evaluation of ore reserves using conditional probability distributions : a case study for the area 5 Prospect, Warren, Maine". Ph. D. Thesis, Stanford University.
33. PARKER HM. (1975) and SWITZER P. : "Use of conditional probability distribution in ore reserve estimation, a case study. Proc. 13th APCOM Symp. Clausthal, W.Germany.
34. PARKER H., JOURNEL A., DIXON W. (1979) : "The use of conditional lognormal probability distribution for the estimation of open-pit ore reserves in strata-bound Uranium deposit".16th APCOM Proceeding AIME, AIME, Publications.
35. WILLIAMSON DR. and MUELLER E (1976) : "Ore estimation at Cyprus Pima Mine". AIME Annual Meeting. Las Vegas (1976).

THE SELECTIVITY OF THE DISTRIBUTIONS AND "THE SECOND PRINCIPLE OF GEOSTATISTICS"

Georges MATHERON

Centre de Géostatistique et de Morphologie Mathématique
ECOLE NATIONALE SUPERIEURE DES MINES DE PARIS, Fontainebleau, France.

FOREWORD (added at the request of the reviewer):

As everyone knows, or should know, thermodynamics is based on two principles: 1) conservation of the energy, and 2) degradation of the energy, or increase of the entropy. In the same way, the most important problem of Geostatistics, that is the change of support, obeys two principles: as the support increases, 1) the mean remains constant and 2) the selectivity is distorted. The present paper is devoted to this "second principle" of Geostatistics.

0. "THE SECOND PRINCIPLE OF GEOSTATISTICS".

The definition of the recoverable reserves is nearly connected with two effects of a purely physical nature: a support effect, and an information effect. For the reserves depend on the support of the selection, that is on the size of the minimal units which can be separately sent either to the mill or to waste. Similarly, they also depend on the ultimate information, that is on the nature of the sampling which will be available when the final destination of each unit will be decided. This is not at all an estimation problem, although, naturally, estimation problems will also arise, but it involves the definition of the "true" or "really" recoverable reserves themselves. In any case, we must expect that an increase of the size of the support, or decrease of the ultimate information will result in a distortion or adulteration of the true grade/tonnage curves. And very often this distortion will be much more important than any estimation error. This "second principle" of Geostatistics seems absolutely general, and it deserves a precise formulation which was given in [3] although the basic idea goes back to D.G. Krige [5].

G. Verly et al. (eds.), Geostatistics for Natural Resources Characterization, Part 1, 421–433.

In the first section, I recall the definition of the "grade/tonnage" curves, represented either in terms of z (cut-off grade) or in terms of T (selected tonnage). The definition of the ordering "F_1 is more selective than F_2" follows, and a powerful criterion is given with the help of a profound theorem due to Cartier. These notions are used to explain the support effect (Section 3) and the information effect (Section 4), and a typical example is examined (Section 5). In appendix, the main properties of a useful selectivity index S/m_o are summarized.

1. THE GRADE/TONNAGE CURVES.

In this first section, we implicitely assume that the true grades of the selection units are (or will be) perfectly known. Of course, this ideal can never be attained in practice, but it provides a good reference for later comparisons: it will be used to gauge the distortion in the grade/tonnage curves due to a change of support (support effect) and/or to a change in the nature of the ultimate information (information effect).

Let F_S, or simply F, be the cumulative distribution function of the grades (defined on a given support S). If we are using a probabilistic model, in which the joint grades z(x) are interpreted as a realization of a random function Z(x), F_S is the c.d.f. of the random variable

$$Z(S) = (1/S) \int_S Z(x)\, dx.$$

It may be better to use a _model-free_ definition: then, the deposit D is imagined as the union $D = \cup S_i$ of disjoint units S_i, and F_S denotes the cumulative histogram of the grades $z(S_i)$. If the units S_i are not equal, it is clear that the frequencies must be weighted by the corresponding tonnages.

The effects of a possible _selection_ can be presented as functions either of the _cut-off grade z_ or of the _selected tonnage T_.

In the first representation, the definitions are:

$$T_-(z) = P(Z \geqslant z) = \int_{z-0}^{\infty} F(du) \quad \underline{\text{or}} \quad T_+(z) = P(Z > z)$$

$$Q_-(z) = \int_{z-0}^{\infty} u\, F(du) \quad \underline{\text{or}} \quad Q_+(z) = \int_{z+0}^{\infty} u\, F(du)$$

$$m_-(z) = Q_-(z)/T_-(z) = E(Z/Z \geqslant z) \quad \underline{\text{or}} \quad m_+(z) = E(Z/Z > z)$$

$$V(z) = Q(z) - z\, T(z) = E\left[(Z - z)_+\right] = \int_z^{\infty} T(u)\, du$$

The functions $T(z)$ (selected tonnage) and $Q(z)$ (the corresponding quantity of metal) are non increasing, and not continuous (except if a density function exists): the left hand values $T_-(z)$, $Q_-(z)$ may be strictly higher than the right hand values T_+, Q_+. The corresponding average grade, $m_\pm(z)$ is non decreasing and not continuous. But the most important function is $V(z)$, because of its economic significance. This function $V(z)$ is non increasing, convex, and, for that reason, always continuous. At each point z, $V(z)$ has a right hand derivative, equal to $-T_+(z)$, and a left hand derivative, equal to $-T_-(z)$.

In the second representation, Q, m, V and z itself are considered as functions of the selected tonnage T. The function $z(T)$ is not increasing and corresponds to the classical $(1-T)$-quantile. Due to the intervals of constancy of $F(z)$, the function $z(T)$ is not continuous, and it has a right hand and a left hand determinations. The definitions are:

$$\begin{cases} z_-(T) = \text{Inf}\left\{z : T(z) < T\right\} \quad ; \quad z_+(T) = \text{Sup}\left\{z : T(z) > T\right\} \\ Q(T) = \int_0^T z(t)\,dt \quad ; \quad m(T) = Q(T)/T \quad ; \quad V(T) : Q(T) - T\,z(T) \end{cases}$$

With the T_- representation, the most interesting function is $Q(T)$, because it is non decreasing, concave and thus continuous. At each point T, $Q(T)$ has a right hand derivative, equal to $z_+(T)$, and a left hand derivative, equal to $z_-(T)$.

In other words, the convex function $V(z)$ and the concave function $Q(T)$ are in duality, in the sense of convex analysis, see [2], pp. 104 & Sq. In particular, they satisfy the reciprocal relations:

$$\text{(a)} \quad \begin{cases} V(z) = \sup\limits_{T\in[0,1]} \left\{Q(T) - z\,T\right\} \\ Q(T) = \inf\limits_{z \geq 0} \left\{V(z) + z\,T\right\} \end{cases}$$

In spite of their simplicity, the relationships (a) are of the greatest importance in understanding grade/tonnage curves. Let us derive a first consequence:

THEOREM 1. Let F_1, F_2 be two distributions, let V_1, V_2 and Q_1, Q_2 be the corresponding V and Q-functions defined as above. Then, we have $V_1(z) \geq V_2(z)$ for any $z \geq 0$ if and only if $Q_1(T) \geq Q_2(T)$ for any $T \in [0,1]$.

Proof. Suppose, for instance, $V_1(z) \geq V_2(z)$ for any z. For a given T and for any z, the second relation (a) implies:

$$Q_2(T) \leq V_2(z) + z\,T \leq V_1(z) + z\,T$$

and thus

$$Q_2(T) \leq \operatorname*{Inf}_z \{V_1(z) + z\,T\} = Q_1(T) \quad \blacksquare$$

2. THE ORDERING: "F_1 IS MORE SELECTIVE THAN F_2".

If $V_1(z) \geq V_2(z)$ for any z, or (which is the same by Th. 1) $Q_1(T) \geq Q_2(T)$ for any T, the distribution F_1 may be considered as "better" than F_2. But this comparison is more instructive if the corresponding expectations m_1 and m_2 are equal, because we chiefly want to compare the distributions defined on various supports inside the same deposit. Hence the following definition:

Definition 1: If F_1 and F_2 are two distributions concentrated on $(0,\infty)$, we say that F_1 is more selective than F_2 if the two following conditions are satisfied:

(i) $m_1 = m_2$ (i.e. $\int z\,F_1(dz) = \int z\,F_2(dz)$)

(ii) $V_1(z) \geq V_2(z)$ for any $z \geq 0$, or, which is the same by Th. 1, $Q_1(T) \geq Q_2(T)$ for any $T \in (0,1)$.

Clearly, the relation "F_1 is more selective than F_2" is a (partial) ordering on the set of the distributions. The following criterion emphasizes the economic importance of our definition:

Criterion. F_1 is more selective than F_2 if and only if we have

$$\text{(b)} \qquad \int \phi(z)\,F_1(dz) \geq \int \phi(z)\,F_2(dz)$$

for any convex function ϕ on R_+.

Proof. Any convex function ϕ on R_+ can be expressed in the form:

$$\phi(z) = a + bz + \int_0^z (z - t)\ \mu(dt)$$

where μ is a positive measure on R_+. If Z is a R.V. and F its d.f., it follows

$$E[\phi(Z)] = a + b\ E(Z) + \int_0^\infty F(dz) \int_0^z (z - t)\ \mu(dt)$$

$$= a + b\ E(Z) + \int_0^\infty \mu(dt)\ V(t)$$

Thus, if F_1 is more selective than F_2, the relation (b) is satisfied.

Conversely, if (b) holds, with $\phi(z) = z$ (resp. $\phi(z) = -z$) we conclude $\int z\ F_1(dz) \geq \int z\ F_2(dz)$ (resp. $\int z\ F_1(dz) \leq \int z\ F_2(dz)$), and thus $m_1 = m_2$. Now, for any $t \geq 0$, the function $\phi(z) = (z - t)_+$ is convex. Then, it follows from (b):

$$V_1(t) = \int (z - t)_+\ F_1(dz) \geq \int (z - t)_+\ F_2(dz) = V_2(t)$$

and F_1 is more selective than F_2. ■

The criterion (b) occurs in a much more general context of functional analysis (see [1], pp. 28). From this literature, we select a definition and a profound theorem due to Cartier.

Definition 2. Let Z_1, Z_2 be two random vectors in $\mathbb{R}^n$, $F_1(z_1)$ and $F_2(z_2)$ their n-variate distributions. We say that F_1 is a dilation of F_2 if there exists a 2n-variate distribution $F(z_1, z_2)$, which has the given marginal distributions $F_1(z_1)$ and $F_2(z_2)$ and such that:

$$\text{(c)} \qquad E(Z_1/Z_2) = Z_2$$

THEOREM 2 (Cartier). With the same notation, F_1 is a dilation of F_2 if and only if we have:

$$\text{(b)} \qquad \int \phi(z)\, F_1(dz) \geq \int \phi(z)\, F_2(dz)$$

for any convex function ϕ on R^n.

The "only if" part is not difficult. In fact, we have $E[\phi(Z)] \geq \phi[E(Z)]$ for any convex function ϕ and whatever be the distribution F of Z. If F is the conditional distribution of Z_1, given that $Z_2 = z_2$, this implies:

$$E[\phi(Z_1)/Z_2] \geq \phi(E(Z_1/Z_2))$$

Now, if (c) holds, we have $\phi(E(Z_1/Z_2)) = \phi(Z_2)$, and we conclude:

$$E(\phi(Z_1)) = E(E(\phi(Z_1/Z_2)) \geq E(\phi(Z_2))$$

On the contrary, the "if" part of the theorem is a very strong result. In our context, the conclusion is that F_1 is more selective than F_2 if and only if the criterion (c) is satisfied.

3. THE SUPPORT EFFECT.

The support effect can be illustrated by two simple examples.

First example. Let $Z(x)$, $x \in R^n$, be a random function, not necessarily stationary. Let $\pi = \cup S_i$ be a panel made up of disjoint blocks S_i. Without loss of generality, we can take the tonnage of π to be 1. Let p_i be the tonnage of the block S_i, so that $\sum p_i = 1$. We define the variables $Z(S_i)$, $Z(\pi)$ in an obvious way, so that we have:

$$\text{(d)} \qquad Z(\pi) = \sum p_i\, Z(S_i)$$

Now, let F_π, F_{S_i} be the distributions of $Z(\pi)$, $Z(S_i)$. Then:

The mixture $F_S = \sum p_i\, F_{S_i}$ is more selective than F_π.

The proof is very simple: the mixture F_S is the distribution of the grade $Z(S)$ of a block S_i chosen at random according to the probability measure p_i. But (d) implies $E(Z(S)/Z(\pi)) = Z(\pi)$, so that F_S is a dilation of F_π. By the Cartier theorem, the conclusion follows.

The <u>second example</u> is a model-free version of the preceding one. Let π_i, $i \in I$ and S_j, $j \in J$ be two <u>partitions</u> of the same deposit D. We take the total tonnage of D to be 1, and denote $p(\pi_i)$, $p(S_j)$ the tonnages of π_i and S_j. We shall say that the S_j (the blocks) constitute a <u>subpartition</u> of the π_i (the panels) if each panel π_i, $i \in I$ is the union of those of the blocks S_j intersecting it, say:

$$\pi_i = \cup \left\{ S_j : j \in J , S_j \cap \pi_i \neq \Phi \right\}$$

The block and panel distributions are the mixtures:

$$F_S = \sum p(S_j) F_{S_j} \quad ; \quad F_\pi = \sum p(\pi_i) F_{\pi_i}$$

(where F_{S_j}, F_{π_i} are Dirac measures located at $z(S_j)$, $z(\pi_i)$ respectively).

Then:

> If the blocks S_j constitute a subpartition of the panels π, the block distribution F_S is more selective than the panel distribution F_π.

The proof is exactly the same.

<u>Comments</u>.

(i) In these examples, we did <u>not</u> assume that the blocks and/or the panels are defined independently of $Z(x)$. In particular, they may be chosen by taking into account the results of certain samples. The basic relation $E[Z(S)/Z(\pi)] = Z(\pi)$ always holds.

(ii) In the <u>polymetallic case</u>, the grade Z becomes a vector $(Z_1, Z_2, \ldots)$ (for instance Z_1 = Pb %, Z_2 = Zn % and so on). If the condition (b) of Th. 2 is chosen as the definition of the ordering "F_1 is more selective than F_2", the conclusions of our two examples remain valid in the polymetallic case.

(iii) It would be nice to be able to state a general rule, for instance:

(R) "If $S \subset S'$, then F_S is more selective than $F_{S'}$".

Unfortunately, this statement is <u>false</u>, although it probably holds in the most part of the practical applications. As a counter example, let us consider the periodic function:

$$Z(x) = a_o + \sum_{n \geqslant 1} A_n \operatorname{Cos} 2n\pi\left(\frac{x}{L} - \phi_n\right)$$

where a_o is a given constant. Then $a_o = m$ is the mean, and each block of size L has the grade $Z(L) = m$: the distribution F_L is concentrated on the single point m and thus is less selective than any other one which has the same mean m. For instance, if $L' = (3/2)L$ the support L' is larger than L, but the distribution $F_{L'}$ is more selective than F_L.

Nevertheless, the rule (R) may be adopted as a <u>heuristic principle</u> for choosing <u>new models of change of support</u>, as it was in the case of the now classical models presented in [4]. Note that the <u>conditional distributions</u> (given such and such sample grades) and not only the "a priori" distributions, must obey the rule (R).

4. THE INFORMATION EFFECT.

Let us now examine the influence of the "ultimate" information, that is the information which will be available when deciding the final destination of a given block. This future information may be more important than the present one, for instance if it involves future blast holes, but it will never be perfect. From time to time, the selection will be wrong, and a really poor block will be sent to the mill if its ultimate samples are rich, and conversely.

Inevitably, this process results in a new distortion of the (true) grade/tonnage curves. This effect is <u>not</u> a consequence of a wrong estimation of the distributions, it would entirely remain even if the true distributions were perfectly known: it is uniquely due to the fact that the final decision will be based upon indirect criteria. Let us examine this point.

Let Z be the R.V. representing the grade of a block π. Let $X = (X_1, X_2, \ldots, X_n)$ be the random vector representing the ultimate samples which will be used when deciding whether or not to mine the block π. Suppose that the cut-off grade is z_o. We need to find a criterion for deciding whether to mine the block, that is we have to choose a Borel set $A \subset R^n$, and to select the block if $X \in A$, and otherwise reject it. Among all possible Borel sets A, for a given cut-off grade z_o, the best ones are the ones which maximize the expectation $E[(Z - z_o)\, I_A(X)]$. Now, if h(X) is the <u>conditional expectation</u> of Z given $X_1, X_2, \ldots, X_n$, i.e.

$$h(X) = E(Z/X)$$

we have $E[(Z - z_o)\, I_A(X)] = E[(h(X) - z_o)\, I_A(X)]$, so that A_{z_o} must be either

$$A^+_{z_o} = \left\{x : h(x) \geq z\right\}, \ \underline{\text{or}}\ A^-_{z_o} = \left\{x : h(x) > z_o\right\}$$

(or any other Borel set between these two).

But it is clear that this comes back to working with the random variable $H = h(X) = E(Z/X)$. In other words, the <u>true</u> grade/tonnage curves depend on the distribution F_H of H and not on the distribution F_π of Z. Even though it might seem paradoxical at first, we may say <u>that we are mining conditional expectations H rather than grades Z</u>.

Now, by its definition, H is the conditional expectation $H = E(Z/X)$. But clearly, the conditional expectation of Z given $H = h(X)$ is again $H = h(X)$ itself, so that we have $E(Z/H) = H$. In other words, F_π is a dilation of F_H, and thus, by Cartier's theorem, <u>F_H is less selective than F_π</u>, i.e.:

$$V_H(z) \leq V_\pi(z) \quad \underline{\text{or}} \quad Q_H(T) \leq Q_\pi(T)$$

This is the <u>information effect</u>: an indirect selection (i.e. one based upon the grades X_i of the ultimate samples and not upon the actual block grade value Z) always has a deleterious effect on the function V or Q.

5. A TYPICAL EXAMPLE.

Let us assume that there will be only one ultimate sample S when selecting the block π. The support S is assumed to be small enough, so that the block π is approximately a union $\pi = \cup S_i$ of disjoint sets S_i equal to S. In this case, F_π is less selective than F_S, but more selective than F_H (see above), that is:

$$V_H \leqslant V_\pi \leqslant V_S \ \underline{\text{or}}\ Q_H \leqslant Q_\pi \leqslant Q_S$$

The first inequality ($V_H \leqslant V_\pi$) represents the information effect, the second one ($V_\pi \leqslant V_S$) corresponds to the support effect.

In order to compare <u>orders of magnitude</u>, let us assume that the grade Z of the block π and the grade X of the sample S have a bivariate lognormal distribution, with the same mean m_o and $\mathrm{Var}(\ell n Z) = \sigma_\pi^2$, $\mathrm{Var}(\ell n X) = \sigma_S^2$. From Cartier's condition $E(X/Z) = Z$, the correlation coefficient $\rho_{\pi S}$ of $\ell n X$ and $\ell n Z$ must be $\rho_{\pi S} = \sigma_\pi/\sigma_S$.

Now the conditional expectation $H = E(Z/X)$ also is lognormal, with the same mean m_o and

$$\mathrm{Var}(\ell n H) = \sigma_H^2 = \rho_{\pi S}^2\ \sigma_\pi^2 = \sigma_\pi^4/\sigma_S^2$$

The various functions T, Q, m, V are easy to calculate from those lognormal distributions. Numerical values are given in Table 1 (with $m_o = 1$, $\sigma_S^2 = 1$, $\sigma_\pi^2 = 1/2$). Note that T_H, Q_H, V_H,... represent the <u>true</u> recoverable reserves. The reserves should be T_π, Q_π,... if there were no information effect. Had we to mine drill cores S instead of blocks π, they should be T_S, Q_S,...

The corresponding <u>selectivity indices</u> $100 \times S/m_o$ (i.e. the mean values of V/m_o, see Appendix below) are:

VARIABLE	SAMPLE	BLOCK	H
SELECTIVITY INDEX	52.1	38.3	27.6

The part of the <u>loss</u> due to the support effect is 52.1 - 38.3 = 13.8, while the part due to the information effect is 38.3 - 27.6 = 10.7.

TABLE I

CUTOFF GRADE z	VARIABLE	T(z)	Q(z)	m(z)	V(z) x 100
z = 0.5	H	.872	.949	1.09	51.3
	π	.735	.909	1.24	54.2
	S	.577	.884	1.53	59.5
z = 1.0	H	.401	.599	1.49	19.7
	π	.362	.638	1.76	27.6
	S	.309	.691	2.24	38.3
z = 1.5	H	.144	.287	1.99	7.1
	π	.177	.413	2.33	14.8
	S	.183	.538	2.94	26.4

APPENDIX. THE SELECTIVITY INDEX S/m_o

If Z is a positive R.V. and F its distribution, the parameter S is defined by

$$S = \int_0^\infty F(z)[1 - F(z)]\, dz$$

S does exist if and only if the expectation $m_o = \int_0^\infty (1 - F(z))dz$ is $< \infty$, and then

$$S/m_o \leqslant 1$$

The index S/m_o is called the <u>selectivity index</u> of the distribution F. The parameter S satisfies the following relations (for the proof, see [7]):

$$\begin{cases} S = E(V(Z)) = \int_0^1 V(T)\, dT \\ \frac{1}{2} S = \int_0^1 [Q(T) - T\, m_o]\, dT \\ \frac{1}{2} S = \mathrm{Cov}\, (z(T), \frac{1}{2} - T) \end{cases}$$

In the last relation, $z(T)$ is the quantile function, as defined above almost everywhere on $(0,1)$, and T is uniformly distributed on $(0,1)$. By Schwarz's inequality, we conclude:

$$S \leqslant \sigma/\sqrt{3}$$

with equality if and only if the distribution F is uniform on a given segment.

In particular, if $Z = \phi(Y)$ where Y is $N\ (0,1)$, the same relation may be rewritten as:

$$\frac{1}{2} S = \int_{-\infty}^{+\infty} \phi(y)[2\ G(y) - 1]dy$$

(G is the standard normal d.f.). If we know an expansion of ϕ in terms of Hermite polynomials $H_n(x) = \exp\left(\frac{x^2}{2}\right) \times \left(\frac{d}{dx^n}\right) \exp\left(\frac{-x^2}{2}\right)$, say

$$\phi(y) = \Sigma \frac{C_n}{n!} H_n(x)$$

S is given by the following expansion

$$S = \frac{-1}{\sqrt{\pi}} \sum_{n=0}^{\infty} \frac{(-1)^n C_{2n+1}}{2^{2n}(2n+1)\ n!}$$

If F is lognormal with the mean m and the (logarithmic) variance σ^2, we find

$$S = m\left[2\,G\left(\frac{\sigma}{\sqrt{2}}\right) - 1\right]$$

If F is $\underline{N}\,(m,\sigma^2)$, we find

$$S = \sigma/\sqrt{\pi}$$

(this value is very near the limit $\sigma/\sqrt{3}$, so that a normal distribution is not very far from a uniform one...).

Finally, if for instance $T < 0.5$, the largest interquantile interval between T and 1-T, which is $z_-(T) - z_+(1-T)$, satisfies a Tchebychev-type inequality, that is:

$$|z_-(T) - z_+(1-T)| \leqslant \frac{S}{T(1-T)}$$

This is the reason why S is called a dispersion parameter.

REFERENCES

1. ALFSEN, G.M.,"Compact Convex Sets and Boundary Integrals", Springer, Berlin, 1971.

2. ROCKAFELLAR, R.T.,"Convex Analysis", Princeton, 1972.

3. MATHERON, G.,"La Sélectivité des Distributions", CGMM, Fontainebleau, 1981.

4. MATHERON, G., "Forecasting Block Grades Distributions: the Tranfer Functions", in Adv. Geostatistics in the Mining Industry, ed. M. Guarascio et al., D. Reidel, 1976, p. 221-236.

5. KRIGE, D.G., "A Statistical Approach to Some Mine Valuations and Allied Problems in the Witwatersrand", Thesis, Un. of the Witwatersrand, 1951.

LOCAL ESTIMATION OF THE RECOVERABLE RESERVES: COMPARING VARIOUS METHODS WITH THE REALITY ON A PORPHYRY COPPER DEPOSIT

Daniel GUIBAL and Armando REMACRE

Centre de Géostatistique et de Morphologie Mathématique
ECOLE NATIONALE SUPERIEURE DES MINES DE PARIS, Fontainebleau, France.

ABSTRACT

The objective of this article is to compare the estimates based on several non-linear methods with the actual figures. The methods tested were disjunctive kriging, multi-Gaussian kriging and a new method (uniform conditioning) which is simpler and computationally quicker than these two and yet still gives comparable results.

INTRODUCTION

The aim of this paper is to apply different non-linear estimation methods under similar conditions where there are enough data to make comparisons and to draw conclusions on the behavior of these methods and their constraints. The methods used are disjunctive kriging, multi-Gaussian and uniform conditioning which is being put into practice for the first time. In addition, the change of support and the bigaussian hypothesis have also been tested, but less thoroughly.

The general problem comes from the fact that the characteristics of many deposits are incompatible with the requirements of non-linear methods (i.e. strictly stationary hypothesis). For instance, in the case of porphyry copper, there is a high grade zone which often leads to preferential sampling at the detriment of the border which is poorer. We are thus confronted with a double problem: presence of a large scale drift and an irregular preferentially sampled grid. We know that these circumstances do not have a great effect on kriging with a univer-

G. Verly et al. (eds.), Geostatistics for Natural Resources Characterization, Part 1, 435–448.

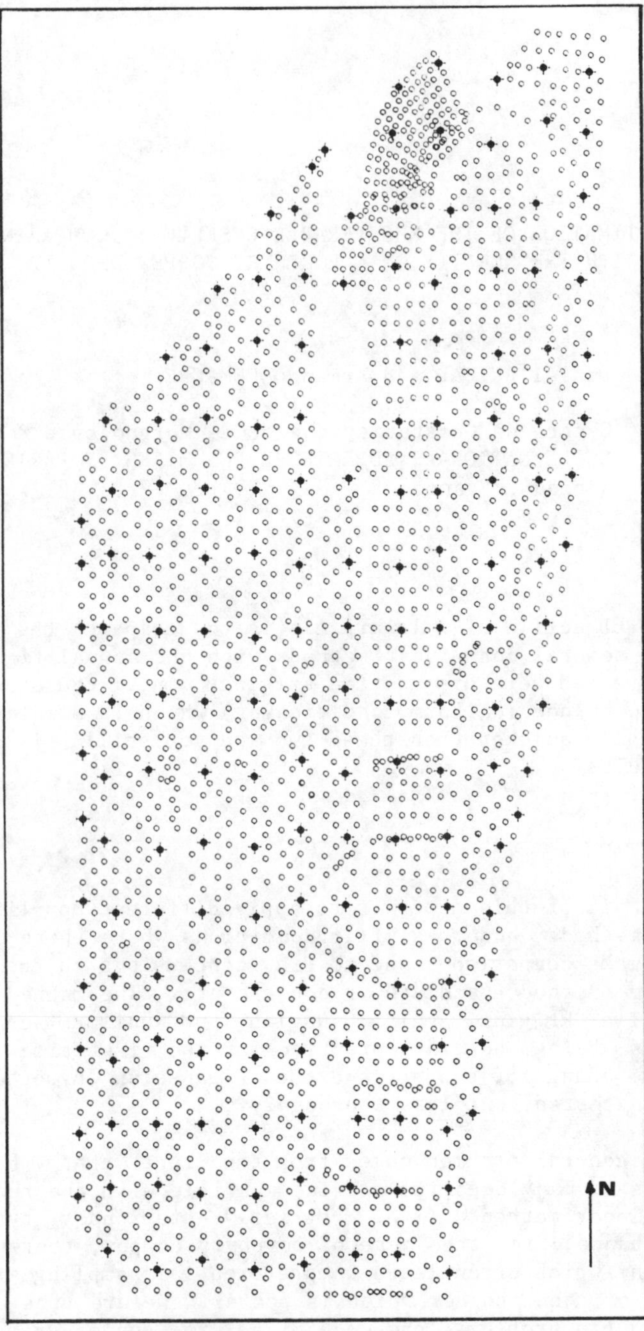

Figure 1. Location of blast-holes and of the data points seleeted.

sality condition. The estimated grades reflect the local means fairly accurately. In this case, the global mean of the kriged grades is close to the mean grade of the data weighted (by the zone of influence or by the kriging weights) and not to their raw mean grade.

The problems with non-linear methods are therefore very delicate. Theoretically the stationary hypothesis plays a fundamental part (can a drift be taken into account?). From a practical point of view, the anamorphosis that precedes almost any of these methods requires the histogram of the data to be representative.

PRELIMINARY CONDITIONS

The data come from one level of a mine where the analysed copper grades of blast-holes were available. They are considered to represent the reality. More precisely, the data correspond to a bench 13 m thick in a porphyry copper type deposit, which is thus reduced to 2 dimensions. The presence of different zones can be observed (in our case, slightly higher grades towards the south of the bench), which confirms the type of deposit mentioned above.

Results from 2 095 blast-holes are available. In order to have basic data for the estimation, these have been classified on a regular grid of panels 30 m x 30 m. By taking the most central samples on each panel, 173 blast-holes were obtained. In order to have the reference "reality", a grid of blocks 10 x 10 x 13 m (= selection unit) has been defined for the 2 095 data. In this grid, blocks with at least 2 blast-holes were taken into account. Then the mean for these blocks was computed, and they were regrouped so as to form panels 30 x 30 x 13 m . Only panels containing more than 5 blocks containing data values were taken into account.

We then have two files: one contains the basic data, the second one the "reality". The basic data are used to compute the recovery functions of blocks 10 x 10 x 13 m using the method mentioned below and the results are then compared to the contents of the second file. We have obtained 173 basic data, 583 blocks and 50 panels with 281 blocks.

REVIEW OF THE MAIN THEORETICAL POINTS

As the theory of non-linear geostatistics has been presented elsewhere (Matheron (1975 (a) (b), 1978)). Only the main points need be mentioned here.

Let Y_v denote the anamorphosed grades of blocks, v, equal in size to the selection unit. The essential problem is to estimate $Q = \Sigma f(Y_{v_i})/L$ where f(.) is a function to be estimated (e.g. the recoverable tonnage or the recoverable metal tonnage) and L is the number of blocks per panel. In cases where the number of blocks per panel is very large, L denotes the number of blocks which are used to represent the panel. These have to be selected to provide a uniform discretization of the panel.

We suppose that the decomposition of f(y) into Hermite polynomials is already known:

$$f(y) = \Sigma \frac{f_n}{n!} H_n(y) \tag{1}$$

Our objective is to estimate $\sum_{i=1}^{L} H_n(Y_{v_i})/L$ using three different sets of assumptions.

a) Multigaussian.

In this model, it is assumed that the distribution of the Y_α (point values) and of the Y_{v_i} is jointly multigaussian. The kriged estimate $Y^k_{v_i}$ (with known mean) of the block v_i represents the conditional expectation $E(Y_{v_i}|Y = y_\alpha)$; it has a standard deviation s_{v_i}.

From the classical relation

$$E\left[H_n(Y)|X\right] = \rho^n H_n(X)$$

valid for the bigaussian variables (X,Y), with correlation coefficient ρ, it is easily shown that the multigaussian estimator (MG) of $H_n(Y_{v_i})$ is given by:

$$\begin{aligned}\left[H_n(Y_{v_i})\right]_{MG} &= E\left[H_n(Y_{v_i})|Y\right] \\ &= E\left[H_n(Y_{v_i})|Y^k_{v_i}/s_{v_i}\right] \\ &= s^n_{v_i} H_n(Y^k_{v_i}/s_{v_i})\end{aligned}$$

As a matter of fact, the correlation coefficient between Y_{v_i} and $Y^k_{v_i}/s$ is equal to s_{v_i}. Therefore the MG estimator of Q can be written as:

$$f_o - f_1 \frac{1}{L} \sum_{i=1}^{L} Y^k_{v_i} + \sum_{n\geqslant 2} \frac{f_n}{n!} \frac{1}{L} \sum_{i=1}^{L} s^n_{v_i} H_n(Y^k_{v_i}/s_{v_i})$$

To get this estimator, L different kriging systems have to be solved.

b) Uniform Conditioning.

To avoid having to solve this number of systems, we use a single linear combination $Y^* = \sum_\alpha \lambda_\alpha Y_\alpha$ for all the v_i within a panel instead of L different ones $Y^k_{v_i}$. We actually take $Y^* = Y^k$, which is ordinary kriging of $1/L \sum Y_{v_i}$ from the Y_α. Let $Y = Y^k/s$, where s is the standard deviation of Y^k and let ρ_{v_i} the correlation coefficient between Y_{v_i} and Y^k/s (this coefficient can be calculated within the frame of the discrete Gaussian model).

The uniform conditioning (UC) estimator of $H_n(Y_{v_i})$ is

$$\left[H_n(Y_{v_i})\right]_{UC} = E\left[H_n(Y_{v_i})\,|\,Y\right]$$

$$= \rho^n_{v_i}\, H_n(Y^k/s)$$

Expanding this, we obtain the UC estimator of Q

$$f_o - f_1 \sum_\alpha \lambda_\alpha Y_\alpha + \sum_{n\geq 2} \frac{f_n}{n!} \frac{1}{L} \sum_{i=1} \rho^n_{v_i}\, H_n(Y^k/s)$$

c) Disjunctive Kriging.

In this case, the disjunctive kriging estimator (DK) of $\sum_{i=1}^{L} H_n(Y_{v_i})/L$ is:

$$\sum_\alpha \lambda_{\alpha,n}\, H_n(Y_\alpha)$$

The coefficients $\lambda_{\alpha,n}$ are obtained by solving the system

$$\lambda_{\alpha,n}\, \rho^n_{\alpha\beta} = \frac{1}{L} \sum_{i=1} \rho^n_{\alpha v_i}$$

Then, the DK estimator can be expressed in the form:

$$f_o - f_1 \sum_\alpha \lambda_\alpha Y_\alpha + \sum_{n\geq 2} \frac{f_n}{n!} \sum \lambda_{\alpha,n}\, H_n(Y_\alpha)$$

We note that for n = 1 all the methods give the same results, which is very interesting in view of the importance of the left-hand term of the development of (1).

In practice, we calculate the point distribution in the

panel and afterwards we introduce the coefficient of change of support r. That is, we calculate the equivalent of $\lambda'_\alpha = \lambda_\alpha / r$ so as to have $H_n^* = 1/L \ \Sigma \ H_n(Y_{x_i})$ for the points, and H_n^*/r^n for the blocks. This makes it possible to easily compute the recovery functions for several supports, if the discretization of the panel can be taken as constant. In terms of grade estimation, it can easily be seen that estimating a panel as:

$$Z_V = 1/L \ \Sigma \ \phi(Y_{x_i}) = \sum_n \frac{C_n}{n!} \left[1/L \ \Sigma \ H_n(Y_{x_i})\right]$$

or

$$Z_V = 1/L \ \Sigma \ \phi_r(Y_{v_i}) = \sum_n \frac{C_n}{n!} r^n \left[1/L \ \Sigma \ H_n(Y_{v_i})\right]$$

which comes back to the same thing.

HISTOGRAMS

Only main histograms obtained are presented here. Figure 2-a shows the histogram of the 173 data together with the fitted distribution and Figure 2-b shows the histogram obtained assuming the permanence of the distribution, i.e. if C_n are the coefficients fitting the point histogram, we calculate $C_n(v) = C_n r^n$ the coefficients of block histograms (10 x 10), from the experimental block histogram.

The table shows the main statistical parameters.

TABLE 1

	ACTUAL	BLOCS	Z	PANELS	BLOCS*
NUMBER OF DATA	2095	583	173	50	281
MEAN	2.315	2.321	2.304	2.453	2.450
S^2	2.755	2.155	2.657	1.713	2.437

The coefficient r of change of support is equal to 0.911 and makes it possible to go from the variance of Z to the block variance. This variance corresponds to the mean value of the covariance in the block $\overline{C}(10,10) = 2.123$ which is fairly close to the experimental variances of the blocks (column 2 of table 1) but rather different from the variance of the blocks used for the comparisons (column 5 of table 1).

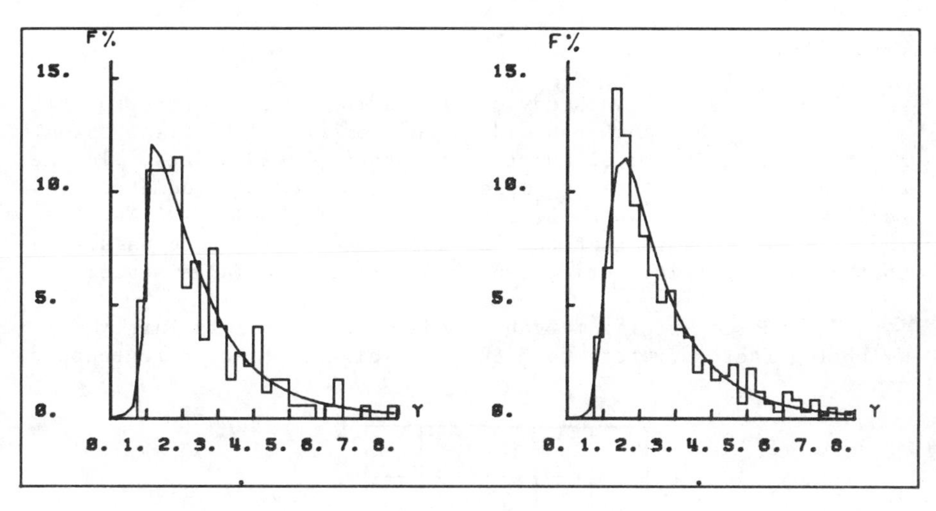

Figure 2a. Experimental histogram (173) and anamorphosis model.

Figure 2b. Experimental histogram of blocks (583) and model of permanence $C_n(v) = C_n . r^n$

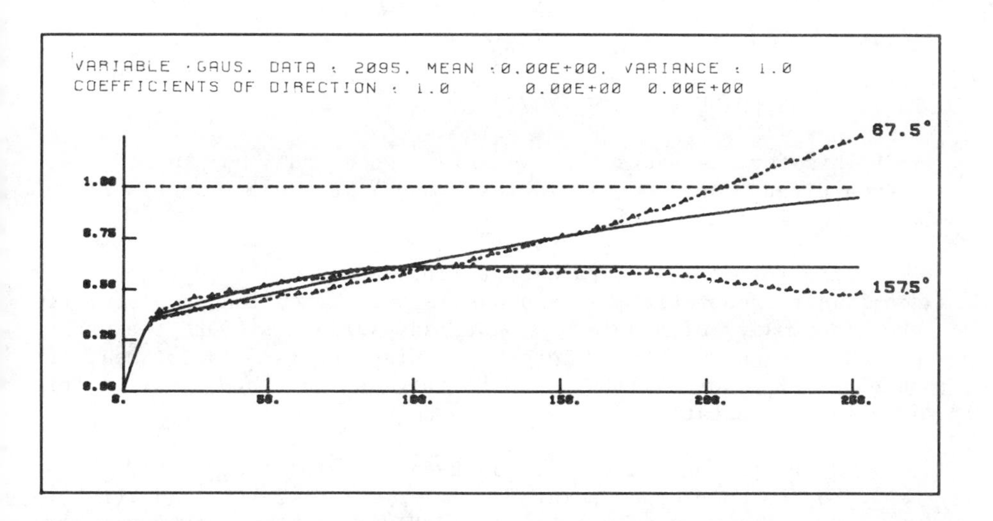

NUGGET: 0.00E+00	NB. STRUCTURE(S): 3		ANISOTROPY	
SPH C: 0.32	A: 12.	X: 1.0	Y: 1.0	Z: 0.00E+00
SPH C: 0.29	A: 0.30E+03	X: 1.0	Y: 3.0	Z: 0.00E+00
SPH C: 0.39	A: 0.35E+03	X: 1.0	Y: 0.00E+00	Z: 0.00E+00

Figure 3. Variograms of blast-hole data and fitted models.

VARIOGRAMS

After an extensive study of the variograms, we chose to fit a model on the anamorphosed values of the 2 095 blast-holes. Figure 3 shows the model adopted for anamorphosed values. In the 2nd and 3rd structures we observed a considerable geometric and zonal anisotropy between the two main directions: 67.5 NNE and 157.5 WNW. We note that the 3rd structure will not be used, and that a non negligible drift can be observed on a large scale.

The chosen kriging neighborhood for 5 x 5 panels involves 13 weighting factors, where each outside weighting factor regroups 3 panels.

		11		
	2	5	7	
10	3	1	8	12
	4	6	9	
		13		

TESTING THE FORMULA: $E[H_n(X)/Y]=\rho^n H_n(Y)$

This formula has to be verified when the couple of two standardized Gaussian variables X and Y is a Gaussian bivariate with correlation coefficient ρ.

This formula is of paramount importance in the theory of non-linear geostatistics and one is very often led to postulate the binormality of couple of Gaussian variables for the sole purpose of using this formula. This test is a useful mean of judging the suitability of the bigaussian hypothesis for non-linear geostatistics.

In the present study, it has been possible to apply this test to the couples of point Gaussian variables Y(x) and Y(x+h). The 2 095 data were anamorphosed and Figure 4 shows the results for $h = 10$ m ± 1 m and n up to 4. On the abscissa we put $-2 < Y < 2$ (which is more than 95% of a Gaussian distribution). The left-hand side of the formula is represented by a continuous line, and the dashed line represented what was expected, i.e. $\rho^n H_n(Y)$.

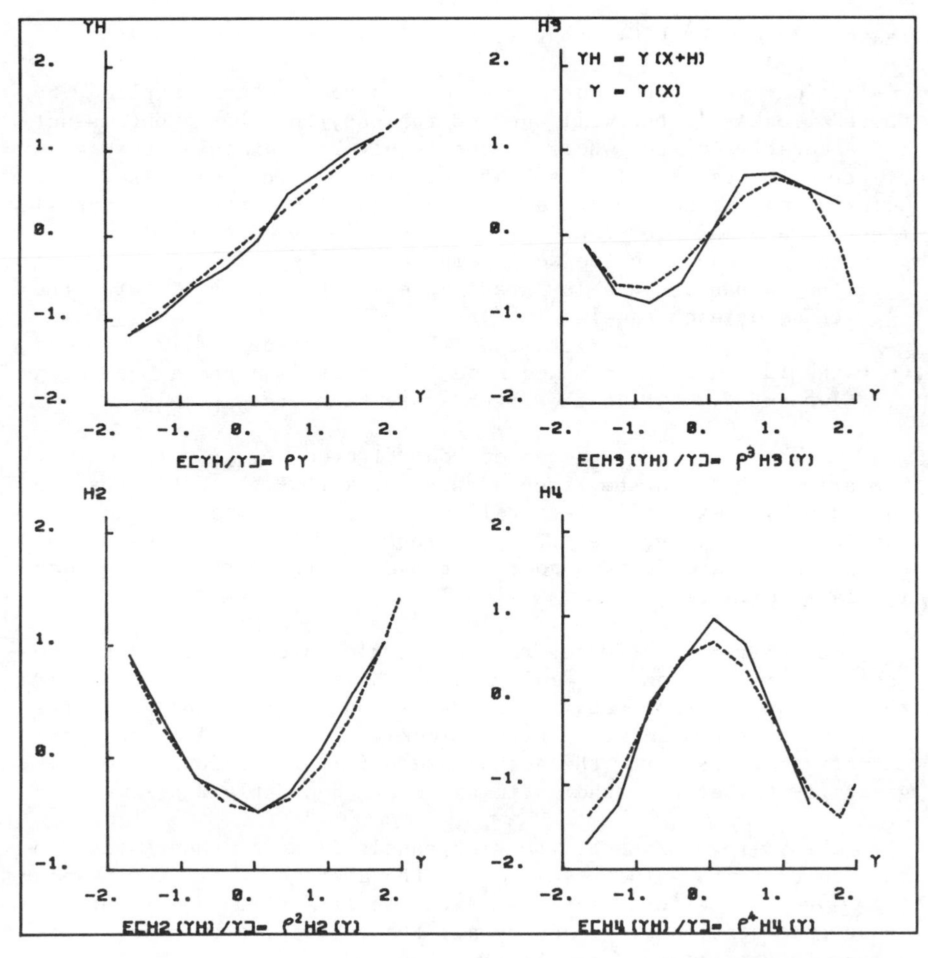

Figure 4. Test of $E[H_n(Y(x+h)|Y(x)] = \rho^n H_n[Y(x)]$.

ESTIMATION OF RECOVERABLE RESERVES

The three methods were used to estimate the proportion P of ore and the metal quantity Q that would be recovered from a set of 30 x 30 m panels with a certain cut-off grade on 10 x 10 selection units. The value of P will be expressed in % while Q will be given using the tonnage of one panel as the unit. Note that when the cut-off grade is zero, Q represents the sum of the quantities of metal in each panel.

Experience has shown that the multigaussian method and disjunctive kriging both give results which are satisfactory globally, i.e. over the whole group of panels in the level where the structural analysis and the anamorphosis are defined. We can reasonably expect U.C. to give equally satisfactory results.

In our study, the cut-off grade has only been applied to those blocks in the well sampled regions, but this group is not representative of the whole of the level. For example the mean of the 50 panels is 2.45% whereas that of the level is 2.30%. Comparisons between the three methods and also with the actual figures were made for four different sets of panels:

- the 50 panels in the well-sampled regions
- the 12 panels in a low grade zone in the north (where the grade of each panel $< 1.5\%$)
- the 6 panels in a rich zone (where the grade $\geq 3.2\%$)
- the 15 panels in a mixed zone (8 panels have grades between 1.5 and 3.2% while the other 7 are high grade panels).

The results are presented on Figures 5, 6, 7 and 8. Comparisons between the three methods show that the estimates are very similar, except for several cases in the last group of panels. This suggests that uniform conditioning gives global results which are just as good as the other two methods, as was indicated earlier.

Moreover the estimated recovery curves are quite close to those obtained from the real values. However it should be noted that the quantity of metal recovered is underestimated when the set of panels contains rich blocks. In particular, when the cut-off grade is zero, the actual grade for the various zones is generally higher than the estimated one. See Table 2 below.

The average grade of the rich panels is more underestimated than in that its value is closer to the average of all the panels (2.30%). This effect is probably due to the strong assumptions about the stationarity made by all the methods, even though this effect is not evident for the poor panels.

TABLE 2. Comparison of actual grades with estimates for a cut-off grade of zero.

	REAL GRADE	ESTIMATED GRADE
All 50 panels	2.45	2.34
The 12 poor panels	1.16	1.12
The 6 rich panels	4.65	4.20
The 15 mixed panels	2.87	2.91

Experience has shown that kriging with a universality condition underestimates the actual grade of rich panels (because of its smoothing influence). But we also know that since the

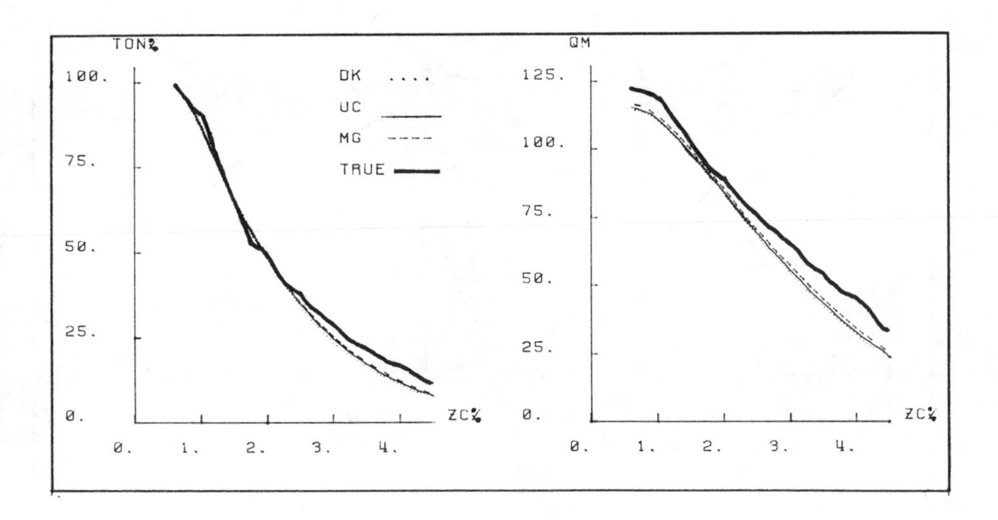

Figure 5. $\frac{\text{Tonage recovered}}{\text{Cut-off grade}}$ and $\frac{\text{Recoverable quantity of metal}}{\text{Cut-off grade}}$ for 50 panels.

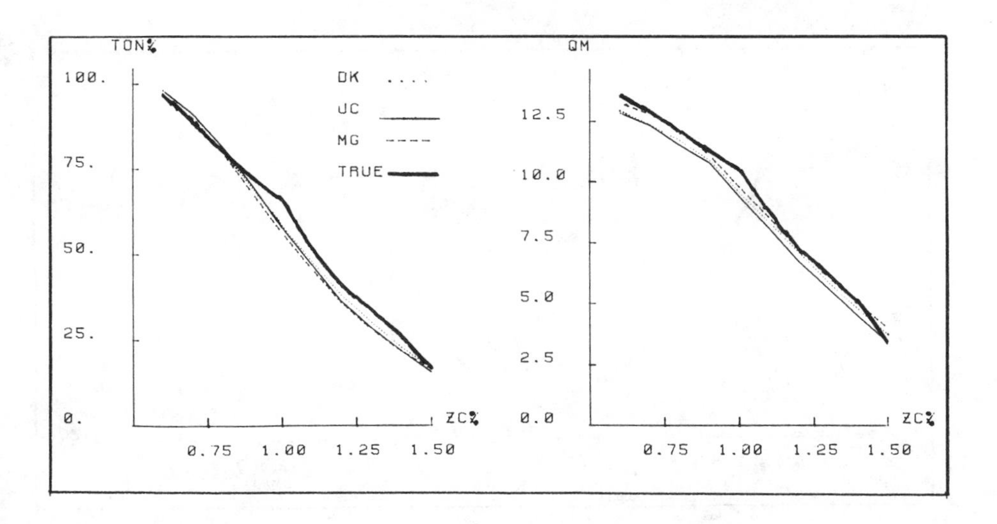

Figure 6. Same as Fig. 5, for 12 panels in a poor area.

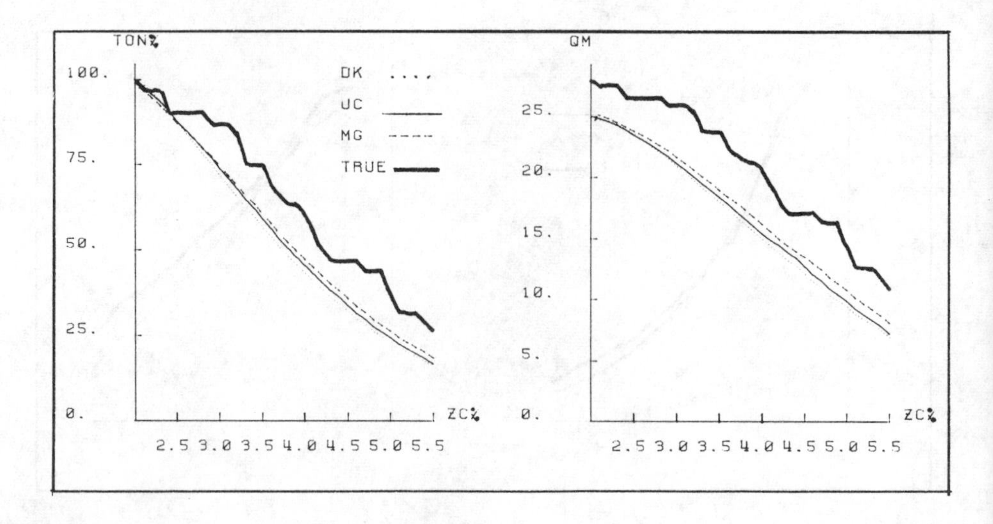

Figure 7. Same as Fig. 5, for 6 panels in a rich area.

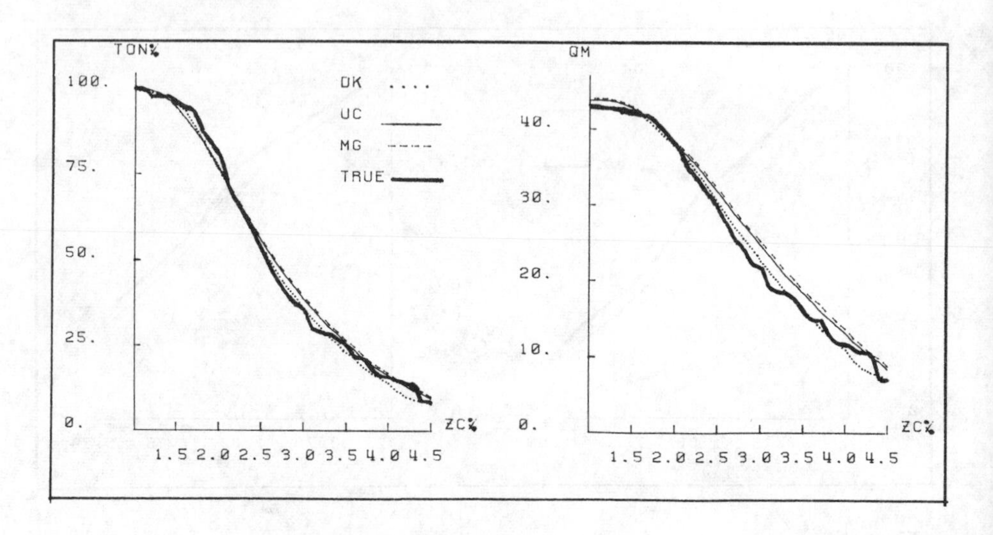

Figure 8. Same as Fig. 5, for 15 panels in a mixed area.

objective in using kriging is to avoid systematic errors when making decisions about the grades, this is of no importance in practice. What is important is that the panels with high estimated grades should be neither over- or under- estimated on average rather than that the truly rich panels be underestimated.

It is well known in ordinary kriging that if a selection is being made on the actual grades of panels (rich ones for example) the average grades and the recoverable reserves are underestimated. The same observation is true in the non-linear case.

This is why, in order to compare estimates and reality, we decided to work on a geographical zone, and not on a population of panels chosen because of their real grades.

Nevertheless, the relatively high discrepancies noticed in the rich area suggest that there is a need for adapting the estimators locally, taking into account local departure from stationarity (such as local trends). Although some preliminary work has been done in that field, it is still open for investigation.

CONCLUSION

This study has made it possible to compare three different non-linear methods both from the theoretical and practical points of view. It has shown that they are very similar.

The uniform conditioning technique first proposed by Matheron [2] in 1975 gave quite satisfactory results and therefore merits further study, since it is easier to use than the other methods. For example it avoids the problems of inacceptable values of the probability density function at high or low grades that sometimes occur with disjunctive kriging. As the results obtained from both methods are very similar, the relative simplicity of uniform conditioning and its greater efficiency in terms of computing requirements come to the fore. The greatest advantage of uniform conditioning is undoubtedly that only one kriging is required to obtain the same results and also that it does not require any additional hypotheses: all that is required is the joint normality of the couples (Y_v, Y_α) and thus of the couples $(Y_v, \sum_\alpha \lambda_\alpha Y_\alpha)$; the "permanence" formulae, i.e. the anamorphosis ϕ_r, is used only at the level of the blocks v, and not of the panels [3, p. 20].

On the negative side, since it depends on only one kriging, two panels with the same kriged estimate and the same kriging neighborhood would then be attributed the same distribution of block grades, whereas one would expect this only if the sample grades were the same.

It is also important to note that although these three methods are fairly sophisticated from the theoretical point of view, they give consistent results and can be used in routine studies.

A current research project, funded by a grant from the French Government concerns the stationarity hypotheses required, with a view to relaxing them whenever possible. The aim of this particular study was to highlight the similarities between the three methods which, up till now, have always been regarded as being quite dissimilar. In fact there is substantial core common to all three, and so rather than trying to find ways of relaxing the stationarity requirements for each of these methods separately, it is better to study them all together.

REFERENCES

1. MATHERON, G., 1976 (a), "Transfert functions and their estimations", Proceeding of NATO A.S.I., D. Reidel, p. 221-236

2. MATHERON, G., 1975 (b), "Les fonctions de transfert de petits panneaux", CGMM, Fontainebleau, France.

3. MATHERON, G., 1978, "Le krigeage disjonctif et le paramétrage local des réserves", Ecole d'Eté, CGMM, Fontainebleau, France.

4. YOUNG, D.S., 1982, "Development and application of disjunctive kriging model; discrete Gaussian model", 17thAPCOM Symp., Colorado School of Mines, EUA.

5. ANDERSON, T.W., 1957, "An Introduction to Multivariate Statistical Analysis", John Wiley and Sons, New York.

6. VERLY, G., 1983, "The Multigaussian Approach and its Applications to the Estimation of Local Reserves", J. of Math. Geol., Vol. 15, n° 2, pp. 263-290.

7. BECKMANN, P., 1973, "Orthogonal Polynomials for Engineers and Physicists", The Golem Press, Golden, Colorado.

ISOFACTORIAL MODELS AND CHANGE OF SUPPORT

Georges MATHERON

Centre de Géostatistique et de Morphologie Mathématique
ECOLE NATIONALE SUPERIEURE DES MINES DE PARIS,
Fontainebleau, France.

ABSTRACT

Disjunctive kriging is a general method, but the usual technique based upon Gaussian anamorphoses and Hermitian expansions become irrelevant in the case of discrete laws or, more generally, if the distribution possesses an atomic part. Isofactorial models with discrete laws, which are relevant in these cases, are suggested.

With the help of of the Markovian semi-groups theory, it is easy to find isofactorial models with symmetric bivariate laws (suitable for the sample/sample or block/block distributions). But it is more difficult to obtain asymmetric bivariate laws compatible with the condition that must be satisfied by a sample/block distribution (one of the marginal laws must be more selective than the other).

As an example, the negative binomial model is examined, and the corresponding model of change of support is compared with true distributions. Another example is given in [2].

G. Verly et al. (eds.), Geostatistics for Natural Resources Characterization, Part 1, 449–467.

The non linear estimation techniques are more difficult when the basic distributions are not continuous. For instance, if Z is a positive variable such that $q = P(Z=0)$ is strictly positive, the "Gaussian" anamorphosis $Z = \phi(Y)$ does exist, but it is not one-to-one. The inverse image of the set $\{Z=0\}$ is not a point but a segment $(-\infty,a)$, so that the variable Y is not really Gaussian. Its real distribution is of the form $q\delta_a + 1_{\{y>a\}}N$. In particular, the Hermite polynomials are no longer orthogonal: disjunctive kriging is still possible, but only via the integral equations of the general theory [1]. In the same way, calculating a conditional expectation like $E[f(Y)/Y_1 \leq a, Y_2 \leq a\ldots]$ becomes fairly tedious. In such a case, it may be better to use an isofactorial model based upon atomic distributions. In practice, it is always possible to discretize a partially continuous distribution, so that we need only examine the purely discrete case. These discrete models seem particularly interesting in the case of diamonds deposits, where one of the basic variables is the number of stones per sample. The following text is only a summary. Proofs and comments are given in the French version [2].

1 - STRUCTURE OF AN ISOFACTORIAL MODEL.

With a view to non linear estimation, three types of bivariate distributions at least are necessary: two symmetric types (sample/sample and block/block) and one asymmetric type (sample block). In the discrete case, we start from two univariate distributions: W_i^v (samples v) and W_i^V (blocks V), and two families of functions $\chi_n^v(i)$ and $\chi_n^V(i)$ called factors, which must be orthonormal and complete in $L^2(\mathbb{N},W^v)$ and $L^2(\mathbb{N},W^V)$ respectively. For two samples v, or two blocks V, located at two points a and b, the symmetric distributions will be of the form:

$$(S)\quad \begin{cases} W_{ij} = W_i^v W_j^v \sum\limits_{n\geq 0} T_n(v_a,v_b)\, \chi_n^v(i)\, \chi_n^v(j) \\ W_{ij} = W_i^V W_j^V \sum\limits_{n\geq 0} T_n(V_a,V_b)\, \chi_n^V(i)\, \chi_n^V(j) \end{cases}$$

where only the coefficients T_n depend on the location. For a sample v_a and a block V_b, the asymmetric distribution will be of the form:

$$(D)\quad W_{ij} = W_i^v W_j^V \sum_{n\geq 0} T_n(v_a,V_b)\, \chi_n^v(i)\, \chi_n^V(j)$$

In the important case of a sample $v \subset V$, the various possible locations of v inside the block V are considered as equivalent, or, which is the same, v is randomly located inside V [1]. The bivariate distribution is of the form:

$$W_{ij}^{vV} = W_i^v W_j^V \sum_{n \geq 0} T_n(v,V) \chi_n^v(i) \chi_n^V(j) \qquad \text{(CS)}$$

It is by itself a model of change of support. Note that this model may be used in two different ways. In the first way, the variable Z_v of interest is the grade of the sample v. It is related to the discrete variable N = i of the model by an anamorphosis $Z_v = \phi_v(i)$, which is known from experimental data. From Cartier's condition $E(Z_v/Z_V) = Z_V$ and the model (CS) above, it follows that the block anamorphosis $Z_V = \phi_V(i)$ is necessarily of the form:

$$\text{(1)} \quad \begin{cases} Z_V = \phi_V(i) = \sum_{n \geq 0} C_n T_n(v,V) \chi_n^V(i) \\ C_n = \sum_j W_j^v \phi_v(j) \chi_n^v(j) \end{cases}$$

In the second case, on the contrary, the variable of interest is the discrete variable N=i itself. It is not a grade, but, for instance, the number of stones in v or in V. In this case, Cartier's condition is:

$$E(N_v/N_V = i) = (v/V)i \qquad (2)$$

Note the difference between these two points of view. In the first case, the model (CS) is given, and then the block anamorphosis ϕ_V is determined by the relation (1). In the second case, there is no anamorphosis, and the relation (2) is a condition which must be satisfied by the model (CS).

Now, our isofactorial model will be entirely defined if we know the coefficients $T_n(v,V)$ of the change of support (CS) and the coefficients $T_n(V_a, V_b)$ of the distribution of two blocks V located at points a and b. Naturally the coefficients $T_n(V_a,V_b)$ depend on the points a and b, or simply on the difference b-a in the stationary case. For, in this model, a sample v_a is randomly located inside its block V_a, so that the variable N_{v_a} may be considered as conditionally independent of the other variables once N_{V_a} is given. It follows that the other T_n coefficients are:

$$\text{(4)} \quad \begin{cases} T_n(v_a,V_b) = T_n(v,V)\, T_n(V_a,V_b) \\ T_n(v_a,v_b) = (T_n(v,V))^2\, T_n(V_a,V_b) \end{cases}$$

so that the model is entirely determined. In particular, if v and v' are two different samples inside the same block, this implies $T_n(v,v') = (T_n(v,V))^2$.

2 - A GUIDELINE.

How to build such a model? Let us first consider the case of the symmetric distributions of type (S). The probability measure W_i and the factors χ_n being given, we must find coefficients T_n such that $W_{ij} \geq 0$. A Markovian point of view will be useful. In fact, let $P_{ij} = W_{ij}/W_i$ be the transition probability from i to j. Then, if T_n and T'_n are admissible coefficients, so are also the products $T''_n = T_n T'_n$. For, the product of the Markovian matrices P and P' associated with T_n and T'_n is the Markovian matrix P" = PP' = P'P associated with the product $T''_n = T_n T'_n$. Moreover, if the distance between two samples is very small, T_n will be almost equal to 1. This suggests an exponential formula $T_n(t)=\exp(-\lambda_n t)$, where t is a parameter without physical significance, but indirectly connected with the distance between samples, λ_n is ≥ 0 and $\lambda_o = 0$.

But the corresponding matrix:

$$P_{ij}(t) = \sum_{n\geq 0} e^{-\lambda_n t} \chi_n(i)\, \chi_n(j)\; W_j$$

describes a Markov process (with continuous time and discrete states). This will be our guideline. The Markovian matrix P(t) always is P(t) = exp(At), where the generator A is a matrix of the form:

$$A_{ij} = - C_i \delta_{ij} + C_i \Pi_{ij}$$

The C_i are positive coefficients, and Π_{ij} is a fixed Markovian matrix.

Moreover, this Markov process must satisfy three conditions. i) W is one of its stationary probability, i.e. W P(t) = W or, which is the same, W A = 0. ii) It must be symmetric with respect to W, i.e. $W_i P_{ij}(t) = W_j P_{ji}(t)$, or, which is the same:

$$C_i W_i \Pi_{ij} = C_j W_j \Pi_{ji} \tag{5}$$

In fact this condition (5) by itself implies $W A = 0$, i.e. the stationarity of W. iii) Finally, the factors χ_n must be eigen functions, i.e.

$$(6) \qquad A \chi_n = - \lambda_n \chi_n$$

In other words, the condition (5) means that the generator A must be a self-adjoint operator on $L^2(\mathbb{N},W)$, and from the condition (6), the spectrum of this operator must be discrete, with eigen values $-\lambda_n \leq 0$ ($\lambda_o = 0$). Moreover, the eigen functions, i.e. the factors χ_n, must form a Hilbertian basis of $L^2(\mathbb{N},W)$.

Explicit Construction.

In the finite case, it is not difficult to build such Markov processes. In fact, let us start from a given probability measure W_i, i=0, 1,... N and an arbitrary symmetric matrix T_{ij} such that $T_{ij} \geq 0$. Put:

$$t_i = \sum_j T_{ij} \quad ; \quad \Pi_{ij} = T_{ij}/t_i \quad ; \quad C_i = t_i/W_i$$

Then, the process defined by the generator $A_{ij} = -C_i \delta_{ij} + C_i \Pi_{ij}$ satisfies all the required conditions. Moreover, the problem of finding the eigen functions and eigen values comes back to diagonalizing the symmetric matrix $- C_i \delta_{ij} + (1/\sqrt{W_i})\, T_{ij} (1/\sqrt{W_j})$.

Naturally, once the matrix P(t) is determined, we may obtain many other models of the form

$$P(\mu) = \int P(t)\, \mu(dt),$$

where μ is a probability on $(0,\infty)$.

So, the first part of the programme, i.e. finding symmetric isofactorial distributions is really easy. The second part, i.e. finding a suitable model of change of support and asymmetric distributions, is more difficult. We shall only examine one example. Another one is given in [2].

3 - THE NEGATIVE BINOMIAL MODEL.

In the discrete version of a diffusion process N(t), the only possible transitions are $i \to i+1$ and $i \to i-1$, so that the generator A is of the form $(Af)_i = -(a_i + b_i) f_i + a_i f_{i+1} + b_i f_{i-1}$. The stationary probability W, if it exists, must satisfy the relations

(7) $$W_i\, a_i = W_{i+1}\, b_{i+1}$$

which are equivalent to relation (5), so that, if W exists, the operator A is necessarily self adjoint. Clearly, W does exist if and only if $\sum_i (a_{i-1}\, a_{i-2} \dots a_o)/(b_i\, b_{i-1} \dots b_1) < \infty$, and in this case $W_i/W_o = (a_{i-1}\, a_{i-2} \dots a_o)/(b_i\, b_{i-1} \dots b_1)$. Note that we always have $b_o = 0$, because the negative values are excluded.

The negative binomial process is of this form with

$$a_i = p(\nu+i) \quad ; \quad b_i = i \quad (\nu > 0,\ 0 < p < 1)$$

The stationary distribution always exists and is defined by $W_i = q^\nu p^i \Gamma(\nu+i)/(\Gamma(\nu) i!)$. This is the negative binomial distribution, and the corresponding generating function is $G(s) = q^\nu/(1-ps)^\nu$.

The conditional generating function $G_i(s,t)$ of $N(t)$ given that $N(0) = i$ is determined by the evolution equation

$$\frac{\partial}{\partial t} G_i(s,t) = A\, G_i = -[\nu\, p + i(p+1)] G_i + p(\nu+i) G_{i+1} + i\, G_{i-1}$$

and the initial condition $G_i(s,0) = s^i$. From the intuitive interpretation of the process, we expect G_i to be of the form

$$G_i(s,t) = H(s,t)\,[\gamma(s,t)]^i$$

where γ describes the descendance of one particle present at time $t = 0$, while H represents the descendance of the new particles born between times 0 and t. This leads to the equations:

$$\begin{cases} \dfrac{1}{H}\dfrac{\partial H}{\partial t} = -\nu\, p(1-\gamma) \quad ; \quad H(s,0) = 1 \\[2ex] \dfrac{\partial \gamma}{\partial t} = (1-\gamma)(1-p\gamma) \quad ; \quad \gamma(s,0) = s \end{cases}$$

The solutions are:

$$\gamma = 1 - \frac{q(1-s)\, e^{-qt}}{1-ps-p(1-s)\, e^{-qt}} \quad ; \quad H = \left(\frac{q}{1-ps-p(1-s)\, e^{-qt}}\right)^\nu$$

The eigen values are $\lambda_n = -nq$ (for the proof, see [2]). Up to a constant factor, the eigen functions $H_n(i)$ are defined by the generating functions:

$$\eta_n(s) = \sum_j W_j \, H_n(j) s^j = q^\nu (1-s)^n / (1-ps)^{n+\nu}$$

Using the notation of the hypergeometric functions, we have:

$$H_n(i) = \frac{q^\nu}{p^n} F(-n, \nu+i, \nu; q) = \frac{q^\nu}{p^n} \sum_k (-1)^k \binom{n}{k} q^k \frac{\Gamma(k+\nu+i)\,\Gamma(\nu)}{\Gamma(\nu+i)\,\Gamma(k+\nu)}$$

These functions $H_n(i)$ satisfy a recurrence relation (with respect to the variable i), i.e. $A H_n = -nq H_n$. But $H_n(i) = H_i(n)$, and this remarkable symmetry implies the very useful recurrence relations (with respect to the variable n):

$$[iq - \nu p - n(p+1)] \, H_n(i) + p(\nu+n) \, H_{n+1}(i) + n \, H_{n-1}(i) = 0$$

with the initial conditions $H_o(i) = 1$, so that the numerical calculations are very easy.

The functions H_n are orthogonal, but not orthonormal. In fact, we have:

$$\sum_i W_i \, H_n(i) \, H_m(i) = \frac{\delta_{nm}}{A_n^\nu p^n} \qquad \left(A_n^\nu = \frac{\Gamma(n+\nu)}{\Gamma(\nu) n!}\right)$$

Moreover, $H_n(i)$ is a polynomial of degree n, so that the factors H_n form a basis of $L^2(\mathbb{N}, W)$, i.e.:

$$\sum_{n \geq 0} A_n^\nu \, p^n \, H_n(i) \, H_n(j) \, W_j = \delta_{ij}$$

The transition matrix of the process is defined by

$$P_{ij}(t) = \sum_{n \geq 0} e^{-nqt} \, A_n^\nu \, p^n \, H_n(i) \, H_n(j) \, W_j \tag{8}$$

and the stationary distribution of N(0) and N(t) has the bivariate generating function:

$$G(s,s';t) = \left(\frac{q}{1-ps}\right)^\nu \left(\frac{q}{1-ps'}\right)^\nu \sum_{n \geq 0} A_n^\nu \, p^n \left(\frac{1-s}{1-ps}\right)^n \left(\frac{1-s'}{1-ps'}\right)^n e^{-nqt} \tag{9}$$

Change of Support.

If we put t=0 in the relation (9), we obtain the generating function of two binomial negative variables N and N', which are almost surely equal:

$$(10) \qquad \left(\frac{q}{1-pss'}\right)^{\nu} = \sum_{n\geq 0} A_n^{\nu}\, p^n \,\frac{q^{\nu}(1-s)^n}{(1-ps)^{n+\nu}}\, \frac{q^{\nu}(1-s')^n}{(1-ps')^{n+\nu}}$$

In order to obtain models of change of support, we will transform this distribution in two different ways. The first transformation is a decimation: if the first variable N (for instance) represents a number of particles, we kill each of them, independently, with the same probability ϖ, i.e. we replace s by $\chi + \varpi s$ ($\chi = 1-\varpi$), so that (10) becomes:

$$\left\{ \begin{array}{l} \left(\dfrac{qq'}{q'-(p-p')s-p'qss'}\right)^{\nu} = \displaystyle\sum_{n\geq 0} A_n^{\nu}\, p'^{\nu}\, \frac{q'^{\nu}(1-s)^n}{(1-p's)^{n+\nu}}\, \frac{q^{\nu}(1-s')^n}{(1-ps')^{n+\nu}} \\[2ex] p' = p\varpi/(1-p\chi) < p \end{array} \right.$$

The corresponding bivariate distribution may be written:

$$(11) \qquad W_{ij}^{\nu}(p',p) = W_i^{\nu}(p')\, W_j^{\nu}(p) \sum_{n\geq 0} A_n^{\nu}\, p'^{\nu}\, H_n^{\nu}(i,p')\, H_n(j,p)$$

with $p' < p$. We have $N \leq N'$ a.s., so that the first variable N must be associated with the sample v, and the second one with the block V.

The second transformation is an addition: the second variable N' is replaced by N' + N", where N" is a negative binomial variable (p,ν'') independent of N and N', so that (10) becomes:

$$(12) \qquad \left(\frac{q}{1-pss'}\right)\left(\frac{q}{1-ps'}\right)^{\nu''} = \sum_{n\geq 0} A_n\, p^n\, \frac{q^{\nu}(1-s)^n}{(1-ps)^{n+\nu}}\, \frac{q^{\nu+\nu''}(1-s)^n}{(1-ps)^{\nu+\nu''}}$$

In other words, the first transformation replaces p by $p' < p$ for the first variable, and the second one replaces ν by $\nu' = \nu + \nu'' > \nu$ for the second variable.

Combining these two transformations, we obtain the following model for the bivariate distribution of N_v and N_V, with $p_v \leq p_V$

and $\nu_v < \nu_V$:

$$(13)\quad W_{ij}^{vV} = W_i^{\nu_v}(p_v)\, W_j^{\nu_V}(p_V) \sum_{n\geq 0} A_n^{\nu_v}(p_v)^n\, H_n^{\nu_v}(i,p_v)\, H_n^{\nu_V}(j,p_V)$$

Clearly, this bivariate distribution exchanges the factors $H_n^{\nu_v}$ and $H_n^{\nu_V}$ according to the rules:

$$(14)\quad \begin{cases} E[H_n^{\nu_v}(N_v;p_v)/N_V = j] = H_n^{\nu_V}(j,p_V) \\[2ex] E[H_n^{\nu_V}(N_V;p_V)/N_v = i] = \dfrac{A_n^{\nu_v}\, p_v^n}{A_n^{\nu_V}\, p_V^n}\, H_n^{\nu_v}(i,p_v) \end{cases}$$

Taking into account that $m_v/m_V = v/V$, the first of the two relations (14), written with n=1, clearly implies $E(N_v/N_V)=v/V\ N_V$. In other words, the distribution (13) satisfies Cartier's condition (2), so that it is an admissible model of change of support.

The two parameters ν_V and p_V must be chosen in such a way that the expectation $E(N_V) = m_V$ and the variance $Var(N_V) = \sigma_V^2$ have the correct values: this is possible, because we have two degrees of freedom. In order to define the model completely, we have to choose the block/block distributions (see Section 2). The simplest possibility is:

$$W_{ij}^{V_aV_b} = \sum_{n\geq 0} (\rho_{ab})^n\, A_n^{\nu_V}(p_V)^n\, H_n^{\nu_V}(i,p_V)\, H_n^{\nu_V}(j,p_V)\, W_j^{\nu_V}(p_V)$$

where ρ_{ab} is the correlation coefficient between N_{V_a} and N_{V_b}. But another model is possible and may be more interesting.

4 - A DISCRETE/CONTINUOUS MODEL.

If the blocks V are large enough, it may be advantageous to replace the discrete distribution of N_V by a continuous one: in practice, this will be the case when the probability of having $N_V = 0$ becomes negligible. This leads to mixed models, where the sample variable N_v remains discrete, while the blocks are associated with a continuous variable X_V.

If N is negative binomial (ν,q), then if $q \to 0$ with ν fixed, the distribution of N_q tends to a limit which is the gamma distribution with the same index ν. More generally, by replacing s by $\exp(-\lambda q)$, the generating function $\eta_n(s)$ of the factors H_n becomes $\lambda^n/(1+\lambda)^{n+\nu}$: this limit is the Laplace transform of the function $g_\nu(x)\ L_n^\nu(x)$, where $g_\nu(x) = (x^{\nu-1}/\Gamma(\nu))\exp(-x)$ is the density function of the gamma distribution of index ν, and the $L_n^\nu(x)$ are the corresponding Laguerre polynomials:

$$L_n^\nu(x) = F(-n,\nu;x) = \sum_k (-1)^k \binom{n}{k} \frac{\Gamma(\nu)}{\Gamma(\nu+k)} x^k$$

These polynomials are orthogonal, but not orthonormal, with respect to the gamma distribution g_ν. Their norms are given by:

$$\|L_n^\nu\|^2 = \int_0^\infty \left(L_n^\nu(x)\right)^2 g_\nu(x)\,dx = \frac{1}{A_n^\nu} = \frac{\Gamma(\nu)n!}{\Gamma(\nu+n)}$$

Moreover they satisfy the useful recurrence relations:

$$(x-\nu-2n)\ L_n^\nu(x) + (\nu+n)\ L_{n+1}^\nu(x) + n\ L_{n-1}^\nu(x) = 0$$

which make the calculations easy, by taking into account the initial condition $L_o(x) = 1$.

In this process (and with the change $t \to t/q$), the transition matrix (8) becomes the density of a transition probability, say:

$$g_t(y/x) = \sum_{n\geq 0} e^{-nt} A_n^\nu\ L_n^\nu(x)\ L_n^\nu(y)\ g_\nu(y)$$

The corresponding Markov process admits the generator $Af = xf'' + (\nu-x)f'$, and as its stationary distribution g_ν. In other words, the discrete negative binomial process is changed into a continuous gamma process, and the Laplace transform of the bivariate stationary distribution is $\Phi_t = [1+\lambda+\mu+\lambda\mu(1-e^{-t})]^{-\nu}$.

Change of Support.

In the same conditions, the asymmetric distribution (11) is changed into a discrete/continuous distribution of the form:

$$(15) \qquad W_j(x) = g_\nu(x)\ W_j^\nu(p_\nu) \sum_{n\geq 0} A_n^\nu(p_\nu)^n\ L_n^\nu(x)\ H_n^\nu(j;p_\nu)$$

This distribution (15) is interesting: Conditionally if N=j is given, the continuous variable X is gamma, with the Laplace transform $(1 + \lambda q)^{-\nu-j}$. Conversely, if X=x is fixed, the discrete variable N is Poisson with E(N/x) = xp/q. Moreover:

$$(16)\qquad \begin{cases} E[H_n(N)/X = x] = L_n(x) \\ E[L_n(X)/N = i] = p_v^n H_n(i) \end{cases}$$

With n=1, the first relationship (16) implies E(N/X = x) = xp/q. Up to a constant factor, this is Cartier's condition. Note that the variable N is integer valued, and does not represent a grade. The grade of a sample v must be defined by $Z_v = N/v$. Let us denote by $m = m_v$ and σ_v^2 the mean and the variance of $Z_v = N/v$ (and not of N_v itself), i.e.: $m = m_v = \nu p/qv$; $\sigma_v^2 = \nu p/q^2v^2$. Then Cartier's condition $E[Z_v/Z_V] = Z_\nu$ is satisfied if we put $Z_V = pX/qv$.

Unfortunately, this relation determines the variance σ_V^2 of Z_V, so that it may describe only one support $V = V_o$ uniquely determined by its variance

$$\sigma_{V_o}^2 = \nu p^2/q^2v^2 = p\,\sigma_v^2$$

In order to get the degree of freedom which is lacking, two ways are possible.

First Model.

In the first way, the continuous variable X_V will be assigned an index $\nu' > \nu$. This comes back to having $q'_V \to 0$ in the relation (13) and leads to the discrete/continuous bivariate distribution:

$$(17)\qquad W_i(x) = W_i^\nu\, g_{\nu'}(x) \sum_{n\geq 0} A_n^\nu\, p^n\, H_n^\nu(i)\, L_n^{\nu'}(x)$$

which exchanges the factors according to the rules:

$$\begin{cases} E[H_n^\nu(i)/X = x] = L_n^{\nu'}(x) \\ E[L_n^{\nu'}(X)/N = i] = \dfrac{A_n^\nu\, p^n}{A_n^{\nu'}} H_n^\nu(i) \end{cases}$$

If n=1, the first of these relations implies $E(i/X=x) = xp\nu/(q\nu')$. Thus, Cartier's condition is satisfied if $Z_V = p\nu X/(qv\nu')$. Naturally, the parameter ν' will be chosen so that it respects the correct variance $\sigma_V^2 = p^2\nu^2/(\nu' q^2 v^2) = p\,\sigma_v^2\,\nu/\nu'$. Note that the condition $\nu' > \nu$ implies $\sigma_V^2 \leq \sigma_{V_o}^2 = p\,\sigma_v^2$, so that this model only holds for supports $V \supset V_o$, i.e. $\sigma_V^2 \leq p\,\sigma_v^2$.

In this model, the grades Z_V and $Z_{V'}$ of two different blocks V and V' are

$$Z_V = p\nu X/(qv\nu') \quad ; \quad Z_{V'} = p\nu Y/(qv\nu')$$

and the density of the two variables (X,Y) is:

$$g_\rho(x,y) = g_{\nu'}(x)\, g_{\nu'}(y) \sum_{n\geq 0} \rho^n A_n^{\nu'} L_n^{\nu'}(x)\, L_n^{\nu'}(y)$$

where $\rho = \rho_{VV'}$ is the correlation coefficient of V and V'. According to Section 2, this completely determines the model. But there exists a second possibility, which is probably better, for reasons which will become apparent in the last section.

Second Model.

In this second model, the parameter ν remains constant. Once the critical support $V = V_o$ as defined above is attained, the grades Z_V have continuous distributions which are defined by anamorphoses $Z_V = \phi_V(X_V)$, where X_V does obey the gamma distribution g_ν with the fixed index ν. If $V = V_o$, we already know this anamorphosis, which is $Z_{V_o} = p\,X_{V_o}/(qv)$. If $V \supset V_o$, Cartier's condition $E(Z_{V_o}/Z_V) = Z_V$ implies $\phi_V(X_V) = (p/(qv))\, E(X_{V_o}/X_V)$. Hence, if we assume that (X_{V_o}, X_V) have the symmetrical density:

$$g_\rho(x_{V_o}, x_V) = g_\nu(x_{V_o})\, g_\nu(x_V) \sum_{n\geq 0} \rho^n A_n^\nu L_n(x_{V_o})\, L_n(x_V)$$

this implies:

$$Z_V = \phi_V(X_V) = (p/qv)\left[\nu + \rho(X_V - \nu)\right]$$

Hence, the anamorphosis is of the "affine correction" type, as is the case whenever the first factor is linear with respect to the variable of interest. Naturally, the parameter ρ will be chosen in such a way that the correct variance will be $\sigma_V^2 = p^2\rho^2\nu/(qv)^2 = p\rho^2\sigma_v^2$. Now, we have $m = E(N_v/v) = \nu p/qv$, so that:

$$(18) \qquad Z_V = \phi_V(X_V) = m[1-\rho + \rho\, X_V/\nu]$$

This leads in an obvious way to a model of change of support. Let v be a sample inside V. Then, the discrete/continuous distribution of (N_v, Z_V) may be obtained by the simple linear transformation (18) from the distribution of (N_v, X_V) which is:

$$(19) \qquad W_i(x) = W_i^\nu\, g_\nu(x) \sum_{n\geq 0} A_n^\nu\, p^n\, \rho^n\, H_n(i)\, L_n(x)$$

Once the model of change of support (19) is known, we complete the determination of the isofactorial model by taking:

$$g(x,x') = g_\nu(x)\, g_\nu(x') \sum_{n\geq 0} r^n A_n^\nu\, L_n(x)\, L_n(x')$$

as the distribution density of $(X_V, X_{V'})$, where r is the correlation coefficient of Z_V and $Z_{V'}$.

5 - COMPARISONS WITH TRUE DISTRIBUTIONS.

Now, we shall try to compare the predictions of our models of change of support with the true distributions which can be obtained in certain cases.

5-1. Continuous Case.

We consider the Markov process, defined by its generator A: $Af = xf'' + (\nu-x)f'$, which has g_ν as its stationary distribution. Let us put $Q_t = \int_o^t X_\tau\, d\tau$ and $\Phi_t(\lambda,\mu;x) = E[\exp(-\lambda Q_t - \mu X_t)/X_o = x]$. Clearly, the function Φ_t is determined by the evolution equation:

$$\frac{\partial \Phi_t}{\partial t} = x \frac{\partial^2 \Phi_t}{\partial x^2} + (\nu - x) \frac{\partial \Phi_t}{\partial x} - \lambda x \Phi_t$$

and the initial condition $\Phi_o = \exp(-\mu x)$. The solution is of the form $\Phi_t = H_t(\lambda,\mu) \exp[-x\psi_t(\lambda,\mu)]$, where the H_t and ψ_t do not depend on x and satisfy the relations:

$$\begin{cases} \dfrac{1}{H_t}\dfrac{\partial H_t}{\partial t} = -\nu\psi_t & ; \quad H_o = 1 \\ \dfrac{\partial \psi_t}{\partial t} = -\psi_t^2 - \psi_t + \lambda & ; \quad \psi_o = \mu \end{cases}$$

If we put, for convenience, $C = \sqrt{1 + 4\lambda}$, the solution is:

$$\begin{cases} H_t(\lambda,\mu) = \left(\dfrac{C \exp(t/2)}{C\,Ch(Ct/2) + (1+2\mu)\,Sh(Ct/2)} \right)^{\nu} \\ \psi_t(\lambda,\mu) = \dfrac{\mu\left[C\,Ch(Ct/2) + Sh(Ct/2)\right] + 2\lambda\,Sh(Ct/2)}{C\,Ch(Ct/2) + (1+2\mu)\,Sh(Ct/2)} \end{cases}$$

We shall examine only the stationary distribution of Q_t, the Laplace transform of which is $\Phi_t(\lambda) = H_t(\lambda,0)/\left[1 + \psi_t(\lambda,0)\right]^{\nu}$, or explicitly:

$$(20) \quad \begin{cases} \Phi_t(\lambda) = \left(\dfrac{C \exp(t/2)}{\left[C\,ch(Ct/4) + Sh(Ct/4)\right]\left[Ch(Ct/4) + C\,Sh(Ct/4)\right]} \right)^{\nu} \\ C = \sqrt{1 + 4\lambda} \end{cases}$$

This is the rigorous solution of the change of support for the gamma Markov process. On the other hand, the (approximate) model, based upon gamma anamorphoses, leads to a prediction of the type "affine correction", i.e. $Q_t = m_t + \sigma_t(X - \nu)/\sqrt{\nu}$ where $m_t = \nu t$, $\sigma_t^2 = 2\nu(\exp(-t)-1+t)$ and X is gamma, ν. By putting $\rho = \rho(t) = \left[2(\exp(-t)-1+t\right]^{1/2}/t$, we have $Q_t = \nu t(1-\rho) + \rho t X$, where X is gamma. Thus, this model leads to the prediction:

$$(21) \qquad \Phi_t^*(\lambda) = \exp\left[-\lambda\nu t(1-\rho)\right]/\left[1 + \lambda\rho t\right]^{\nu}$$

From a numerical point of view, the agreement between the rigorous formula (20) and the approximate one (21) seems very good. But the significance of this agreement is not easy to interpret. A more tractable criterion is given by the order-3 cumulant $\chi_3 = (\partial^3/\partial\lambda^3) \log \Phi_t(\lambda)$. For the exact distribution (20) we find:

$$(22) \qquad \frac{\chi_3(t)}{(\sigma_t)^3} = \frac{6(t + t\exp(-t) - 2 + 2\exp(-t))}{\sqrt{2}\,(\exp(-t) - 1 + t)^{3/2}} \frac{1}{\sqrt{\nu}}$$

On the other hand, the model (21) leads to the same value as the gamma distribution itself, i.e. $2/\sqrt{\nu}$, which coincides with the value of (22) at t=0. But it turns out that $\chi_3/(\sigma_t)^{3/2}$ remains almost constant if t is not too large:

t	$\chi_3\sqrt{\nu}/\sigma_t^3$	t	$\chi_3\sqrt{\nu}/\sigma_t^3$
0	2	1.5	1.9384
0.1	1.99967	2	1.8985
0.5	1.99215	3	1.8056
1	1.97060	4	1.7071

Thus, we may expect a good agreement between the two distributions, at least for $t \leq 2$.

In fact, an explicit inversion of (20) is possible at least in the case $\nu=1$, and makes a direct comparison possible. From the general results presented in [3] we know that $\Phi_t(\lambda)$ is of the form:

$$\Phi_t(\lambda) = \prod_{n\geq 0} \left(\frac{1}{1 + \lambda\lambda_n}\right)^{\nu}$$

so that Q_t is a sum $\Sigma\, \lambda_n S_n$, where the S_n are independent ν gamma variables, and the λ_n are the eigen values of the covariance considered as a kernel operator on the segment $(0,t)$. In the present case, the quantities $b_n = 1/\lambda_n$ are the roots of the function $[\mathrm{Ch}(Ct/4) + (1/C)\,\mathrm{Sh}(Ct/4)][\mathrm{Ch}(Ct/4) + C\,\mathrm{Ch}(Ct/4)]$, $C = \sqrt{1 + 4\lambda}$. Thus, by putting $\lambda = -(1/4 + 4\omega^2/t^2)$, we have to find the roots of $f(\omega) = [\cos\omega + (t/4\omega)\sin\omega][\cos\omega - (4\omega/t)\sin\omega]$. These roots are $\pm\omega_n$, $n = 0, 1\ldots$ with $\omega_n = n(\pi/2) + \varepsilon_n$ and $\tan\varepsilon_n = t/2[n\pi + 2\varepsilon_n]$ $(0 < \varepsilon_n < \pi/2)$. The numerical calculation is very easy. Hence we find the ω_n, and the $b_n = 1/\lambda_n = 4\omega_n^2/t^2 + 1/4$. For n large, $b_n = n^2\pi^2/t + 2/t + 1/4 + 0(t/n)$.

TABLE 1

True and approximate values of 1-F and K, for t=1

x	1-F / 1-F*	K / K*
x = 0.14	1-F = 0.96775	K = 0.99653
	1-F* = 1	K* = 1
x = 0.16	0.95399	0.99447
	0.97950	0.99690
x = 0.25	0.87873	0.97893
	0.88194	0.97698
x = 0.5	0.66911	0.89919
	0.65896	0.89472
x = 1	0.36995	0.68562
	0.36788	0.68343
x = 2	0.11460	0.32697
	0.11466	0.32766
x = 3	0.03550	0.13678
	0.03573	0.13785
x = 5	0.00341	0.01994
	0.00347	0.02033
x = 7	0.00033	0.00257
	0.00034	0.00265

In the case $\nu=1$, the Laplace transform $\Phi_t(\lambda)$ is

$$\Phi_t(\lambda) = \frac{\exp(t/2)}{f(\omega)} = \prod \frac{\omega_n^2 + t^2/16}{\omega_n^2 - \omega^2}$$

This is a Weierstrass type expansion in infinite product, and classically we have $1/f(\omega) = \Sigma C_n/(\omega_n^2 - \omega^2)$ with $C_n = -2\omega_n/f'(\omega_n)$. After some calculations, we conclude:

$$\phi_t(\lambda) = \Sigma B_n/(b_n + \lambda) \quad ; \quad B_n = (-1)^n \frac{4b_n - 1}{2(1 + b_n t)} \exp(t/2)$$

so that the density of Q_t is $f_t(x) = \Sigma B_n \exp(-b_n x)$, this expression being convergent for $x > 0$, and:

$$(23) \quad \begin{cases} 1 - F_t(x) = \Sigma (B_n/b_n) \exp(-b_n x) \\ K_t(x) = \Sigma (B_n/b_n)(x + 1/b_n) \exp(-b_n x) \end{cases}$$

where $1 - F_t$ is the tonnage $\geq x$, and $K_t(x) = \int_x^\infty x f_t(x)\, dx$ is the quantity of metal $\geq x$.

Table 1 presents, in the case $t=1$, the exact values of $1 - F_t$ and K_t, computed from (23), and the approximate values predicted by the model (21). The model (21) predicts a density $f_t(x) = 0$ for $x \leq 0.15$: the real density is > 0, but very small, so that the agreement remains admissible. If $x > 0.15$, the agreement seems extraordinarily good, and it is always the case for $t \leq 2$ or 2.5.

5-2. The Discontinuous Case.

If $N(t)$ is the negative binomial Markov process considered in Section 3, we may consider various variables: $Q_t = \int_0^t N(\tau)\, d\tau$, $N^+(t)$: number of positive transitions $i \to i+1$ and $N^-(t)$: number of negative transitions between the times 0 and t. The conditional 4-variate distribution of Q_t, $N^+(t)$, $N^-(t)$ and $N(t)$ given that $N(0) = i$ is given in [2]. The most interesting distribution is the (discrete) stationary distribution of $N^+(t)$. Its generating function is:

$$
(24)\quad \begin{cases} G^{+}(s,t) = \\ \left[\dfrac{q\ C\ \exp(qt/2)}{\left[q\ \mathrm{Ch}(Ct/4) + C\ \mathrm{Sh}(Ct/4)\right]\left[C\ \mathrm{Ch}(Ct/4) + q\ \mathrm{Sh}(Ct/4)\right]} \right]^{\nu} \\ C = \sqrt{(1+p)^2 - 4ps} = \sqrt{q^2 + 4p(1-s)} \end{cases}
$$

If $\nu \to 0$ and $t \to \infty$, in such a way that $\nu t = b = C^{ste}$, we find:

$$
G^{+}(s,t) \to \exp\left[\frac{bq}{2}\left(1 - \sqrt{1 + \frac{4p}{q^2}(1-s)}\right)\right]
$$

This limit is the SICHEL-distribution. Note also that the distribution (24) may be obtained from (20) by replacing t by qt and λ by $(4p/q^2)(1-s)$: the distribution (24) is a Poisson mixture directed by the distribution (20).

By using the results of Section 5-1, it is possible to calculate the numerical values of the probabilities associated with the generating function (24), and to compare them with the predictions provided by the two models: the first one is the model (13), which predicts for $N^{+}(t)$ a negative binomial distribution (ν',p'), with ν' and p' determined by the mean and the variance. The second model is obtained by replacing λ by $(4p/q^2)(1-s)$ in the approximate continuous model (21): if $\nu=1$, this implies that $N^{+}(t)$ is the sum of a Poisson and a Pascal independent variables. Numerical values are given in [2]. It turns out that the two models are good, but the second one is by far the better.

REFERENCES.

Isofactorial models originate from Quantum Mechanics, where they have been widely and systematically used since the twenties. See for instance [4]. In particular, Hermite polynomials appear as eigen functions in the case of the one dimensional harmonic oscillator. Among statisticians, this Hermitian model and some others are mentioned by Cramer [5] as early as 1945. Isofactorial models were also used in the field of Markov processes, see [6]. In data analysis, they constituted the starting point of correspondence analysis, [7], [8]. In Geostatistics, they have been used since 1973, [9], [1]. Note well that in general the factors are not polynomials. But, at the request of the reviewer, I also mention the fact that reference [10] discusses orthogonal polynomials.

1. MATHERON, G., "A Simple Substitute for Conditional Expectation: the Disjunctive Kriging", Proc. NATO ASI, "Advanced Geostatistics in the Mining Industry", D. Reidel, p. 221-236. 1976.

2. MATHERON, G., "Modèles Isofactoriels et Changement de Support", CGMM, Fontainebleau, 1983.

3. MATHERON, G., "Remarques sur les Changements de Support", CGMM, Fontainebleau, 1981.

4. LANDAU, L.D., and LIFSCHITZ, E.M., "Kvantovaïa Mekanika", Mir, Moscow, 1974.

5. CRAMER, H., "Mathematical Methods of Statistics", Princeton, 1945.

6. FELLER, W., "An introduction to Probability Theory and its Applications", Vol. 1, J. Wiley and Sons, 1957.

7. BENZECRI, J.P., "L'Analyse des Données", Dunod, Paris 1973.

8. NAOURI, "Analyse Fonctionnelle des Correspondances Continues", Thesis, Paris, 1972.

9. MATHERON, G., "Le Krigeage Disjonctif", CGMM, Fontainebleau, 1973.

10. BECKMANN, P., "Orthogonal Polynomials for Engineers and Physicists", The Golem Press, Boulder, Colorado, 1973.

GEOSTATISTICAL APPLICATION IN TABULAR STYLE LEAD-ZINC ORE AT PINE POINT CANADA

Gary F. Raymond

Senior Mine Project Engineer, Cominco Ltd.

ABSTRACT. Tabular style lead-zinc ore at Pine Point, Canada, occurs as a number of small, flat lying, very erratically mineralized orebodies. This paper describes geostatistical studies based on back analysis of exploration and blasthole samples from one of the first of these orebodies mined by open pit methods. Conditional probability ore estimates from exploration drillhole assays are compared to kriged estimates from blasthole assays and to ore actually mined. Improvements to both local exploration ore estimation and pit design are indicated. As a result of this study, the conditional probability method is now being used for pit design in tabular ore at Pine Point, although the more familiar manual estimates are presently being retained for ore reserves and as a check on pit design.

1. INTRODUCTION

Tabular style lead-zinc ore at Pine Point exists as a number of small, flat lying, very erratically mineralized orebodies surrounded by barren waste rock. Both exploration and production ore estimates are a problem. To examine the potential for applying geostatistics in tabular ore, a detailed back analysis was made of a 3.2 million tonne (3.5 million ton) mined out orebody based on exploration drillhole (DH) grades, production blasthole (BH) grades, and actual mined ore. Ore/waste boundaries estimated using ordinary kriging of BH grades were compared to actual mined limits based on visual estimation during mining. Exploration conditional probability estimates were calculated using ordinary kriging, relative variograms, and a 3 parameter lognormal approximation to conditional distributions. Comparisons were

G. Verly et al. (eds.), Geostatistics for Natural Resources Characterization, Part 1, 469–483.

made to present exploration estimates based on a modified polygon technique, and to milled ore. The effect on economic pit design was examined.

Indicated improvements in profitability over the original pit design provided the incentive to extend the conditional probability method to exploration estimates in new orebodies. Further refinements to the method included correction for sample bias and limiting estimates by kriging variance. Variogram calculation from exploration DH's in new orebodies remains an unsolved problem. Instead, the relative variogram shape observed from BH studies is assumed fundamental, and the magnitude of estimation variances is modified by observed errors from jackknife analysis of exploration DH's in the new area.

2. THE MINE

Pine Point Mines Ltd., owned 69% by Cominco Ltd., is located in Northwest Territories, Canada. From the start of production in 1964 to the end of 1981, the mine had produced about 50 million tonnes (55 million tons) of ore averaging about 3.0% Pb and 6.5% Zn, mainly using open pit methods with large equipment - generally 11.3 m^3 (15 cu. yd.) loaders. The property which is about 55 km long by 20 km wide (about 35 by 12 miles), includes several dozen known orebodies and is still only partly explored. Exploration is geared to maintaining a suitable ore reserve ahead of production.

Geologically (5), the orebodies are of Mississippi Valley-type, deposited in very gently, west-dipping carbonates along a Middle Devonian barrier reef complex. Development of high porosity by dolomitization and subsequent karsting provided depositional sites for sphalerite, galena, and pyrite-marcasite mineralization. Geometry of the ore is controlled mainly by geometry of the voids prior to mineralization. Orebodies at Pine Point are classified by geometry as either "prismatic" or "tabular".

Prismatic ore occurs within collapse breccias cutting stratigraphy as roughly spherical, massive, high grade orebodies with well-defined boundaries. These orebodies can be located by geophysical methods and have been the major ore producers to present time. In contrast, tabular mineralization was more lithologically controlled resulting in lower grade, erratically mineralized, thin and horizontally extensive orebodies. Tabular orebodies are generally elongated along the trend of the barrier, but vary dramatically in local horizontal and vertical extent. With the depletion of prismatic orebodies, tabular mineralization is becoming increasingly important as an ore source at Pine Point. The geostatistical approach to both exploration and production grade estimates in tabular ore appeared to be a logical extension of proven geostatistical applications, by this author, in erratically mineralized copper orebodies (3,4).

Traditionally, Pine Point ore reserve estimates in tabular ore have been done manually by compositing DH's in bench heights centered on the mineralization and using a modified polygon technique and dilution factors. Polygons are modified by grouping together higher grade exploration intersections which appear to be familiar shapes geologically. For example, a row of higher-grade composites aligned along the trend of the barrier is a probable indication of solution channelway mineralization. Internal waste may be included with groups of higher-grade intersections. For pit design, groups of composites are assigned their average grade to reduce the local influence of unusually high grade intersections. At the production stage, sorting of ore and waste is done by visual inspection of the mining face using BH assays as an initial guide only. Where there is a sharp contrast between dark ore and light-coloured host rock, visual sorting is very effective; however, this is not always the case.

Of the tabular orebodies, L37 orebody was one of the largest and was one of the first to be mined by open pit methods. Initial geostatistical study involved an analysis of L37 grade estimates from exploration through to produced mill head grades. If good estimates could be obtained in a known orebody, the method could be extended with confidence to new orebodies.

3. THE L37 BLASTHOLE STUDY

Because of anticipated problems in calculating geostatistical parameters from the limited exploration DH assays, geostatistical study of L37 was begun from BH assays. An equally important objective of the BH study was to determine the relationship between mined ore and ore estimated by kriging BH assays. This would indicate the potential for using BH kriging to determine production ore-waste boundaries, and would provide production grade estimates on a regular grid for exploration comparisons. Pb, Zn and Fe grades from 6200 DH's, taken on about 6 m (20 ft.) centres from upper and lower halves of each of the two major 7.6 m (25 ft.) mining benches were used in the study. Initial study was confined to composited 7.6 m (25 ft.) benches which represented the normal mining height.

Figure 1 shows BH assays and mined ore from the second (lower) 7.6 m (25 ft.) bench in L37. BH grades are spotty, producing a salt-and-pepper pattern with more chance of ore samples in high grade areas than in low grade areas. However, visual estimates (shaded in Figure 1) show the ore to be much more continuous, at least on a scale that can be mined. Geologically, the elongated northerly and southerly portions of the orebody

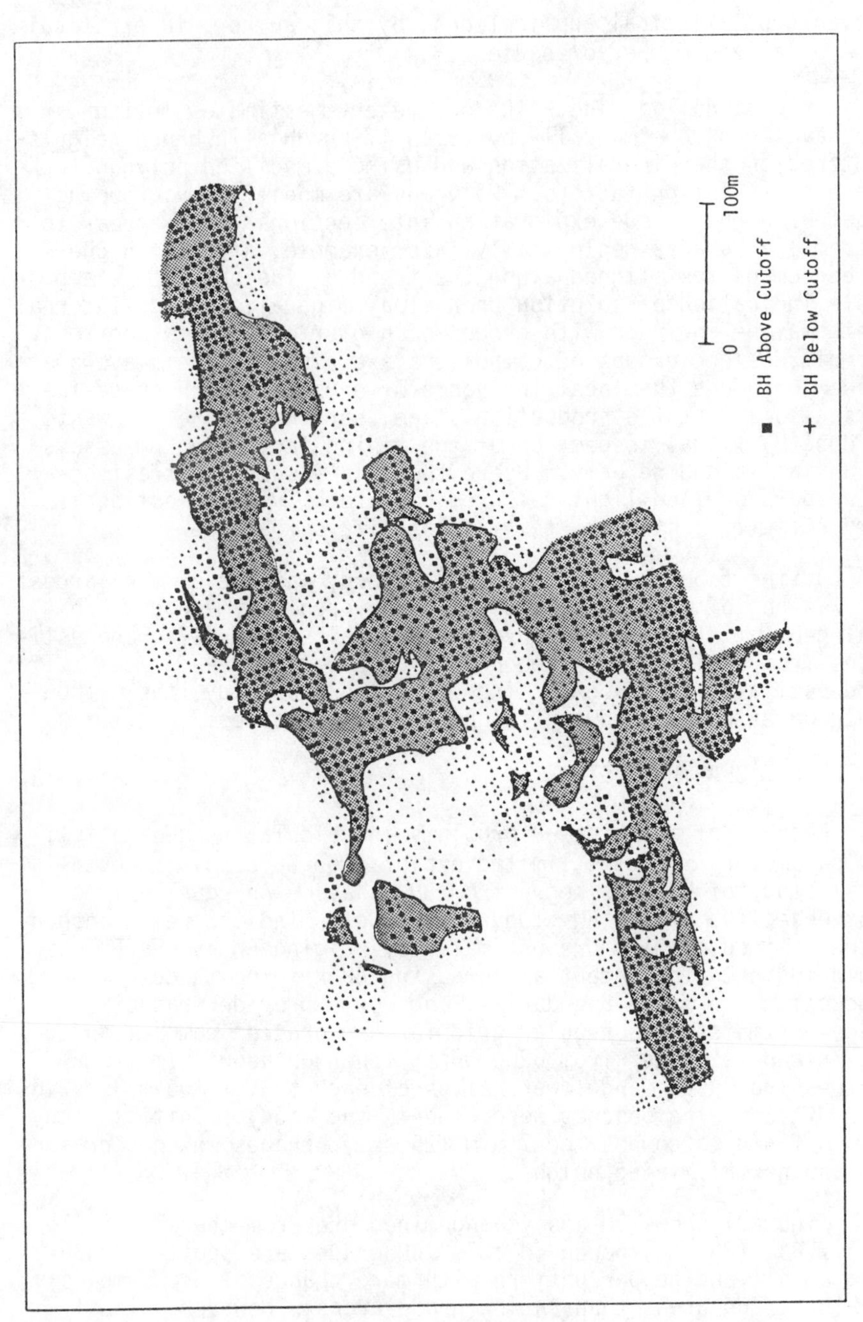

Figure 1. Ore Limits Mined By Visual Estimates

represent high grade solution channelway ore. The central area represents lower grade bedding-replacement ore, disseminated ore and a few local collapses. There is an obvious influence of the crescent shape of the mining face on these ore/waste boundaries.

Geostatistical study on BH grades involved calculating relative variograms in various directions, jackknife analysis, and contouring ore/waste boundaries using ordinary kriging. Relative variograms were calculated by dividing variogram values at individual distance lags by (grade + constant) squared. This approach will be further discussed under conditional probability. Jackknife analysis involved removing BH's one-by-one, estimating a grade at their locations using ordinary kriging of nearby BH's, and comparing estimates to actual. Since the mining cutoff is determined by combined Pb+Zn grades, and since initial studies showed similar variogram shapes for Pb, Zn and Fe, detailed studies as described in this paper, were based on combined Pb+Zn grades. However, final methods have since been extended to include individual Pb, Zn and Fe estimates. Because of grade-density dependency, all statistics and estimates were density-weighted. Sample density is calculated from Pb, Zn and Fe grades assuming pure minerals and constant porosity. Because of the limited vertical extent of tabular ore, only 2 dimensional (horizontal) estimates were used in this study.

Figure 2 shows L37 BH variograms relative to $(\text{grade} + 0.5\%\ \text{Pb+Zn})^2$ calculated in along-trend and cross-trend directions ($\pm$ 22.5 degrees). The first distance lag represents only a few sample pairs. Succeeding lags represent 5,000 - 15,000 pairs. A spherical variogram model with a 2:1 geometric anisotropy has been fitted to the plot as shown. Relative to grade squared (no additive constant), the nugget effect of the variogram would be 1.4.

Despite the high variability among BH assays, the BH jackknife estimate was nearly perfectly conditionally unbiased. In practical terms this means that, <u>if we were to mine to kriged ore/waste boundaries from BH assays, we would obtain the estimated tons and grade on average</u>. This conclusion, proven elsewhere by long-term comparisons to mill head grades (3), is valid only if BH assays are unbiased and if kriging variance is not too large. Using the BH variogram and the nearest (corrected for anisotropies) 10 BH samples per estimate, ore/waste boundaries shown in Figure 3 were calculated using ordinary kriging. Kriged boundaries are very similar to those obtained by visual estimates (Figure 1). Comparisons between BH kriging and milled ore from the portion of L37 pit studied are shown in Table 1.

Note that milled ore itself is not an exact figure, being back-calculated through several stockpiles based on mill head grades and truck counts. Reconciled milled ore tonnage from L37 is thought to be an over-estimate of actual production. From

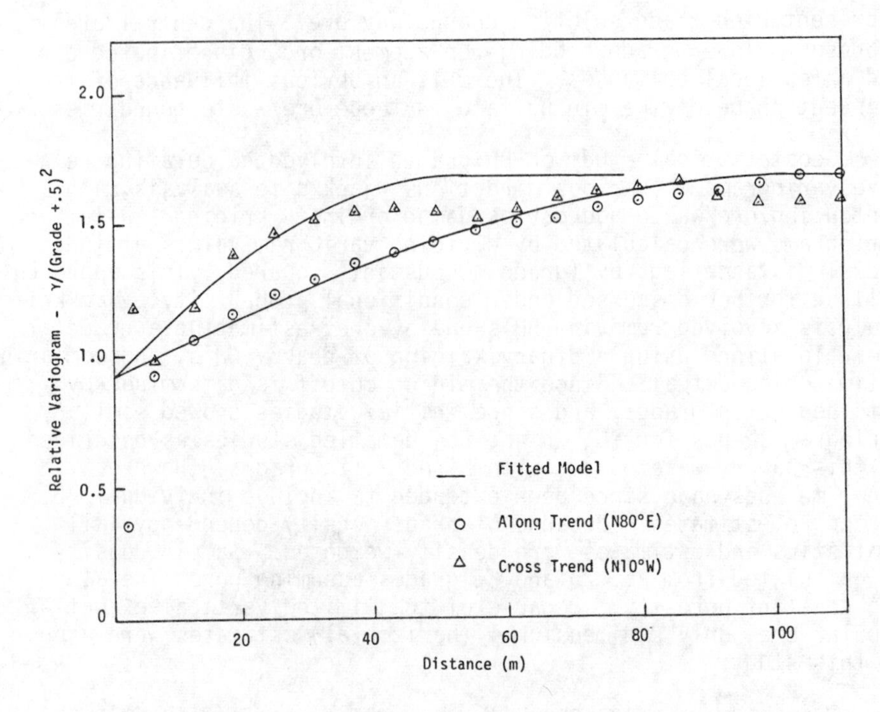

Figure 2. BH Assay Relative Variograms

	Ore Tonnes	Ore Grade	Ore Metal Tonnes
BH Kriging	2,215,000	4.42 (%Pb+Zn)	97,900
Milled Ore	2,559,000	4.54 (%Pb+Zn)	116,200
Milled/Kriging	+15.5%	+2.7%	+18.7%

Table 1. Milled Ore Versus BH Kriged Estimates

these comparisons, it was concluded that BH kriged estimates are a reasonable approximation to ore actually mined by visual sorting in L37 and are, therefore, valid for exploration back analysis. Although visual estimation methods presently used are very effective in sorting ore and waste, there may be potential to improve ore recovery, particularly in low-grade areas, by combining BH kriging and visual estimates.

4. THE L37 EXPLORATION DRILLHOLE STUDY - CONDITIONAL PROBABILITY

First, a note on the exploration sampling. Because of the large volume of exploration drilling, DH core is first visually inspected for sulphides and only potentially interesting samples are assayed. Samples rejected by visual inspection are assigned

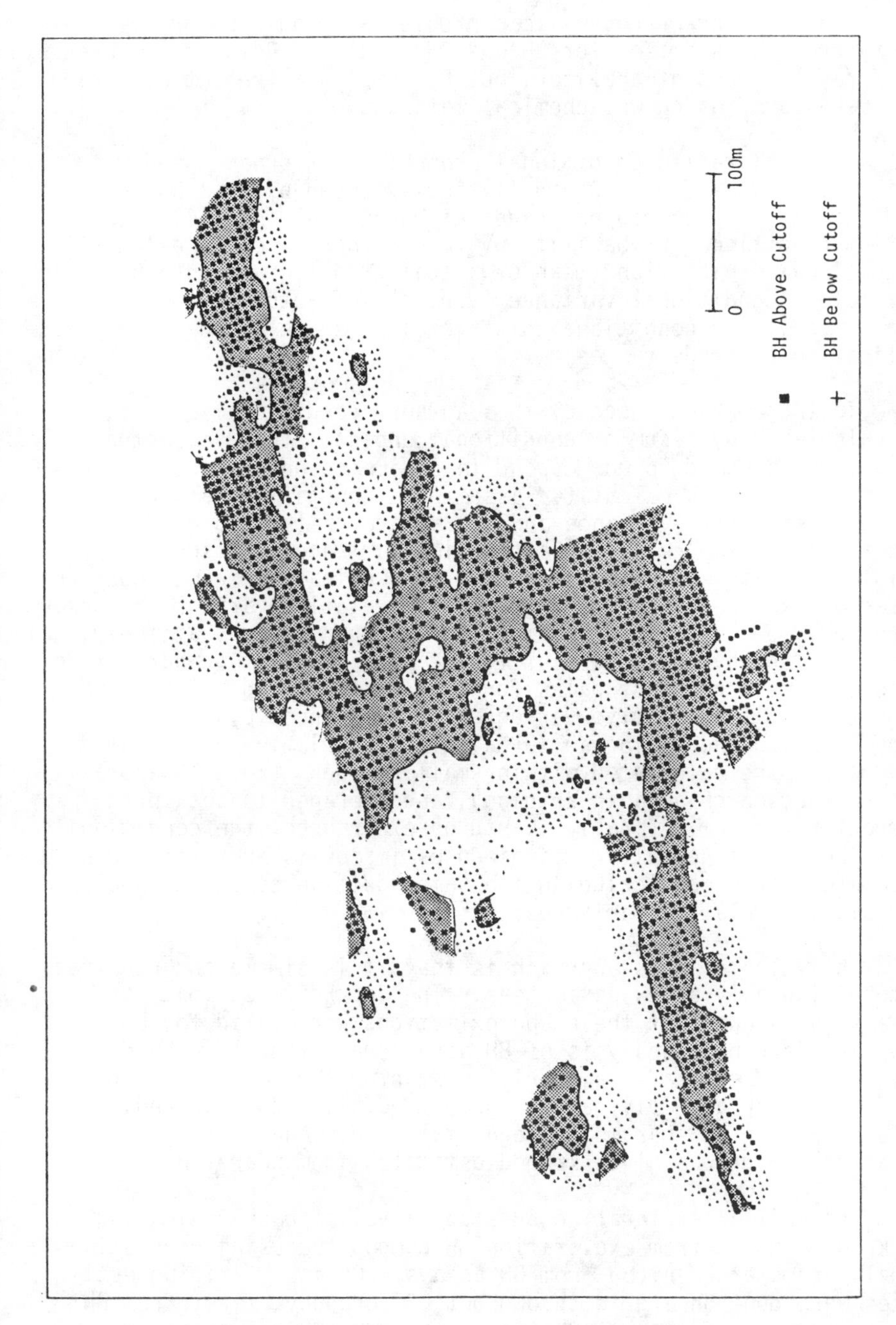

Figure 3. Ore Limits By Kriging BH Assays

zero grades. Core recovery is a problem and could be grade-related. Assay procedures differ for BH and DH samples. Both are assayed by an X-ray fluorescent analyzer, but DH samples above cut-off grade are re-assayed using wet chemical methods.

The exploration conditional probability approach used was originally suggested by David (1) and extended by this author in a detailed study of copper grades at Mount Isa Mines (4). Calculating conditional probability by this method involves calculating conditional expectation (mean of actual grades given an exploration estimate), conditional variance (variance of actual grades about their mean), and conditional distribution (shape of the grade classification by grade ranges).

In the approach used by this author, conditional expectation is calculated by assuming conditional unbias: that is, actual grades are assumed to correspond to estimates on average. Conditional variance is calculated assuming a constant relationship between estimation variance and the square of (kriged grade + constant). This "relative" estimation variance corresponds to kriging variance as calculated from the ordinary kriging equations based on the "relative" variogram. The relative variogram as used here would be calculated by dividing variogram values at individual distance lags by the square of (grade plus constant), where grade refers to mean grade of pairs for that distance lag. Since the magnitude of the relative variogram is influenced by the size of area in which the variogram is calculated, relative variograms may have to be calculated within smaller blocks (roughly approximating the search area in kriging), and averaged to obtain the correct variogram. Finally, based on experience, the conditional distribution of actual grades given estimates is approximated as 3 parameter lognormal including the same additive constant used in calculating relative variances.

The value of this approach is that it is simple and requires determining only a few parameters. The jackknife estimate of BH assays had shown that these approximations were valid for L37. That is, jackknife analysis of BH assays had produced a conditionally unbiased estimate, a nearly constant ratio between observed estimation variance divided by (grade + 0.5% Pb+Zn) squared, and an approximately 3 parameter lognormal conditional distribution of actual BH grades given kriged estimates from nearby BH's.

The initial exploration DH study involved back-estimating BH kriged grades from exploration DH composites using geostatistical parameters derived from BH assays. Ordinary kriging estimates were done on a grid throughout L37 orebody, first from BH assays and then from exploration DH assays. BH estimates were averaged within ranges of DH estimates.

Resulting comparisons showed a regression effect: high grade exploration estimates tended to over-estimate BH estimates, and vice versa. There was also an apparent bias in total Pb+Zn estimates: the mean of exploration estimates was 7% lower than the mean of BH estimates. As a preliminary measure, the regression effect was corrected for by fitting a curve to the observed relationship. Since observed estimation variances (errors) were nearly constant when divided by (grade + 0.5% Pb+Zn) squared, a constant relationship was assumed and the observed magnitude of relative estimation variances was used. Conditional distributions of BH kriged estimates given exploration kriged estimates were assumed 3 parameter lognormal with a 0.5% Pb+Zn additive constant.

This semi-empirical approach to conditional probability gave comparisons to milled ore as shown in Table 2. Also shown in Table 2 are comparisons to original polygon exploration estimates, and to exploration kriging estimates (regression effect corrected) without conditional probabilities. The conditional probability approach gave a reasonable estimate of minable ore and was a great improvement over both ordinary kriging from exploration DH's and the original polygons.

	Ore Tonnes	Ore Grade (% Pb+Zn)	Ore Metal Tonnes
Milled Ore	3,250,000	4.45	144,600
Conditional Probability (Conditional Bias Corrected)	2,859,000	4.24	121,200
Original Polygons	2,114,000	4.96	104,900
Exploration Kriging (Conditional Bias Corrected)	2,475,000	3.67	90,800

Table 2. Milled Ore Versus Exploration Estimates

Although obtaining a closer local exploration ore estimate would certainly improve mine scheduling problems, it was felt that the major benefit of applying geostatistics would come from improved pit design. Therefore, L37 pit was redesigned based on conditional probability exploration estimates and using the same economic criteria as was used in the original pit design. However, taking advantage of the increased number of estimates, the new pit design was an incremental marginal profit design done manually from computer-produced section and plans of profits (or losses) expected by mining individual 15x15x7.5 m (50x50x25 ft.) blocks. Block profitability was calculated as probable ore tonnage and grade times ore profitability factors, less probable waste tonnage times waste costs.

Table 3 shows exploration conditional probability estimates

of ore, and waste plus overburden within the pit actually mined, and within the pit as redesigned based on conditional probabilities.

	Ore Tonnes	Ore Grade (% Pb+Zn)	Waste + OB Tonnes
Actual Mined Limits	2,859,000	4.24	12,445,000
Conditional Probability Pit	2,985,000	4.22	12,311,000

Table 3. Conditional Probability Estimates In L37 Pit

Despite large errors in the original polygon exploration estimates, a reasonably good pit design had been achieved by the Pine Point practice of averaging groups of DH composites, rather than considering individual polygons. The resulting pit design had encompassed most of the mineralized DH's. This approach had given the flexibility to further expand pit limits during mining where ore was encountered in unexpected areas. As a result, total mined tonnage was significantly more than had been anticipated from the original polygon exploration design. On the other hand, the conditional probability pit design was much closer to pit limits finally mined.

Compared even to the mined pit, the conditional probability design still gave an estimated 4.4% increase in ore tonnes and a 1.1% decrease in stripping. Although this may seem to be a small gain, in 1976 dollars this difference would have represented a $700,000 increase in profit from the pit.

5. REFINING THE L37 EXPLORATION ESTIMATES

Having justified the project, the next step was to identify the causes of the regression effect, and to relate observed estimation variances to calculated relative kriging variances. First, two sources of sample bias were identified and eliminated. Vertical variogram studies showed very similar BH assays from upper and lower halves of the same bench. This was interpreted as a sample preference for the upper half of the bench as a result of contamination of the lower sample. For this reason, and because of increased emphasis on selectivity in mining, further studies were limited to 3.8 m (12.5 ft.) BH composites from upper halves of L37 benches. Figure 4 shows probability distributions of these BH assays versus nearest (corrected for anisotropies) 3.8 m (12.5 ft.) exploration DH composites from L37. Differences between the two distributions in low grade assays appear to be an exploration sampling bias which results from the practice of assigning zero grades to samples visually estimated to be low in sulphides. About 45% of samples within L37 pit limit were assigned zero grades in this manner. Upgrading these zero grade samples to 0.5% Pb+Zn gave the same total mean grade estimate from exploration kriging as from BH assays and increased exploration conditional probability ore tonnage estimates by 10% without reducing ore grade estimates.

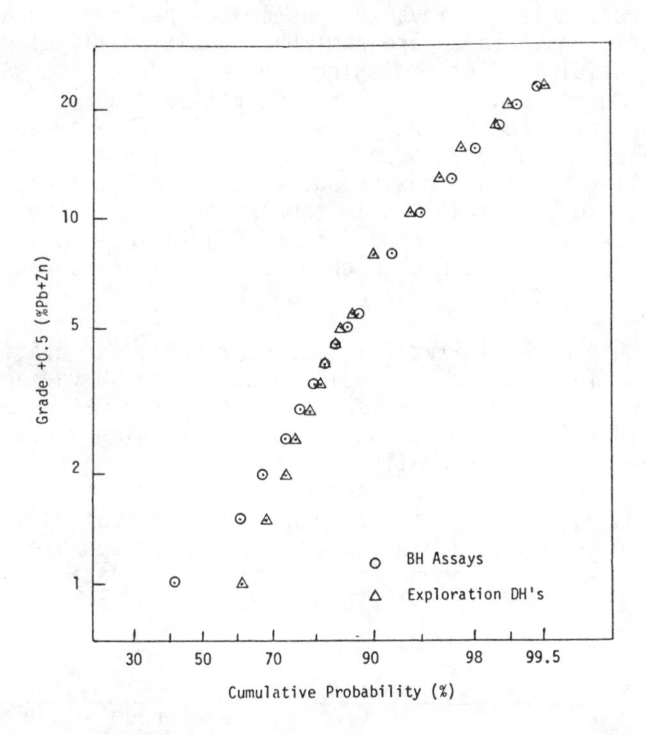

Figure 4. BH vs DH Grade Distribution

Using this new data set, kriged estimates from exploration DH's at individual BH locations were compared to actual BH assays. The variogram used was derived from the 3.8 m (12.5 ft.) BH assays, and was again relative to (grade + 0.5% Pb+Zn) squared. Comparisons were grouped by increments of exploration relative kriging variance as shown in Table 4.

Relative Kriging Variance	Total Grade (% Pb+Zn)		Above 2(% Pb+Zn)	
	Estimate	Actual BH	Estimate	Actual BH
< 0.50	2.23	2.38	4.11	3.91
0.50 - 0.68	2.29	2.02	4.16	3.28
> 0.68	2.01	1.30	3.86	1.47

Table 4. BH Assays Given Exploration Kriging - Grouped By Kriging Variance

Estimates with relative kriging variance below 0.5 are very nearly conditionally unbiased. Those above 0.68 relative kriging variance are completely invalid. Estimates between 0.5 and 0.68 relative kriging variance are somewhat conditionally biased, but are still useful in distinguishing higher grade areas. As a result of these comparisons, only exploration estimates below 0.68 relative kriging variance are presently used, and all are assumed conditionally unbiased. However, to ensure conditional unbias, final exploration drilling grids are now being designed to limit relative kriging variance to less than 0.5 in tabular ore. Limiting estimates by maximum kriging variance has an added advantage that the abrupt termination of ore contours quickly points out areas requiring more drilling.

From this more limited set of comparisons, exploration kriging variance was found to be as predicted using the BH variogram. Selectivity in mining is assumed to be as approximated by BH kriging. That is, as an approximation, BH kriging variance is subtracted from the exploration kriging variance of true grades to obtain the estimation variance of minable grades given exploration estimates. A 3 parameter lognormal distribution of actual grades given estimates again appears to be a reasonable approximation. Table 5 shows comparisons between exploration conditional probabilities and blasthole kriging using these more refined estimates.

	Ore Tonnes	Ore Grade (% Pb+Zn)	Ore Metal Tonnes
Conditional Probability			
1st Bench Upper Half	209,000	3.75	7,800
2nd Bench Upper Half	776,000	5.23	41,300
BH Kriging			
1st Bench Upper Half	233,000	4.01	9,300
2nd Bench Upper Half	764,000	4.92	37,600

Table 5. Exploration Conditional Probabilities Versus BH Kriging

Figure 5 shows resulting exploration conditional probability contours as limited by relative kriging variance, from the upper 3.8 m (12.5 ft.) of the second bench of L37.

We have as yet, been unsuccessful in deriving variograms from exploration DH sampling, partly because of the general lack of close-spaced drilling, and partly because of the fringe of zero-grade DH's surrounding every mineralized area (the resulting variogram changes with the size of area considered). Instead, the relative shape of the L37 BH variogram is assumed to be fundamental to

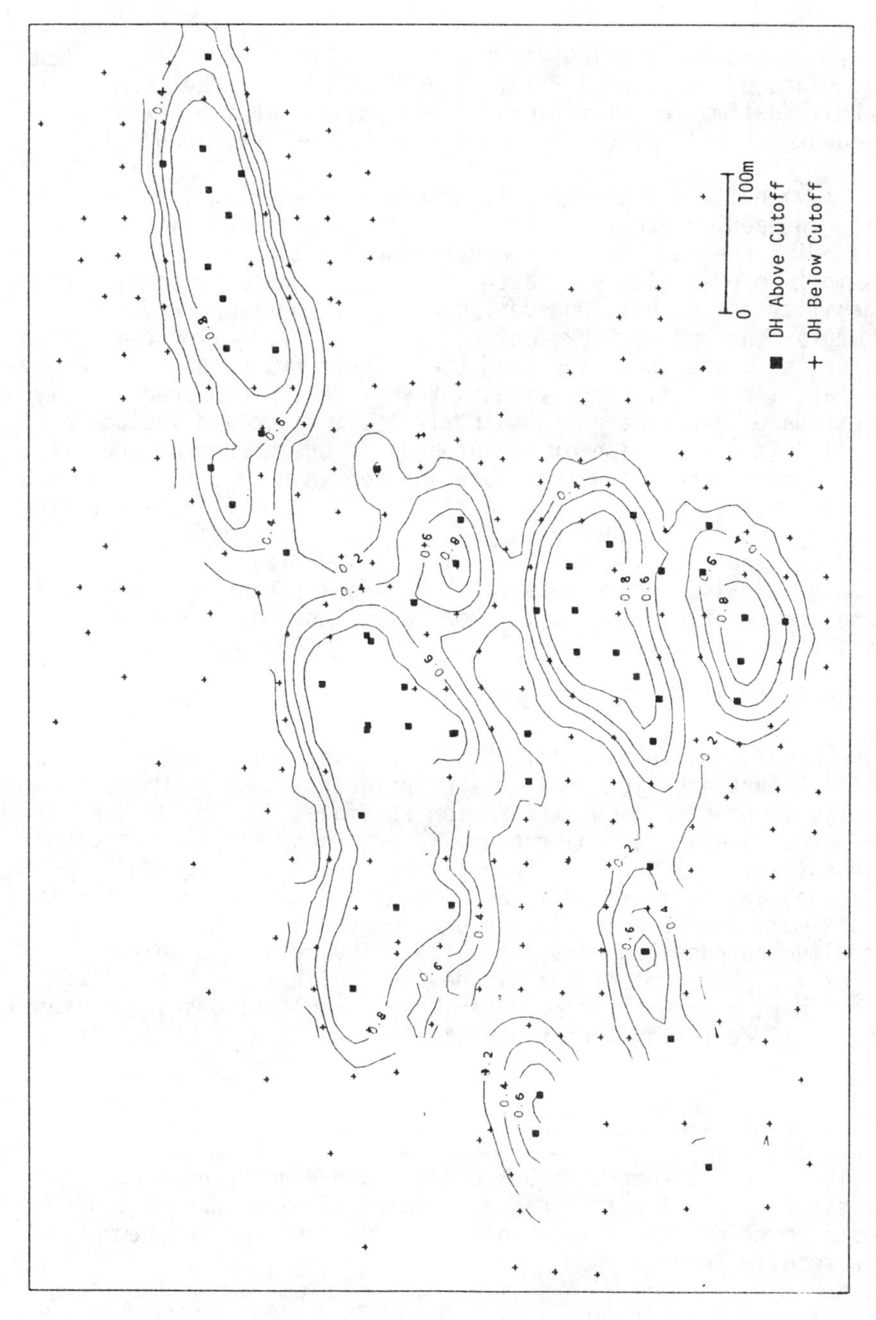

Figure 5. Exploration Conditional Probability Of Ore

tabular orebodies. The magnitude of the variogram is obtained for new orebodies by calculating a ratio of observed to estimated (using the L37 variogram) relative estimation variances from jackknife estimation of exploration composites in the new area. This approach will be checked as new orebodies are mined out.

The conditional probability method is now being used for pit design in tabular ore at Pine Point, although design limits are still being checked locally against manual estimates. To date, pits in five orebodies have been re-designed based on conditional probability estimates. However, because of the temporary mine closure at the end of 1982, only preliminary production comparisons are available and these are from the highest grade tabular orebody known at Pine Point. Improvements to pit design compared to polygon ore estimates are evident in this first pit mined and include a 40 m (130 ft.) extension of the orebody in one corner of the pit. However, mined ore grades have been higher than conditional probability estimates, apparently as a result of better visual selection in mining the high-grade ore. The necessity for considering structural and lithological controls to minerlization in the estimate has also become apparent. Polygons are being retained as the ore reserve method until more experience has been gained with the new estimate.

6. CONCLUSION

A conditional probability model based on relative variograms and the 3 parameter lognormal distribution has been shown by back analysis to provide good exploration estimates in tabular ore at Pine Point Mines. The major economic advantage of this approach is an estimated 5% increase in ore tonnage recovery for the same overburden and waste removal, as a result of pit design closer to the optimum. Improvement in exploration ore estimation for scheduling purposes is also indicated. The method is now being used as the primary tool for pit design in tabular ore at Pine Point Mines, although the more familiar manual methods are presently being retained for ore reserve estimates and as a check on pit designs.

ACKNOWLEDGEMENTS

The author wishes to thank Cominco and Pine Point Mines, for permission to publish this paper. The continuing support and encouragement of Pine Point Geology and Engineering Departments is gratefully acknowledged.

REFERENCES

1. David, M.: 1977, "Geostatistical Ore Reserve Estimation". Elsevier, 364 pp.

2. David, M.: 1969, The Notion Of Extension Variance And Its Application To The Grade Estimation Of Stratiform Deposits. In: "A Decade Of Digital Computing In The Mineral Industry." AIME, pp. 63-81.

3. Raymond, G.: 1979, Ore Estimation Problems In An Erratically Mineralized Orebody. CIM Bulletin, Vol. 72, No. 806, pp. 90-98.

4. Raymond, G.: 1982, Geostatistical Production Grade Estimation Using Kriging In Mount Isa's Copper Orebodies. Proc. Aust. IMM, No. 284, pp. 17-39.

5. Skall, H.: 1975, The Paleoenvironment Of Pine Point Lead-Zinc District. Economic Geology, Vol. 70, pp. 22-47.

LOCAL ESTIMATION OF THE BLOCK RECOVERY FUNCTIONS

Amilcar Soares

Research Fellow CVRMUTL, Mining Engineer FERROMINAS,E.P.

Methods for local estimation of block recovery functions - b.r.f., based on bigaussian and multigaussian hypotheses, are analysed. The estimation methods were performed on a simulated deposit in order to compare their results.

A new development of M.G. approach applied to b.r.f. estimation is proposed.

INTRODUCTION

In mine planning, a problem that often arises is the estimation, in local panels, of the recoverable reserves (tonnage or quantity of metal remaining above a given cutoff grade) based on the selective mining units (the smallest block which may be separated in a mining operation as ore and waste).

The estimation of block recoverable reserves requires a block distribution model.

MATHERON (1974) proposed the discrete gaussian model (or hermitian model) based on a generalized permanence of law hypothesis (bigaussianity). In particular these models have been applied to Disjunctive Kriging (D.K.).

Recently a much simpler post-kriging Affine Correction of variance (A.C.) was introduced by JOURNEL (1983).

For point probability density function estimation, the Multigaussian Approach (M.G.) (MATHERON, 1974), although based on stronger hypotheses (Multigaussianity) than D.K., avoids,conversely the estimation problems of the latter (VERLY, 1983).

In this paper, the M.G. approach is proposed to handle the problem of block recovery functions estimation. Carrying out, ab initio, an affine correction of variance and assuming an extra

G. Verly et al. (eds.), Geostatistics for Natural Resources Characterization, Part 1, 485–493.

hypothesis of point-block multigaussianity the problem is solved in a straighforward way.

As it is well known, all above mentioned methods are highly distribution dependent and based on constitutive hypotheses difficult to check. Hence, a certain trade-off must be accomplished, balancing the accuracy of each method (in terms of theoretical estimation variance) with its actual performances.

In this paper, in order to compare the a.m. extended M.G. approach with D.K. and A.C. and assess their pros and cons, block values from an iron ore deposit were simulated.

BLOCK-RECOVERY FUNCTIONS

Defining:

- the indicator function $I_{X_o}(X_i)$

$$I_{X_o}(X_i) : \begin{cases} 1 & \text{iff } X_i \geq X_o \\ 0 & \text{iff } X_i < X_o \end{cases}$$

- the transform function ϕ_v

$$Z_{v_i} = \phi_v(X_i), \text{ with:}$$

X_i being a standard gaussian distributed variable

Z_{v_i} being the block - v value,

the block-recovery functions - tonnage and quantity of metal above a given cut-off z_o in a local panel A. can be defined as

$$T_v(z_o) = \int_{-\infty}^{+\infty} I_{X_o}(X) \cdot f_{X_{i}/y_\alpha}(X)\, dX$$

$$Q_v(z_o) = \int_{-\infty}^{+\infty} I_{X_o}(X) \cdot \phi_v(X) \cdot f_{X_{i}/y_\alpha}(X)\, dX$$

$z_o = \phi_v(X_o)$ being the cutt-off grade and y_α the data conditioning the panel A.

The estimate of those quantities - $T_v(z_o)$ and $Q_v(z_o)$ - can be deduced from the estimate of the conditional density function

f_{X_i/y_α} (X), that is from the composite distribution of block-v grades within the panel A.

MULTIGAUSSIAN APPROACH

The variables $X_i = \phi_v^{-1}(Z_{v_i})$ and $y_i = \phi^{-1}(zi)$ are univariate normal. Assuming an aditional hypotheses of multigaussianity of X_i and y_i, the composite conditional density function f_{X_i/y_α} is fully determined by its conditional expectation $E\{X_{i/y_\alpha}\}$ and conditional variance VAR $\{X_{i/y_\alpha}\}$(MATHERON, 1974).

Hence $E\{X_i/y_\alpha\} = \sum_{\alpha=1}^{N} \lambda_\alpha(i)\, y_\alpha$

The weights λ_α are given by a simple kriging system.

VAR $\{X_{i/y_\alpha}\}$ is the simple kriging variance.

The correlation coefficient point-point $E\{y_i \,.\, y_j\}$ is determined from the transformed point variogram.

The correlation coefficient point-block $E\{X_i \,.\, y_\alpha\}$ is obtained from the point correlation coefficients, as it is deduced in the sequel.

Assuming that, given a value X_i, y_x (x randomized in v_i) is conditionally independent of y_j with $i \neq j$. This conditional independence assumption is rather strong and can only be stated if the support v is small. Than it can be written the following equality:

$$E\{y_j \,.\, Y_{v_i}\} = E\{E\{y_j/X_i\} \,.\, E\{Y_{v_i}/X_i\}\} \qquad (1)$$

where $Y_{v_i} = \frac{1}{v_i} \int_{v_i} y_x \, dx$

The two terms of (1) are written:

$$E\{y_j/X_i\} = r_{j,i} \,.\, X_i$$

$$E\{Y_{v_i}/X_i\} = \frac{1}{v_i} \int_{v_i} r_{x,i} \, dx \,.\, X_i$$

where $E\{y_j \,.\, X_i\}$ is denoted by $r_{j,i}$.

The expression (1) comes:

$$E\{y_j \,.\, Y_{v_i}\} = r_{j,i} \,.\, \frac{1}{v_i} \int_{v_i} r_{x,i} \, dx \qquad (2)$$

The integral $\frac{1}{v_i} \int_{v_i} r_{x,i}\, dx$ does not depend on i, ie, it is constant for all blocks. Then from (2) :

$$r_{j,i} = \left(\frac{1}{v_i} \int_{v_i} \rho_{j,x}\, dx\right)/S \qquad (3)$$

where $S = \frac{1}{v} \int_{v} r_{x,i}\, dx$

$$E\{Y_j \cdot Y_{v_i}\} = \frac{1}{v_i} \int_{v_i} \rho_{j,x}\, dx$$

$E\{Y_i \cdot Y_x\}$ is denoted by $\rho_{i,x}$

Integrating both terms of (2) in v_i, the parameter S comes:

$$S = \frac{1}{v_i} \sqrt{\left(\int_{v_i} \int_{v_i} \rho_{x,y}\, dx\, dy\right)} \qquad (4)$$

All this formalism can be visualized as an approximation of the discrete gaussian model. Then the following warning has to be considered: the consistence between the correlation coefficient $r_{i,j}$, as calculated by (3) and (4) , and the bigaussian law of (X_i, y_j) is assured if the block size is small.

CASE STUDY ON A SIMULATED IRON ORE DEPOSIT

Characteristics of the data

From a 1 x 1 m grid simulated deposit, a "slice" which covers a global area of 500 x 100 meters (the third dimension is ignored) was choosen.

This "slice" was divided in ten small panels of 50 x 100 meters.

Inside each one of these panels - P_1 to P_{10}, the spatial distribution of a 5 x 5 meters block (v.d. Fig. 1) was estimated by the previous methods.

To construct the "real" block-v histograms it was considered that a "real" block value is the average of the simulated point values within each small block (Fig. 1).

The estimation of each panel is based on 6 samples on a 50 x 50 m grid.

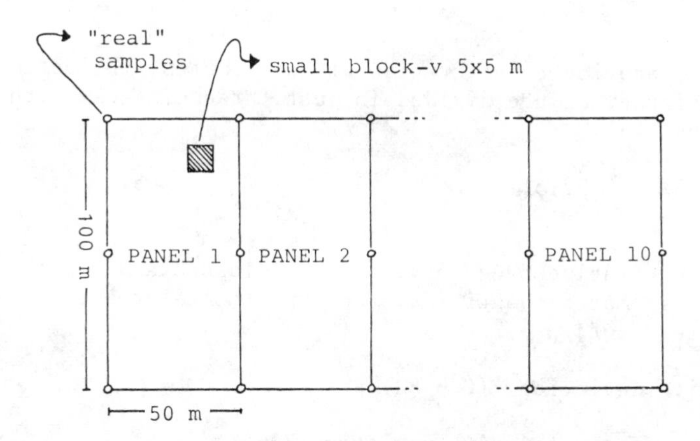

Fig. 1: Configuration of the simulated deposit

For the point variable z_i the following statistics were estimated:

mean - 41.2% variance - $41.11\%^2$

Structural analysis

The variogram model is isotropic, with two nested spherical structures:

$$a_1 = 12\text{m} \qquad C_1 = 16.4\%^2 \qquad C_o = 4.1\%^2$$

$$a_2 = 50\text{m} \qquad C_2 = 20.6\%^2 \qquad r^2 = h_x^2 + h_y^2$$

$$\gamma(h_x, h_y) = C_o + C_1 \ \text{Sph}(r) + C_2 \ \text{Sph}(r)$$

Application of the methods

- Using the "D.K." method, a good hermitian expansion fit with a five order polynomial was found, and the "hermitian model" was used to define the block distribution model.

- The point spatial distribution in each panel was estimated by the Multigaussian Approach, then corrected for block variance (A.C. method).

- For the extended M.G. approach it was achieved a quite satisfactory anamorphosis ϕ_v. The block size was considered sufficiently small (compared with the structural distances) to make the block distribution model consistent.

Results

In order to provide a better view of the results, the estimated and real histograms are divided in just three classes with limits 37.5% and 45%:

$$Z_{v_i} \leq 37.5\% \qquad 37.5\% < Z_{v_i} \leq 45\% \qquad Z_{v_i} > 45\%$$

F being the relative frequency of these classes and F^* the corresponding estimated quantities, the experimental results are presented in terms of:

a - relative deviation $(F - F^*)/F$

b - modulus of the relative deviation $|F - F^*|/F$

c - quadratic deviation $(F - F^*)^2$

For each class the relative deviation for the 3 methods are shown in Fig. 2, 3 and 4. These relative deviation are then average over the 10 panels considered, c.f. Table I.

From Fig. 2 to 4 and Table I it can be withdrawn the following non trivial conclusions:

. In terms of deviation values per panel, M.G. stands systematically in the "middle" position, being also relevant a bias on the A.C. second class (Fig. 2 to 4). These features are linked to diferences on the dispersion variances of estimated histograms ($\sigma^{*2}_{DK} > \sigma^{*2}_{MG} > \sigma^{*2}_{AC}$).

. In terms of relative deviation and respective modulus for the average of the ten panels, D.K. produces the best statistics for the second class, and the worst for the tail classes(Table I).

A.C. presents the higher quadratic deviation values for the three classes, neverthless a low average relative deviation combined with a high average quadratic deviation is a common characteristic of the three methods (Table I).

For the upper tail class and for the average of the three classes M.G. provides better results than D.K. and A.C..

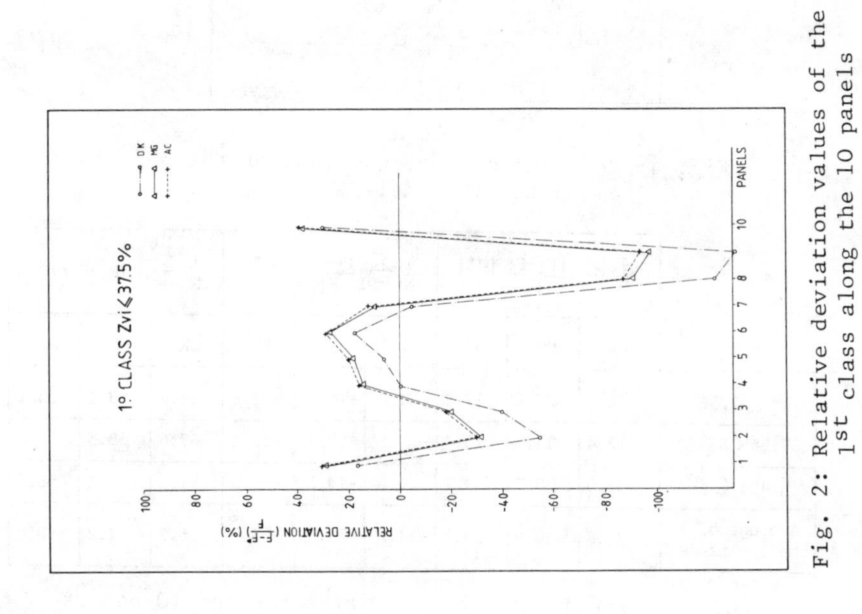

Fig. 2: Relative deviation values of the 1st class along the 10 panels

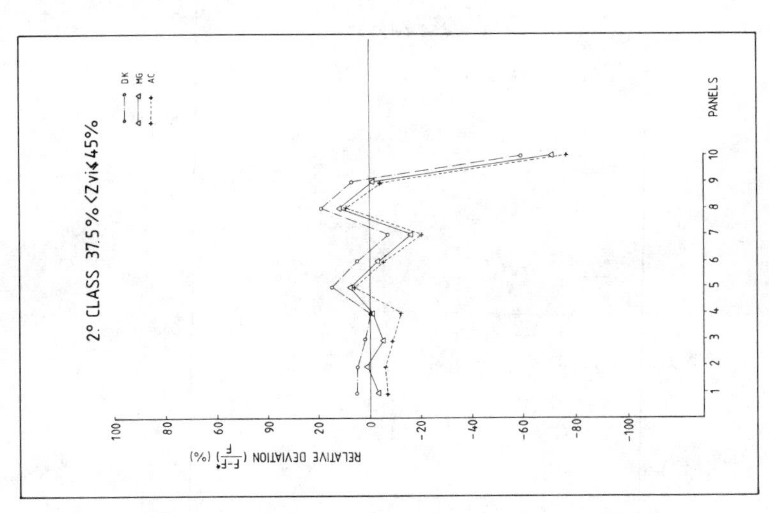

Fig. 3: Relative deviation values of the 2nd class along the 10 panels

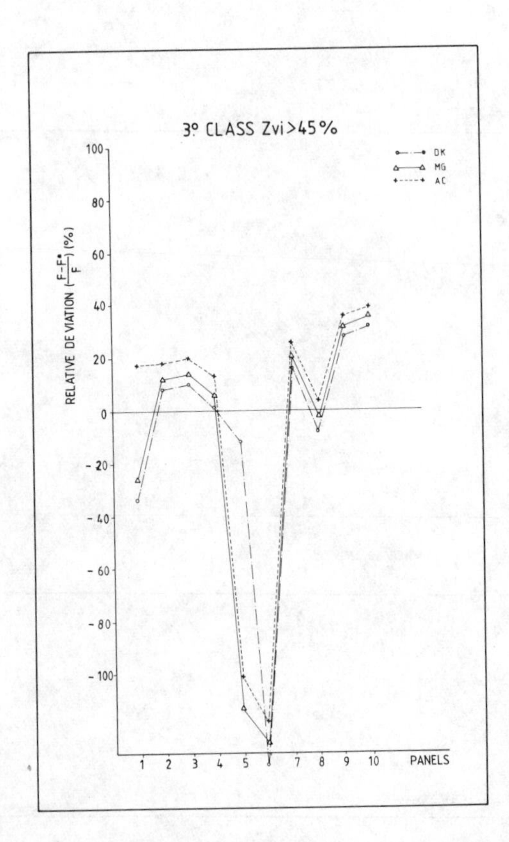

Fig. 4: Relative deviation values of the 3rd class along the 10 panels

	$\frac{1}{N}\sum_{i=1}^{10} \frac{\lvert F_i-F_i^*\rvert}{F_i}$ (%)			$\frac{1}{N}\sum_{i=1}^{10} \frac{F_i-F_i^*}{F_i}$ (%)			$\frac{1}{N}\sum_{i=1}^{10} (F_i-F_i^*)^2$		
	DK	MG	AC	DK	MG	AC	DK	MG	AC
$Z \leqslant 37.5\%$	42.9	37.9	37.8	-21.2	-10.3	- 8.6	48.4	53.1	55.1
$37.5\% < Z \leqslant 45\%$	12.4	13.	15.4	- .8	- 9.4	-12.4	69.6	59.5	92.4
$45\% < Z$	39.5	38.7	39.1	-20.6	-14.8	- 4.5	51.	50.1	59.0
Average of the 3 classes	31.6	29.9	30.1	-14.2	-11.5	- 8.5	56.3	54.2	68.8

TABLE I - Average of the statistics for the 10 panels

CONCLUSIONS

Although M.G. is based on very strong constitutive hypotheses, they are not invalidated by the experimental results for this particular case, and so, as theoretically expected, M.G. provides the lowest deviation parameters.

However this can not be stated as general rule and special care must be taken when dealing with both heavy theoretical assumptions (such as multigaussianity) and performing approximations.

In any case it is important to emphasize the performing simplicity of the A.C. or M.G. approaches (mainly A.C. method) over D.K..

ACKNOWLEDGEMENTS

The management of Ferrominas, E.P. is greatefully acknowledged for having supported this study and allowing its publication.

The colaboration of A.P.Braga in the computational area was indispensable to the achievment of this work.

The encouragement of Prof. Q.Rogado was deeply appreciated.

This paper was induced by helpfull suggestions and comments made by Prof. A.Journel, during the author's geostatistical training in Stanford University.

REFERENCES

- Journel, A. (1983) - Non-Parametric Estimation of Spatial Distribution. Math-Geology, Vol. 15, no. 3, pp. 445-468.

- Matheron, G. (1974) - Les Fonctions de Transfert des Petits Panneaux. Les Cahiers du Centre de Morphologie Mathématique de Fontainebleau, N-395, 73 p.

- Verly, C. (1983) - The Multigaussian Approach and its Applications to the Estimation of Local Recoveries, in Math-Geology, Vol. 15 no. 2, pp. 263-290.

THE BLOCK DISTRIBUTION GIVEN A POINT MULTIVARIATE NORMAL DISTRIBUTION

Georges Verly

Applied Earth Sciences Departement
Stanford University

ABSTRACT

The linear properties of the multivariate normal distribution make it the most straighforward model for obtaining conditional distributions. Problems of multiple integration have been solved through a Monte-Carlo type approach, thus rendering the multigaussian approach fully operational for the estimation of local reserves and associated spatial distributions.

KEYWORDS: Geostatistics, Conditional Distribution, Multigaussian, Local Recoveries, Monte Carlo Method.

I - INTRODUCTION

The estimation of local recoveries in a selective mining operation is one of the most challenging problems in mining geostatistics. This problem amounts to calculating uni- and bi-variate conditional distributions of grades, from which the conditional expectations and variances of local recoveries are evaluated.

Several krigings have been developed to provide a solution

G. Verly et al. (eds.), Geostatistics for Natural Resources Characterization, Part 1, 495–515.

to this problem: Disjunctive (DK), Lognormal (LnK)*, Multigaussian (MG), Indicator (IK), and Probability (PK) Krigings [6 to 11, 14 to 16]

As long as the selective mining units (smu's) are of point support, or more precisely of the same support as the data, then DK, LnK, IK, and PK need only a small number of hypotheses to come out with an estimate of the sought conditional univariate grade distribution. Unfortunatly, the smu's are most often not of data size support and each of these krigings requires additional and in some cases inconsistent hypotheses about the grade distribution of the smu's. Moreover, none of these methods gives access to the conditional bivariate distribution of grade, thus not allowing the inference of the conditional estimation variance of the local recoveries.

The Multigaussian approach is the only method presently giving access to both uni- and bi-variate conditional distributions whatever the size of the smu's. The solution is not only complete but also optimal, provided the multigaussian hypothesis is correct. The method also has the advantage of not requiring any orthogonal polynomial expansion, which makes it easy to work with.

II - THE THEORY OF THE MULTIGAUSSIAN APPROACH

II.1 - THE LOCAL RECOVERY FUNCTIONS

Given a panel of size V and smu's of size v ($v < V$), the basic recovery functions associated with the panel V are:

- The recovered tonnage of ore, $t(z_c)$, proportional to the number of smu's within the panel with grade higher than the cut-off grade z_c.
- The recovered quantity of metal, $q(z_c)$, or quantity of metal in the recovered smu's.

The recovered average grade is then defined as the ratio $q(z_c)/t(z_c)$.

---

* The Lognormal Kriging as described in [6] and [14] is considered here. LnK is then a particular case of MG if the selective mining units (smu's) are of point size support. However, it differs from MG in the non-point support case.

Introducing the indicator function

$$i_c[z_v(\underline{x})] = \begin{cases} 1 \text{ if } z_v(\underline{x}) > z_c \\ 0 \text{ if not} \end{cases} \quad (1)$$

where $z_v(\underline{x})$ is the grade of an smu located at $\underline{x}$, the recovery function can be expressed as

$$t(z_c) = \frac{To}{L} \sum_{l=1}^{L} i_c[z_v(l)] \quad (2)$$

$$q(z_c) = \frac{To}{L} \sum_{l=1}^{L} z_v(l) \cdot i_c[z_v(l)] \quad (3)$$

where To is the total tonnage within panel V constituted of a set of L smu's with grades $Z_v(l)$, l=1 to L.

Note that both the recovered tonnage and quantity of metal have the same general expression

$$r(z_c) = \frac{To}{L} \sum_{l=1}^{L} h[z_v(l)] \quad (4)$$

where $h[z_v(l)]$ is some function of the smu grades. From now on, this general expression will be used.

II.2 - ESTIMATION OF THE LOCAL RECOVERIES

Geostatistics interprets the grade $z(\underline{x})$ as a realization of a random variable (RV) $Z(\underline{x})$. The set of all these RV's in the deposit $\{Z(\underline{x});\ \underline{x} \in \text{Deposit}\}$, constitutes a random function (RF) $Z(\underline{x})$.

Within this new context, the previous recovery function, $r(z_c)$, is considered as a RV, $R(z_c)$, and its best estimator is the conditional expectation

$$E_n R(z_c) = E\{R(z_c) \mid Z(\alpha) = z(\alpha),\ \alpha \in (N)\} \quad (5)$$

where $R(z_c)$ is the RV whose realization is $r(z_c)$ and $Z(\alpha)=z(\alpha)$, $\alpha \in (N)$, are the N drill-hole samples informing the panel.

Combining the equations (4) and (5) gives

$$E_nR(z_c) = \frac{To}{L}\sum_{l=1}^{L} E_n\{h[z_v(l)]\}$$
$$= \frac{To}{L}\sum_{l=1}^{L}\int_{-\infty}^{+\infty} h(z)\bullet f_{v(l)}(z|(N))\bullet dz \quad (6)$$

where $f_{v(l)}(z|(N))$ is the conditional probability density function (pdf) of $z_v(l)$ given $Z(\alpha)=z(\alpha)$, $\alpha \in (N)$.

The conditional variance of $R(z_c)$ is

$$V_nR(z_c) = Var\{R(z_c) \mid Z(\alpha) = z(\alpha),\ \alpha \in (N)\}$$

$$= E_n\{[R(z_c)]^2\} - [E_nR(z_c)]^2$$
$$= \frac{To^2}{l^2}\sum_{l=1}^{L}\sum_{l'=1}^{L}\int_{-\infty}^{+\infty}\int_{-\infty}^{+\infty} h(z)\bullet h(z')\bullet f_{v(l)v(l')}(z,z'|(N))\bullet dzdz' - [E_nR(z_c)]^2 \quad (7)$$

where $f_{v(l)v(l')}(z,z'|(N))$ is the conditional bivariate pdf of $Z_v(l)$ and $Z_v(l')$ given $Z(\alpha)=z(\alpha)$, $\alpha \in (N)$.

Thus it appears that the problem of estimating local recoveries within a panel is solved provided the conditional uni- and bi-variate grade pdf's of the smu's within the panel are known.

II.3 - THE MULTIGAUSSIAN APPROACH

The derivation of conditional distributions such as those appearing in relations (6) and (7) is greatly simplified when the RF is multinormal. Indeed in this case, all conditional distributions distributions are normal. The conditional means depend linearly on the conditioning data. The conditional variances and covariances are independent of the values of the conditioning data. Hence the idea, rather classical in statistics, of transforming the original RF $Z(\underline{x})$ into a RF $Y(\underline{x})$ standard multinormally distributed and carrying out all calculations with this new RF. The following notation is used

$$Z(\underline{x})=\phi[Y(\underline{x})] \quad \text{or} \quad Y(\underline{x})=\phi^{-1}[Z(\underline{x})] \quad (8)$$

where ϕ is a strictly increasing function.

In practice, the gaussian transform ϕ is defined graphically by a one-to-one correspondence between the cumulative frequency distribution (cdf) of the RF $Z(\underline{x})$ and a standard normal cdf [5,p 478]. The normal score $Y(\underline{x})$ then obtained is guaranteed to have a normal marginal distribution but the multinormality of its multivariate pdf has to be hypothesized. This hypothesis is very strong and must be assumed with caution.

The conditional distributions defined in the previous section (eqs. 6 and 7) can now be developed with further details. For the sake of simplicity, only the conditional univariate distribution of $Z_{v(l)}$ is considered. The developments relative to the conditional bivariate distribution of $Z_{v(l)}$ and $Z_{v(l')}$ are similar in every respect.

The conditional cdf of $Z_{v(l)}$ is

$$F_{v(l)}(a|(N)) = P\{Z_v(l) \le a \mid Z(\alpha) = z(\alpha),\ \alpha \in (N)\} \tag{9}$$

Approximating $Z_v(l)$ by the average of M point-support values within v

$$Z_v(l) \simeq \frac{1}{M}\sum_{i=1}^{M} Z(i) = \frac{1}{M}\sum_{i=1}^{M} \phi[Y(i)] \tag{10}$$

the following expression is obtained

$$F_{v(l)}(a|(N)) \simeq P\left\{\sum_{i=1}^{M} \phi[Y(i)] \le Ma \mid Y(\alpha) = y(\alpha),\ \alpha \in (N)\right\} \tag{11}$$

The RF $Y(\underline{x})$ being standard multinormal, the pdf of $[Y(i), Y(\alpha), i \in (M), \alpha \in (N)]$ is (M+N) standard multivariate normal and the conditional pdf of $[Y(i), i \in (M) \mid y(\alpha),\ \alpha \in (N)]$ is M multivariate normal (M-normal). The conditional mean vector is [11,p 28]

$$E\{Y(i),\ i \in (M) \mid Y(\alpha) = y(\alpha),\ \alpha \in (N)\}$$

$$= E\{\underline{Y}_i \mid \underline{y}_\alpha\} = \Sigma_{i\alpha} \cdot \Sigma_{\alpha\alpha}^{-1} \cdot \underline{y}_\alpha \quad \text{and is denoted by} \quad \underline{\nu}_{i/\alpha} \tag{12}$$

The conditional covariance matrix is

$$E\{[\underline{Y}_i - \underline{\nu}_{i/\alpha}][\underline{Y}_i - \underline{\nu}_{i/\alpha}]' \mid \underline{Y}_\alpha = \underline{y}_\alpha\}$$

$$= \Sigma_{ii} - \Sigma_{i\alpha} \cdot \Sigma_{\alpha\alpha}^{-1} \cdot \Sigma_{\alpha i} \quad \text{and is denoted by} \quad \Sigma_{ii/\alpha} \tag{13}$$

where

$$\begin{aligned} &\text{the vector } \underline{Y}_i = [Y(i),\ i \in (M)]' \\ &\text{the vector } \underline{Y}_\alpha = [Y(\alpha),\ \alpha \in (N)]' \\ &\text{the covariance matrix } \Sigma_{ii} = E\{\underline{Y}_i \cdot \underline{Y}_i'\} \\ &\text{the covariance matrix } \Sigma_{\alpha\alpha} = E\{\underline{Y}_\alpha \cdot \underline{Y}_\alpha'\} \\ &\text{the covariance matrix } \Sigma_{i\alpha} = \Sigma_{\alpha i}' = E\{\underline{Y}_i \cdot \underline{Y}_\alpha'\} \end{aligned} \tag{14}$$

Note that the conditional mean vector $\nu_{i/\alpha}$ (12) is none other than the SK estimate of the Y(i)'s.

The relation (11) then becomes

$$F_{v(1)}(a|(N)) \simeq \int_D g(\underline{y}_i|(N))d\underline{y}_i \tag{15}$$

where D is the set of vectors $\underline{y}_i$ such that $\{\phi[y(1)]+...+\phi[y(M)] \leq Ma\}$ and $g(\underline{y}_i|(N))$ is the conditional normal pdf of $[\underline{Y}_i \mid \underline{Y}_\alpha = \underline{y}_\alpha]$ fully characterized by its two moments given in relations (12) and (13).

II.4 - THE MONTE CARLO METHOD AS A SOLUTION TO MULTIPLE INTEGRALS

There exists no simple analytical or numerical procedure to evaluate a multiple integral such as of relation (15) [4,p 43], particularly when the dimension of the integral is high and its domain of integration complex as in D. A convenient alternative is to estimate it using a Monte Carlo method. Such a method consists in a simulation of a certain variable followed by an estimation of the integral through a statistical treatment of the simulated values.

Several Monte Carlo methods can be applied to estimate multiple integrals [3, pp. 50-65]. The "Importance Sampling" method has been chosen for it has been shown (cf. [2]) to provide a reasonable solution to the present integration problem. The Importance Sampling method consists in simulating J vectors M-normal having for mean vector $\underline{\nu}_{i/\alpha}$ and covariance matrix $\sum_{ii/\alpha}$ those given in relations (12) and (13). Such a simulation can be done at very low cost by using subroutines of the International Mathematical and Statistical Library (IMSL). An unbiased estimate of the multiple integral (15) is then

$$^{*}F_{v(1)}(a|(N)) = \frac{J_a}{J} = \theta \tag{16}$$

where J is the total number of simulated vectors and J_a is the number of those belonging to the domain of integration D defined in relation (15).

The number J of simulated vectors necessary to achieve an acceptable precision on the estimator θ is currently under investigation. This number is not expected to be very large for the following two reasons:

- The precision of the estimator θ, measured through its standard deviation, is of order $J^{-1/2}$, whatever is the dimension M of the integral. This means that the number of vectors to be simulated is independent of the dimension of the integral.

- Considering the final precision associated with the estimated recoveries, a 1% precision at most is needed on the estimator of the integral. Possibly a 5% or even a 10% precision would be sufficient thus allowing a dramatical reduction of the number of simulated vectors.

II.5 - SOME APPROXIMATIONS

It can be argued that the computation required to estimate local recoveries, and particularly the Monte Carlo simulation part of it, is tedious. Indeed it is, but this is the price one has to pay to strictly respect the multigaussian hypothesis, which is the only one giving direct access to the exact conditional expectation. There exist approximations, with various degree of theoretical consistency, that would dramatically simplify these computations. These approximations fall into two categories: construction of a model for the smu grade distribution, and post-kriging correction.

One possible model for the smu grades is the Discrete Gaussian model [11].

Another one, more straightforward, consists in defining, in addition to the transform ϕ associating $Z(\underline{x})$ and $Y(\underline{x})$ (eq. 8), a second strictly increasing function ϕ_v transforming the RF $Z_v(\underline{u})$ (smu grade) into a new RF $B(\underline{u})$, then hypothesizing that $B(\underline{u})$ and $Y(\underline{x})$ are jointly multinormal.

$$B(\underline{u}) = \phi_v^{-1}[Z_v(\underline{u})] \quad \text{or} \quad Z_v(\underline{u}) = \phi_v[B(\underline{u})]$$

$$\text{with } Z_v(\underline{u}) = \frac{1}{v}\int_{v(u)} Z(\underline{x})d\underline{x} \tag{17}$$

The cdf of the RF $Z_v(\underline{u})$ is itself deduced from the RF $Y(\underline{x})$ by the relation:

$$F_v(a) = P\{Z_v(\underline{u}) \leq a\} \tag{18}$$

Using relation (8) and approximating $Z_v(1)$ with the average of M point-support values within v, the previous relation becomes

$$F_v(a) \simeq P\left\{\sum_{i=1}^{M} \phi[Y(i)] \leq Ma\right\} \tag{19}$$

where $[Y(1),\ldots,Y(M)]'$ is a M-normal vector with a zero mean vector and a covariance matrix fully characterized by the variogram of $Y(\underline{x})$. The cdf (19) is then obtained by using a Monte Carlo method.

It is worth noting that the above method can be added to the list of techniques used to obtain a block grade cdf from a point-support grade cdf.

The inconsistency of the model is that if the first part of the hypothesis is true, more precisely if the RF $Y(\underline{x})=\phi^{-1}[Z(\underline{x})]$ is multinormal, then the RF $B(\underline{u})=\phi_v^{-1}[Z_v(\underline{u})]$ cannot be multivariate normal unless the transforms ϕ and ϕ_v are linear. This inconsistency is shared by both the discrete gaussian model and the model for smu grades used in Lognormal Kriging [6, 14].

The conditional cdf of $Z_v(l)$ used in eqs. (5) and (9), can be expressed as

$$\begin{aligned} F_{v(l)}(a|(N)) &= P\{Z_v(l) \leq a \mid Z(\alpha), \alpha \in (N)\} \\ &= P\{\phi_v[B(l)] \leq a \mid Y(\alpha), \alpha \in (N)\} \end{aligned} \tag{20}$$

where the univariate conditional distribution of $[B(l) \mid y(\alpha), \alpha \in (N)]$ is normal with mean and variance obtained from the Simple Kriging system

$$\begin{aligned} &\sum_{i=1}^{N} \lambda_i \bullet r(i,\alpha) = R(\alpha,l) \qquad \alpha = 1 \text{ to } N \\ &\text{where } r(i,\alpha) = E\{Y(i)\bullet Y(\alpha)\} \text{ and} \\ &\qquad R(\alpha,l) = E\{Y(\alpha)\bullet B(l)\} \end{aligned} \tag{21}$$

The covariance $R(\alpha,l)$ is itself deduced from the multinormality of $[Y(\underline{x}),B(\underline{u})$. Using relations (8) and (17), and discretizing the smu into M points, we have

$$R(\alpha,l) \simeq E\left\{Y(\alpha)\bullet\phi_v^{-1}\left[\frac{1}{M}\sum_{i=1}^{M} \phi(Y_i)\right]\right\} \tag{22}$$

where $[Y(\alpha),Y(1),\ldots,Y(M)]'$ is a (M+1)-normal vector with a mean vector $\underline{0}$ and a covariance matrix fully characterized by the variogram of $Y(\underline{x})$. The expectation in (22) has to be obtained once again by a Monte Carlo method. In practice, these expectations need not be computed for each configuration (α,l). It is sufficient that a limited number of configurations be considered in a pre-kriging step.

The post-kriging correction method consists of evaluating all conditional distributions within the panels at the point level, for which no multiple integration of the type (15) is required , cf. reference [16]. The point-conditional distributions are then corrected for support effect. Various possibilities for such corrections are discussed in Journel [8].

III - THE PRACTICE OF MG

A mathematical model is most often a simplistic image of the reality and the MG approach is no exception. The pratice of MG requires a series of careful checks and occasionally some adaptations. This section is a review of the main steps in an MG study with a discussion of the problems that can occur, together with their practical solutions.

III.1 - DATA ANALYSIS

This most important step is required in every geostatistical study. Its importance is due to the fact that any undetected anomaly in the data will be carried on throughout the entire estimation procedure with possibly severe consequences.

The data must be clean, that is free of measurement errors. Various statistics should be computed globally, and also locally within moving cells over the deposit. An inspection of these statistics will point out outliers and also problems of non-stationarity.

When the hypothesis of strict stationarity of the RF $Z(\underline{x})$ seems to be unreasonable, the MG approach and all the other methods for local recovery estimation need to be adapted locally.

The presence of outliers will not have any serious effect on the variogram of the transformed y values since these values depend on the ranks and not on the values of the original z data. However, if outliers are included in the definition of the gaussian transform ϕ (eq. 8), they can affect the estimates at the time of back-transforming from the y's to the z's (eqs. 6, 11, and 15). The decision wether to include or not outliers in the construction of ϕ is subjective and depends on the objective of the study.

III.2 - THE "DESPIKING" PROCEDURE

The occurrence of a spike at the origin of the z values histogram can have serious consequences if not handled adequately. Consider for example an histogram of 500 z values showing a 30% proportion of zero values. Each of those 150 zeros will correspond later to a different normal score y value according to its ranking. The question is how to assign these rankings which vary from 1/500 to 150/500.

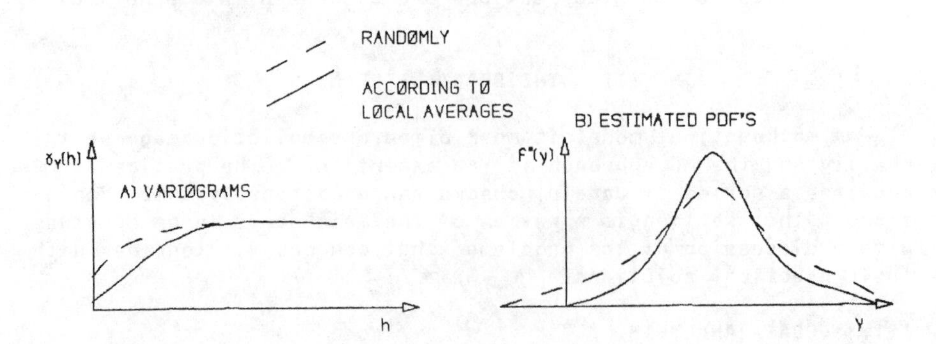

Fig. 1. Effect of two 'despiking' procedures on the variogram and the estimated pdf's.

If the ranking assignment is done randomly, it would be equivalent in this example to attributing random locations to 150 y's whose values vary from $G(1/500)=-2.88$ to $G(150/500)=-.53$. The consequence of such a procedure is the input of an artificial noise in the y's that would be reflected in the variogram (Figure 1a) and could jeopardize the later estimation procedures (Figure 1b).

One practical and very simple solution that would restore some continuity among the y's is first to rank order the zero z values according to their local averages. These local averages are calculated within moving windows centered on the zero z values. The corresponding normal scores are then computed using the obtained rankings. This solution has been applied on various case studies with satisfactory results (Figure 1).

III.3 - THE GAUSSIAN TRANSFORM ϕ

The gaussian transform ϕ is graphically obtained from a one-to-one correspondence between the cdf of $Z(\underline{x})$ and a standard normal cdf. A declusterized histogram of the z values is used as an estimate of the cdf of $Z(\underline{x})$.

The transform ϕ need not be approximated by a series of Hermite polynomials. The graphical transform is defined through a series of bounds

$$\left\{ z_k, y_k \; : \; y_k = G^{-1}\left[\frac{F(z_k)}{1+\epsilon}\right]\right\} \qquad k=1,NB \tag{23}$$

where F is the cumulative frequency of z_k, G the standard normal cdf, and ϵ a small number (<<1) in order to avoid the problem of $G^{-1}(1)=\infty$

Note that since the original z values have been "despiked", the z_k's in relation (23) are all different, so the series of bounds is indeed an approximation of an increasing function.

In theory, the function ϕ is defined for $y \in]-\infty,+\infty[$. However, since the cdf of $Z(\underline{x})$ is inferred from a limited number of z values, the function ϕ is approximated for y's within a finite interval. Taking again the example of an histogram of 500 z values, the corresponding y values will vary from -2.87 to +2.87. At the time of backtransforming the y's to the z's (eqs. 6, 11, and 15), integrals involving $\phi(y)$ have to be computed from $-\infty$ to $+\infty$. Most of the time these will be correctly estimated within a bounded interval,[-2.87,+2.87]. However, particularly for rich panels, a better solution consists of artificially extending the approximation of ϕ out to two extreme bounds

$$(z_{min},y_{min}) \text{ and } (z_{max},y_{max}) \quad \text{where } G(y_{min})=F_z(z_{min})=\delta \quad \text{and } G(y_{max})=F_z(z_{max})=1-\delta \tag{24}$$

If $\delta=10^{-6}$, then $y_{min}=-5.2$ and $y_{max}=+5.2$. The values of z_{min} and z_{max} are unknown but can be inferred from the histogram of $Z(\underline{x})$. Most of the time, z_{min} is set to zero and z_{max} to a physical absolute maximum or to a value slightly higher than the maximum histogram value.

III.4 - THE MULTIGAUSSIAN HYPOTHESIS

The hypothesis of multinormality of $Y(\underline{x})=\phi^{-1}[Z(\underline{x})]$ is very severe, hence worth some checks. It implies first the strict stationarity of the phenomenon, already inspected at the time of data analysis. The multinormality itself can be tested through various properties [4,p 59-61]. One of them, stating that any linear combination of the components of a multinormal vector is itself normal [13,p 578] is now being used to develop some checks that are easy to implement in a geostatistical study [12].

In a case where the hypothesis of multinormality of $Y(\underline{x})$ would be unreasonable but not the hypothesis of binormality, then the technique of Disjunctive Kriging [10] might be applied. If the binormal hypothesis is itself inapropriate, then the only choice left would appear to be the non-parametric approach of Indicator Kriging [7, 8].

III.5 - STRUCTURAL ANALYSIS

The structural analysis of the RF $Y(\underline{x})$ generally is not difficult. The variograms of the y's are well behaved, even if the variograms of the z's are quite erratic. This is due to the fact, already mentioned, that the y's are deduced from the rankings and not the values of the z's.

It is essential to estimate correctly the short scale variability for the covariance matrix of the smu's depends heavily on it, cf relation (13)

As an additional check on binormality, the variogram calculation from the original z values can be compared with the same variogram derived from the y's by means of the following relation

$$\gamma_z(h) = \tfrac{1}{2}\int_{-\infty}^{+\infty}\int_{-\infty}^{+\infty} (\phi(y)-\phi(y'))^2 \bullet g(y,y') \bullet dydy' \qquad (25)$$

where $g(y,y')$ is the standard binormal density with covariance $\sigma(h)=1-\gamma_y(h)$. In practice, the integrals are approximated by a discrete series. In fact relation (25) could be used to evaluate the variogram of the z's, under a binormal hypothesis, when a direct computation of it gives too erratic results.

III.6 - COMPUTATION OF LOCAL RECOVERIES

The evaluation of the local recoveries within a panel involves three steps:

1. Evaluation of the conditional mean vector and conditional covariance matrix correponding to each smu (eqs. 12 and 13)
2. Using the results in 1, simulation of multinormal vectors and calculation of the conditional grade distributions (eq. 16).
3. Using the results in 2, compositing the conditional distributions of the smu's within the panel and evaluation of the recoveries (eq. 6).

The first step is very similar to a simple kriging and therefore does not involve much computer time. The last step consists of a few do-loops. The middle step, however, requires some careful planning. For example, one should take advantage of any symmetry in the data configuration that would allow using the same simulated vectors for several smu's, thus shortcutting step 2. Also random kriging approximations may be designed to limit the number of data configurations to be considered in step 2.

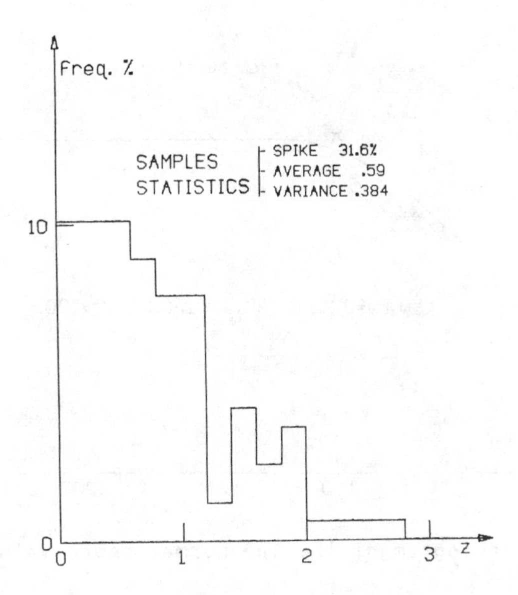

Fig. 2. Histogram of the 168 original z values.

IV - CASE STUDY

The deposit used for the case-study is a two-dimensional simulation obtained by the turning band method [5, p. 498]. This simulated deposit, called Stanford II, has already been used in a comparative study of MG and DK for the estimation of local recoveries given point-support smu's [16]. Stanford II is 550 m in length by 110 m in width, and is perfectly known through 60,500 point values.

The available information consists of 168 data located on a regular 20 by 20 m grid. The objective of the study is to estimate local recoveries within 104 panels. Each panel is 20 by 20 m and contains sixteen 5 by 5 m smu's. The panel size has been chosen in order to have a sufficient statistical mass when comparing the estimated and true recoveries. The z-values histogram shows 31.5% zero values and a strong asymmetry in the distribution of positive grades (Figure 2). Since by construction, the simulation has a multigaussian "flavor", no special statistics were computed to check either the strict stationarity of the RF $Z(\underline{x})$, or the multinormality of its normal score $Y(\underline{x})$.

The 168 z values were "despiked" and combined with their normal scores to define the gaussian transform ϕ. Directional variograms were computed from the 168 y values. Some fences of close data helped to evaluate the short-scale variability. The

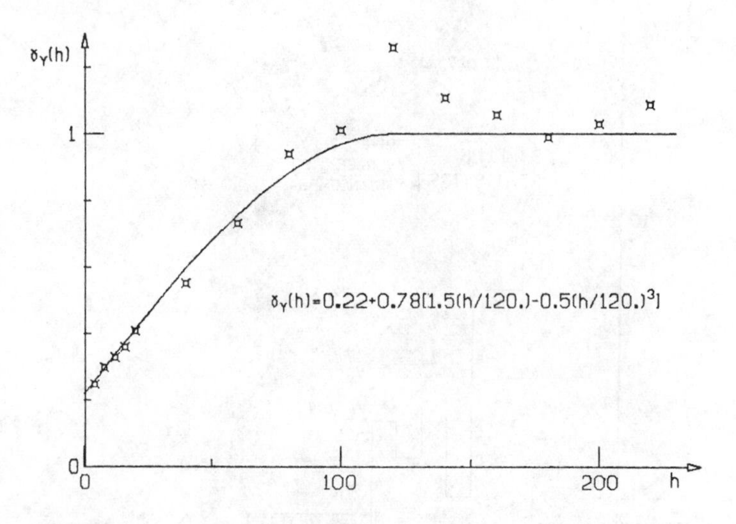

Fig. 3. Variogram of the 168 normal score values.

chosen model is an isotropic spherical structure plus a small nugget effect (Figure 3).

Thanks to the regular grid of data, the recoveries within each panel were estimated with an identical and symmetrical configuration of data: one datum at the center of the panel and eight in the periphery. As a result, the computation costs were negligible.

The local recoveries have been computed for several cut-off grades whose values, together with their respective percentiles for the histogram of 168 data, are given in the table below:

#	cutoff value	percentile
1	0	0
2	0.001	31.5
3	0.170	41.1
4	0.360	51.2
5	0.579	60.1
6	0.989	78.0
7	1.423	88.7

Scattergrams showing the true versus the estimated local recoveries given cutoffs #1, 3, 5, and 7 are shown in Figures 4

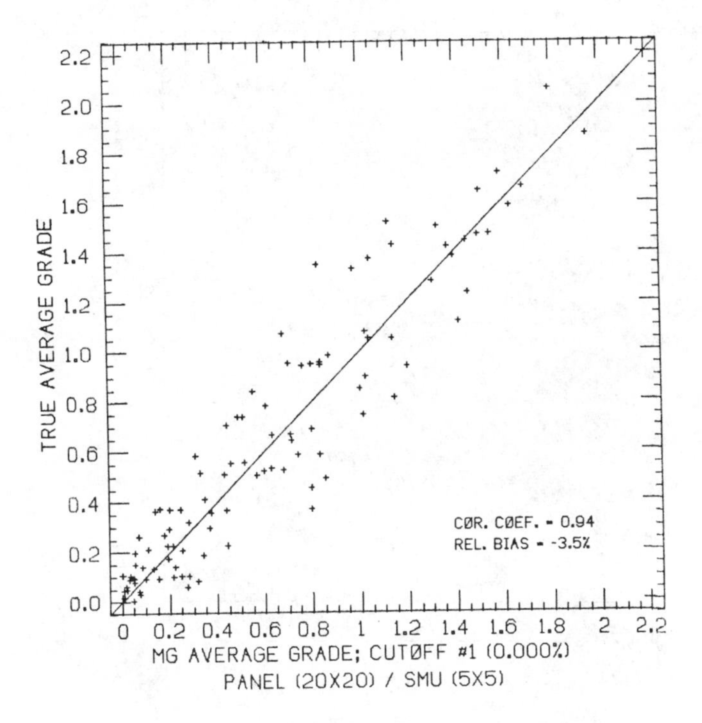

Fig. 4. Average grades, true versus estimated, within each panel. The panel size is 20x20 m. The SMU size is 5x5 m. A negative bias corresponds to underestimation.

through 7. The global recovery functions are shown in the Figure 8.

The performance of MG in estimating the average grade within each panel (or recovered quantity of metal given no cutoff) is excellent (Figure 4). The scattergram shows no significant conditional bias. The correlation coefficient between the true and estimated values is high (0.94). The size of the global relative bias, -3.5%, compares favorably with the relative estimation standard deviation of 5.5%.

The same comments can be applied to the estimated recovered quantities of metal (Figures 5 and 6) although there is a degradation of the estimation for the last cutoff (Figure 7). More details about these results are given in the next paragraph.

The scattergrams of the recovered tonnages (Figures 5, 6, and 7) show sometimes large discrepancies between the true and estimated values. Still the results are considered satisfactory for the reasons below:

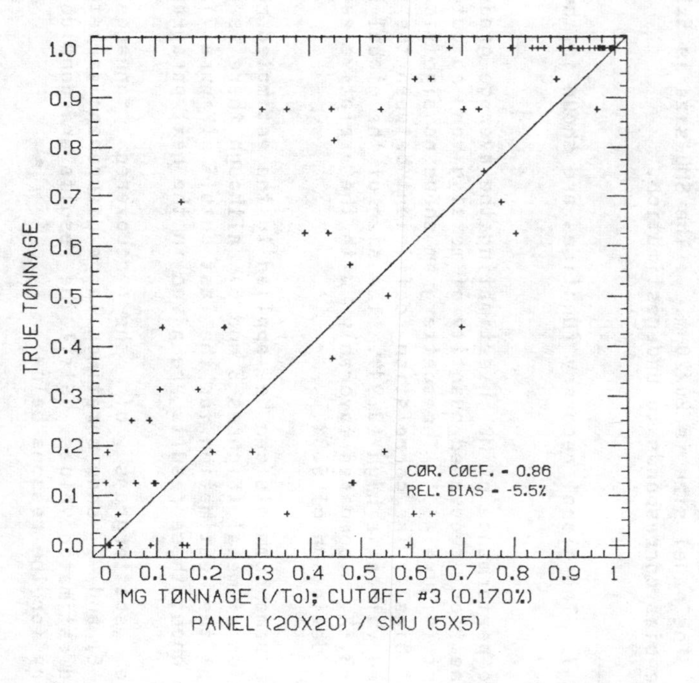

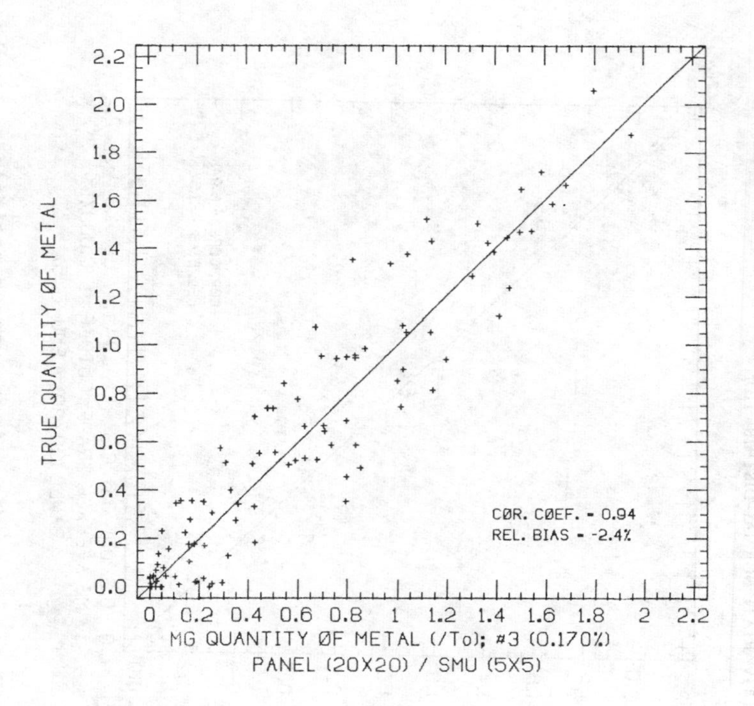

Fig. 5. Recovered tonnages and quantities of metal, true versus estimated, given the 3rd cutoff (0.170). This cutoff corresponds to the 41th percentile of the distribution.

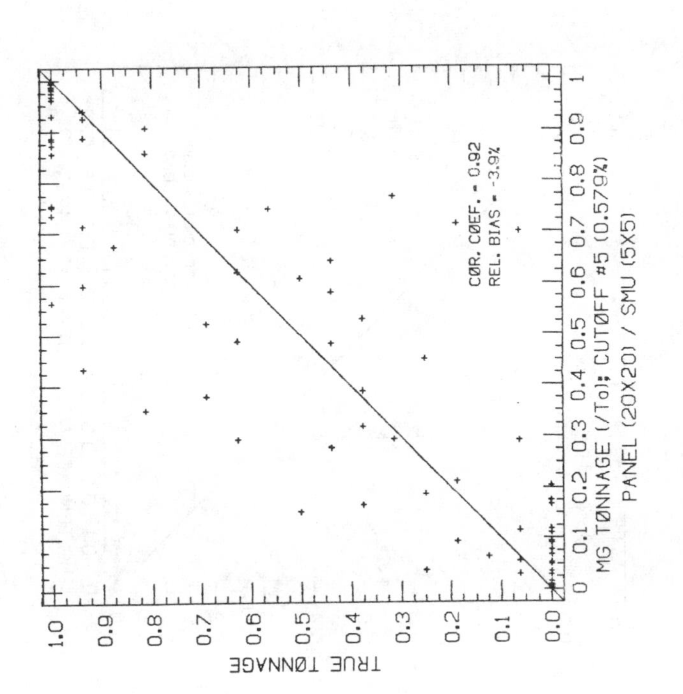

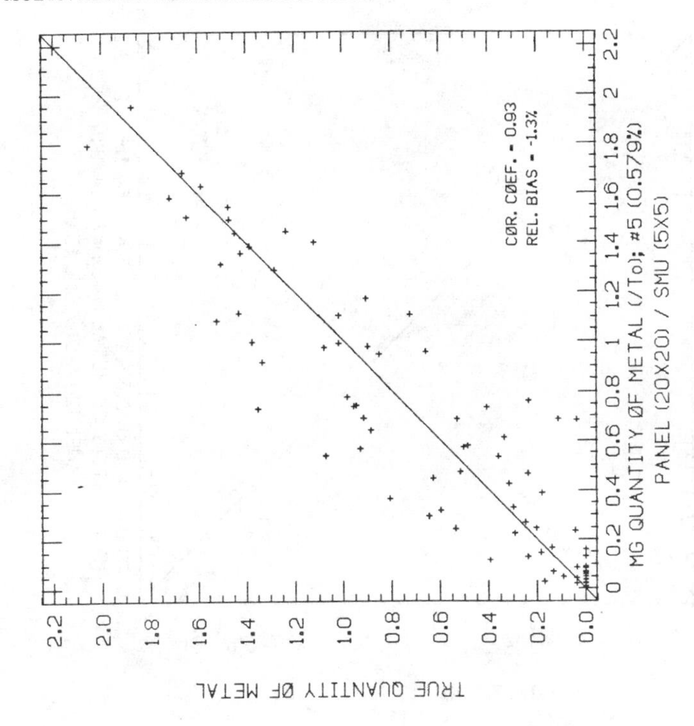

Fig. 6. Recovered tonnages and quantities of metal, true versus estimated, given the 5th cutoff (0.579). This cutoff corresponds to the 60th percentile of the distribution.

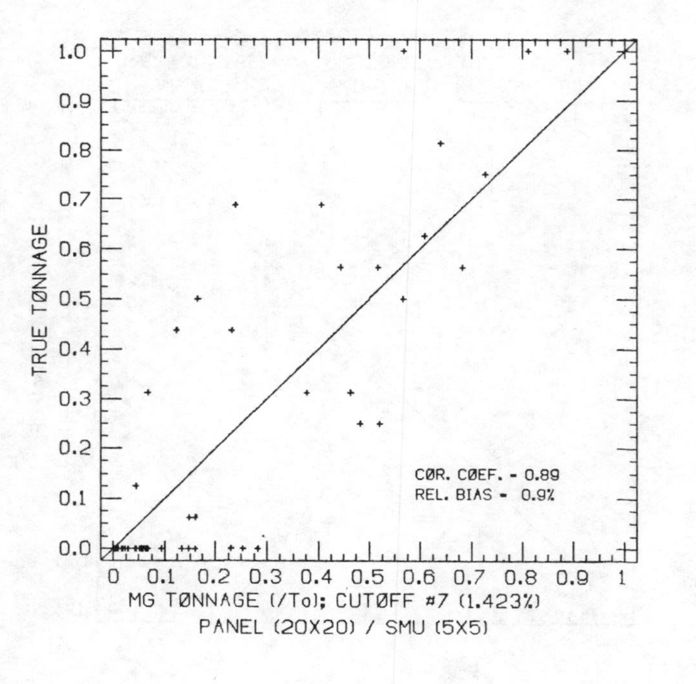

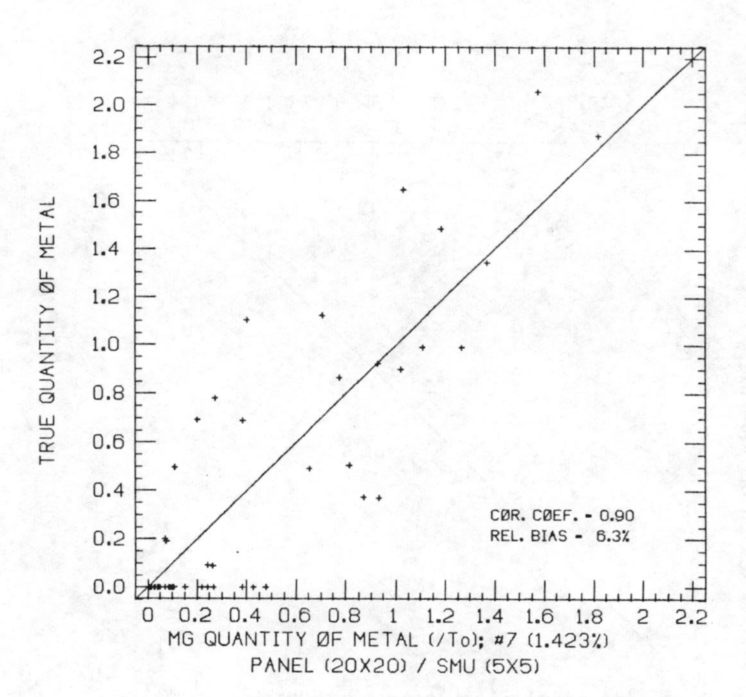

Fig. 7. Recovered tonnages and quantities of metal, true versus estimated, given the 7th cutoff (1.423). This cutoff corresponds to the 89th percentile of the distribution.

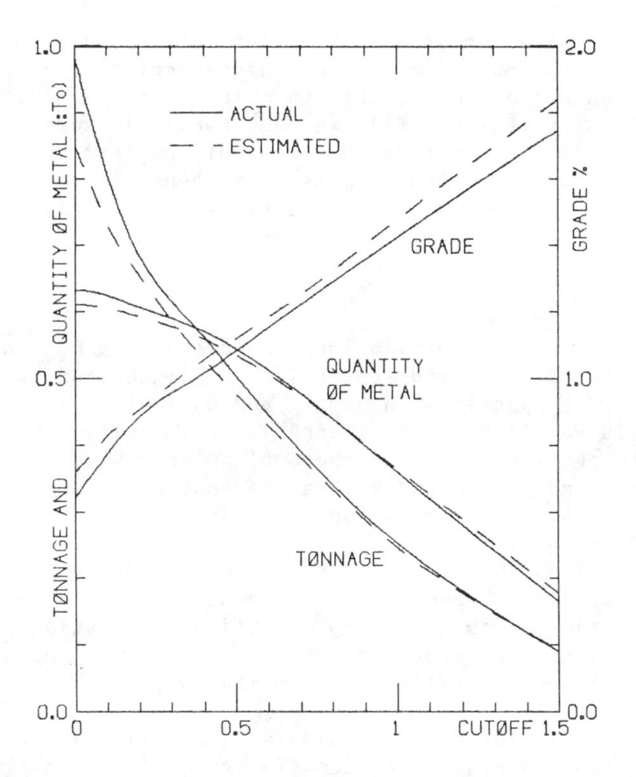

Fig. 8. Global true and estimated recovery functions.

- The global relative biases, although higher than the ones observed for the estimated quantities of metal, are still reasonably low.
- Most of the points (≃ 70%) are concentrated in the upper right and/or lower left corners of the plots and therefore correspond to correctly estimated tonnages. This fact is confirmed by the high correlation coefficients: ρ between .86 and .92.
- The bulk of points (≃ 30%) for which large discrepancies between the true and estimate tonnages are observed, correspond to quantities of metal with values between 0.0 and 0.8. The relative precisions on these quantities of metal, as it can be seen from the scattergrams, are not better than those on the corresponding tonnages. This lack of precision is not due to the method itself but rather to the small size of the panels.

The global performance of the method, illustrated in the Figure 8, is considered excellent. With the exception of the recovered tonnages and grades at very low cutoffs, the size of the relative bias never exceed 6% in absolute value, and most of the time is less than 3.5%. The higher relative bias on the recovered tonnages and grades at low cutoffs (respectively -13.8% and 12% given the .001 cutoff) is due partly to some panels for which the available information does not reflect the content. However, the grade of these panels is generally low and the amplitude of this bias quickly decreases as the cutoff increases.

V - CONCLUSIONS

The problem of estimating local recoveries on non-point support has been fully solved under the multigaussian hypothesis. The solution is perfectly general and optimal given that the hypothesis is verified. The theory is relatively simple and at no time requires the use of orthogonal polynomials. The necessary computations unfortunately can be costly. Approximations are currently under investigation in order to reduce these costs.

Some problems that occur in practice have been discussed and practical solutions proposed. The existence of these problems should be considered as a warning: a blind application of the MG approach may lead to severe errors, although this point can be made for any other geostatistical application.

The results obtained on a simulated deposit are promising. A simulated multinormal type deposit is certainly a favorable case to test the MG approach; however, it is the only case for which a comparison of true and estimated recoveries can be obtained on sufficient statistical masses. Implementation of this MG approach on a real deposit (gold) is currently being done with satisfactory results.

REFERENCES

1. Anderson, T.W., (1957). "An Introduction to Multivariate Statistical Analysis". John Wiley and Sons Inc., New York.

2. Escoufier, Y., (1967). "Calculs de probabilites par une methode de Monte Carlo pour une variable p-normale". Revue de Statistique Appliquee, Vol. 15, No. 4, pp. 5-15.

3. Hammersley, J., and Handscomb, D., (1964). "Monte Carlo Methods". Methuen, London.

4. Johnson, N.L., and Kotz, S., (1972). "Distributions in Statistics: Continuous Multivariate Distributions". John Wiley and Sons, Inc., New York.

5. Journel, A.G., and Huijbregts, Ch.J., (1978). "Mining Geostatistics". Academic Press, London, 2nd printing.

6. Journel, A.G., (1980). "The lognormal approach to predicting local distributions of selective mining units grades". Journal of Mathematical Geology, Vol. 12, No. 4, pp. 283-301.

7. Journel, A.G., (1983). "Non-parametric estimation of spatial distribution". Journal of Mathematic Geology, Vol. 15, No. 3, pp. 445-468.

8. Journel,A.G., (1983). "The place of non-parametric geostatistics". NATO Avanced Study Institute, Tahoe - 1983, D. Reidel Pub. Co., this volume, p. 307.

9. Marechal, A., (1975). "The practice of transfer functions: numerical methods and their applications. NATO Advanced Study Institute, D. Reidel Publishing Company, Dordrecht (Holland), p. 253 - 276.

10. Matheron, G., (1975). "A simple substitute for conditional expectation: disjunctive kriging". NATO Advanced Study Institute, D. Reidel Publishing Company, Dordrecht (Holland), p. 221-236.

11. Matheron, G., (1975). "Forecasting block grade distributions: the transfer functions". NATO Advanced Study Institute, D. Reidel Publishing Company, Dordrecht (Holland), p. 237-251.

12. de Oliveira-Leite, S., (1983). "Checks for multinormality". Unpublished Research Note, Stanford University, Departement of Applied Earth Sciences.

13. Rao, C.R., (1965). "Linear Statistical Inference and its Applications". John Wiley and Sons, Inc., New York.

14. Rendu, J.M., (1979). "Normal and lognormal estimation". Journal of Mathematical Geology, Vol. 11, No. 4, pp.407-422.

15. Switzer, P., and Parker, H., (1975). "The problem of ore versus waste discremination for individual blocks: the lognormal model". NATO Advanced Study Institute, D. Reidel Publishing Company, Dordrecht (Holland), p. 203-218.

16. Verly, G., (1983). "The multigaussian approach and its applications to the estimation of local reserves". Journal of Mathematical Geology, Vol. 15, No. 2, pp. 263-290.

[illegible] Academic Press, [illegible]

THREE-DIMENSIONAL, FREQUENCY-DOMAIN SIMULATIONS OF GEOLOGICAL VARIABLES

Leon Borgman
University of Wyoming, Laramie, Wyoming

M. Taheri
INTEVEP, Caracas, Venezuela

R. Hagan
Anaconda Minerals, Denver, Colorado

ABSTRACT

Layered three-dimensional, frequency-domain computer simulations have many advantages an an alternative to the turning band method, for certain geological problems. Computer speed and even, accurate duplication of statistical properties are the main advantages. Limitations in the permitted extent in the third dimension are the primary disadvantages of the method.

The symmetries implicit in the discrete Fourier transform are exploited to gain computer speed. An extensive mathematical theory for one-, two-, and three-dimensional spectra corresponding to common variograms is developed for use in the frequency-domain simulations.

A numerical example of the computational procedures is presented.

G. Verly et al. (eds.), Geostatistics for Natural Resources Characterization, Part 1, 517–541.

1. INTRODUCTION

Let W(x) be a function such that

$$\int_{-\infty}^{\infty} |W(x)|^2 \, dx < \infty \qquad (1.1)$$

Such a function is said to be square integrable, or belong to L_2 space. Then the Fourier transform, A(f), defined by

$$A(f) = \int_{-\infty}^{\infty} W(x)[\exp(\overline{i2\pi fx})]dx \qquad (1.2)$$

forms a dual function to W(x) in the sense that

$$W(x) = \int_{-\infty}^{\infty} A(f) \exp[(i2\pi fx)] df \qquad (1.3)$$

(Here i is the imaginary unit in complex-variable theory).

The generalization to two-dimensional space is straightforward.

Two-Dimensions:

$$A(f_1,f_2) = \int_{-\infty}^{\infty}\int_{-\infty}^{\infty} W(x_1,x_2)\exp[-i2\pi\ (f_{1x1}+f_{2x2})]dx_1dx_2 \qquad (1.4a)$$

$$W(x_1,x_2) = \int_{-\infty}^{\infty}\int_{-\infty}^{\infty} A(f_1,f_2,f_3)\exp[i2\pi(f_{1x1}+f_{2x2})]df_1df_2 \qquad (1.4b)$$

Analogous formulas may be written for three or more dimensions.

Corresponding discrete summation dualities also hold. Let $\{W_n,\ n=1,2,3,\dots,N-1\}$ be a sequence of numbers conceptually representing $W(n\Delta x)$ where Δx is a fixed increment in the x-direction. A sequence of equally spaced coordinates along the f-axis is defined by $f=m\Delta f$, such that

$$\Delta f = (N\Delta x)^{-1} \tag{1.5}$$

With these definitions, the natural numerical integral approximation to (1.2) and (1.3) is

$$A_m = \Delta x \sum_{n=-N/2+1}^{N/2} W_n \exp[-i2\pi nm/N] \tag{1.6a}$$

$$W_n = \Delta f \sum_{m=N/2+1}^{N/2} A_m \exp[i2\pi mn/N] \tag{1.6b}$$

where W_n is repeated periodically (with period N) for subscripts outside the range (0,1,2,...,N-1). A_m similarly has period N.

The periodicity allows the summations to be shifted to (0,1,2,...,N-1) to obtain

$$A_m = \Delta x \sum_{n=0}^{N-1} W_n \exp[-i2\pi mn/N] \tag{1.7a}$$

$$W_n = \Delta f \sum_{m=0}^{N-1} A_m \exp[i2\pi mn/N] \tag{1.7b}$$

This is called the discrete or finite Fourier transform pair and is exact for any finite sequences of complex numbers [4].

What does all of this have to do with the computer simulation of geological variables? Two relevant properties make the dual relation very useful in certain applications. These are as follows.

1. If $\{W_n, n=0,1,2,\ldots,N-1\}$ is a periodic, correlated, stationary, real-valued, random sequence, then $\{A_m, m=0,1,2,\ldots,N/2\}$ are uncorrelated, complex-valued, random variables having a changing variance [2]. Also A_{N-m} is the complex conjugate of A_m.

2. The computational operations in (1.7) can be performed very rapidly using the fast Fourier transform algorithm[4].

The variances of the coefficients are related to the spectra, S_m,

$$S_m = \Delta x \sum_{k=0}^{N-1} C_k \exp[-i2\pi mk/N] \qquad (1.8)$$

where

$$C_k = \text{statistical expectation of } [(W_n-\mu)(W_{n+k}-\mu)] \qquad (1.9)$$

For stationary, periodic, mean zero, random, Gaussian sequences, the real and imaginary parts of A_m are independent of each other and each have variance equal to $N\Delta x S_m/2$ for $0<m<N/2$. The coefficients A_0 and $A_{N/2}$ are both real-valued and have variances equal to $N\Delta x S_0$ and $N\Delta x S_{N/2}$ respectively. The coefficients A_{N-m} are the complex- conjugate of A_m for $0<=m<N/2$. All coefficients are Gaussian.

Simulation of the sequence proceeds as follows. The real and imaginary parts of A_m for $0<=m<=N/2$ are produced by multiplying independent, standard normal, pseudo-random numbers by the square root of the appropriate variances. The rest of the coefficients are developed by conjugation. The sequence A_m is inverted with the FFT to give X_n. A constant mean is added if desired. A nugget effect can be introduced by simulating the sequence without the nugget effect, and then adding white noise with the nugget-effect variance.

The last terms in the X_n sequence are correlated with the first terms by the artificially enforced periodicity with stationarity. If this last-to-first correlation is a problem, it can be eliminated by deleting a "zone of infuence" of terms from the end of the X_n sequence.

The discrete Fourier transform can be easily generalized to two-dimensions.

$$A_{m_1,m_2} = \Delta x_1 \Delta x_2 \sum_{n_1=0}^{N_1-1} \sum_{n_2=0}^{N_2-1} W_{n_1,n_2} \exp[-i2\pi(\frac{m_1 n_1}{N_1} + \frac{m_2 n_2}{N_2})] \qquad (1.10)$$

$$W_{n_1,n_2} = \Delta f_1 \Delta f_2 \sum_{m_1=0}^{N_1-1} \sum_{m_2=0}^{N_2-1} A_{m_1,m_2} \exp[i2\pi(\frac{m_1 n_1}{N_1} + \frac{m_2 n_2}{N_2})] \qquad (1.11)$$

In a similar way, the discrete Fourier transform can be generalized to three or more dimensions.

2. SIMULATION OF CORRELATED RANDOM VECTORS

Several techniques can be used to develop correlated normal vectors of moderate dimension, say less than 100 [3]. A method which has many advantages is based on an eigenvector analysis of the covariance matrix (idem, p. 391). Let C be the covariance matrix for the random vector. Let L be the diagonal matrix with the eigenvalues down the main diagonal and zeros elsewhere, and define B as the matrix whose j-th column is the j-th eigenvector. The set of n eigenvector relations can be written also as an equivalent single matrix equation

$$CB = BL \qquad (2.1)$$

A simulation of the n-dimensional vector, $\underline{W}$, having the covariance matrix C can be produced by

$$\underline{W} = \underline{\mu} + BL^{\frac{1}{2}} \underline{Z} \qquad (2.2)$$

where $\underline{Z}$ is a vector of independent, normal, zero mean, unit variance random numbers [10] and $\underline{\mu}$ is the vector mean to be imposed on $\underline{W}$. The resulting vector, $\underline{W}$, will have the appropriate mean vector and covariance matrix and will be multivariate normal.

The eigenvalues are typically reported out by the computer in descending order,and it is usually the case that the eigenvalues are almost zero if $j>k$, for some k. For this situation, let L_1 be the kxk upper left square submatrix of L which involves the first k eigenvalues and define B_1 as the nxk submatrix of B which is composed of the first k columns of B. A simulation of $\underline{W}$ with almost the same statistical properties of (2.2) can be obtained from

$$\underline{W} = \underline{\mu} + B_1 L_1^{\frac{1}{2}} \underline{Z}_1 \tag{2.3}$$

$\underline{Z}_1$ is a k-component vector of independent standard normals. This last simulation equation is a very parsimonious procedure which minimizes computer time.

3. TWO-DIMENSIONAL RELATIONS

Let $\{W_{n1,n2}\}$ be a 2-D field of values for $0<=n_1<N_1$ and $0<=n_2<N_2$, which are stationary and repeated periodically along both axes. The Fourier coefficients $A_{m1,m2}$ defined in (1.11) may be written as

$$A_{m_1,m_2} = U_{m_1,m_2} - i\, V_{m_1,m_2} \tag{3.1}$$

The pairs of random variables $(U_{m1,m2}, V_{m1,m2})$ are uncorrelated with each other within certain sub-domains and related by complex conjugation to corresponding pairs in other sub-domains. The relevant sub-domains in two-space are shown in Fig. 3.1. These are classified into zones A, B, C_1, and C_2, each with their own statistical characterization.

The statistical properties in each zone [8] may be summarized. All $U_{m1,m2}$ and $V_{m1,m2}$ have theoretical mean zero and are normally distributed if the $\{W_{n1,n2}\}$ are multivariate normal with components having zero means. Let

$$T_1 = N_1 \Delta x_1$$
$$T_2 = N_2 \Delta x_2 \qquad (3.2)$$

Zone A:

$$\mathrm{Var}(U_{m_1,m_2}) = T_1 T_2 S_{m_1,m_2}$$
$$V_{m_1,m_2} = 0 \qquad (3,3)$$

Zone B and C_1:

$$\mathrm{Cov}\begin{pmatrix} U_{m_1,m_2} \\ V_{m_1,m_2} \end{pmatrix} = \frac{T_1 T_2}{2} \begin{pmatrix} S_{m_1,m_2} & 0 \\ 0 & S_{m_1,m_2} \end{pmatrix} \qquad (3.4)$$

Zone C_2:

$$\mathrm{Cov}\begin{pmatrix} U_{m_1,m_2} \\ V_{m_1,m_2} \end{pmatrix} = \frac{T_1 T_2}{2} \begin{pmatrix} S_{m_1,N_2-m_2} & 0 \\ 0 & S_{m_1,N_2-m_2} \end{pmatrix} \qquad (3.5)$$

The $S_{m1,m2}$ spectral set is obtained from

$$S_{m_1,m_2} = \Delta x_1 \Delta x_2 \sum_{k_1=0}^{N_1-1} \sum_{k_2=0}^{N_2-1} C_{k_1,k_2} \exp[-i2\pi(\frac{m_1 k_1}{N_1} + \frac{m_2 k_2}{N_2})] \qquad (3.6)$$

$C_{k1,k2}$ is the 2-D covariance set, taken as periodic with period T_1 and T_2.

The Fourier coefficients for the other subdomains of the plane may be filled in by the symmetries for $0<m_1<N_1/2$ and $0<m_2<N_2/2$ given by

$$A_{n_1-m_1,N_2-m_2} = \text{complex conjugate of } A_{m_1,m_2}$$
$$A_{n_1-m_1,m_2} = \text{complex conjugate of } A_{m_1,N_2-m_2} \quad (3.7)$$

The two-dimensional simulation proceeds as follows.

1. The Fourier coefficients for zones A, B, C_1, and C_2 are produced by multiplying standard normal random numbers by the square root of the variance for that frequency.

2. The rest of the Fourier coefficients are obtained by symmetry.

3. The 2-D inverse fast Fourier transform is applied to the Fourier coefficients to obtain the simulation of the 2-D set of geologic variables, $W_{n1,n2}$.

4. TWO-DIMENSIONAL SPECTRA

The discrete spectra in (1.8) and (3.6) are analogies to the corresponding integral relations

$$S(f) = \int_{-\infty}^{\infty} C(h) \exp[-i2\pi fh]dh \quad (4.1)$$

$$S(f_1,f_2) = \int_{-\infty}^{\infty}\int_{-\infty}^{\infty} C(h_1,h_2) \exp[-i2\pi(f_1h_1+f_2h_2)]dh_1dh_2 \quad (4.2)$$

It is convenient to evaluate the continuous spectral functions by calculus and then discretize them to get the functions for (3.6).

A number of different two-dimensional cases were studied by Taheri [9] and Hagan [6]. They concentrated on two- (and three-) dimensional covariance functions which may be described as "radial covariances with elliptical (or ellipsoidal) bases." These terms will be defined as follows, at least for two-dimensional space.

Let

$$h_0 = h_0\ (\theta) \tag{4.3}$$

be the equation of an ellipse in polar form where the long axis of the ellipse is a, the short axis is b, the long axis has an angle with the x-axis, and the ellipse is centered over the origin.

Definition: Let (h_1, h_2) be expressed in polar coordinates with radius h and argument θ.

A covariance function which can be expressed as

$$C(h_1,h_2) = g[h/h_0(\theta)] \tag{4.4}$$

for some univariate function g(x) will be said to be a radial covariance function with elliptical base.

The spectral functions for radial covariance functions with elliptical bases has a very interesting relation to the Hankel transform in special function theory [5], p. 3ff)

The Hankel transform of order nu of a function g(x) is defined as

$$H(y;\nu) = \int_0^\infty g(x)\ J_\nu(xy)\ \sqrt{xy}\ dx \tag{4.5}$$

The inverse relation is

$$g(x) = \int_0^\infty H(y;\nu)\ J_\nu(xy)\ \sqrt{xy}\ dy \tag{4.6}$$

Here, $J_\nu(z)$ is the Bessel function of order nu [8].

THEOREM 4.1: Let $C(h_1, h_2)$ be a two-dimensional radial covariance function with elliptical base given by (4.3). Let the elliptical base have long axis a, short axis b, angle of long axis with h_1-direction equal to α, and be centered over the origin. Define r by

$$r = 2\pi[a^2 (f_1 \cos\alpha + f_2 \sin\alpha)^2 + b^2(f_1 \sin\alpha - f_2 \cos\alpha)^2]^{\frac{1}{2}} \quad (4.7)$$

and let

$$\rho = h/h_0 \quad (4.8)$$

and radial covariance function be based on $g(\rho)$ where

$$C(h_1,h_2) = g(\rho) \quad (4.9)$$

Then $S(f_1, f_2)$ is related to the Hankel transform of $g(\rho)$ as follows.

$$S(f_1,f_2) = \frac{2ab\pi}{\sqrt{r}} \int_0^\infty \sqrt{\rho}\; g(\rho)\; J_0(r\rho)\; \sqrt{r\rho}\; d\rho \quad (4.10)$$

$$= S_*(r)$$

where $J_0(z)$ is a Bessel function of order zero. Note that $S(f_1, f_2)$ is purely a function of r which is a function of f_1, f_2 and the geometric parameters of the base ellipse.

Proof: The proof follows from a mapping of the elliptical base onto the unit circle, followed by integration with respect to angle over the full circle [6], [9].

The advantage of theorem (4.1) is that it permits the use of published tables of Hankel Transforms ([5],vol.2, for example) to study common variogram and covariance models. Several common examples are as follows [6], [9].

(a) Simple transitive model

$$g(\rho) = \begin{cases} 1-\rho, & \text{for } 0 \leq \rho \leq 1 \\ 0, & \text{for } \rho > 1 \end{cases} \tag{4.11}$$

$$S_*(r) = \frac{2\pi ab}{\sqrt{r}} \int_0^1 \sqrt{\rho}\,(1-\rho)\; J_0(r\rho)\; \sqrt{r\rho}\; d\rho \tag{4.12}$$

(b) Spherical model

$$g(\rho) = \begin{cases} 1-(3\rho/2)+(\rho^3)/2, & \text{if } 0 \leq \rho \leq 1 \\ 0, & \text{if } \rho > 1 \end{cases} \tag{4.13}$$

$$S_*(r) = \frac{2\pi ab}{\sqrt{r}} \int_0^1 \sqrt{\rho}\,\left(1 - \frac{3\rho}{2} + \frac{\rho^3}{2}\right) J_0(r\rho)\; \sqrt{r\rho}\; d\rho \tag{4.14}$$

(c) Exponential model

$$g(\rho) = \exp(-\rho) \tag{4.15}$$

$$S_*(r) = 2\pi ab(1+r^2)^{-3/2} \tag{4.16}$$

(d) Gaussian model

$$g(\rho) = \exp(-\rho^2) \tag{4.17}$$

$$S_*(\rho) = \pi ab\pi \exp(-r^2/4) \tag{4.18}$$

(e) "Hole effect" model

$$g(\rho) = (\sin\rho)/\rho \tag{4.19}$$

$$S_*(r) = \begin{cases} 2\pi ab(1-r^2)^{-\frac{1}{2}}, & \text{if } 0 \leq r < 1 \\ 0, & \text{if } r > 1 \end{cases} \tag{4.20}$$

Tables and graphs of these spectral functions are given by [6] for the spherical, transitive, exponential, and Gaussian cases. The transitive spectral function goes slightly negative over a small range of argument r, thus verifying the known fact that it is not positive definite in two-space. However the departure into negative values is so slight, that the $S_*(r)$ function can be set to zero for those values of r with no effect on practical problems.

The two-dimensional spectra provide basic tools for the $S_{m1,m2}$ needed in the variances of the frequency-domain Fourier coefficients. Here

$$S_{m_1,m_2} \cong S(m_1 \Delta f_1, m_2 \Delta f_2) \tag{4.21}$$

where

$$f_j = (N_j \Delta x_j)^{-1} \tag{4.22}$$

and (N_1, N_2) are the number of grid points in the x_1 and x_2 directions respectively, while $(\Delta x_1, \Delta x_2)$ are the spacing increments in the two directions. The value of r is computed for the two arguments of $S(f_1, f_2)$ and finally the value is computed from the listed formulas or read from the tables. The ease of computation with the exponential and Gaussian models suggest strongly that further research in Hermite or Laguerre expansions of covariance functions would be useful.

5. THREE-DIMENSIONAL SPECTRA

Taheri [9] and Hagan [6] investigate in some detail properties of three-dimensional spectral functions for radial covariance functions. The base here is an ellipsoid with long axis a, short axis c, and intermediate axis b, centered over the coordinate origin, and with arbitrary orientation of ellipsoidal axes. As before, the formulas typically reduce to a Hankel transform. Many theoretical questions remain in the evaluations of the transform for selected covariance models. However, the Gaussian curve reduces to a closed form formula and appears to be a good starting point for the practical representation of covariance models in 3-space.

In applications, (4.10) must often be evaluated by numerical integration. If the covariance function is truly non-negative definite as theory requires, then the spectral function should be everywhere positive. However, in the round off errors of numerical integration, some oscillations plus and minus will be seen in regions where the spectral functions are almost zero. These are spurious oscillations and not real. They do not affect computations in most cases and the spectra may usually be set to zero arbitrarily in these regions. If the simple transitive variogram or covariance model in 3-D space is used, then since it is known to be not non-negative definite, the negative values give a measure of how serious the lack of this property is in the use of the model.

Consider two planes perpendicular to the x_3-axis, separated by a lag or spacing of h_3. Let $C(h_1, h_2, h_3)$ be the covariance function and $S(f_1, f_2, f_3)$ be the spectral function. What is the 2-D spectra for the field of values in each plane and what is the cross-spectra between the fields of values in the two planes? The required relations are

$$\int_{-\infty}^{\infty} S(f_1,f_2, f_3)df_3 = S_{11}(f_1,f_2) = S_{22}(f_1,f_2) \tag{5.1}$$

$$\begin{aligned} S_{12}(f_1,f_2) &= \int_{-\infty}^{\infty} S(f_1,f_2, f_3) \exp[i2\pi f_3 \Delta h_3)df_3 \\ &= C_{12}(f_1,f_2)-i\, q_{12}(f_1,f_2) \end{aligned} \tag{5.2}$$

The functions, $c_{12}(f_1,f_2)$ and $q_{12}(f_1,f_2)$ are called the co- and quad-spectral densities respectively.

6. THREE-DIMENSIONAL SIMULATIONS WITH LAYERED TWO-DIMENSIONAL COMPUTATIONS

Two-dimensional Fourier coefficients are uncorrelated with other coefficients within the A, B, C_1, and C_2 zones of Fig. 3.1. The same thing is true with the Fourier coefficients for values in two or more two-dimensional parallel planes. These Fourier coefficients in zones A, B, C_1, and C_2 at the same m_1 and m_2 are correlated with each other, but not with any other coefficients in the four zones.

Let (U,V) denote the Fourier coefficients on the first plane and (U_*,V_*) be the coefficients on the second plane, both at the same 2-D frequency. Also let c and q denote the corresponding co- and quad-spectral densities and S the spectral density. If the random field is statistically stationary and Gaussian with mean zero, then the expected value of all coefficients is zero and the covariance matrix is given by

$$\mathrm{Cov}\begin{pmatrix} U \\ V \\ U_\star \\ V_\star \end{pmatrix} = aT_1T_2 \begin{bmatrix} S & 0 & c & q \\ 0 & S & -q & c \\ c & -q & S & 0 \\ q & c & 0 & S \end{bmatrix} \tag{6.1}$$

The value of a is 1.0 in zone A and 0.5 in the other zones. Also only the first and third rows and columns of the matrix are used in zone A because the imaginary parts of the coefficients are zero.

Three-dimensional simulation for several parallel mining horizons can, thus, be reduced to two-dimensional operations separately for each plane of values. Suppose that J planes are involved. Considering real and imaginary parts, there will be 2J real random variables at each (m_1, m_2) frequency in zones A, B, C_1, and C_2. Standard Normal random numbers are used to develop the 2J correlated normals having the 2Jx2J covariance matrix of the form symbolized by (6.1) and mean zero. Usually the method of eigenvalues is the most efficient procedure saving the most computer time. The set of J complex Fourier coefficients at each frequency in (A,B,C_1,C_2) can be developed separately since they are independent of each other. The rest of the Fourier coefficients are developed from symmetry relations within each plane. Then the 2-D fast Foiurier transform is used to revert the Fourier coefficients in each plane to real space to obtain the actual simulations of the geological variables.

The main advantage of the layered two-dimensional methods is that of computer-time savings. The disadvantage is, of course, the mathematical complexity and the inability to work with more than 10 or 20 layers in any practical way. However, for most three-dimensional spectral functions, large regions of the frequency spaces contain Fourier coefficients which are essentially zero. Only very small portions of the (f_1, f_2) space require simulation of Fourier coefficients. The rest of space may be filled with zeros for a particular simulation.

7 ILLUSTRATIVE EXAMPLE

Suppose it is desired to simulate simultaneously two intercorrelated mean zero Gaussian processes each representing one level of a given mine. Let the first process represent the surface of mine (z=0) and the second process represent the level one meter below the surface (z=1). Let the fluctuations in the given mine be represented by a three dimensional Gaussian covariance function. Then the corresponding three-dimensional spectral density is

$$S(r) = abc(\pi)^{3/2} \exp\,[-r^2/4] \tag{7.1}$$

where a, b, and c are the major, minor and vertical axes of the ellipsoid respectively, and

$$r = 2\pi(a^2f_1{}^2 + b^2f_2{}^2 + c^2f_3{}^2)^{\frac{1}{2}} \tag{7.2}$$

Now let the discrete version of the two processes be represented by $X_{n1,n2}$ and $Y_{n1,n2}$ for $0 <= n_1 < 16$; $0 < = n_2 < 16$, and the FFT coefficients corresponding to these processes by $A_{m1,m2}$ and $B_{m1,m2}$.

Step 1: The covariance matrix for the FFT coefficients in zone A and zones B and C were given previously. The components of these matricies are the spectral densities for the processes X and Y, and the cross-spectral densities between the processes X and Y. These spectras and cross-spectras are computed using the formulas given in Section 5. For the given example, four 2x2 covariance matrices in zone A, and 126 4x4 covariance matrix in zones B and C are developed.

Step 2: In this step, four two-component multivariate normal random vectors of Fourier coefficients for the frequencies in zone A, and 126 four-component multivariate normal vectors of the Fourier coefficients corresponding to the frequencies in zones B and C are generated and the covariance matrices developed above. The components of each generated vector have mean zero and are correlated according to the given covariance matrix.

Step 3: This step completes the determination of the Fourier coefficients in the entire Fourier transform planes. Up to now the Fourier coefficients for each process in zones of determination are generated. Then the conjugate symmetries are applied to determine the Fourier coefficients in entire transform plane for each process. The real and imaginary parts of the generated Fourier coefficients for the first process are shown in Tables 7.1 and 7.2 and for the second process are given in Tables 7.3 and 7.4.

Step 4: The Fourier coefficients in each transform plane separately are inverted to the space domain using the FFT. The generated processes in space domain are shown in Tables 7.5 and 7.6.

The covariances corresponding to the simulated processes shown in Tables 7.5 and 7.6 are computed and listed on Tables 7.7 and 7.8. The zone of influence and the variances of the simulated processes are almost the same as the theoretical ones.

8. FREQUENCY-DOMAIN CONDITIONAL SIMULATION

Conditional simulations are generally performed by the space-domain methods given by Journel [6]. It is interesting to see if it is possible to carry out such operations in frequency domain. Three theorems may be stated which show that under some circumstances, frequency- domain simulations are feasible and useful. This is basically when values are to be simulated on a large grid of values, conditional on known values at a relatively small set of locations.

The first two theorems are [3], p. 406-407). Let $\underline{W}$ be a normal random (column) vector with n components, which has mean vector $\underline{\mu}$ and covariance matrix C. Let $\underline{W}$ be partitioned into two vectors $\underline{W}_1$ and $\underline{W}_2$ with n_1 and n_2 components respectively. The vector $\underline{\mu}$ and the matrix C are similarly partitioned. Thus

$$n = n_1 + n_2 \tag{8.1}$$

$$\underline{W} = \begin{bmatrix} \underline{W}_1 \\ \underline{W}_2 \end{bmatrix}, \ \underline{\mu} = \begin{bmatrix} \underline{\mu}_1 \\ \underline{\mu}_2 \end{bmatrix}, \ C = \begin{bmatrix} C_{11} & C_{12} \\ C_{12}^T & C_{22} \end{bmatrix} \tag{8.2}$$

where the superscript T denotes the matrix transpose.

Theorem 8.1. The conditional probability law for $\underline{W}_2$, given $\underline{W}_1 = \underline{w}_1$, is multivariate normal with conditional mean of

$$E[\underline{W}_2 | \underline{W}_1 = \underline{w}_1] = \underline{\mu}_2 + C_{12}^T C_{11}^{-1} (\underline{w}_1 - \underline{\mu}_1) \tag{8.3}$$

conditional covariance matrix

$$\mathrm{Cov}\, [\underline{W}_2 | \underline{W}_1 = \underline{w}_1] = C_{22} - C_{12}^T C_{11}^{-1} C_{12} \tag{8.4}$$

Proof. (Anderson, [1], pp. 27-29).

Theorem 8.2. Let $\underline{W}$ be an unconditional simulation of the random vector. That is, $\underline{W}$ follows a multivariate normal probability law with mean $\underline{\mu}$ and covariance matrix C. The vector $\underline{\tilde{W}}_2$ defined by

$$\underline{\tilde{W}}_2 = C_{12}^T C_{11}^{-1} (\underline{w}_1 - \underline{W}_1) + \underline{W}_2 \tag{8.5}$$

will be a conditional simulation of $\underline{W}_2$, given $\underline{W}_1 = \underline{w}_1$. The mean vector and covariance matrix for $\underline{W}_2$ are the same as the conditional mean and covariance relations specified in Theorem 8.1 and $\tilde{W}_2$ is a multivariate normal random vector.

Proof. Verified by direct computation.

The third theorem is a new result. The separation of the values at grid locations into those known w_1 and those to be simulated W_2, produces a reordering of the normal grid order. Let n_j be the vector of location integers for the j-th component of W. Let m_j be the same reordering of the vector of frequency integers. Finally define E as the matrix of exponentials such that the (i,j) element of E is

$$(E)_{ij} = \exp(i2\pi \underline{m}_i^T \underline{n}_j) \tag{8.6}$$

That is, rows correspond to reordered frequency vectors and columns to reordered location vectors. Let E = (E_1, E_2) be the same partitioning of locations as used in (8.2). Also let $\underline{\Delta x}$ be Δx (if one-dimension), $\Delta x_1 \Delta x_2$ (if two-dimensional), and $\Delta x_1 \Delta x_2 \Delta x_3$ (if three-dimensional). The discrete Fourier transform may be written

$$\underline{A}_m = \underline{\Delta x}\ E\ \underline{W} \tag{8.7}$$

Theorem 8.3. The notation from Theorem (8.2) will be used here. Let A be the Fourier coefficients for the array of values

$$\tilde{\underline{W}} = \begin{pmatrix} \underline{w}_1 \\ \tilde{\underline{W}}_2 \end{pmatrix} \tag{8.8}$$

where $\tilde{W}_2$ is the conditional simulation of W_2, given W_1 = w_1. Let A be the Fourier coefficients for an unconditional simulation, W. Then

$$\tilde{\underline{A}} = \underline{A} + E \begin{pmatrix} C_{11} \\ C_{21} \end{pmatrix} C_{11}^{-1} (\underline{w}_1 - \underline{W}_1) \tag{8.9}$$

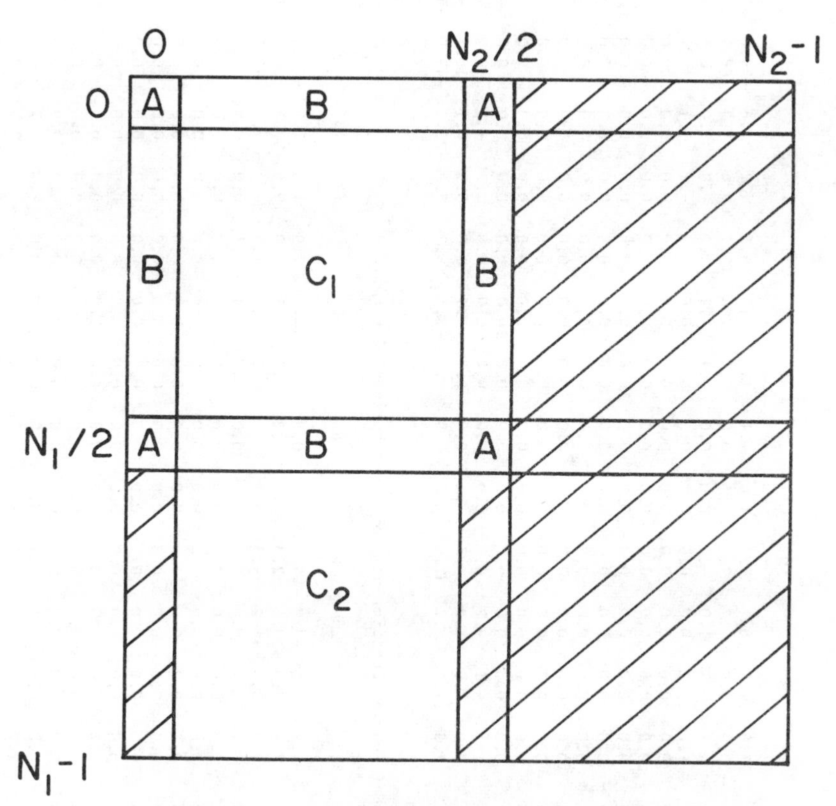

Figure 3.1 Geometric arrangement of 2-D Frequency domain structure

TABLE 7.1 Real part of Fourier coefficients for first horizon of Example B.

M1= M2	0	1	2	3	4	5	6	7	8	9	10	11	12	13	14	15
0	-19.6	-12.4	-59.6	27.6	1.2	6.7	-.7	-1.6	-.1	1.6	.7	6.7	1.2	27.6	-59.6	-12.4
1	-22.3	-10.5	-49.0	-20.9	5.9	.7	-3.5	-.5	0.0	.2	2.4	.7	2.4	1.3	-68.1	-32.6
2	9.4	7.6	-16.1	4.1	3.7	1.0	-.1	0.0	0.0	-.1	.7	.2	-.9	7.0	8.6	-3.3
3	-.9	-.1	0.0	-.2	.2	0.0	0.0	0.0	0.0	0.0	0.0	.2	-.1	1.1	.1	.3
4	0.0	-.1	0.0	0.0	0.0	0.0	0.0	0.0	0.0	0.0	0.0	0.0	0.0	0.0	0.0	0.0
5	0.0	0.0	0.0	0.0	0.0	0.0	0.0	0.0	0.0	0.0	0.0	0.0	0.0	0.0	0.0	0.0
6	0.0	0.0	0.0	0.0	0.0	0.0	0.0	0.0	0.0	0.0	0.0	0.0	0.0	0.0	0.0	0.0
7	0.0	0.0	0.0	0.0	0.0	0.0	0.0	0.0	0.0	0.0	0.0	0.0	0.0	0.0	0.0	0.0
8	0.0	0.0	0.0	0.0	0.0	0.0	0.0	0.0	0.0	0.0	0.0	0.0	0.0	0.0	0.0	0.0
9	0.0	0.0	0.0	0.0	0.0	0.0	0.0	0.0	0.0	0.0	0.0	0.0	0.0	0.0	0.0	0.0
10	0.0	0.0	0.0	0.0	0.0	0.0	0.0	0.0	0.0	0.0	0.0	0.0	0.0	0.0	0.0	0.0
11	0.0	0.0	0.0	0.0	0.0	0.0	0.0	0.0	0.0	0.0	0.0	0.0	0.0	0.0	0.0	0.0
12	0.0	0.0	0.0	0.0	0.0	0.0	0.0	0.0	0.0	0.0	0.0	0.0	0.0	0.0	0.0	-.1
13	.9	.3	-.1	1.1	-.1	-.2	0.0	0.0	0.0	0.0	0.0	0.0	.2	-.2	0.0	-.1
14	9.4	-3.3	8.6	7.0	-.9	.2	.7	-.1	0.0	0.0	-.1	1.0	3.7	4.1	-16.1	-7.6
15	-22.3	-32.6	63.1	1.3	2.6	.7	2.4	.2	0.0	-.5	-3.5	.7	5.9	-20.9	-49.0	-10.5

TABLE 7.2 Imaginary part of Fourier coefficients for first horizon of Example B.

M1= M2	0	1	2	3	4	5	6	7	8	9	10	11	12	13	14	15
0	0.0	-39.3	25.5	-1.9	-21.0	-7.9	-14.2	-.1	0.0	.1	14.2	7.9	21.0	1.9	-25.5	39.3
1	.2	-52.2	-48.9	7.9	-8.3	-13.5	2.3	-.6	.2	-.3	-2.8	6.0	19.5	-57.3	26.2	-68.9
2	-4.7	1.4	7.7	-9.3	3.3	-1.2	-.3	-.1	-.1	.2	-2.1	-1.4	-.9	-12.2	-4.1	2.4
3	.4	-1.1	.2	-.7	.4	0.0	.1	0.0	0.0	0.0	-.1	.1	0.0	0.0	0.0	-.6
4	0.0	0.0	0.0	0.0	0.0	0.0	0.0	0.0	0.0	0.0	0.0	0.0	0.0	0.0	0.0	0.0
5	0.0	0.0	0.0	0.0	0.0	0.0	0.0	0.0	0.0	0.0	0.0	0.0	0.0	0.0	0.0	0.0
6	0.0	0.0	0.0	0.0	0.0	0.0	0.0	0.0	0.0	0.0	0.0	0.0	0.0	0.0	0.0	0.0
7	0.0	0.0	0.0	0.0	0.0	0.0	0.0	0.0	0.0	0.0	0.0	0.0	0.0	0.0	0.0	0.0
8	0.0	0.0	0.0	0.0	0.0	0.0	0.0	0.0	0.0	0.0	0.0	0.0	0.0	0.0	0.0	0.0
9	0.0	0.0	0.0	0.0	0.0	0.0	0.0	0.0	0.0	0.0	0.0	0.0	0.0	0.0	0.0	0.0
10	0.0	0.0	0.0	0.0	0.0	0.0	0.0	0.0	0.0	0.0	0.0	0.0	0.0	0.0	0.0	0.0
11	0.0	0.0	0.0	0.0	0.0	0.0	0.0	0.0	0.0	0.0	0.0	0.0	0.0	0.0	0.0	0.0
12	0.0	0.0	0.0	0.0	0.0	0.0	0.0	0.0	0.0	0.0	0.0	0.0	0.0	0.0	0.0	0.0
13	-.4	.6	0.0	0.0	0.0	-.1	.1	0.0	0.0	0.0	-.1	0.0	-.4	.7	-.2	1.1
14	4.7	-2.4	4.1	12.2	.9	1.4	2.1	-.2	.1	.1	.3	1.2	-3.3	9.3	-7.7	-1.4
15	-.2	68.9	-26.2	57.3	-19.5	-6.0	2.3	.3	-.2	.6	2.3	13.5	8.3	-7.9	48.9	52.2

TABLE 7.3 Simulated values for first horizon of Example B

M2 \ M1=	0	1	2	3	4	5	6	7	8	9	10	11	12	13	14	15
0	-.8	-.5	-1.0	0.0	1.2	.7	-.4	-.8	-.2	.8	.9	0.0	-.6	-.4	-.7	-1.2
1	-.1	.4	.3	.3	.9	.3	-.7	-1.4	-.8	.6	1.2	.5	-.5	-.8	-1.0	-1.2
2	.5	1.2	.4	.5	.5	-.1	-1.0	-1.9	-1.4	.3	1.2	.7	-.4	-.9	-1.2	-1.0
3	.9	1.7	.8	.5	.2	-.5	-1.0	-2.1	-1.8	-.2	.8	.7	-.3	-.9	-1.0	-.8
4	1.0	1.9	1.0	.5	0.0	-.5	-.7	-1.9	-2.0	-.7	.2	.5	0.0	-.5	-.7	-.5
5	1.0	1.7	.9	.4	-.1	-.2	-.1	-1.4	-2.0	-1.3	-.5	.3	.3	-.1	-.2	-.1
6	.8	1.2	.5	.3	.1	.4	.7	-.7	-1.9	-1.7	-1.0	.2	.6	.2	.1	.1
7	.5	.6	.1	.2	.3	1.1	1.5	0.0	-1.6	-1.9	-1.3	.2	.9	.5	.3	.2
8	.1	-.1	-.3	.2	.6	1.6	2.2	.5	-1.2	-1.8	-1.3	.3	.9	.5	.2	.1
9	-.4	-.7	-.7	.1	.9	2.0	2.5	.9	-.7	-1.5	-1.2	.2	.8	.5	.1	-.2
10	-.9	-1.1	-1.1	0.0	1.0	2.1	2.5	1.1	-.3	-1.1	-1.1	.1	.6	.5	.1	-.6
11	-1.5	-1.5	-1.3	-.2	1.1	2.0	2.1	1.1	.2	-.7	-1.1	-.5	.2	.5	.1	-.9
12	-1.9	-1.8	-1.6	-.3	1.2	1.7	1.6	.9	.6	-.2	-1.0	-1.0	-.1	.5	.2	-1.2
13	-2.1	-1.9	-1.3	-.4	1.3	1.5	1.0	.7	.8	.3	-.8	-1.2	-.4	.5	.2	-1.3
14	-1.9	-1.8	-1.8	-.4	1.4	1.3	.5	.3	.8	.6	-.3	-1.1	-.6	.4	.1	-1.3
15	-1.4	-1.3	-1.5	-.2	1.4	1.1	.1	-.2	.4	.8	.3	-.6	-.6	.1	-.2	-1.3

TABLE 7.4 Real part of Fourier coefficients for second horizon of Example B.

M2 \ M1=	0	1	2	3	4	5	6	7	8	9	10	11	12	13	14	15
0	-20.2	24.0	-63.7	26.3	-2.9	5.1	-.4	-.7	.4	-.7	-.4	5.1	-2.9	26.3	-63.7	24.0
1	-49.5	-15.2	-54.2	-25.3	1.3	2.4	-3.3	-.2	-.2	0.0	1.3	1.3	-2.7	-2.7	69.6	-17.1
2	5.3	7.1	-14.0	4.5	1.7	.9	0.0	0.0	0.0	.1	.6	.3	-1.0	4.5	6.5	.1
3	.8	.3	.3	-.2	.1	0.0	-.1	0.0	0.0	0.0	0.0	-.2	-.1	1.1	.3	.6
4	0.0	0.0	0.0	0.0	0.0	0.0	0.0	0.0	0.0	0.0	0.0	0.0	0.0	0.0	0.0	0.0
5	0.0	0.0	0.0	0.0	0.0	0.0	0.0	0.0	0.0	0.0	0.0	0.0	0.0	0.0	0.0	0.0
6	0.0	0.0	0.0	0.0	0.0	0.0	0.0	0.0	0.0	0.0	0.0	0.0	0.0	0.0	0.0	0.0
7	0.0	0.0	0.0	0.0	0.0	0.0	0.0	0.0	0.0	0.0	0.0	0.0	0.0	0.0	0.0	0.0
8	0.0	0.0	0.0	0.0	0.0	0.0	0.0	0.0	0.0	0.0	0.0	0.0	0.0	0.0	0.0	0.0
9	0.0	0.0	0.0	0.0	0.0	0.0	0.0	0.0	0.0	0.0	0.0	0.0	0.0	0.0	0.0	0.0
10	0.0	0.0	0.0	0.0	0.0	0.0	0.0	0.0	0.0	0.0	0.0	0.0	0.0	0.0	0.0	0.0
11	0.0	0.0	0.0	0.0	0.0	0.0	0.0	0.0	0.0	0.0	0.0	0.0	0.0	0.0	0.0	0.0
12	0.0	0.0	0.0	0.0	0.0	0.0	0.0	0.0	0.0	0.0	0.0	0.0	0.0	0.0	0.0	0.0
13	.8	.6	.3	1.1	-.1	-.2	0.0	0.0	0.0	0.0	-.1	0.0	.1	-.2	.3	.3
14	5.3	.1	6.5	4.5	-1.0	.3	.6	-.1	0.0	0.0	0.0	.9	1.7	4.5	-14.0	-7.1
15	-49.5	-17.1	69.6	-2.7	-2.7	1.3	1.3	0.0	-.2	-.2	-3.3	2.4	1.3	-25.3	-54.2	15.2

TABLE 7.5 Imaginary part of Fourier coefficients for second horizon of Example B.

M1= M2	0	1	2	3	4	5	6	7	8	9	10	11	12	13	14	15
0	0.0	-58.9	2.3	-13.7	-22.3	-10.5	-13.2	-.4	0.0	.4	13.2	10.5	22.3	13.7	-2.3	53.9
1	1.9	-40.9	-57.5	9.8	-4.7	-13.6	4.2	-.3	.1	-.3	-2.3	3.5	22.4	-53.5	33.2	-71.6
2	-8.9	5.4	7.0	-8.9	5.4	-1.5	-.1	-.2	-.1	.1	-2.3	-1.5	0.0	-11.4	-3.3	2.0
3	.5	-1.4	.2	-.8	.3	0.0	0.0	0.0	0.0	0.0	-.1	.1	.1	0.0	-.1	-.8
4	0.0	0.0	-.1	0.0	0.0	0.0	0.0	0.0	0.0	0.0	0.0	0.0	0.0	0.0	0.0	0.0
5	0.0	0.0	0.0	0.0	0.0	0.0	0.0	0.0	0.0	0.0	0.0	0.0	0.0	0.0	0.0	0.0
6	0.0	0.0	0.0	0.0	0.0	0.0	0.0	0.0	0.0	0.0	0.0	0.0	0.0	0.0	0.0	0.0
7	0.0	0.0	0.0	0.0	0.0	0.0	0.0	0.0	0.0	0.0	0.0	0.0	0.0	0.0	0.0	0.0
8	0.0	0.0	0.0	0.0	0.0	0.0	0.0	0.0	0.0	0.0	0.0	0.0	0.0	0.0	0.0	0.0
9	0.0	0.0	0.0	0.0	0.0	0.0	0.0	0.0	0.0	0.0	0.0	0.0	0.0	0.0	0.0	0.0
10	0.0	0.0	0.0	0.0	0.0	0.0	0.0	0.0	0.0	0.0	0.0	0.0	0.0	0.0	0.0	0.0
11	0.0	0.0	0.0	0.0	0.0	0.0	0.0	0.0	0.0	0.0	0.0	0.0	0.0	0.0	0.0	0.0
12	0.0	0.0	0.0	0.0	0.0	0.0	0.0	0.0	0.0	0.0	0.0	0.0	0.0	0.0	.1	0.0
13	-.5	.8	.1	0.0	-.1	-.1	.1	0.0	0.0	0.0	0.0	0.0	-.3	.3	-.2	1.4
14	8.9	-2.0	8.9	11.4	0.0	1.5	2.3	-.1	.1	.2	.1	1.5	-5.4	8.9	-7.0	-5.4
15	-1.9	71.6	-38.2	53.5	-22.4	-3.5	2.3	.3	-.1	.3	-4.2	13.6	4.7	-9.8	57.5	40.9

TABLE 7.6 Simulated value for second horizon of Example B.

M1= M2	0	1	2	3	4	5	6	7	8	9	10	11	12	13	14	15
0	-.9	-.1	-.3	.2	.3	.1	-1.1	-1.5	-.9	.3	.7	-.2	-.3	-.5	-.8	-1.5
1	-.2	.8	.5	.6	.6	-.2	-1.3	-2.6	-1.5	.2	1.1	.2	-.9	-1.0	-1.2	-1.5
2	.4	1.6	1.2	.9	.4	-.5	-1.3	-2.3	-1.9	0.0	1.1	.4	-.8	-1.2	-1.4	-1.3
3	.3	2.2	1.6	1.1	.3	-.6	-1.2	-2.3	-2.2	-.4	.7	.5	-.6	-1.2	-1.4	-1.1
4	1.1	2.4	1.9	1.1	.3	-.4	-.8	-2.0	-2.3	-.9	.1	.3	-.2	-.8	-1.0	-.7
5	1.2	2.3	1.6	1.0	.3	0.0	-.1	-1.4	-2.2	-1.5	-.6	.2	.2	-.4	-.5	-.3
6	1.1	1.8	1.2	.9	.5	.6	.8	-.7	-2.0	-2.0	-1.2	.2	.6	.1	-1	.1
7	.9	1.2	.7	.7	.3	1.4	1.7	0.0	-1.7	-2.2	-1.5	.2	.9	.4	.3	.3
8	.6	.5	.1	.5	1.0	2.0	2.4	.7	-1.2	-2.0	-1.5	.3	1.0	.6	.4	.2
9	.1	0.0	-.3	.4	1.2	2.4	2.7	1.1	-.7	-1.7	-1.4	.2	.9	.6	.4	0.0
10	-.4	-.5	-.6	.2	1.3	2.4	2.6	1.2	-.2	-1.3	-1.3	-.1	.6	.6	.3	-.3
11	-.9	-.9	-.9	0.0	1.3	2.1	2.0	1.0	.1	-.8	-1.2	-.4	.3	.6	.2	-.7
12	-1.5	-1.2	-1.2	-.1	1.2	1.6	1.3	.6	.3	-.5	-1.2	-.8	-.1	.6	.2	-1.1
13	-1.8	-1.4	-1.3	-.2	1.1	1.1	.4	.2	.3	-.1	-.9	-1.1	-.3	.5	.1	-1.4
14	-1.8	-1.3	-1.3	-.2	1.1	.7	-.2	-.4	.1	.1	-.5	-1.0	-.5	.3	-.1	-1.5
15	-1.5	-.9	-.9	-.1	1.0	.4	-.7	-.9	-.3	.3	.1	-.7	-.7	0.0	-.4	-1.5

TABLE 7.7 Data covariance computed from the simulation shown in Table 8.7. mean = -.076, variance = .985

M1= M2	0	1	2	3	4	5	6	7	8	9	10	11	12	13	14	15
0	1.0	.7	.1	-.3	-.4	-.3	-.1	0.0	0.0	0.0	-.1	-.3	-.4	-.3	.1	.7
1	.9	.7	.2	-.3	-.4	-.2	-.1	0.0	0.0	0.0	-.2	-.3	-.4	-.3	.1	.6
2	.7	.6	.2	-.2	-.3	-.2	0.0	.1	0.0	-.1	-.2	-.3	-.3	-.3	0.0	.4
3	.5	.5	.2	-.1	-.2	-.1	0.0	.1	.1	-.1	-.2	-.2	-.2	-.2	-.1	.2
4	.2	.3	.2	0.0	-.1	0.0	.1	.1	.1	-.1	-.1	-.1	-.1	-.1	-.1	0.0
5	-.1	.1	.1	.1	0.0	0.0	.1	.1	.1	0.0	-.1	-.1	0.0	0.0	-.2	-.2
6	-.3	-.1	0.0	.1	.1	.1	.1	.1	.1	0.0	0.0	0.0	.1	0.0	-.2	-.3
7	.-4	-.3	0.0	.1	.1	.1	.1	.1	.1	.1	0.0	.1	.1	.1	-.1	-.4
8	-.5	-.4	-.1	.1	.2	.1	.1	.1	.1	.1	.1	.1	.2	.1	-.1	-.4
9	-.4	-.4	-.1	.1	.1	.1	0.0	.1	.1	.1	.1	.1	.1	.1	0.0	-.3
10	-.3	-.3	-.2	0.0	.1	0.0	0.0	0.0	.1	.1	.1	.1	.1	.1	0.0	-.1
11	-.1	-.2	-.2	0.0	0.0	-.1	-.1	0.0	.1	.1	.1	0.0	0.0	.1	.1	.1
12	.2	0.0	-.1	-.1	-.1	-.1	-.1	-.1	.1	.1	.1	0.0	-.1	0.0	.2	.3
13	.5	.2	-.1	.-2	-.2	-.2	-.2	-.1	.1	.1	0.0	-.1	-.2	-.1	.2	.5
14	.7	.4	0.0	-.3	-.3	-.3	-.2	-.1	0.0	.1	0.0	-.2	-.3	-.2	.2	.6
15	.9	.6	.1	-.3	-.4	-.3	-.2	0.0	0.0	0.0	-.1	-.2	-.4	-.3	.2	.7

TABLE 7.8 Data covariance computed from the simulation shown in Tables 8.10 mean = -.079, variance = 1.146

M1= M2	0	1	2	3	4	5	6	7	8	9	10	11	12	13	14	15
0	1.1	.8	.2	-.3	-.4	-.3	-.1	.1	.1	.1	-.1	-.3	-.4	-.3	.2	.8
1	1.1	.8	.2	-.2	-.4	-.2	0.0	.1	.1	0.0	-.1	-.3	-.4	-.3	.1	.7
2	.9	.7	.2	-.2	-.3	-.2	0.0	.1	.1	0.0	-.2	-.3	-.3	-.2	.1	.6
3	.6	.5	.2	-.1	-.2	-.1	0.0	.1	0.0	-.1	-.2	-.3	-.2	-.2	0.0	.3
4	.3	.3	.2	0.0	-.1	0.0	0.0	0.0	0.0	-.1	-.2	-.2	-.1	-.1	-.1	.1
5	0.0	.1	.1	.1	0.0	0.0	0.0	0.0	-.1	-.1	-.2	-.1	0.0	0.0	-.1	-.1
6	-.3	-.1	.1	.1	.1	-.1	0.0	0.0	-.1	-.2	-.1	0.0	.1	0.0	-.1	-.3
7	-.4	-.3	0.0	.2	.2	.1	0.0	-.1	-.1	-.1	-.1	0.0	.1	.1	-.1	-.4
8	-.5	-.4	-.1	.1	.2	.1	-.1	-.1	-.1	-.1	-.1	-1	.2	.1	-.1	-.4
9	-.4	-.4	-.1	.1	.1	0.0	-.1	-.1	-.1	-.1	0.0	-1	.2	.2	0.0	-.3
10	-.3	-.3	-.1	0.0	.1	0.0	-.1	-.2	-.1	0.0	0.0	-1	.1	.1	.1	-.1
11	0.0	-.1	-.1	0.0	0.0	-.1	-.2	-.1	-.1	0.0	0.0	0.0	0.0	.1	.1	.1
12	.3	.1	-.1	-.1	-.1	-.2	.2	-.1	0.0	0.0	0.0	0.0	-.1	0.0	.2	.3
13	.6	.3	0.0	-.2	-.2	-.3	-.2	-.1	0.0	.1	0.0	-.1	-.2	-.1	.2	.5
14	.9	.6	.1	-.2	-.3	-.3	-.2	0.0	.1	.1	0.0	-.2	-.3	-.2	.2	.7
15	1.1	.7	.1	-.3	-.4	-.3	-.1	0.0	.1	.1	0.0	-.2	-.4	-.2	.2	.8

9. REFERENCES

1 Anderson, T. W. (1958), An Introduction to Multivariate Statistical Analysis, John Wiley, New york.

2 Borgman, L. E. (1973), "Statistical Properties of Fast Fourier Transform Coefficients," Res. Paper 23, College of Commerce and Industry, University of Wyoming, Laramie, Wyoming.

3 Borgman, L. E. (1982), "Techniques for Computer Simulation of Ocean Waves," Advanced Topics in Ocean Physics, Italian Physical Society, Bologna, Italy.

4 Brigham, O. C. (1974), The Fast Fourier Transform, Prentice-Hall, Englewood Cliffs, N. J.

5 Erdelyi, A., Magnus, W., Oberhettinger, F., and Tricomi, F. G. (1954), Tables of Integral Transforms, vol. II, McGraw-Hill Book Co., New York.

6 Hagan, R. L. (1982), Application of Spectral Theory and Analysis in Mining Geostatistics and Statistical Linear Wave Theory, Ph.D. Thesis, Statistics Department, University of Wyoming, Laramie, Wyoming.

7 Journel, A. G. (1974), "Geostatistics for Conditional Simulation of Ore Bodies," Economic Geology, vol. 69, pp. 673-687.

8 Oliver, F. W. J. (1964), Bessel Functions of Integer Order," Handbook of Mathematical Functions, (Abramowitz, M. and Stegun, I. A.: Editors) pp. 355-433, U. S. Government Printing Office, Washington, D. C.

9 Taheri, S. M. (1980), Data Retrieval and Multidimensional Simulation of Mineral Resources, Ph.D. Thesis, Statistics Department, University of Wyoming, Laramie, Wyoming.

10 Zelen, M. and Severo, N. C. (1964), Probability Functions," Handbook of Mathematical Functions, (Abramowitz, M. and Stegun, I. A.:Editors), pp. 925-995, U. S. Government Printing Office, Washington, D. C.

FACTORIAL KRIGING ANALYSIS : A SUBSTITUTE TO SPECTRAL ANALYSIS OF MAGNETIC DATA

Alain GALLI

Centre de Géostatistique, Ecole des Mines de Paris
35 rue Saint Honoré - 77305 FONTAINEBLEAU - FRANCE

Françoise GERDIL-NEUILLET, Claire DADOU

Compagnie Française des Pétroles
39-43 Quai André Citroën - 75015 PARIS - FRANCE

ABSTRACT

Factorial kriging analysis is a new method which combines kriging analysis and principal component analysis into the framework of geostatistics. The variables are split into principal components corresponding to different frequency ranges.

It was tempting to apply this method to magnetic data, since the main objective of the study of these data is to determine the local structures that correspond to certain frequency ranges. In this paper, we present a few ways of applying this method to magnetism, and show the first results.

1. INTRODUCTION

Using structural analysis on the studied variables (covariances and cross covariances) factorial kriging analysis consists in splitting these variables into components corresponding to particular frequency bands. In each of these bands, each main component is estimated from the raw variables by cokriging, after factorial analysis has been performed on the corresponding part of the covariance matrix.

This method incorporates the regionalized nature of the variables into the study via their covariances. Despite the problems inherent to estimating the covariances (and more particularly the cross-covariances), the model in general reflects the main scales of variability and the spatial correlation between the variables.

G. Verly et al. (eds.), Geostatistics for Natural Resources Characterization, Paty 1, 543–557.

This method therefore makes it possible to incorporate the spatial character of the phenomenon into data analysis, and to give us a type of "spectral analysis". While this would not be quite as precise as traditional harmonic analysis, as far as the identification of the frequencies is concerned, it nevertheless gives a reliable indication of the main frequency ranges as well as of the spatial regionalization of the variables.

Given these possibilities, it therefore seems very appropriate for gravimetric and magnetic data processing.

Gravity and magnetic anomalies come from variations of density and susceptibility of the earth material. Magnetic and gravity surveys have an economically important application in the exploration of new areas. They can determine the local relief of the basement and then basinal areas.
Each gravity and magnetic measurement determines the sum of all effects from the ground. A magnetic or a gravity map is almost never a simple picture of a simple source but often is a combination of sharp anomalies which must be of shallow origin and of broad anomalies of a regional nature. Therefore gravity and magnetic interpretation begins with some transformations which separate the anomalies from superficial sources on one hand and deep sources on the other.

The results obtained by kriging analysis can subsequently be compared with those obtained by FOURIER analysis which is frequently used with data of this type.

In the sequel we will illustrate the factorial kriging analysis on magnetic data only.
With this limitation, the problem becomes a univariate one. The relevant theory which comes to a type of spectral analysis, will be presented briefly before going on to the case study of the magnetic data.

2. IMPLEMENTATION

2.1 Factorial kriging analysis in the univariable case

The general theory of factorial kriging analysis has been developped by G. MATHERON [1]. In what follows we will briefly present the theory in the stationary univariate case (for a more detailed account of this particular case see [4]).

The model is :

$$Z(x) = \sum_{i=1}^{n} a_i \, Y^i(x) \qquad (1)$$

Z(x) is the field measured at points Z (x_j), the Y^i (x) are mutually orthogonal random functions representative of differents scales of variability.

In the first place, after structural analysis of the Z (x_j), and modelling their experimental covariance (let C (x-y) this covariance), we determine the number n of components Y^i , their normed covariance C^i (x-y) and the coefficients ai. Then the Y^i are estimated by cokriging, using the Z (x_j), ie :

$$(2) \quad Y^{i*}(x) = \sum \lambda^{ij} Z(x_j)$$

It is obvious that relation (1) does not specify the expectation of the Y^i. It can be easily seen, for instance by writing the non bias condition, that if :

1) E (Z(x)) = 0 the cokriging of the Y^i will be performed without universality condition

2) E (Z(x)) = m $\neq$ 0, Y^i must be cokriged with the condition $\Sigma\lambda^{ij} = 0$, which imposes E (Y^i) = 0, and the model is thus :

$$(3) \quad Z(x) = \sum ai\, Y^i(x) + m$$

In case 1), the expression (3) is still valid, <u>but</u> the cokriging is performed without conditions on the λ^{ij}.

It can then easily be shown that, if Z^* (x) is the kriging estimate of Z (x), Y^{i*} (x) the cokriging estimate of Y^i (x), m^* the kriging estimate of m, we have :

$$(4) \quad Z^*(x) = \sum ai\, Y^{i*}(x) + m^*$$

On the other hand, we know from the dual formalism (cf. [2], [3]) that Z^* (x) can be written as :

$$(5) \quad Z^*(x) = \sum b^j\, C(x-x_j) + m^*$$

It can be shown, because of the linearity (cf. [1]), that :

$$(6) \quad Y^{i*}(x) = ai^2 \sum b^j\, C^i(x-x_j)$$

As the dual system does not use the coordinates of the point to be estimated, this system has only to be solved once (in unique neighbourhood), to obtain the b^j and m^*. Then the Y^{i*} (x) <u>whatever i and x</u> are calculated by a simple scalar product.

2.2 Unique or moving neighbourhood

The question is whether one should work in unique or moving neighbourhood.

The arguments for and against each method are as follows :

Unique neigbourhood

For :
- Univocal determination of the Y^{i*} and m^*
- Simplicity due to the dual system

Against :
- Need for statistical inference valid on the whole field, and inadequacy of polynomial drift for a large neighbourhood
- Problems of solving a very large system

Moving neighbourhood

For :
- Easier statistical inference
- Numerous programs

Against :
- Dependency of the Y^{i*} on the number of points used for their estimation
- If $E(Z(x)) = m \neq 0$, it is no longer m^* which is estimated, but $m^*(x)$, that is a local mean. The model thus becomes

$$Z(x) = \Sigma \; a_i Y^i(x) + m(x)$$

therefore

$$Z^*(x) = \Sigma \; a_i Y^{i*}(x) + m^*(x)$$

In our case, we choose to work with a moving neighboorhood. As the maps provided by factorial kriging analysis are to be used as a basis for the interpretation, it is essential that the estimations of the Y^{i*} be little affected by the size of the neighbourhood. To put it in a more realistic terms, is it possible to determine a neighbourhood size beyond which the estimators are stable ? In the sequel we will pay special attention to this point.

3. MAGNETIC DATA

In a classical study, complex transformations are applied to data in order to isolate the effect of particular sources.

The classical transformations computed on magnetic data are the reduction to the pole, derivation, and continuations, without forgetting the separations "Regional - Residual". These transformations can be calculated in the spacial or frequency domain.

The reduction of data to the pole enables to bring anomalies upright the originating structures. Vertical gradient emphasizes shallow sources effects, and increases focusing of magnetic effects plumb with their sources. Upward and downward continuation is also frequently used as a smoothing (upward) or focusing transformation (downward).

Then, from the different transformations computed, structural maps of magnetic sources can be established.

4. THE CASE STUDY

4.1 Presentation of the data

We worked on magnetic data reduced to the pole, available at the node of a regular 76 x 96 grid (with equal spacings along both axes). In fact values existed for 5879 of the 7296 points. The mean is -1050. and the variance $.13 \times 10^7$.

A contour map of the values is shown on figure 1. Two essential features of the data are seen : the overall smoothness (regularity) of the phenomenon associated, nevertheless, with important variations in the levels. The structural analysis indicates that the phonomenon is stationary in the 4 directions tested (i.e. OX, OY, and the 45° bisectors). There is a very slight geometric anisotropy. However since this is indistinguishable at the scale used in practice, the variogram model was fitted to the average experimental variogram. It consisted of two nested cubic models with ranges of 6. and 20. respectively and sills of 140,000 and 950,000.

$$\gamma(h) = \begin{cases} c\,(7\,h^2/a^2 - 8.75\,h^3/a^3 + 3.5\,h^5/a^5 - .75\,h^7/a^7) & \text{if } h < a \\ c & \text{if } h \geqslant a \end{cases}$$

It would also have been possible to use a combination of a cubic and a spherical variogram (with similar ranges as the others) as the model, but it seemed preferable to use the two cubic models which are much more regular. This seems more appropriate to this type of data which derive from a potential field and is therefore regular. The mean variogram of raw data is shown on figure 2.

We also worked on anamorphosed data. The model fitted was very similar to the one for raw data. It consists of 2 cubic models with ranges 9. and 27. and sills 0.2 and 0.8

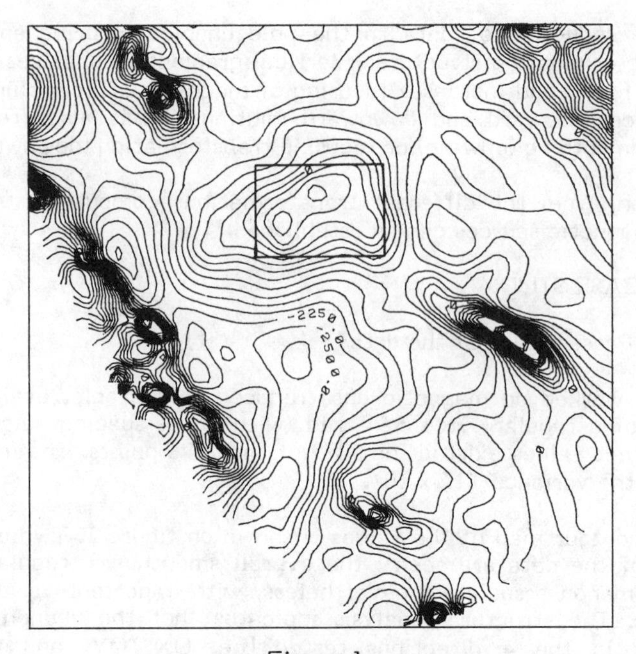

Figure 1

Magnetic values
(The rectangle shows the test zone for stability study)

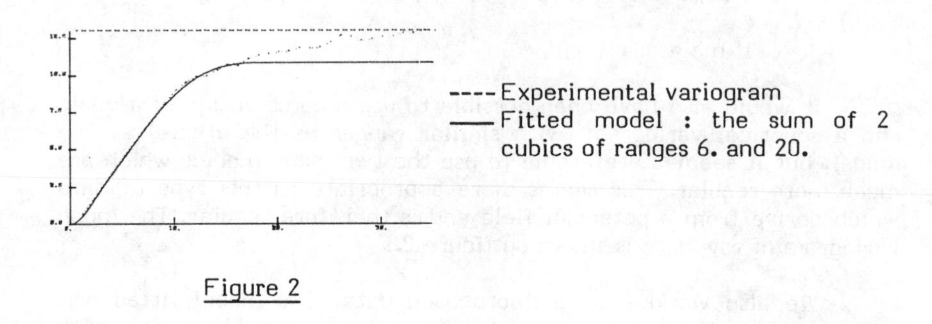

Figure 2

4.2 Stability of the components estimation

By structural analysis, it was found that the phenomenon could be modeled, in the form :

$$Z(x) = Y^1(x) + Y^2(x) + m(x)$$

where the components Y^1 and Y^2 have cubic variograms as described in the previous chapter. Y^1 corresponds to a small range, Y^2 , to a large range.

Note : Y^1 and Y^2 incorporate the terms a_1 and a_2 .

There are two possibilities for the term m(x) : to work either on anamorphosed data with a zero mean, or on raw data with the arithmetic mean removed. In this case we have no universality condition and m(x) = 0, or else we can work with a non-zero mean. In this case m(x) is the constant term of the drift, similar to a local mean, and is estimated by kriging on the same neighbourhood as the 2 components.

Intensive tests were performed on the 400 points zone indicated on figure 1.

a) On raw data kriged with a universality condition, we estimated Y^{1*} , Y^{2*} and m^* with 8 then 30, 50, 70, 90 neighbourhood points succesively. Contour levels of the estimate with 8 neighbourhood points of Y^{1*} and m^* are shown on figures 3a and 3b. While on figures 4a and 4b we see Y^{1*} and m^* estimated with 90 neighbourhood points. It is immediately clear that the estimate of Y^{1*} is hardly affected by the number of points ; while that of m^* is much more sensitive to this number.

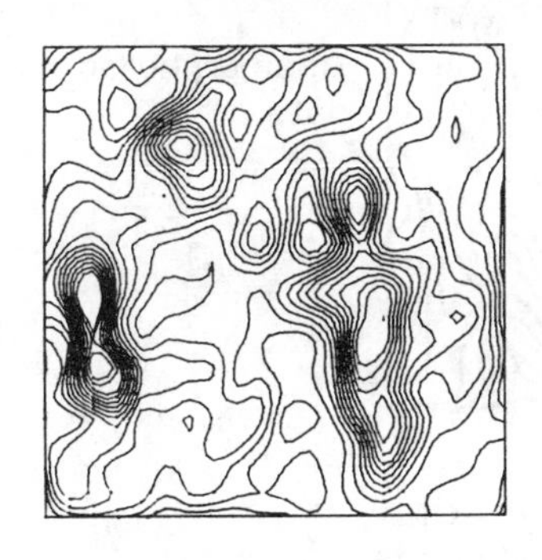

3a
Y^{1*} estimated with
8 neighbourhood points

3b
m^* estimated with
8 neighbourhood points

Raw data

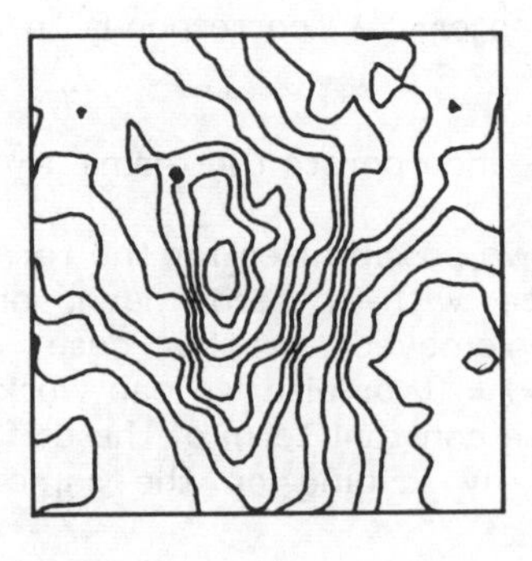

Raw data

4a
$Y1^*$ estimated with
90 neighbourhood points

4b
m^* estimated with
90 neighbourhood points

b) the same estimates were made on the anamorphoses, kriged with 1 universality condition and 16, 30, 50, 70, 90 , neighbourhood points. Similar results were obtained as shown on figures 5a, 5b, for 16 neighbourhood points, and on figures 6a, 6b for 90 points.

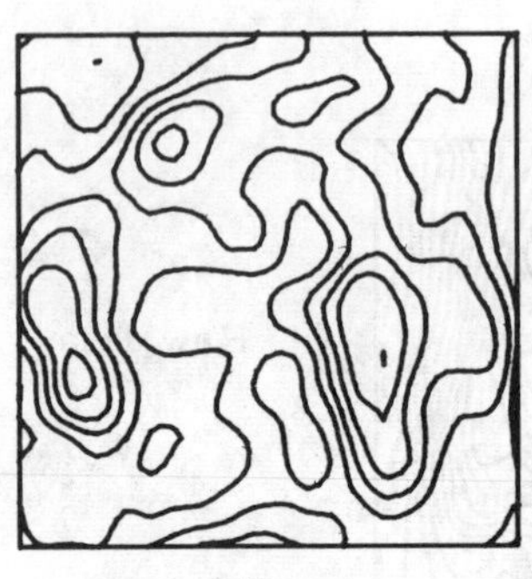

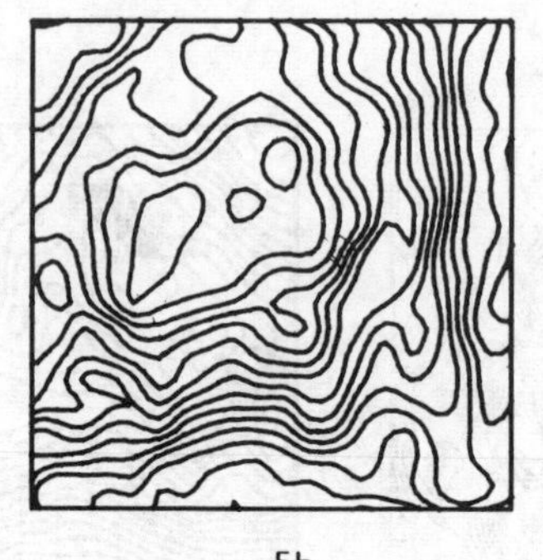

Anamorphosed data

5a
$Y1^*$ estimated with
16 neighbourhood points

5b
m^* estimated with
16 neighbourhood points

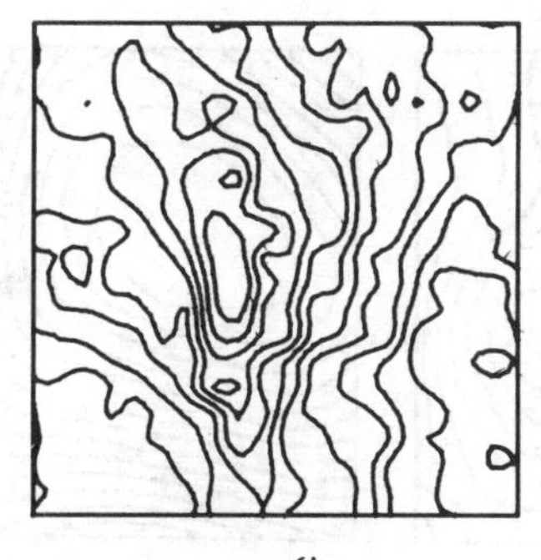

Anamorphosed data

6a
Y^{1*} estimated with
90 neighbourhood points

6b
m^* estimated with
90 neighbourhood points

c) We performed the same tests on anamorphosed data, this time without a universality condition, which is the right method for data with a zero mean. We again considered 16, 30, 50, 70, then 90 neighbourhood points. Figures 7a, 7b show Y^{1*} and Y^{2*} estimated with 16 neighbourhood points, while figures 8a, 8b show Y^{1*} and Y^{2*} estimated with 90 points.

Conclusions on Stability

The following conclusions can be drawn from this stability study :

1) There are no problems concerning the estimation of the low range factor in any of the tested cases (including the raw data reduced to a zero mean and kriged without universality condition, which is not discussed here). (Note that this factor does not seem sensitive to the presence or absence of universality conditions).

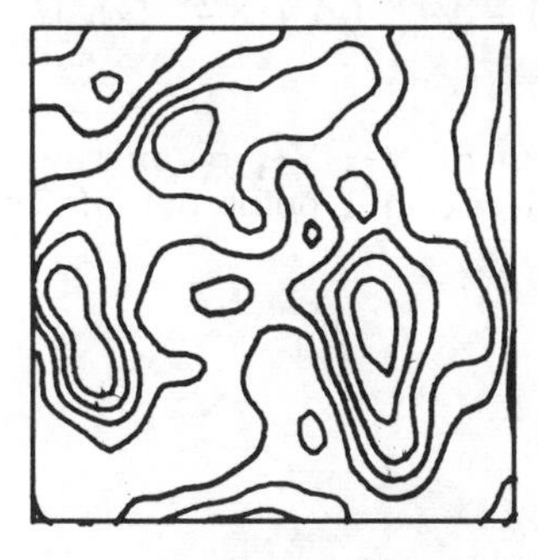

Anamorphosed data

7a
Y^{1*} estimated with
16 neighbourhood points

7b
Y^{2*} estimated with
16 neighbourhood points

Anamorphosed data

8a
Y^{1*} estimated with
90 neighbourhood points

8b
estimated with
90 neighbourhood points

2) An additional problem linked to the neighbourhood size arises for the high range factor and the local mean in the case of raw magnetism. With 70 neighbourhood points (which is a large number) we cover an 8 x 8 square and the 11.3 maximal distance is quite inferior to the range of the second factor. Therefore, the estimated range of the factor is not 20, but about half this.

Hence, we should either increase the number of neighbourhood points (which is rather difficult) or use 1 point out of 2 or out of 4. This elimination makes sense, since it filters the high frequencies, which hardly play a part in the estimation of a high range phenomenon. Experience shows that the best compromise is to take 1 point out of 2 to estimate the factor Y^2 and 1 point out of 4 to estimate the mean.

3) It is worthwhile to note that it is not necessary to increase the neighbourhood size when estimating a high range factor with the anamorphoses because of the relation $Z^*(x) = Y^{1*}(x) + Y^{2*}(x)$. If Y^{1*} is stable so is Y^{2*}.

Figures 9a, 9b, 9c show the estimation of Y^1, Y^2, m on raw magnetism, using 70 neighbourhood points and taking 1 point out of 2 for Y^2, and 1 point out of 4 for the mean.

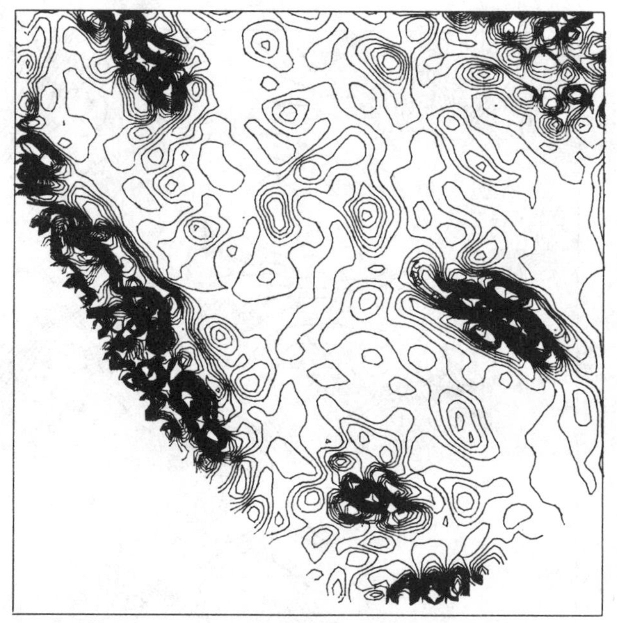

Figure 9a
Raw data : Y^{1*} estimated with 70 neighbourhood points

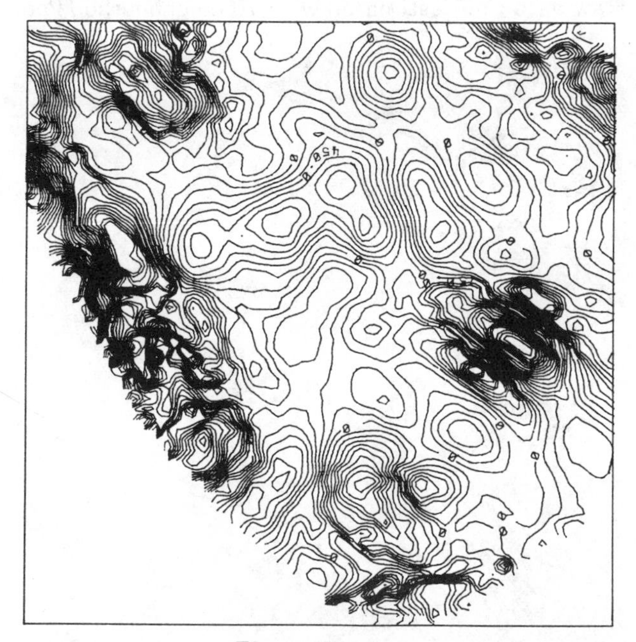

Figure 9b
Raw data : Y^{2*} estimated with 70 neigbourhood points

Figure 9c
Raw data : m^* estimated with 70 neighbourhood points

Figures 10a and 10b show the estimation of factors Y^1 and Y^2 on the anamorphoses (hence without universality condition) with 70 neighbourhood points.

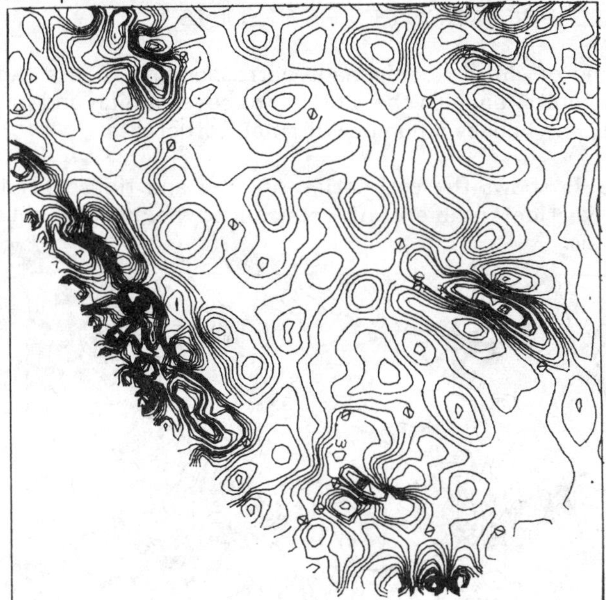

Figure 10a : Anamorphosed data : Y^{1*} with 70 neighbourhood points

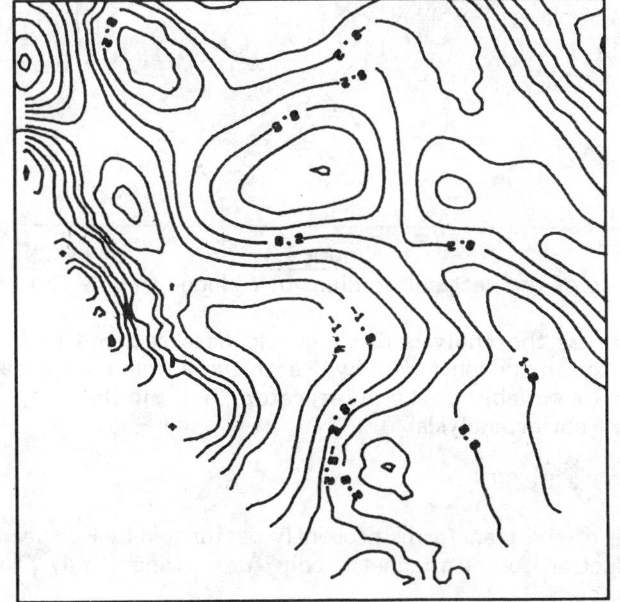

Figure 10b : Anamorphosed data : Y^{2*} with 70 neighbourhood points

Note : All the figures have been drawn at the "Centre de Géostatistique" by the program CARTOLAB 5 .

5. INTERPRETATION OF THE RESULTS

The map of Y^{1*} (componant corresponding to a small range), points up the faults, enables the plotting of several high and low axes of the basement, and bounds a number of small structures.

Figure 9a shows the estimation of Y^1, and figure 11 shows the map of the vertical gradient of reduced to pole field obtained by Fourier analysis.

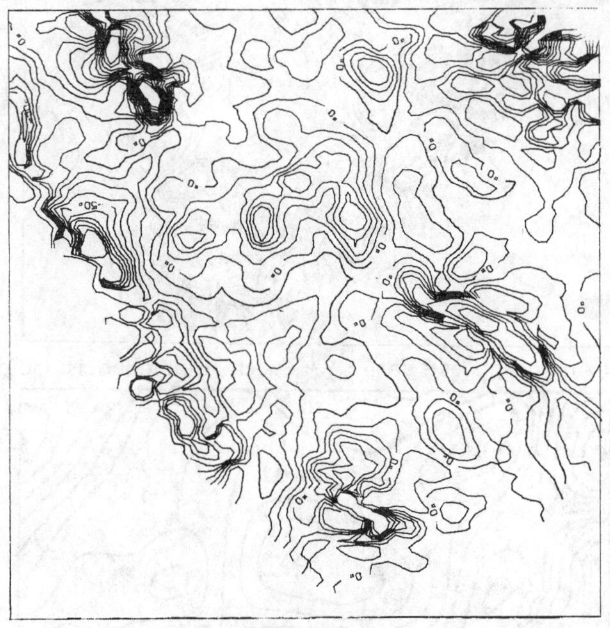

Figure 11

Map of the vertical gradient of reduced to pole field

As far as the analysis of magnetic data reduced to the pole is concerned, the maps obtained by Factorial Kriging Analysis at this stage seem to be able to be interpreted in a similar way to those produced by Fourier analysis.

6. CONCLUSION

Some of the transforms presently performed by Fourier methods, such as reduction to the magnetic pole for instance, may probably be done by kriging.

If further research confirms this impression and if, moreoever, the method works just as well with non reduced magnetism, it would have the tremendous advantage over existing methods of working directly with the data values, which would avoid the problems associated with using an intermediate regular grid as well as eliminating the filtering, smoothing and extrapolating inherent in Fourier methods.

REFERENCES

1 Matheron, g., 1982 : "Pour une analyse krigeante des données régionalisées". Internal Report n° N-732, Centre de Géostatistique, Fontainebleau.

2 Matheron, G., 1981 : "Splines and Kriging : their Formal Equivalence". In "Down to Earth Statistics - Solutions looking for Geological Problems", pp. 77-95, D.F. Merriam Editor, Syracuse University Geology Contributions.

3 Galli, A., Murillo, E., Thomann, J. : "Dual Kriging : Its properties and its uses in direct contouring". (1984) D. Reidel, p. 621.

4 Luc Sandjivy. "The factorial kriging analysis of regionalized data - Its application to geochemical prospecting" (1984) D. Reidel p. 559.

5 JL. Mallet - "Programmes de cartographie automatique : présentation de la bibliothèque CARTOLAB" - Sciences de la Terre, série informatique géologique n° 7, Nancy

THE FACTORIAL KRIGING ANALYSIS OF REGIONALIZED DATA. ITS APPLICATION TO GEOCHEMICAL PROSPECTING.

Luc SANDJIVY

Centre de Géostatistique et de Morphologie Mathématique
ECOLE NATIONALE SUPERIEURE DES MINES DE PARIS, Fontainebleau, France.

ABSTRACT

Uni and/or multivariate structural analyses lead to modelling the variability of the phenomenon under study. Factorial Kriging Analysis (F.K.A.) assigns to each of the structures in the models a "fictitious variable" which corresponds to a certain frequency level of the phenomenon. These can then be estimated by means of cokriging. The theory is reviewed for the stationary case and a practical application to geochemical prospecting is presented. This gives a geostatistical way of defining the concepts of "anomaly" and "regional background".

INTRODUCTION

The standard data analysis procedures used for handling multivariate data fail to take into account the spatial features of a regionalized phenomenon. There is no simple way of incorporating the location of the data points into the analysis. The aim of this paper is to present a new geostatistical method, named Factorial Kriging Analysis, which makes this possible.

The geostatistical approach to this question begins with the results of the structural analysis which points out the structure of a (co-) regionalisation, the presence or absence of a "drift", and quantifies them by modelling the (cross-) variograms or generalized covariances. These spatial structures on different scales which are identified through their covariances can be considered as "harmonics" or "frequency ranges" of the phenomenon. It then appears natural to apply data analysis methods to each of these

G. Verly et al. (eds.), Geostatistics for Natural Resources Characterization, Part 1, 559–571.

"harmonics" separately.

The purpose of the Factorial Kriging Analysis (F.K.A.) models is to make it possible to decompose a phenomenon, whether stationary or not, into its own different characteristic "frequencies" as shown by the structural analysis. In theory there is no need for a regular sampling pattern, and the information can be uni- or multivariate. In the latter case, they could have been collected at the same places or not.

Even though the method is always the same, the simplest way to understand it and to see its usefulness is to begin with a univariate study in the stationary case.

A UNIVARIATE STUDY.

Theory

Let $Z(x)$ be a regionalized variable on a field V, we consider it to be the realization of a weakly stationary random function $Z(x)$. Under this hypothesis $E(Z(x)) = m$ exists and is independent of x, and the covariance function, $Cov(x,y) = E\,(Z(x)-m)(Z(y)-m) = C(h)$ exists and depends only on h.

Let us now consider the following decomposition of $Z(x)$:

$$Z(x) = \sum_u a_u Y_u(x)$$

where the $Y_u(x)$ are weakly stationary random functions which are mutually orthogonal:

$$Cov(Y_u(x),\ Y_v(x+h)) = C_{uv}(h) = 0 \quad \forall u \neq v$$

$C_u(h)$ denotes the covariance function associated with $Y_u(h)$. Then $C(h)$, the covariance function of $Z(x)$ is:

$$C(h) = \sum_u a_u^2\, C_u(h)$$

We want to estimate $Y_u(x)$ by cokriging, using the data values Z_α; that is, we are looking for:

$$Y_u^*(x) = \sum_\alpha \lambda_\alpha Z_\alpha \quad \text{where } \lambda_\alpha \text{ are the cokriging weights}$$

The "non-bias" condition is $E(Y^*(x) - Y(x)) = 0$.

If $E(Z(x)) = m$ is not zero, it is not possible to find a unique decomposition of m into the different $Y_u(x)$.

We must study Z(x)-m if we know m or otherwise impose a condition on the λ_α's.

In the stationary case, m corresponds to the "drift" and Z(x)-m to the "residual", and <u>we estimate the "residuals" of Y(x) related to the "drift" m.</u>

We can then write: $E(Y_u(x)) = 0$ and the non bias condition is $\sum_\alpha \lambda_\alpha = 0$.

$$E\left(\sum_\alpha \lambda_\alpha Z_\alpha - Y_u(x)\right) = 0 \quad \text{if} \sum_\alpha \lambda_\alpha = 0 \text{ and } E(Y_u(x)) = 0.$$

Note that when m is equal to 0 or is known "a priori" we do not need any condition on the λ_α's. This is similar to kriging with known mean whereas $\sum_\alpha \lambda_\alpha = 0$ is the corresponding universality condition for kriging with unknown mean.

The minimisation of the quantity $E(Y_u^*(x) - Y_u(x))^2$ leads to the following system :

$$\begin{cases} \sum_\beta \lambda_\beta \, C(\alpha,\beta) = a_u \, C_u(\alpha,x) + \mu \quad \forall \alpha = 1 \dots n_\alpha \\ \sum_\alpha \lambda_\alpha = 0 \end{cases}$$

when m is unknown

μ : Lagrange coefficient

or

$$\left\{ \sum_\beta \lambda_\beta \, C(\alpha,\beta) = a_u C_u(\alpha,x) \right.$$

when m is known.

The corresponding variance $\sigma^2 = C_u(o) - a_u \sum_\alpha \lambda_\alpha C_u(\alpha,x)$ is not a real estimation variance of $Y_u(x)$; but of its projection on the space generated by the $\lambda(x)Z(x)$.

We have thus shown that in the stationary model, if we know the covariance function associated with a regionalized variable Z(x), and if its function can be expressed as the sum of basic covariance models, we can estimate the different components $Y_u(x)$ corresponding to each covariance at every point x in the field V.

Moreover the following consistency relation is verified:

$$\forall x \in V \quad Z^*(x) = \sum_u a_u Y_u^*(x) + m^*$$

where

Z^* : kriged value of Z at point x

$Y_u^*(x)$: cokriged value of Y_u at point x

m^* : known expectation $E(Z(x))$ or its kriged estimate in V.

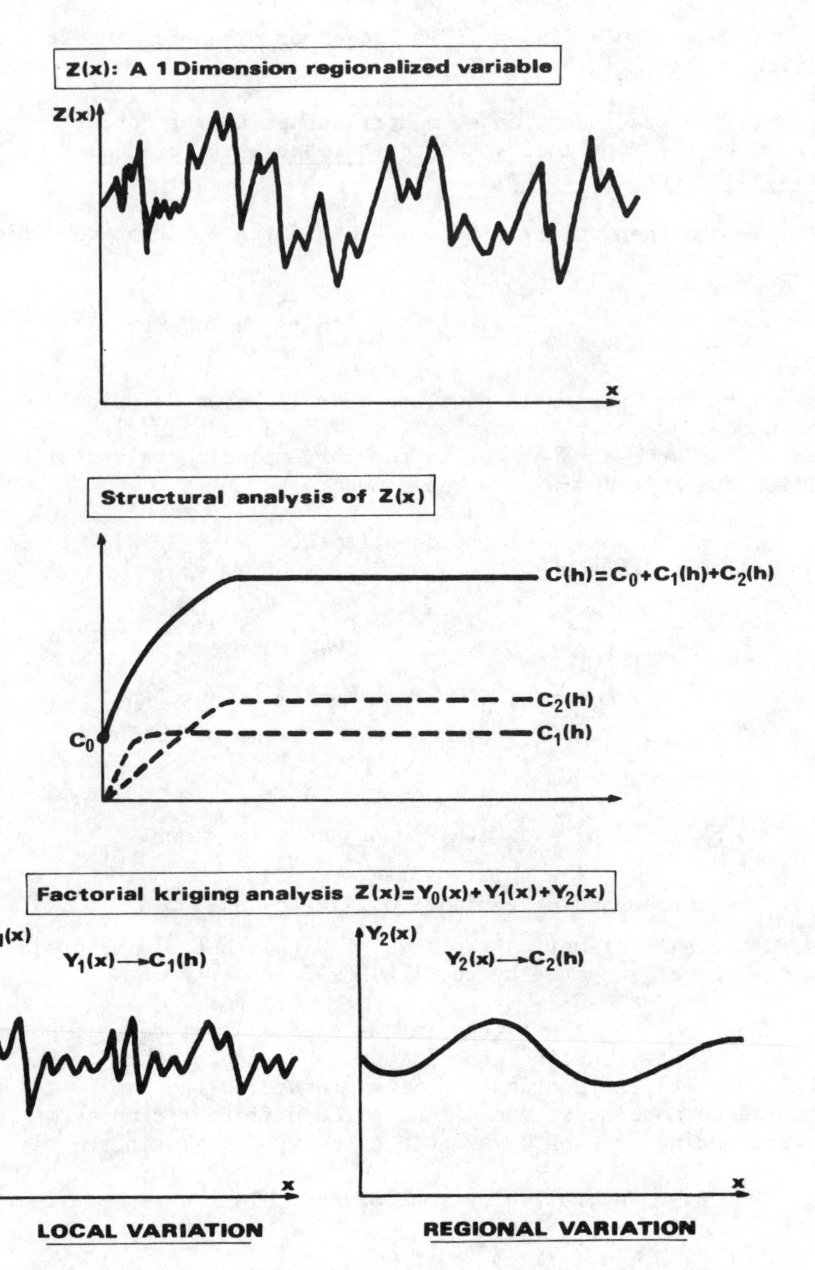

Figure 1

Illustration of F.K.A. procedure on a 1-Dimension example.

This procedure is illustrated in Figure 1.

A GEOCHEMICAL CASE STUDY

An interesting implication for geochemical prospecting is that F.K.A. can help defining the notions of "regional background" and "anomaly". The following example came from the study of the distribution of Cu in soil samples collected during a large grid prospection in the MUNSTER area (France).

977 soil samples were analyzed for 22 chemical element contents. We have first studied the variability of Cu, Zn, Pb and we give the results of the F.K.A. of Cu :

The distribution of Cu has a unimodal histogram of "lognormal" shape with m = 19.7 ppm and σ^2 = 747. To map Cu using kriging, the Gaussian anamorphosis (equivalent to the log transformation in the lognormal model) is applied before performing the structural analysis. The regionalization of Cu is modelled with an isotropic stationary covariance model :

$$C(h) = a_o^2 C_o + a_1^2 C_1(h) + a_2^2 C_2(h)$$

C_o : nugget effect $a_o^2 = .30$

C_1 : spherical model range 2 units $a_1^2 = .15$

C_2 : spherical model range 9 units $a_2^2 = .60$

We can associate these three basic models with three different "harmonics" of the regionalization :

C_o with its very erratic aspect, C_1 with the local structures, C_2 with the regional background. In other words, C_o corresponds to the high frequencies, C_1 to the medium ones and C_2 to the low ones.

We estimate the corresponding three components Y_o, Y_1 and Y_2 of the decomposition of Z(x) :

$$Z(x) = a_o Y_o(x) + a_1 Y_1(x) + a_2 Y_2(x)$$

and the coefficients of correlation (Z(x), $Y_i(x)$ of the form :

$$r_i(x) = \frac{a_i \|Y_i(x)\|}{\sqrt{\sum_i a_i^2 \|Y_i^*(x)\|^2}}$$

Comparing the maps of Cu, Y_1 and Y_2 , (not Y_o , because $Y_o = 0$ except at sampling points) shows that we have filtered the Cu map :

Y corresponds to the regional background, linked to the lithology seen on the geological map, while Y can be considered as the local fluctuation map, an anomaly map.

This is shown on maps 2, 3 and 4.

For every statistically anomalous value (say samples with Cu > 50 ppm), the associated coefficients of correlations r_o , r_1 and r_2 with Y_o , Y_1 , Y_2, help to decide whether these values are regional or anomalous (see Table 5). A high correlation with the nugget effect (r_o) shows an anomalous point ; a high correlation with the continuous component (r_2) indicates a background value. Note that the decision is clear in most cases.

The geochemist will decide on the usefulness of this procedure but our first trial has shown results in agreement with those previously established at the end of the campaign.

A MULTIVARIATE STUDY IN THE INTRINSIC CORRELATION MODEL

Theory :

In multivariate analysis, the general procedure is the same but is not so simple to apply since a number of (cross)-variograms have to be computed and modelled and large co-kriging programs used. This has not been done so far, but there is a very interesting case that leads back to the univariate one ; that is, when the variables have an intrinsic correlation. Then, all the (cross)-variograms are proportional to the same variogram function.

If $Z_i(x)$ are the different variables and $C_i(h)$ their associated covariances, then $C_{ij}(h) = e_{ij}\, C(h)$ where the e_{ij}'s are the coefficients of the correlation matrix and C(h) is the basic variogram model $C(h) = \sum_u a_u\, C_u(h)$.

The $Z_i(x)$ are usually not orthogonal and cannot be expressed directly as a sum of $Y_u(x)$ corresponding to $C_u(h)$, but the factors $F_i(x)$ resulting from the Principal Component Analysis of matrix e_{ij} are orthogonal and can be decomposed as follows :

$F_i(x) = \sum_j \alpha_{ij}\, Z_j(x)$, $F_i(x)$ being the corresponding factor to eigenvalue v_i.

$F_i(x) = \sum_u b_u Y_u(x)$ decomposition of $F_i(x)$ into the different $Y_u(x)$ corresponding to the $C_u(h)$

$G_i(h) = \sum_u b_u^2 C_u(h)$ covariance of $F_i(x)$

$$b_u^2 = a_u^2 v_i$$

Given the change of variable $Z_i(x) \rightarrow F_i(x)$, we are back to the univariate case and we proceed to the F.K.A. of the F_i's.

It is very interesting to notice that the F_i's have an intrinsic character in this model, as they do not rotate with h. Otherwise, the F_i's basis changes with the distance h, and we are led to use other models for F.K.A., as the linear coregionalisation one, where the P.C.A. is performed not directly on the e_{ij} matrix but in the e^u_{ij} matrix of the coefficients of each $C_u(h)$.

Case Study

To illustrate the intrinsic model, we consider the value of Cu, Pb and Zn analyzed in our previous example. In a first approximation (small variation of $r_{ij}(h) = C_{ij}(h) / \sqrt{C_{ii}(h) C_{jj}(h)}$, we can choose an intrinsic correlation model with the following parameters :

$C_{ij}(h) = e_{ij}\ C(h)$

$$e_{ij} = \begin{pmatrix} 1. & .48 & .42 \\ .48 & 1. & .63 \\ .42 & .63 & 1. \end{pmatrix}$$

$C(h) = a_o^2 C_o + a_1^2 C_1(h) + a_2^2 C_2(h)$

C_o = nugget effect $\quad a_o^2 = .35$

$C_1(h)$ = spherical model range 2. units $\quad a_1^2 = .30$

$C_2(h)$ = spherical model range 9. units $\quad a_2^2 = .40$

Eigenvalues v_i :

$v_1 = 2.0254 \quad v_2 = .61031 \quad v_3 = .3643$

We study factor F_1 (67% of variance) corresponding to v_1: $F_1 = .528\ Pb + .61\ Zn + .59\ Cu$. represents the overall content

Its covariance function is $G_1(h) = 2.0254\ C(h)$

There again, we obtain synthetic maps of the regional background and highly variable areas, the study of coefficients of correlation, performed as before for $C_u(h)$, help to isolate anomalous values.

The results of F.K.A. of F_1 are given through maps and tables (6, 7, 8, 9) as before.

CONCLUSION

Although there is no theoretical obstacle to the application of F.K.A. to non-stationary phenomena and the use of multivariate data, it still requires more research to define a satisfactory routine procedure and corresponding computer programs. Nevertheless, with F.K.A. a link has been established between purely statistical data analysis and the geostatistical approach. Its future liesin the hands of potential users of this technique and in their interpretation of the results.

The first case study in geochemical prospecting illustrates a possible geostatistical definition of anomaly and background in the interpretation of a sampling campaign, which is in accordance with the actual results already known.

Further collaboration with geochemists will show the usefulness of this new tool and the scope of its possible applications in applied earth sciences and industry.

REFERENCES

(1) Matheron, G., 1982 : "Pour une analyse krigeante des données régionalisées". Note n° N-732, Centre de Géostatistique, Fontainebleau.

(2) Sandjivy, L., 1982 : "Etude statistique et structurale de données géochimiques". Note n° N-747, Centre de Géostatistique, Fontainebleau.

(3) Sandjivy, L., 1982 : "Cartographie des données géochimiques". Note n° N-768, Centre de Géostatistique, Fontainebleau.

(4) Journel, A.G. , 1978 : "Mining Geostatistics" , Academic Press, London.

PRESENTATION OF RESULTS

The results of F.K.A. can be presented on maps and tables. The F.K.A. of copper leads to the following decomposition of the anamorphosed copper values (ANA) :

$$ANA = \sqrt{.30}\ PEP + \sqrt{.15}\ SP1 + \sqrt{.60}\ SP2$$

In the same way in the multivariate case, FACT1 can be written

$$FACT1 = \sqrt{.35}\ PEP + \sqrt{.30}\ SP1 + \sqrt{.40}\ SP2$$

PEP, SP1, SP2 being the variables corresponding to the nugget effect and spherical models (small (SP1) and large (SP2) ranges) identified by the structural analysis.

The mapping of ANA and FACT1 along with SP1 and SP2 shows the local and regional features of the regionalisation. See maps 2,3,4 for Cu and 6,7,8 for FACT1.

As all the variables are centered, we represent the negative values by dotted lines and the positive values by solid lines.

One can decide whether a sample is anomalous by looking at the corresponding coefficients of correlation of the variable with its components. This is shown for samples containing more than 50 ppm of Cu (Table 5) or showing values of FACT1 superior to 3 (Table 9).

NUM is the sample number
ANA : anamorphosed values of Cu
FACT1 : PCA factor
PEP, SP1, SP2 : values of the F.K.A. components
R0, R1, R2 : coefficients of correlations with PEP, SP1, SP2

R0 is high : the point is anomalous
R2 is high : the point belongs to the regional background.

F.K.A OF COPPER

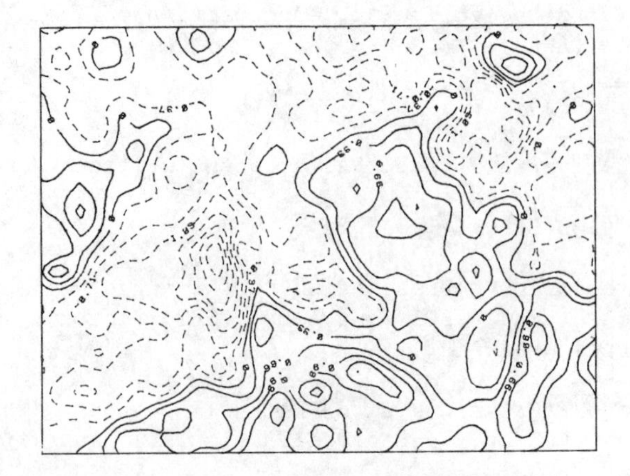

Map 2 ANA

Distribution of Cu

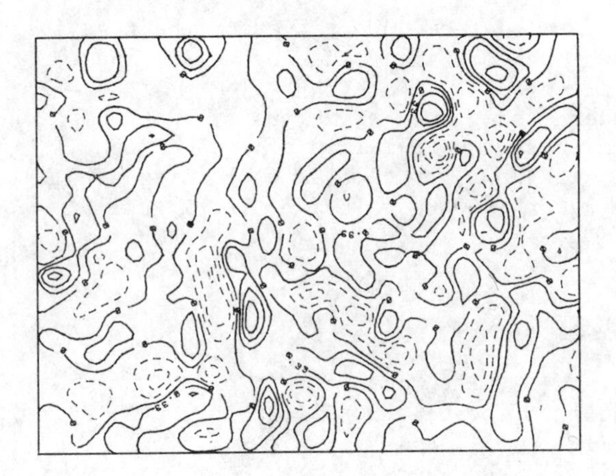

Map 3 SP1

Local Variations

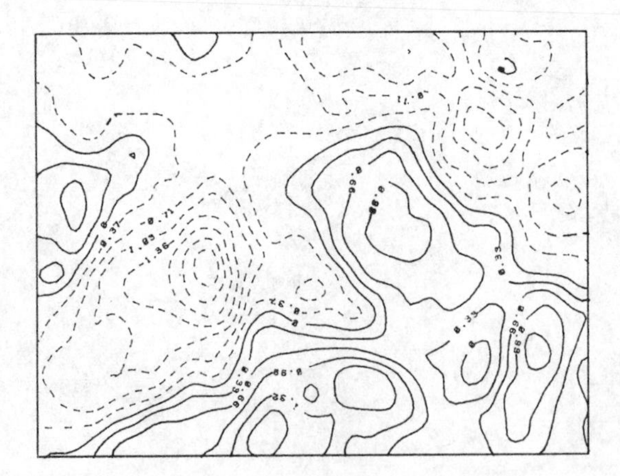

Map 4 SP2

Regional Variations

VARIABLE CUIVRE
****

NUM	PPM	ANA	PEP	SP1	SP2	R0	R1	R2
169	61.0	1.942	2.228	0.596	0.633	0.914	0.173	0.367
201	122.0	2.536	3.317	1.586	0.136	0.946	0.320	0.055
235	55.0	1.849	1.807	0.636	0.792	0.832	0.207	0.515
264	63.0	1.959	3.211	0.908	-0.195	0.977	0.195	-0.084
301	74.0	2.123	2.695	0.920	0.376	0.955	0.230	0.188
318	60.0	1.909	2.343	1.172	0.222	0.935	0.331	0.126
582	117.0	2.477	3.315	1.928	-0.110	0.924	0.380	-0.043
654	70.0	2.077	2.634	1.297	0.171	0.941	0.328	0.086
679	107.0	2.425	2.723	1.418	0.497	0.912	0.336	0.235
683	82.0	2.203	3.472	0.554	0.112	0.993	0.112	0.045
750	56.0	1.864	3.510	0.990	-0.571	0.957	0.191	-0.220
793	89.0	2.300	2.776	0.694	0.559	0.939	0.214	0.268
814	67.0	2.015	1.948	0.584	0.931	0.816	0.173	0.552
834	52.0	1.635	0.723	0.539	1.589	0.302	0.159	0.940
838	82.0	2.233	0.885	1.202	1.656	0.335	0.321	0.886
839	86.0	2.265	0.829	0.956	1.860	0.292	0.238	0.926
841	51.0	1.822	0.445	0.656	1.709	0.178	0.186	0.966
850	329.0	2.977	2.347	1.067	1.651	0.691	0.222	0.688
852	184.0	2.806	2.641	0.635	1.438	0.785	0.133	0.605
857	51.0	1.795	1.120	0.703	1.175	0.542	0.241	0.805
925	60.0	1.925	3.923	1.269	-0.923	0.927	0.212	-0.308
980	67.0	2.034	2.778	0.772	0.276	0.972	0.191	0.137
987	71.0	2.099	1.678	1.186	0.789	0.803	0.358	0.477
1001	51.0	1.808	0.530	0.858	1.531	0.229	0.263	0.937
1002	50.0	1.783	0.496	0.564	1.669	0.203	0.163	0.966
1177	105.0	2.379	2.405	1.049	0.847	0.863	0.266	0.430
1182	74.0	2.148	1.871	1.013	0.944	0.777	0.298	0.555
1195	651.0	3.084	3.513	1.119	0.937	0.916	0.206	0.345
1212	63.0	1.977	1.184	1.007	1.212	0.538	0.323	0.778
1228	161.0	2.693	2.121	1.093	1.430	0.700	0.255	0.667
1256	56.0	1.878	1.012	0.862	1.279	0.468	0.282	0.837
1293	95.0	2.338	1.309	1.201	1.492	0.499	0.324	0.804
1296	67.0	2.055	1.396	0.767	1.283	0.593	0.231	0.771
1325	133.0	2.606	1.425	1.421	1.647	0.490	0.345	0.801
1334	63.0	1.995	0.972	0.683	1.547	0.398	0.198	0.896
1337	82.0	2.175	1.125	0.904	1.560	0.440	0.250	0.863
1338	59.0	1.894	0.922	0.886	1.349	0.417	0.283	0.863

TABLE 5

F.K.A. of Copper - Decomposition of sample values > 50 ppm (See presentation of results for comments).

F.K.A. OF THE FIRST FACTOR OF THE PCA OF Cu, Pb, Zn.

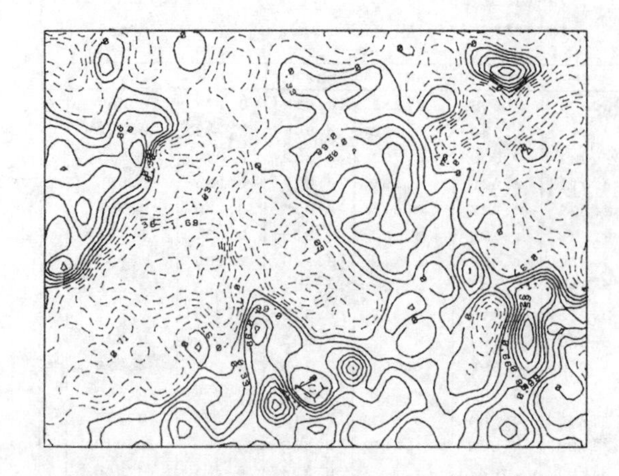

Map 6 FACT1

Distribution of the first factor issued from the PCA of Pb, Zn, Cu.

Map 7 SP1

Local Variations

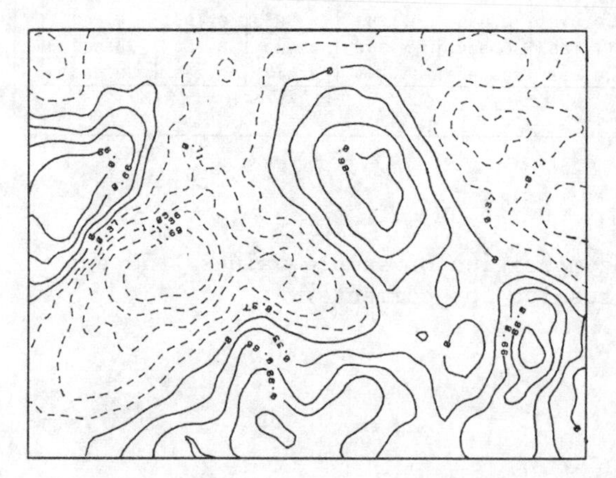

Map 8 SP2

Regional Variations

VARIABLE FACTEUR1

NUM	FACT1	PEP	SP1	SP2	R0	R1	R2
201	4.242	3.286	1.848	0.032	0.887	0.462	0.009
235	2.903	1.570	0.617	1.220	0.741	0.270	0.615
251	3.324	1.619	1.218	1.121	0.701	0.489	0.519
264	2.761	2.755	1.062	-0.432	0.930	0.332	-0.156
301	4.332	2.912	1.383	0.687	0.877	0.386	0.285
305	3.210	2.353	1.016	0.482	0.910	0.364	0.199
508	2.615	1.175	0.768	1.140	0.640	0.388	0.663
553	2.655	1.132	1.009	1.014	0.620	0.513	0.594
582	3.734	2.718	1.904	-0.048	0.839	0.545	-0.016
583	2.717	2.011	1.326	-0.015	0.853	0.522	-0.007
638	2.691	1.786	1.438	0.069	0.801	0.598	0.033
654	3.850	2.957	1.690	0.042	0.884	0.468	0.013
679	4.550	3.005	1.940	0.558	0.846	0.506	0.168
683	3.219	2.959	0.665	0.230	0.976	0.203	0.081
750	2.555	2.710	1.047	-0.607	0.919	0.329	-0.220
838	3.876	1.224	2.138	1.305	0.451	0.730	0.514
839	4.079	1.469	1.873	1.532	0.524	0.620	0.584
850	4.159	2.153	1.320	1.460	0.736	0.418	0.533
980	4.022	2.743	1.456	0.637	0.876	0.431	0.217
998	2.883	1.649	1.409	0.437	0.765	0.606	0.217
1001	3.328	0.984	1.564	1.419	0.425	0.626	0.654
1002	2.500	1.197	0.498	1.225	0.653	0.252	0.714
1037	2.878	2.104	0.831	0.506	0.913	0.334	0.235
1080	2.724	1.936	0.924	0.412	0.895	0.396	0.204
1139	2.599	1.838	0.822	0.454	0.898	0.372	0.237
1173	3.213	2.748	0.704	0.387	0.963	0.228	0.145
1177	3.369	1.612	1.574	0.869	0.682	0.617	0.393
1182	3.669	1.875	1.631	0.907	0.722	0.582	0.373
1184	3.360	2.225	1.276	0.543	0.860	0.458	0.224
1195	5.182	4.070	1.576	0.580	0.932	0.334	0.142
1211	2.812	0.171	1.837	1.370	0.076	0.756	0.650
1212	3.809	1.442	1.801	1.319	0.551	0.638	0.538
1219	3.130	1.351	0.551	1.734	0.575	0.217	0.789
1222	3.468	1.836	1.410	0.911	0.748	0.532	0.396
1228	4.722	2.218	1.811	1.598	0.680	0.514	0.523
1238	2.926	1.566	0.930	0.979	0.756	0.416	0.505
1277	3.067	1.543	1.056	1.046	0.720	0.457	0.522
1310	2.851	1.496	0.875	1.008	0.743	0.403	0.535
1337	3.067	1.426	1.286	0.956	0.672	0.562	0.482
1338	3.189	1.613	1.424	0.798	0.716	0.586	0.378
1348	2.715	1.275	0.971	0.980	0.679	0.479	0.557

TABLE 9

F.K.A. of P.C.A. factor F1 - Decomposition of sample values > 2.5
(See presentation of results for comments)

THE ANALYSIS OF SECOND-ORDER STATIONARY PROCESSES: TIME SERIES ANALYSIS, SPECTRAL ANALYSIS, HARMONIC ANALYSIS, AND GEOSTATISTICS

Andrew R. Solow

Stanford University

ABSTRACT

Time series analysis, spectral analysis, and geostatistics are all tools designed for analyzing second-order stationary stochastic processes. This paper compares these three approaches at a broadly theoretical level.

INTRODUCTION

Historically, the study of second-order stationary processes originated in the study of Gaussian processes, for which second-order stationarity entails strict stationarity. This is not surprising, considering how much of applied probability and statistics has its origins in the study of Gaussian random variables. To be sure, the application of Gaussianbased techniques need not require the Gaussian assumption. Rather, in the absence of better methods, these often nonrobust procedures will have to suffice. Of course, there now exist many robust and non-parametric techniques which often provide excellent alternatives to their more classical counterparts.

The notion of modelling a process as stationary is not without theoretical justification. Many processes -- positive recurrent Markov chains and certain diffusion processes among them --

G. Verly et al. (eds.), Geostatistics for Natural Resources Characterization, Part 1, 573–585.

exhibit limiting, stationary behavior. That is, regardless of initial conditions, if such a process is allowed to evolve through time (or over space), the resulting probability structure will reach and remain at a stationary state.

The notion of modelling a process as Gaussian is also not without theoretical justification. A number of Central Limit Theorems ensure that under certain conditions processes which in some sense represent sums of random variables will have limiting normal distributions. Donsker's Theorem, in particular, states that the asymptotic distribution of such a process does not depend on the underlying distribution. When the conditions of this theorem are met, the limiting probability structure may be approximated by that of a suitably scaled Gaussian process.

From the theory of stationary processes have come the important techniques collectively known as time series analysis. Time series analysis may be approached through the frequency domain in which case it is called Fourier, harmonic, or spectral analysis, or through the time domain in which case the BoxJenkins methodology is commonly applied. Geostatistics may be viewed as the application of the time domain approach to multi-dimensional problems. This paper discusses at a broadly theoretical level the similarities and differences between these approaches.

THE GOALS OF TIME SERIES ANALYSIS

The techniques that have been developed to analyze stationary processes have arisen in response to certain classes of problems and also as a consequence of the physical models favored by the analysts. Three important problems that are often addressed in time series analysis are prediction, modelling, and exploratory data analysis.

In prediction, the object is to interpolate or extrapolate a sampled realization. That is, the unknown value of the discrete (or discretely sampled) realization at time t, z(t), is estimated from the data z(t-1), z(t-2),... (and including z(t+1), z(t+2),... in interpolation). What is important in prediction is the value generated by the process. If this can be estimated without explicit modelling, as in linear geostatistics, that is fine.

In modelling, an explicit parametric expression of the process is found. Typically, the form of the model is chosen on the basis of some structural characteristics of the process from among a family of models, and the model parameters are chosen by some fitting procedure. The goals of modelling include:

understanding the nature of the process, controlling the process by sequential decisions, filtering the process to rid it of noise, and predicting unknown values of the process, which need not require modelling. Harmonic analysis and Box-Jenkins analysis are both modelling procedures. Harmonic analysis models the process as the sum of periodic terms, while Box-Jenkins analysis models it as a so-called autoregressive moving average.

Exploratory data analysis is often useful in deciding what sort of approach is appropriate to more fully analyze a process. It is also useful in questions of classification and recognition, as when a process is catalogued according to some summary characteristic. In geostatistics, the semi-variogram provides such a characteristic. In spectral analysis the spectral density plays this role, while in Box-Jenkins analysis the autocorrelation function is used.

WOLD'S DECOMPOSITION THEOREM

Before discussing the different approaches to time series analysis, it is useful to introduce an important theorem due to Wold. Consider a sequence of linear autoregressions:

$z(t)$ on $z(t-1), z(t-2), \ldots$

$z(t)$ on $z(t-2), z(t-3), \ldots$

$z(t)$ on $z(t-3), z(t-4), \ldots$

Each autoregession has the form:

$$z(t) = R(t) + a(t)$$

where $R(t)$ is a linear combination of past values and $a(t)$ is the residual process. Let $s^2(q)$ be the variance of the qth autoregression and let S^2 be the variance of the process itself. Three cases may occur:

$$\lim s^2(q) = 0$$

$$\lim s^2(q) = S^2$$

$$0 < \lim s^2(q) < S^2$$

where in all cases the limit is taken as q approaches infinity. In the first case, the process is called purely deterministic because knowledge of the remote past allows perfect prediction. In the second case, the process is called purely indeterministic because such knowledge is useless for prediction. The third case is intermediate.

Wold's Decomposition Theorem states that any stationary process may be decomposed into the sum of a purely deterministic and a purely indeterministic process. Moreover, the purely indeterministic component may be written as a linear combination of uncorrelated random variables with zero mean and equal variance. Such a stationary sequence is often called an innovation process. The power of this theorem is that it allows any purely indeterministic stationary process to be written:

z(t) = c0a(t) + c1a(t-1) + c2a(t-2) + ...

where a(t) is the innovation process. As will be seen, such a representation corresponds to Yule's model of a stationary process which is the basis of the Box-Jenkins approach.

THE FREQUENCY DOMAIN APPROACH

The earliest work done in the frequency domain approach was by Schuster and others in the last century. This work is perhaps best termed Fourier analysis since it consists of fitting a model composed of sinusoids to the process, although it is ususally called harmonic analysis. It was thought at the time that many physical processes were truly composed of complex sums of such terms, and this became known as the model of hidden periodicities, because the terms may interact to obscure the contributions of individual components.

A representation theorem due to Wiener (sometimes attributed to Karhunen) states that any stationary process has the form:

$$z(t) = \int_0^{\pi} \cos \omega t \, du(\omega) + \int_0^{\pi} \sin \omega t \, dv(\omega)$$

where u(w) and v(w) are uncorrelated processes with independent increments. This is the theoretical justification for harmonic analysis which fits models of the form:

$$z(t) = \sum_k a_k \cos \omega_k t + \sum_k b_k \sin \omega_k t$$

The second main approach in the frequency domain is spectral analysis. The seminal work in this field was done by Kolmogorov and Wiener in the 1930's and 1940's. Spectral analysis consists of estimating and interpreting a characteristic of a stationary process called the spectral density or the spectrum. The spectrum is a measure of the contribution to the total variance of the process that is made by terms at each frequency, w. The sampling properties of the spectrum are briefly discussed later in this paper.

The frequency domain approach is useful in three particular instances. In analyzing processes that are truly composed of sums of periodic components, harmonic analysis may provide a good-fitting model involving a small number of parameters to be estimated. Spectral analysis is sometimes useful in exploratory data analysis. The spectrum in some cases is quite striking, exhibiting one or more sharp peaks. Such behavior will ordinarily have physical significance, thus aiding structural analysis. Finally, it has sometimes proved easier to carry out theoretical analysis in the frequency domain, although this may be an analyst-specific advantage.

On the other hand, the frequency domain approach suffers from certain shortcomings. While Wiener's theorem ensures that any stationary process may be expressed as a Fourier series, it does not ensure that the estimation of this representation from limited data will be easy or accurate. It is a theme of this paper that the analyst should be guided by what seems the most natural. In many cases, the exceptions being obvious upon encounter, the frequency domain approach will not seem natural. Periodic behavior will often best be treated as a deterministic component and handled accordingly.

THE TIME DOMAIN APPROACH

The Box-Jenkins approach to time series analysis is based on Yule's model of a stationary process. This model has the schematic representation:

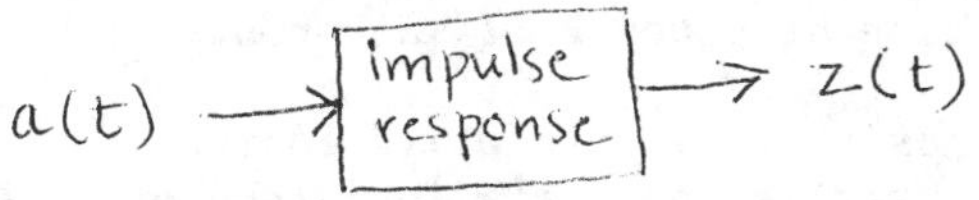

a(t) → [impulse response] → z(t)

That is, this model represents a possibly highly correlated process, z(t), as a moving average of past and present values of an uncorrelated innovation process, a(t). This model is written:

$$z(t) = a(t) + b1a(t-1) + \ldots + bpa(t-p)$$

Such a model is called a moving average of order p, or MA(p). It can be shown that this process has the equivalent representation:

$z(t) = a(t) + a1z(t-1) + a2z(t-2) + \ldots$

which is an autoregressive model of infinite order, or AR( ). This equivalence is, in fact, only true if the process is invertible. Throughout this discussion, all processes are assumed to be invertible.

The family of models composed of moving averages, autoregressions, and mixtures of the two -- autoregressive moving average, or ARMA -- forms the basis of the Box-Jenkins methodology.

The modelling procedure consists of estimating the autocorrelation structure of the process and comparing it to certain canonical autocorrelation functions to identify the type and order of the process. Apart from Gaussian processes, the autocorrelation structure does not determine a unique process. Therefore, it may be necessary to apply a further criterion to uniquely define a model. The criterion suggested by Box and Jenkins is parsimony. That is, broadly speaking, in choosing between models that reproduce the autocorrelation structure equally well the choice should be that model which minimizes the number of parameters to be estimated. The necessity for such an additional rule arises from the lack of sufficient data to decisively distinguish between ARMA processes with similar autocorrelation functions. Indeed, the use of mixed ARMA-type models is designed to economize on parameters. Other criteria besides parsimony have been suggested and applied with some success. Notable among these is Akaike's Information Criterion, or AIC.

To deal with periodocity and trend, Box and Jenkins suggest differencing the process at the appropriate lag and order to produce an aperiodic stationary process which may then be modelled as an ARMA process. Because the final model is then reached by de-differencing, it is called an autoregressive integrated moving average, or ARIMA. This is closely related to the intrinsic random function of order k of Matheron.

The Box-Jenkins approach has several attractive features. It is easy to implement and its results are easy to interpret. Also, the ARIMA family of models is extremely rich, certainly richer than the harmonic representation of equal order.

It is interesting to note that the role of periodic components in a process has been reduced to that of a deterministic component that is factored out by differencing. This attitude stems from the fact that Box-Jenkins analysis was first developed for the analysis of economic time series, in which periodicity usually reflects fairly regular seasonal influences. Also, it should be noted how Yule's model fits with Wold's Decomposition Theorem. In Yule's model, periodicity is treated as a purely deterministic component and the purely indeterministic component is modelled as a moving average.

THE LINKAGE BETWEEN THE TIME AND FREQUENCY DOMAINS

The differences between the time domain and the frequency domain approaches are clear. The frequency domain approach models a process as the sum of sinusoids, while the time domain approach models a process as a parsimonious ARMA process. While both models are theoretically equally general, the harmonic model presumes a greater understanding of the physical process. It is linked to the model of hidden periodicities. The family of ARMA models, while equally general, is less bound to a particular philosophy. In terms of structural analysis, the differences are perhaps even sharper. The spectrum is designed for decomposition of variance according to frequency. The autocorrelation function, on the other hand, is designed for the study of the pattern of correlation.

On a theoretical level, the two approaches are strongly linked. The Wiener-Khinchine Theorem shows that the autocovariance function is the Fourier transform of the spectral density in one dimension. Similarly, the representation theorem of Wiener and Wold's Decomposition Theorem ensure that any stationary process may be expressed by both a harmonic expansion and a moving average. Thus, the two approaches are mathematically isomorphic.

The choice between analysis in the frequency domain and time domain must be made on more practical grounds. First, it should depend on the physical nature of the process under study. In those applications where the spectrum has physical meaning, as is often the case in electrical engineering, the frequency domain approach may be best. On the other hand, many real processes in economics and other fields are by nature autoregressive or Markovian and in that case the time domain approach is indicated. Again, broadly speaking, the ARMA family is richer and less bound to a particular physical model, and so it has superseded the harmonic representation to a great extent. Second, it is important to keep the goal of analysis in mind. For example, as was mentioned before, if this goal entails decomposition of variance among frequencies then spectral analysis should be used, while the time domain approach is better suited to studying the pattern of correlation. Third, numerical matters may need to be considered. For example, the question of the behavior of sample measurements is quite important, and though little can be said of this at a general level it is taken up in the next section.

A BRIEF LOOK AT SAMPLING CONSIDERATIONS

It is of cold comfort to the practicing analyst that the spectrum, correlogram, and variogram are mathematically equivalent. This is because none of these functions is actually available to the analyst: they must be estimated from often limited data. It is therefore of primary interest how accurately these functions can be estimated. Of course, the quality of the estimation will depend on the quality, number, and configuration of the data, as well as on the strength and pattern of the correlation. If the dependence (not only correlation) between observations vanishes after a certain lag -- say, m -- the process is said to be m-dependent. If the domain of observation is large compared to m then it is possible to invoke central limit theorems and the sample variogram and correlogram values will be consistent (asymptotically unbiased) and normally distributed with calculable variance. The rate of convergence to normality is probably too slow in most practical cases to be of much use. Apart from this, however, general results have been hard to come by. Some broad conclusions should be noted.

With regard to the sample spectrum, the most natural estimator -- the periodogram -- is not consistent and indeed may fluctuate quite wildly. If the process is noise-like, the sample spectrum is distributed chi-square with 2 degrees of freedom, entailing a coefficient of variation of 1. Several techniques have been developed by Tukey, Bartlett, and others starting in the 1950's to handle this. Pre-whitening of the data, smoothing the estimated spectrum, and truncating the spectrum are examples. While the sampling properties of the pre-whitened, smoothed, and truncated estimator are nicer than those of the periodogram, obviously a loss of information is involved. Other problems also occur. As in any smoothing, a window must be chosen and this involves a trade-off between variance and bias. Also, smoothing the spectrum imparts correlation to successive values, which hinders interpretation and statistical testing.

With regard to the sample correlogram, the theoretical work has taken a slightly different approach. In this case, not-large sample results have been worked out for certain types of processes. Circular processes have been perhaps the most intensively studied. On a broader level, it is known that the classical estimator of covariance has a bias of order 1/N. This bias is not very important and it can be reduced in order to 1/N by Quenouille's jackknife. The sample correlogram, being the ratio of two random quantities, has properties more difficult to determine. It has a bias, also reducible in order, and it has an asymptotic variance that depends on all values of the correlogram and that increases with lag. The distribution of correlogram values under the null hypothesis that all values are zero can be found, allowing a test of the hypothesis to be made.

The variogram estimator appears to be unbiased, but apart from asymptotics little can be said about its sampling distribution in the general case. Almost certainly, it behaves in a similar way as the sample correlogram.

Overall, these results suggest that it is useless to attempt to discern fine structure in these functions. In particular, the fact that successive values are correlated means that, first, structures are spread out over many lags and, second, the influence of outliers is also spread out. It means also that statistical tests of individual values may not be done independently. At best, it can be hoped that the sample functions accurately reproduce the gross structures in the underlying function.

Finally, the influence of outliers on all these functions can be disastrous. Also, departures from normality make the classical estimators non-optimal. These issues are the subjects of other papers on the robust estimation of the variogram.

It is interesting to note that because the spectrum is primarily a tool for structural analysis, smoothing is routinely done to reduce noise and accentuate gross structure. Because the correlogram and variogram are used also for prediction, smoothing (as opposed to fitting) is not desirable.

GEOSTATISTICS AND TIME SERIES ANALYSIS

Linear geostatistics -- variography and kriging -- is clearly closely linked to the time domain approach to time series analysis. In geostatistics, the explicit modelling stage is omitted, and prediction is based directly on the correlation structure. The lack of a modelling stage is due to the types of problems geostatistics was designed to solve. In mining, it is the value of the ore grade that is most interesting and not a parametric model of the process that produced it. Similarly, since Box-Jenkins analysis was designed to model economic and industrial processes, for which understanding and controlling the process is important, an explicit model is an important goal of the analysis.

The greatest difference between geostatistics and the time domain approach is that geostatistics is applicable in more than one dimension and the time domain approach in its present form is not. The difficulty in applying the time domain approach to multiple dimensions addresses a basic property of time. Time can, essentially without disagreement, be ordered. It is meaningful to say that t-1 precedes t. It is possible to write models of the form:

$$z(t) = f(z(t-1), z(t-2), \dots)$$

and to comfortably predict z(t) from the already realized values of z(t-1), z(t-2), ... In multiple dimensions, however, the notion of ordering space is not a meaningful one.

To say that geostatistics "works" in multiple dimensions and that the time domain approach does not is an oversimplification. Geostatistics is a good deal less ambitious than the time domain approach, in that it does not seek to model the process, per se. Incidentally, the limitation of applying time series analysis to regularly spaced data also stems from the modelling role. That is, it is easiest to model a process as being discrete through time, the regular spacing of the data corresponding to the unit time interval. Geostatistics avoids this problem by treating the process as continuous and modelling the variogram as a continuous function.

Although the problems faced by the time domain approach in multiple dimensions and with irregularly spaced data both arise from the modelling aspect, they are still distinct problems. The matter of dimensionality is more profound, as it addresses a fundamental property of space. The matter of data spacing is concerned with the distinction between discrete and continuous processes. That is, while it is possible to model a process as continuous -- although not within the ARMA scheme -- it is not possible to avoid the dimensionality problem.

A second difference between geostatistics and the time domain approach lies in the form of the estimators of unknown values. The autoregressive form of the linear kriging estimator is linked to the multi-Gaussian case for which the conditional expectation -- which is "best" in the classical sense -- of the unknown value given the data is precisely the kriging estimator. This is, of course, the theoretical justification for the multi-Gaussian approach in geostatistics. The ARMA estimator, on the other hand, is more general, based as it is on Yule's model. Since any invertible ARMA process has an AR( ) representation, the limitation of the form of the kriging estimator is not important. By delimiting a kriging neighborhood that excludes data which would have near-zero kriging weights, kriging is using an implicit AR( ) model.

CONCLUSIONS

Geostatistics is at a stage now where it is encountering many problems, both practical and theoretical, familiar to statisticians in general. The number of papers at this Institute on robust estimation of the variogram is testimony to this. Because of the particularities of the geostatistical problem -- significant correlation, multiple dimensions, and irregularly spaced data -- it is not possible to adopt techniques which have proved successful elsewhere without intelligently adapting them. Much of the work done on processes defined over space, and not just time, by mainstream statisticians attempts to define suitable models for such processes. There are at least three problems with this approach. First, the number of such models is limited. Second, such modelling is out of step with the current emphasis on robust, nonparametric approaches. Third, from a practical point of view, modelling becomes essentially impossible in the face of irregularly spaced data. Part of the problem seems to be that these statisticians view the goals of this analysis as the same as the analysis of time series. It is here that the geostatistician may make an important contribution. Being acquainted with practical problems in mining and other fields, he is able to derive techniques which provide accurate, relevant results, even if they do not rival the modelling capability of time series analysis.

As the territories of geostatistics and statistics come to overlap, communication becomes more and more important. Statisticians do not have solutions to all of our problems, or for that matter to any. But they certainly have insight and experience that may help guide us. This paper is an attempt to sketch out the place of geostatistics in the classical statistical framework.

As applied probability and statistics have evolved and as they have come to address a wider class of problems in the analysis of physical processes, techniques have been developed that at first glance appear unrelated. The theme of the first part of this paper is that in the analysis of stationary processes the main techniques are closely related, even isomorphic. The differences between them result from the particularities of the problems which they were designed to solve and from the attitudes of the analysts towards the physical phenomenon. The choice between applying one or the other approach should depend on the ease of application, and this in turn depends on which approach best captures the nature of the process. Because the time domain approach is based on a more general underlying model than the frequency domain approach, with the exception of processes with marked periodicity and striking spectral behavior, it has tended to supplant the frequency domain approach. Even in the case of marked periodicity, modern analysts have tended to treat the periodicity as deterministic and factor it out by differencing.

When comparing geostatistics to the time domain approach, as the second part of this paper suggests, it is important to specify the application in mind. To say that geostatistics out-performs Box-Jenkins analysis in multiple dimensions and with irregularly spaced data is not sensible. Geostatistics was designed to be applied under such conditions, and under such conditions, without making unreasonable assumptions, variography and kriging are about the most sophisticated things that can routinely be done. Under the conditions presented by a typical economic time series -- a large, regularly spaced data set in one dimension -- it is possible to go beyond structural analysis and prediction and to produce an explicit model of the process.

LITERATURE

There is an enormous body of literature on the analysis of stationary processes, both in book form and in journal articles. Much of the best theoretical work was done by Russian statisticians and is not readily available. The book by Cox and Miller contains an excellent review of the theory in both domains. Yaglom's book also provides an excellent review of the theory. Hannan's book contains both theoretical and practical material. For the time domain approach, the book by Box and Jenkins is a classic, while Jenkins and Watts present a thorough introduction to the frequency domain approach. The articles by Harris and Tukey are of historical interest. For discussion of the sampling theory of the autocorrelation function, see Kendall and Kendall and Stuart, as well as many of the works mentioned above. All of these books contain excellent bibliographies, although none is recent. The books by Ripley and Bartlett represent attempts by classical statisticians to address the problem of spatial stochastic processes.

REFERENCES

Bartlett, M.S. (1975). The Statistical Analysis of Spatial Pattern. Chapman and Hall, London.

Box, G.E.P. and Jenkins, G.M. (1970) Time Series Analysis.

Holden-Day, San Francisco.

Cox, D.R. and Miller, H.D. (1968). The Theory of Stochastic Processes.. Chapman and Hall, London.

Feller, W. (1966). An Introduction to Probability Theory and its Applications. Wiley, New York.

Hannan, E.J. (1960). Time Series Analysis. Methuen, New York.

Harris, B. "Introduction to the Theory of Spectral Analysis of Time Series" in Spectral Analysis of Time Series. (1967). Wiley, New York.

Jenkins, G.M. and Watts, D.G. (1968). Spectral Analysis and its Applications. Holden-Day, New York.

Kendall, M. (1973). Time Series. Griffin, London.

Kendall, M. and Stuart, A. (1968). Advanced Theory of Statistics, vol. 3. Griffin, London.

Ripley, B.D. (1981). Spatial Statistics. Wiley, New York.

Tukey, J.W. "An Introduction to the Calculations of Numerical Spectrum Analysis" in Spectral Analysis of Time Series. (1967). Wiley, New York.

Yaglom, A.M. (1962). Stationary Random Functions. Prentice-Hall, Englewood Cliffs, N.J.